Enhancing Autonomous and Adaptive Systems With AI and IoT

Ben Othman Soufiene
King Faisal University, Saudi Arabia

IGI Global
Scientific Publishing
Publishing Tomorrow's Research Today

Vice President of Editorial	Melissa Wagner
Director of Acquisitions	Mikaela Felty
Director of Book Development	Jocelynn Hessler
Production Manager	Mike Brehm
Cover Design	Jose Rosado

Published in the United States of America by
IGI Global Scientific Publishing
701 East Chocolate Avenue
Hershey, PA, 17033, USA
Tel: 717-533-8845 | Fax: 717-533-7115
Website: https://www.igi-global.com E-mail: cust@igi-global.com

Library of Congress Cataloging-in-Publication Data

LCCN: 2025013367 (CIP Data Pending)
ISBN13: 979-8-3373-3146-1
Isbn13Softcover: 979-8-3373-3147-8
EISBN13: 979-8-3373-3148-5
British Cataloguing in Publication Data
A Cataloguing in Publication record for this book is available from the British Library.

Table of Contents

Detailed Table of Contents

S. Anusha, Easwari Engineering College, India
S. Vidyasagar, SRM Institute of Science and Technology, India
V. Kalyanasundaram, SRM Institute of Science and Technology, India
K. Saravanan, SRM Institute of Science and Technology, India
L. A. Anto Gracious, R.M.K. College of Engineering and Technology, India
M. Nalini, SA Engineering College, India

This chapter explores the fundamental components of the Internet of Things (IoT) within the context of an AI-driven intelligent ecosystem. It discusses the role of IoT sensors, connectivity, and communication protocols in enabling seamless interaction between devices, networks, and intelligent systems. The integration of AI into IoT systems is highlighted, emphasizing its potential in enhancing data analytics, decision-making, and automation. The chapter also addresses key challenges in scalability, interoperability, security, and privacy, while outlining emerging trends such as quantum computing and next-generation communication technologies like 5G. The future of IoT and AI convergence offers significant opportunities for creating smarter, more efficient environments in various industries, from healthcare to smart cities. This chapter concludes by discussing the transformative potential of IoT and AI for shaping the future of intelligent ecosystems.

practices, and ethical AI deployment. Using interdisciplinary research and practical examples, it guides researchers, engineers, and industry professionals in advancing intelligent IoT systems.

The integration of Artificial Intelligence (AI) and the Internet of Things (IoT) has the potential to revolutionize urban planning, fostering the development of smart cities. AI enhances urban management through predictive analytics, decision-making, and automation, while IoT enables real-time data collection and monitoring of city infrastructure. Together, they optimize transportation, energy consumption, waste management, and public safety, contributing to more sustainable, efficient, and liveable urban environments. However, the implementation of these technologies faces challenges, including data privacy, security concerns, and interoperability issues. Addressing these barriers is crucial for unlocking the full potential of smart cities. This chapter explores the applications, challenges, and future directions of AI and IoT in urban planning, offering insights into the transformative role of these technologies in shaping the cities of tomorrow.

This chapter discusses the colliding spaces between artificial intelligence (AI) and the Internet of Things (IoT) to form intelligent cities, i.e., cities that are not only smart but also resilient, adaptive, and human-centered. It offers an imaginary and technological vision of the process of the AI-IoT convergence shaping real-time governance, infrastructure optimization, environmental oversight, and participatory urban planning. The chapter encompasses global case studies and ethical analysis, putting forward the necessity of inclusive, secure, and sustainable digital transformation. We also suggest policy directions and implementation strategies that can promote equitable and sustainable urban ecosystems.

In the transport industry, such a revolution revolves around the autonomous vehicles (AVs) industry with a particular unmatched nature, and it can be attributed to breakthroughs in the realms of artificial intelligence (AI) and the Internet of Things (IoT). Appropriate navigation technologies such as the global positioning system (GPS) and inertial navigation system (INS) are reported as some of the most extensive ones in driverless vehicle applications, whereas SLAM is introduced as one of the primary AI-based technologies that allows performing real-time mapping and positioning. Moreover, in the chapter, AI-controlled safety systems which would avoid collisions and help to respond to the emergency are also brought up, finally resulting in smart travel management and the quality of life. The latter discusses the emerging patterns, technological limitations, and ethics on which the future research and implementation processes will develop. The chapter published as a book can be considered referential and practical to researchers, engineers, and policymakers who have the interest of developing the next level of self-driving cars.

The overlap of real-time IoT systems, quantum computing, and the ethical dimensions of autonomous AI is reshaping intelligent automation. This work outlines key technologies like edge computing, low-latency communication, and distributed algorithms that support real-time responses in critical areas such as healthcare, autonomous vehicles, and smart infrastructure. It explores how AI methods, machine learning, computer vision, and reinforcement learning enable systems to adapt and act autonomously. Quantum computing's potential to enhance processing, optimization, and security is also examined, with a focus on its integration into AI-IoT systems. The study addresses challenges like algorithmic bias, opacity, accountability, and the societal impacts of automation. Emphasizing Explainable AI (XAI), it advocates for transparent, trustworthy decisions in high-stakes applications. The chapter promotes inclusive dialogue and responsible governance to align innovation with ethical standards.

The integration of AI and IoT is transforming autonomous systems, enabling data-driven solutions in industries like precision farming, smart irrigation, utility maintenance, healthcare, manufacturing, and transportation. This work presents a framework explaining AI-powered IoT networks and their role in industrial automation. It highlights AI techniques like machine learning and real-time analytics to enhance efficiency. The research examines challenges in AI-IoT integration, such as energy management, security, interoperability, infrastructure limits, and data privacy. Implementation examples show how these technologies transform smart cities and industries, improving decision-making and adaptability. It also explores the societal and ethical impacts of AI-driven innovation, emphasizing fairness and sustainability. Additionally, the text outlines future directions in emerging agriculture technologies, and government policies in farming. By fostering collaboration, it provides insights for experts and engineers, equipping them with tools to develop efficient AI-driven frameworks.

A profound transformation in agriculture is being realized through the widespread adoption of precision farming and smart irrigation systems, leading to a fundamental enhancement in energy management. This chapter provides an insight into modern, smart irrigation technologies, including soil moisture sensors, weather-based controllers, and Internet of Things (IoT) based systems, all of which have been developed to advance water management through improved accuracy in irrigation. Recent advancements have highlighted the crucial role of ICT technologies in fostering the development of smart and sustainable agriculture. Within this context, an overview of contemporary smart irrigation software solutions, including Dacom Irrigation Advice and Irris Scheduling Software, is provided in this chapter. Additionally, the SWAMP project's view, architecture, pilots, and its scenario-based development process are presented. The microgrid is overseen by an energy management system (EMS), the foundation of which is model predictive control (MPC).

Industrial Internet of Things (IIoT) generates vast volumes of data from interconnected sensors and devices, creating opportunities for predictive analytics to enhance operational efficiency and reliability. This chapter reviews key predictive analytics models and techniques applied in IIoT, including traditional statistical methods, machine learning, and deep learning approaches. It explores essential IIoT data sources, infrastructure, and applications such as predictive maintenance, process optimization, and safety management. The chapter also discusses critical challenges including data quality, scalability, security, model interpretability, and integration with legacy systems. Finally, future research directions highlight advancements in edge AI, digital twins, explainable AI, and sustainable IIoT practices.

High-Tech Manufacturing (HTM) is a capital-intensive sector requiring substantial national investments. Integrating Multi-Agent Technology (MAT) into HTM presents new growth opportunities by optimizing processes and enhancing decision-making. This study investigates the impact of MAT on HTM using Multi-Criteria Decision-Making (MCDM) techniques, specifically the Analytical Hierarchical Process (AHP) and Decision-Making Trial and Evaluation Laboratory (DEMATEL). A conceptual framework comprising eight factors and forty-five variables is established. AHP identifies the "high-tech manufacturing process" as the most critical factor, followed by product features, market analysis, and marketing decisions. DEMATEL reveals strong interdependencies, with the manufacturing process exerting the most influence, while quality management, maintenance integration, and organizational factors

N. Jagadish Kumar, SRM Institute of Science and Technology, India
Chin-Shiuh Shieh, National Kaohsiung University of Science and Technology, Taiwan
Mong-Fong Horng, National Kaohsiung University of Science and Technology, Taiwan
G. Elangovan, SRM Institute of Science and Technology, India
K. Rajkumar, SRM Institute of Science and Technology, India

The amalgamation of wearable IoT sensors with QSVM technology is transforming the early identification and ongoing surveillance of Parkinson's Disease (PD). This chapter examines the enhancement of PD diagnostics through quantum machine learning by evaluating real-time kinematic data from various sensors. Traditional models such as Classical SVM, RF, and DNN encounter difficulties due to the intricacy and high dimensionality of time-series sensor data. QSVM, employing Variational Quantum Circuits, optimizes feature extraction and classification, thereby diminishing computational requirements while improving accuracy. In benchmark evaluations, QSVM attained an accuracy of 97.2%, exceeding that of Classical SVM, RF, and DNN, and concurrently decreasing computational time by 35%. The efficacy of QSVM, in conjunction with edge computing and federated learning, has the potential to revolutionize personalized, privacy-centric healthcare. Notwithstanding the constraints of quantum hardware, QSVM presents significant promise in enhancing PD diagnosis and advancing intelligent healthcare.

Sonam Gour, Department of Computer Engineering, Poornima College of Engineering, Jaipur, India

The concept of Internet-connected devices has been around for a while. New research fields, ranging from industrial control to environmental monitoring, have been established as a result of market trends and innovations in the electronics sector. The amount of data travelling worldwide is increasing at an exponential rate due to the growing number of users and connected devices. Mobile phones, computers, and numerous commonplace appliances like air conditioners, TVs, refrigerators, tablets, PDAs, and smart sensors are all immediately connected to the Internet thanks to the widespread adoption of wireless communication. A big, intricate cable network needs a high throughput. New design problems arise from the exponential growth in the number of connected devices and the mixed use of wired and wireless network components. Service providers are searching for ways to enhance

The union of Blockchain and Federated Learning (FL) technologies in the Internet of Vehicles (IoV) domain has ushered in new possibilities for privacy-preserving, decentralized, and cyber-resilient financial use cases. As autonomous and connected vehicles (CAVs) are integrated with edge computing and 5G/6G networks, these smart nodes increasingly engage in real-time financial transactions spanning from usage-based insurance, automated tolls, electric vehicle charging fee payments, to intelligent vehicle leasing and decentralized vehicular identity management. Such a dynamic scenario is best addressed by Blockchain technology, which provides a decentralized ledger system that promises immutability, transparency, and trust, which are perfect for authenticating transactions like insurance claims, micropayments, and contractual fulfilment in mobility finance. This enables IoV stakeholders to build predictive systems with data sovereignty and low latency. Even so, the joint employment of blockchain and FL in financial environments poses important regulatory and ethical challenges.

Preface

The rapid advancement of Artificial Intelligence (AI) and the Internet of Things (IoT) has profoundly transformed the way intelligent systems perceive, learn, and interact with their environments. The convergence of these technologies has paved the way for autonomous and adaptive systems capable of operating with minimal human intervention while continuously improving through experience and contextual awareness. This book, Enhancing Autonomous and Adaptive Systems With AI and IoT, explores the scientific foundations, engineering methodologies, and emerging applications that define this transformative paradigm.

The chapters in this volume bring together contributions from leading researchers and practitioners who address critical aspects such as distributed intelligence, real, time decision, making, edge computing, self, organization, and intelligent sensing. Collectively, these works demonstrate how AI, driven IoT architectures, are shaping the next generation of autonomous systems in domains ranging from healthcare and transportation to smart cities, industry, and environmental monitoring. Beyond technical innovation, the book also emphasizes issues of reliability, scalability, and ethical design, key factors in developing sustainable intelligent ecosystems. By bridging theory and practice, it provides readers with both conceptual understanding and practical insights into designing systems that are not only smart but also resilient and context, aware.

The target audience for *Enhancing Autonomous and Adaptive Systems With AI and IoT* includes researchers, graduate students, and industry professionals working in artificial intelligence, the Internet of Things, robotics, cyber, physical systems, and intelligent automation. It also serves as a valuable reference for engineers, data scientists, and system architects involved in developing adaptive and self, governing solutions across sectors such as healthcare, manufacturing, smart cities, transportation, and environmental monitoring. Policymakers and technology strategists interested in the societal and ethical implications of autonomous systems will also find the discussions in this volume relevant and insightful.

COVERAGE

The many academic areas covered in this publication include, but are not limited to:

1. Artificial Intelligence for Autonomous Systems
2. Adaptive and Self, Learning IoT Architectures
3. Edge and Fog Computing for Real, Time Decision Making
4. Machine Learning and Deep Learning for Dynamic Environments
5. Cognitive IoT and Context, Aware Computing
6. Multi, Agent Systems and Distributed Intelligence
7. Reinforcement Learning for Control and Optimization
8. Human–Machine Collaboration and Interaction
9. Cyber, Physical Systems and Digital Twins
10. Intelligent Sensors and Embedded AI
11. Security, Privacy, and Trust in Autonomous IoT Networks
12. Energy, Efficient and Sustainable AIoT Systems
13. Explainable and Ethical AI in Adaptive Systems
14. Applications in Smart Cities, Healthcare, and Industry 5.0
15. Future Trends in AI, Enabled Autonomy and Self, Organization

TABLE OF CONTENTS

Chapter 1 — Fundamentals of IoT: Sensors, Connectivity, and Protocols in the AI-Driven Intelligent Ecosystem

This foundational chapter provides a comprehensive introduction to IoT architecture, including sensors, communication protocols, and connectivity technologies. It highlights how AI enhances IoT by enabling advanced analytics, automation, and decision-making within intelligent ecosystems. The authors address key technical and societal challenges such as scalability, interoperability, privacy, and security, while examining next-generation technologies such as 5G and quantum computing.

Chapter 2 — Artificial Intelligence: Exploring Machine Learning, Deep Learning, and Neural Networks

Focusing on the pillars of modern AI, this chapter presents an in-depth study of machine learning, deep learning, and neural network architectures. It explores their evolution, applications, and prevailing challenges such as data quality, explainability, and computational efficiency. Emerging paradigms, including Explainable AI and

Federated Learning, are examined as pathways to more transparent, secure, and sustainable AI systems.

Chapter 3 — Transfer Learning for IoT Applications: Explainable AI for Adaptive Systems Integration

The authors investigate how Transfer Learning optimizes IoT applications by reducing data and computational requirements across domains such as smart cities, energy management, and healthcare. The chapter emphasizes the role of Explainable AI in ensuring transparency, fairness, and ethical compliance. It bridges theory and real-world practice, offering guidance for building reliable, adaptive AI-IoT systems.

Chapter 4 — AI and IoT for Urban Planning and Management: Towards Smart and Sustainable Cities

This chapter examines how AI and IoT technologies are transforming urban environments. Real-time monitoring, predictive analytics, and automation support improvements in transportation, public safety, waste management, and energy consumption. The authors address critical barriers including security, privacy, and interoperability and outline future directions for building resilient, sustainable cities.

Chapter 5 — Leveraging AI and IoT for Smart Urban Planning and Management: Intelligent Cities

Expanding on the smart-city theme, this chapter explores the concept of intelligent cities that are adaptive, human-centered, and resilient. Through global case studies and ethical evaluations, the authors propose policy recommendations and strategic frameworks for equitable and sustainable urban transformation. Emphasis is placed on participatory planning, environmental oversight, and real-time governance.

Chapter 6 — Autonomous Vehicles: AI and IoT for Navigation and Safety

This chapter delves into the technological backbone of autonomous vehicles, including machine learning, sensor fusion, IoT connectivity, GPS, INS, and SLAM. The authors discuss safety systems, communication protocols, and real-time decision-making mechanisms. They also explore emerging research directions and ethical considerations critical to the future of self-driving mobility.

Chapter 7 — Real-Time IoT, Quantum Advances, and Ethical Implications in Autonomous AI Systems

Here, the authors examine the intersection of real-time IoT computing, quantum technologies, and ethical governance in autonomous systems. Innovations such as edge computing and low-latency communications are reviewed alongside quantum-enhanced optimization and security. Ethical issues including bias, transparency, and accountability are addressed with a strong emphasis on Explainable AI and responsible deployment.

Chapter 8 — Green Intelligence in Agriculture: Integrating AI-Driven Irrigation, Energy Management, and Sustainability

The chapter outlines how AI-enabled IoT frameworks support eco-smart farming, precision agriculture, and energy optimization. It discusses real-time analytics, automation, and policy considerations, while addressing key challenges such as interoperability, energy constraints, and data security. The work emphasizes sustainability and ethical innovation in agricultural transformation.

Chapter 9 — Precision Farming and Smart Irrigation Systems for Energy Management

Focusing specifically on irrigation technologies, this chapter explores soil moisture sensing, weather-based controllers, and IoT-enabled water management. It presents an overview of leading smart irrigation software and the SWAMP project, detailing its architecture and microgrid-based energy management. The work highlights how precision farming can significantly enhance sustainability and energy efficiency.

Chapter 10 — Predictive Analytics in Industrial IoT (IIoT): Enhancing Efficiency and Reliability

This chapter reviews predictive analytics techniques—statistical models, machine learning, and deep learning—applied within IIoT environments. The authors cover predictive maintenance, process optimization, and safety management, while addressing challenges such as data quality, scalability, and legacy-system integration. Future trends include digital twins, edge AI, and sustainable IIoT frameworks.

Chapter 11 — Impact of Multi-Agent Technology on High-Tech Manufacturing Using MCDM Analysis

Applying AHP and DEMATEL methodologies, this chapter evaluates how Multi-Agent Technology influences high-tech manufacturing. The study identifies key factors affecting production processes, quality management, and organizational performance. The findings show that effective MAT implementation enhances automation, product targeting, and overall manufacturing efficiency.

Chapter 12 — Artificial Intelligence in Detecting and Diagnosing Microbiome-Driven Cancer

This chapter explores how AI technologies analyze microbiome data to support early cancer detection, diagnosis, and treatment planning. Machine learning and deep learning models enable identification of cancer-associated signatures and integration of multi-omics datasets. The authors discuss non-invasive diagnostic tools, challenges in microbiome-based precision medicine, and opportunities for AI-enabled immunotherapy.

Chapter 13 — Quantum SVM for Parkinson's Disease Prediction Using Wearable IoT Sensors

This chapter evaluates how QSVM models, powered by quantum machine learning, analyze high-dimensional sensor data from wearable IoT devices to improve Parkinson's disease prediction. Benchmark comparisons demonstrate significantly higher accuracy and lower computational costs relative to classical models. The authors discuss the promise of QSVM in advancing privacy-centric, personalized healthcare despite current hardware limitations.

Chapter 14 — Real-Time Monitoring of Patients Using IoT Protocols and AI

The chapter discusses the architecture, protocols, and challenges of real-time patient monitoring systems. As connected devices proliferate, issues related to data throughput, scalability, and mixed wired-wireless environments become critical. The authors highlight the need for robust design frameworks to ensure reliable, high-quality medical data transmission in AI-driven healthcare networks.

Chapter 15 — A Deep Learning-Based Lip Reading Module for Forensic Analysis and Silent Speech Decoding

This chapter proposes a computer-vision-based lip-reading system to support forensic investigations, especially when audio data is missing or unreliable. Using deep learning techniques, the system extracts speech information from visual footage, addressing challenges in contextual understanding, privacy, and ethical use. The work offers a powerful tool for silent speech decoding and digital forensics.

Chapter 16 — Policy and Regulatory Considerations for Blockchain and Federated Learning in the Internet of Vehicles (IoV)

The final chapter explores the policy landscape surrounding blockchain and federated learning in vehicular ecosystems. As connected vehicles engage in financial transactions—insurance, tolling, charging, leasing—secure, decentralized solutions become essential. The authors discuss privacy, ethical challenges, and regulatory requirements necessary for deploying blockchain-FL systems in IoV environments.

CONCLUSION

Enhancing Autonomous and Adaptive Systems With AI and IoT has explored the rapidly evolving intersection of Artificial Intelligence (AI) and the Internet of Things (IoT), highlighting how these technologies collectively enable intelligent, adaptive, and autonomous systems across diverse domains. The book began with foundational concepts, including IoT sensors, connectivity protocols, and AI techniques such as machine learning, deep learning, and neural networks. It then progressed to advanced topics such as transfer learning, explainable AI, and real-time data processing, emphasizing their role in adaptive decision-making and resilient system design.

Subsequent chapters illustrated practical applications, from smart urban planning, healthcare, and autonomous vehicles to precision farming, energy management, and industrial IoT. Specialized discussions on safety monitoring, predictive analytics, gesture control, microbiome-driven cancer detection, Parkinson's disease prediction, and silent speech decoding demonstrated the transformative potential of AIoT solutions in real-world scenarios. The book also addressed emerging technologies, including quantum computing, federated learning, and blockchain, alongside ethical, regulatory, and policy considerations crucial for responsible deployment.

Throughout, emphasis has been placed on integrating theory with practice, showcasing how AI and IoT frameworks improve efficiency, safety, sustainability, and human-centric design. By uniting cutting-edge research, technological innovation, and application-driven insights, this book serves as a comprehensive resource for researchers, engineers, and practitioners aiming to develop the next generation of autonomous and adaptive systems. Ultimately, it underscores the promise of AI and IoT as catalysts for intelligent, resilient, and ethical solutions across industries and society.

Acknowledgment

This work was supported by the Deanship of Scientific Research, Vice Presidency for Graduate Studies and Scientific Research, King Faisal University, Saudi Arabia [Grant No. 254442]

Chapter 1

Fundamentals of IoT: Sensors, Connectivity, and Protocols in the AI-Driven Intelligent Ecosystem

S. Anusha
Easwari Engineering College, India

S. Vidyasagar
SRM Institute of Science and Technology, India

V. Kalyanasundaram
SRM Institute of Science and Technology, India

K. Saravanan
SRM Institute of Science and Technology, India

L. A. Anto Gracious
https://orcid.org/0000-0003-2386-9182
R.M.K. College of Engineering and Technology, India

M. Nalini
https://orcid.org/0000-0001-8987-3120
SA Engineering College, India

ABSTRACT

This chapter explores the fundamental components of the Internet of Things (IoT) within the context of an AI-driven intelligent ecosystem. It discusses the role of IoT sensors, connectivity, and communication protocols in enabling seamless interaction between devices, networks, and intelligent systems. The integration of AI into IoT systems is highlighted, emphasizing its potential in enhancing data analytics, decision-making, and automation. The chapter also addresses key challenges in scalability, interoperability, security, and privacy, while outlining emerging trends such as quantum computing and next-generation communication technologies like 5G. The future of IoT and AI convergence offers significant opportunities for creating

DOI: 10.4018/979-8-3373-3146-1.ch001

smarter, more efficient environments in various industries, from healthcare to smart cities. This chapter concludes by discussing the transformative potential of IoT and AI for shaping the future of intelligent ecosystems.

INTRODUCTION

1.1 Background of IoT

The IoT can be described as the use of technology to connect objects, systems and devices in the world around us in such a way that they can all share information and in some cases, make decisions on their own. This is not just limited to the machine type communication but is used in various fields such as health, agriculture, manufacturing and smart city technologies(Deshmukh et al., 2023). The best definition to describe IoT is the integration of sensors analysis, processors, and software into physical objects to support the exchange of data in real-time, with the view of improving decision-making the efficiency of a business operation(Khang, 2023; MAP et al., 2025).

The development of IoT technologies has been significantly linked to the progress in several fields such as wireless communication, sensors, and cloud computing. Initially, IoT was used only for basic control and monitoring of machines and devices. Yet as the internet access was no longer rare and the prices for sensors and computational capabilities were brought down, IoT started to grow. Current IoT systems are enabled to handle big data and deliver information that was not easily conceivable before. With the advancement in technology, numerous devices that can sense changes in the environment such as humidity, motion, among others, have become part and parcel of human life and business activities.(Kashyap, 2019)

IoT has been supported by several technological factors such as the wireless communication standards (e.g., Wi-Fi), the cloud computing for the storage of large data, and the AI for smart data processing. These technologies have made the development of advanced, effective and manageable IoT systems, which are considered as the key enabler for digital transformation in various sectors(Kumar et al., 2020).

1.2 Importance of IoT in the AI-Driven Intelligent Ecosystem

In the intelligent ecosystem that is characterized by the use of AI, IoT serves as a key component by feeding data into the AI systems. AI entails the use of large datasets to train algorithms and make forecasts and improve the organizational workflow. IoT devices due to the fact that they are constantly gathering information from the physical environment act as a constant and real-time data feed for AI systems.

It was considered as raw data such as GPS, motion, humidity, pressure, light, and temperature, which are the base of AI algorithms(Karthik et al., 2023).

Among all the benefits IoT brings to the AI-powered systems, one can pinpoint the key aspect of predictive analytics. In different fields, IoT sensors can detect the status of the machines, vehicles, or structures, and their condition in real-time. The data can also be used by AI algorithms to forecast future occurrences such as equipment breakdown, traffic jam and so on and thus necessary actions like equipment repair, avoiding traffic congestion can be taken. This change from reactive to predictive and subsequently adopting preventive type decisions is fastest changing industries through cost cutting, optimization and safer operations.

However, IoT and AI are not only used as complementary in the context of data acquisition and processing. AI provides an added value to IoT systems by making them more self-sufficient and responsive. This is because IoT devices are capable of advanced processing and analyze Real-time data and perform anomaly detection and even make intelligent decisions without the assistance of humans by the application of ML models and AI techniques. For instance, in the smart homes, variable like thermostats and lighting systems through implementing AI can learn the usage inclination of dwellers and adjust accordingly thus providing a more personalized environment(Krishnamurthi et al., 2019).

The relationship between IoT and AI does not end with smart devices, but it goes further to the intelligent environment where various devices and systems work in unison. For example, in a smart city, the IoT, machines, and devices such as the traffic light, road, and vehicles can inform on traffic density, air quality, and transport services. This data can be processed by the AI systems in order to control traffic lights, recommend the most effective route to drivers, or regulate the consumption of energy in order to minimise emissions. In healthcare, smart devices like health monitoring devices can be used to monitor the health of the patients while artificial intelligence can be used to study the data collected by the health monitoring devices in order to diagnose the health complications at an early stage and advise the patients on the best way to handle the situation.

This integration makes a system more effective, flexible and sensitive to the needs of the people and the quality of services offered to the society.

1.3 Objective of the Chapter

This chapter aims to explore the IoT basics by identifying the most important aspects of the IoT ecosystem and the elements of sensors, connectivity, and communication protocols, as well as their importance in an AI-based intelligent environment. The chapter also has the goal of offering detailed information on the integration of IoT technologies to support intelligent systems that can analyze data and make decisions in different fields.

In order to address these issues, the chapter will discuss the different categories of IoT sensors, their function in data acquisition, and their input to the development of value-added information for AI systems. Sensors are the entry point of the data acquisition process in IoT and it's important to consider the various forms of IoT sensors as well as their functions to properly assess their role in the system. Moreover, the chapter will explore the different connectivity solutions of IoT devices covering short-range communication as well as long-range communication protocols including Bluetooth and 5G respectively. Connectivity is one of the most important factors that influence the transmission of data from one device or platform to another.

Also, the chapter will discuss various communication standards that can be adopted in IoT systems to enable the devices to interconnect & communicate seamlessly in order to exchange information in a secure and efficient manner. From the analysis of these aspects the chapter will elucidate the relations between IoT devices as they are integrated in an AI environment, allowing intelligent systems to work and bring value in various application domains.

In conclusion, this chapter will discuss about the importance of IoT in the new intelligent ecosystems based on AI, the problems, the prospects and the future trends in the field. The amalgamation of IoT with AI is expected to change industries and open new horizons for enhancing efficiency, developing new technologies and better enhanced living and working conditions.

2.IoT SENSORS

2.1 Introduction to IoT Sensors

Sensors are defined as the most basic components of the IoT as they are the devices that capture information from the physical environment and encode it into electronic signals that can be processed by IoT networks. The primary purpose of IoT sensors is to record environmental or physical features like temperature, humidity, movement, pressure, light, etc(Solanki et al., 2019).

These sensors are the receptors of the devices and systems by which the devices and systems are able to feel and communicate with the environment. Sensors are therefore a crucial component of the IoT since it would be impossible for the devices to sense the environment in which they are placed.

It is imperative to acknowledge the need for sensors in IoT. They underpin the first building block of data acquisition as the key activity in smart environment decision-making. For example, temperature sensors are used in smart homes for control of the heating and cooling systems whereas the motion sensors are used for the detection of occupants which in turn can control the lighting or security systems. Also, the IoT sensors are widely used in the predictive maintenance, monitoring and analyzing the environmental conditions, diagnosing the health of patients and many others applications that necessarily base on the accurate and real time data collection.

There are different types of IoT sensors as they are categorized based on the physical or environmental value that they are meant to track. Some of these sensors are temperature sensors, which are used to measure temperature changes, humidity sensors that are used to measure the moisture content of the environment, pressure sensors that are used to measure the pressure that is exerted on surfaces, motion sensors that are used to measure motion and proximity sensors that are used to detect the presence of objects. There are also specialized sensors for gases, air, pH, sound and others. All type of sensors plays a specific part in IoT applications and make a significant contribution to the IoT system intelligence(Singh & Singh, 2023; Sathya et al., 2024).

2.2 Sensor Technologies

There are several types of sensors in IoT systems and the most basic classification is based on the analog and digital sensors. Analog sensors provide output signals that are smooth functions of time and can have any value within a range. These sensors create an analog voltage or current signal and their output directly corresponds to the physical quantity to be measured. For instance, a specific temperature sensor can creates steady voltage signal that increases proportional to the temperature. However, the analog signals are sensitive to noise and therefore need to be converted before they can be fed to digital systems(Tripathy & Anuradha, 2017).

On the other hand, digital sensors generate digital signals that are most often analog to digital where the output is in binary form and can either be on or off or 1 and 0 respectively. Digital sensors are normally more accurate and dependable because they have less noise compared to analog devices. Also, these digital sensors can be easily connected to microcontrollers or computers because they do not require conversion circuit after sensing the signal or data.

Figure 1. **IoT Sensors**

One of the most commonly used sensors in IoT is MEMS which stands for Micro-Electro-Mechanical Systems. MEMS sensors are small devices that contain mechanical components, sensors, actuators and electronics on a single chip.

These sensors are small in size and also relatively cheaper and consume less power; a factor that makes them suitable for IoT devices. There are different types of MEMS sensors, such as acceleration, pressure, temperature, and motion sensors. Due to their small size, they can be installed in various types of devices, from wearables to industrial devices, thus, making it possible to create compact IoT systems.

Other important innovation in sensors is smart sensors. Smart sensors have a form of local processing where they are able to process and analyze the collected data to some extent before forwarding it to the main system. They are capable of functions such as signal conditioning, self-calibrating and in some cases, simple data analysis. Smart sensors also incorporate intelligence so that it reduces the amount of computation that requires other computing devices and also the amount of data transmitted to other devices hence increases efficiency of the system. Moreover, they may include self-testing functions that will enable the identification of faults and malfunctions, increasing dependability and minimizing the expenses on maintenance (Venkateswaran et al., 2023).

2.3 Sensor Data Collection and Processing

After IoT sensors gather data, the next important stage is data acquisition as well as data processing. Data acquisition is the method of collecting data from the sensors and converting them to binary form if required and then sending them to a processing unit like microcontroller or cloud server. This must be very efficient to enable fast and real-time data capturing in the case where actions have to be taken immediately like in the case of self-driving cars or industrial use of the technology.

One of the challenges that can be encountered in many IoT systems is the fact that the data provided by the sensors can be immense especially when the system is connected to numerous IoT devices. Thus, the processing of the sensor data is one of the critical issues in the IoT systems. One way of addressing this problem is through edge computing that involves computation of data near its source. Edge computing also makes it possible to minimize the amount of data that is sent through the network and minimize latency and bandwidth. Also, it promotes real-time decision making because data collected can be processed and decisions made at the grassroots level without having to wait for instructions from a central server.

Edge computing is most effective in cases where IoT devices have to work independently or in the cases where data processing cannot wait, for example in autonomous cars, smart factories and health monitoring systems.

For instance, in smart factory, there are sensors installed to monitor various aspects of a machine, through edge computing the data that indicate the possibility of breakdown can be predicted and maintenance request made before the machinery fails.

2.4 Applications of IoT Sensors

Internet of Things sensors can be applied in many sectors, as they are mainly used for developing new products and enhancing productivity. In the health care system, Wearable technology encompasses the devices placed on the human body that gather data of vital metrics like pulse rate, etc. This data is transmitted to the healthcare givers who in turn monitor the health status of the patient in real-time and identify any irregularities that may require change in treatment. For instance, IoT sensors in diabetic patient's glucose monitor keep monitoring the blood sugar level to avoid complications and enhance the quality of the care delivered.

In the farming industry, smart sensors are applied in matters to do with moisture, temperature and weather conditions that are ideal for plant growth. These devices assist farmers in decisions relating to the use of water in irrigation, the use of fertilizers and pest control and thereby increase productivity and efficiency. In the same way, the intelligent cities have incorporated the use of sensors in traffic, air quality and public infrastructure. For instance, intelligent traffic lights employ motion and occupancy sensors to manage the traffic light as per the movement of cars, on one hand, and air quality sensors to monitor the level of pollution and to notify officials of the change, on the other hand.

In industries, IoT sensors are useful in predicting equipment failure and management of assets. Smart sensors installed in the machines and equipment help in the tracking of its performance and signs of failure, to notify the operators to take necessary action that prevents breakdowns, hence increase the productivity. Also in supply chain, IoT sensors are applied to monitor inventory, maintain temperature-sensitive products, and guarantee the product quality.

Overall, sensors play a crucial role in IoT systems being the source of data on which decisions are made in various industries. As the IoT sensors continue to develop new sensors, data acquisition, and data analysis technologies, it remains evident that IoT sensors are the driving force behind connected systems that offer more intelligent and efficient solutions in different sectors.

3. IoT CONNECTIVITY

3.1 Introduction to IoT Connectivity

The most crucial component in an IoT system is the connection which facilitates the flow of information between devices over a network. In IoT systems, the data exchange between the devices, sensors and servers is very important for monitoring, analysis and decision making.

Communication is at the core of IoT as it enables sensors to send their collected information to the processing centers or the cloud, and IoT devices to act on inputs received based on the processed information. Without proper connectivity solutions, IoT systems would barely operate efficiently, and the horizons of those technologies would remain unutilized.

Connectivity is a crucial aspect in IoT since it defines the way data is exchanged between devices, different systems and users. In most IoT systems, data has to be transmitted over different distance some of which require high speed, some require a large range and some require higher bandwidth. The IoT applications can be of high bandwidth kind for tasks like video surveillance or low bandwidth and low power for monitoring temperature and motion. Therefore, the connectivity solutions that can be employed in IoT devices are important in addressing the need for different applications regarding IoT such as smart homes & industrial processes.

Figure 2. **IoT Connectivity**

3.2 Types of IoT Communication Networks

The most suitable choice of communication network in an IoT system depends on the range, energy consumption, the rate of data transfer, and reliability required for communication in an IoT system. These may be categorized as short-range and long-range communication technologies, although they have their strengths and weaknesses.

Wireless short-range communication networks are applicable in IoT applications that involve short distances and need high data rate. Popular examples of these are Wi-Fi, Bluetooth, Zigbee and Z-Wave. Wi-Fi is one of the most commonly used technologies in IoT devices for home automation since it is fast and stable but consumes relatively more power than other technologies making it less appropriate for use in devices with batteries. Bluetooth is widely used in IoT devices especially Bluetooth Low Energy (BLE) because of its power efficiency and ease of implementation and this makes it suitable for applications such as fitness monitors, smart watches and home automation. Zigbee and Z-Wave are two technologies used in home and building automation, low power and low data rate, which supports mesh network, to extend the coverage of the network.

On the other hand, low power wide area networks are intended for devices to have long-range communication that may function at low power and low data rates. These networks are LoRa (Long Range), NB-IoT (Narrowband IoT), and 5G. LoRa is widely used for IoT applications in rural and outdoors as it has long range and low power consumption and is ideal for smart agriculture, smart cities, and environment monitoring.

NB-IoT is a cellular technology most suitable for low data rate, large number of connections that need coverage in areas where connectivity is a challenge, such as smart metering and asset tracking. There are expectations from the 5G networks to provide ultra-reliable, low-latency and high capacity IoT networks which in turn will popularize applications such as autonomous automobile and smart manufacturing.

Apart from the terrestrial networks, satellite and LEO networks are useful for IoT applications especially in the regions that are not easily accessible. These networks are suitable for industries such as agriculture, logistics, and environmental monitoring since the IoT devices should work in areas that do not have conventional connectivity. Satellite IoT solutions are used in applications such as monitoring of structures, tracking and monitoring of wildlife or vehicles, to mention but a few, where communication does not require physical connection.

3.3 IoT Network Architectures

Architecture of an IoT network is the manner in which the devices are connected and the pattern of communication between these devices. The type of the network affects aspects being data flow, dependability, expandability, as well as energy utilization efficiency.

Some of the general network topologies used in IoT systems are star, mesh, and hybrid networks. It is a type of network topology in which all the units are connected to a central point, which is a hub and can transmit data between devices. This is a straightforward and deprived kind of architecture made use of usually in home automating systems in which light bulbs, thermostats and many others are self-governing and connect specifically to the master control panel. However, if the central hub is lost then the network of devices is unable to communicate and this hampers the network.

Mesh networks are slightly more flexible and reliable than other networks, and can also be used for communication. In this configuration, the devices are connected where one device can forward the data to other connected devices. This enhances the reliability and the coverage of the network as the data will find its way through the different paths. Mesh networks are especially applied in such large-scale IoT applications as smart cities and industrial automation, as devices have to cover a vast area and remain connected even if some of them are out of range or malfunctioning. Two of these technologies include Zigbee and Z-Wave, which employ mesh networking for applications that require such features as reliability and scalability.

A hybrid topology is a combination of star and mesh topology and it has the advantages of both the topology types; the organization of devices is maintained in a central point while the data can travel through several paths.

Many IoT applications may require a combination of these features, and as such, hybrid networks are quite versatile.

Other important factors in developing IoT network architecture include the device-to-device communication and the device-to-cloud communication. In device-to-device communication, the devices are connected directly with one another and is often carried out through short range technologies such as Bluetooth or Zigbee. This type of architecture is well fitting for the cases when devices have to transfer data or act in parallel, for example, in smart homes where lights, thermostats, and security systems should function as separate entities but be interconnected and work in parallel. On the other hand, device-to-cloud communication means that the data from the devices is sent to a cloud server for processing and analysis. This model is often used in the application such as supervisory control and data acquisition, condition monitoring, and data gathering, analysis and processing where large volumes of data are generated with high volumes and need to be collected, analyzed and forwarded to the central stations.

3.4 Challenges in IoT Connectivity

Although IoT connectivity has brought significant improvements in different sectors, there are several issues related to the development of stable IoT connections. The first one is the bandwidth issue. Due to this, IoT devices produce large volumes of data and transmitting these data in the network can be very resource intensive especially in the likes of video streaming or real time data processing. A major challenge in dealing with large-scale IoT systems is to provide enough bandwidth to allow for the implementation of the large number of devices which does not overload the network.

Two of the fundamental issues are latency and reliability. When it comes to IoT solutions, many of them especially in such fields as healthcare and automotive need to either update or collect information in near real time. When the latency or reliability of the communication is poor, it means that decision-making is also affected, which is disastrous for certain applications such as emergency response systems or monitoring of critical infrastructure.

Cost is another issue in Internet of Things connectivity especially on the energy required by battery driven devices. LoRa and NB-IoT are expected to be low power technologies but managing power consumption for IoT devices when deployed in large quantities and still ensure that they are connected is challenging.

In summary, security and privacy of communication in IoT are emerging challenges. Since IoT devices are designed to gather and transmit data, they are at high risk of being hacked. Encryption, authentication & access control are some of the ways of ensuring that transmission of data is secure and unauthorized access is prevented. IoT networks should also have measures in place to counter threats of denial of service, data theft and other acts that may lead to system compromise and interruption of service.

In this context, the selection of the communication networks, network architecture to be used, and how the band-width, latency and power consumption issues as well as security will be addressed will define the success and mass adoption of IoT across industries.

4.1 Introduction to IoT Protocols

Since the IoT systems involve the integration of various devices and appliances, communication protocols are crucial to facilitate the flow of information between the devices. This is because they provide guidelines on how devices should communicate, transfer data and even interact over networks. These protocols are crucial in the ability of the devices, that may be developed by different manufacturers and used in various conditions, to interconnect. In addition, there are certain communication protocols that are employed regarding to Internet of Things; they are used not only data transmission, but also for interoperability, security, and data flow management in IoT application. This is because as more IoT systems are developed with billions of connected devices, it is essential to have well-established and efficient communication protocols for IoT systems.

IoT protocols can be divided according to the OSI model of networking at the application layer, the transport layer and the network layer. Each of them has been developed to address the needs of IoT applications, for example, low energy consumption, low bandwidth, timely data transfer and security. These protocols are therefore important in defining the functionality, reliability and security of IoT systems and they are important in making sure that IoT networks can support various applications that range from smart homes to industrial IoT.

4.2 Standard IoT Communication Protocols

There are several communication protocols that have been adopted due to their effectiveness in IoT systems, flexibility, and reliability in handling IoT devices. The most widely used protocol is MQTT or Message Queuing Telemetry Transport, lightweight messaging protocol. MQTT is intended for use in such conditions as low bandwidth and high latency, so it is perfect for IoT devices that have to send small and frequent messages. The publish-subscribe model benefits the system because it only delivers messages that the clients need, and thus cuts down on excessive traffic. MQTT is particularly used in the following applications: remote monitoring, smart homes, and industrial IoT systems.

Another protocol is CoAP, which is a web transfer protocol that is used in low power and low bandwidth devices. CoAP is similar to HTTP but has been designed for use in resource-scarce devices and networks.

It uses UDP (User Datagram Protocol) instead of TCP, and this makes it suitable for IoT devices that have limited computational power as well as memory. CoAP is particularly applied in places where devices should exchange the information in real-time, for instance, smart cities, healthcare, or environment monitoring.

HTTP/HTTPS the standard for web communication is also used in IoT systems especially when a device is required to connect with cloud service or web applications. Despite the fact that HTTP/HTTPS is not particularly efficient for IoT when compared to MQTT or CoAP, the latter is still used due to its universality and seamless integration with web technologies. In IoT systems where security is an issue, the use of HTTPS the secure version of HTTP has the capability to encrypt data during transfer.

Figure 3. **IoT Protocols**

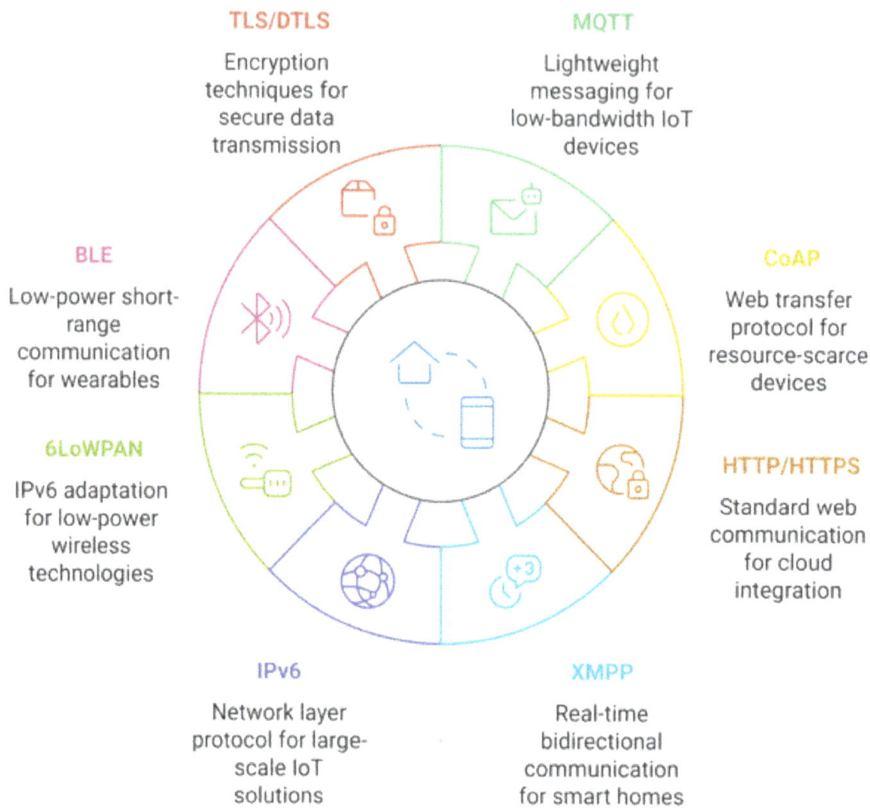

Another protocol that has also been adopted for use in the IoT applications is known as the XMPP (Extensible Messaging and Presence Protocol). Developed primarily as an application for instant messaging, XMPP is easily adaptable to IoT applications because it supports real time bidirectional communication between devices and users. XMPP is used in smart home automation, remote control of devices & monitoring systems where real time updates or status are critical.

4.3 Network Layer Protocols

At the network layer, IoT protocols that are involved in managing the connectivity of the devices, addressing of the devices as well as the routing of the data within the network are key important. IPv6 (Internet Protocol version 6) is the current network layer protocol used in IoT mostly because of the increased number of devices to be connected. IPv6 also has the capability to address individual devices around the world, this is important for IoT solutions at a large scale. It makes sure that number of IoT devices can be assigned unique IP address, so that they can directly communicate with each other and also enhances the scalability of IoT networks.

One of the technologies based on IPv6 is 6LoWPAN, a compression and adaptation layer that can be used to allow IPv6 to be used over the low power, low bandwidth wireless technology, and it is optimized for Zigbee and Bluetooth Low Energy (BLE) technologies, and it is used to support IoT devices with limited resources to be able to communicate over large distance while at the same time consuming minimal power. This makes it suitable for smart homes, industrial IoT, and agriculture where devices should work independently and cover a large area with low power consumption. Another network layer protocol in IoT is BLE.

BLE is designed for low power applications for short range where the devices involved have to exchange small packets of data with little energy. It is mostly applied in wearables, smart home devices, and health monitoring systems where longevity of the battery and stable connectivity are the priorities. The fact that BLE can sustain connections for extended durations while consuming little power makes it suitable in IoT systems in health and fitness monitoring and home automation.

4.4 Security Protocols in IoT

Security is one of the most important concerns of IoT communication since IoT is vulnerable to different threats like unauthorized access, interception of received data, and attack on the device's integrity. To this end, several measures have been put in place to enhance security of data exchanged through IoT networks with regards to confidentiality, integrity and authenticity.

Encryption is one of the most basic mechanisms that are applied in an attempt to secure data in transit. It makes certain that any intercepted data is not comprehensible to unauthorized persons. TLS and DTLS are the most commonly used encryption techniques in IoT. TLS is mainly applied to TCP-based connection, for instance, HTTP or MQTT, while DTLS is employed to UDP-based connection, which is widely used in IoT applications like CoAP.

Another critical aspect of security in IoT systems is usually authentication. It also checks that devices connected to the network are genuine and are allowed to be connected to the network. Some protocols such as the Internet Protocol Security (IPSec) is used for authenticating and encrypting the IP packets in the IoT networks so as to have security at the network level.

Other security measures that are important in IoT include access control protocols. These policies limit the access of confidential data and resources to certain devices and users in accordance with the set policies. Security policies are usually implemented in IoT systems by two or more methods like the role-based access control or attribute-based access control.

4.5 Emerging Protocols and Trends

The IoT systems are growing more complicated and the variety of IoT devices is expanding, that is why new protocols are being developed in the world. One of them is the appearance of artificial intelligence and machine learning to create the AI-enhanced protocols that can flexibly adjust the communication process according to the current circumstances(Ponmalar et al., 2024). These can include protocols that can enhance the flow of data, the utilization of bandwidth and the flow of the network by analyzing traffic and changes in the environment.

Another trend in IoT communication is the integration of 5G technology 5G is expected to deliver ultra-reliable, low-latency, high-speed connectivity which is a key requirement to support many new IoT applications such as automobile industry, factory automation, smart city, etc. It has the potential of handling a large number of connected devices with low latency which will help IoT to expand more by adding new applications which require high volume and real-time data transmission.

In conclusion, it can be stated that the communication protocols are the core of the IoT infrastructure that allows devices to connect and communicate in a wide range of applications. From the application layer extending to the network layer, there is a range of protocols that have been developed to suit IoT devices and applications. With the evolution of IoT systems, new trends and protocols that are being developed include; AI based protocols and IoT along with 5G.

5.1 The Role of AI in IoT

AI is becoming more prominent in the IoT landscape, as it pertains to data processing and utilization in particular. AI with IoT can be integrated to develop smart systems to take decisions, to perform various functions with specific intelligence in numerous sectors. AI has been also incorporated into IoT through data analytics and machine learning. AI is used in IoT to process large amounts of data from different sensors or systems of IoT devices. Such Machine learning algorithms are capable of pattern recognition and analysis, that is, they can predict and find out more information from data which is raw information(Prabhu et al., 2025). This makes IoT systems more efficient and intelligent to adapt to the data they have collected, and give users better understanding and details of the users, the environment and the operations.

Decision making and automation are inherent to IoT systems, especially in areas that require timely actions and prevention measures. In industrial setting, AI can assist in carrying out routine tasks out on the basis of data gathered from the IoT sensors in order to monitor and manage equipment or facility. This can result in improved efficiency, time-saving and safety. In smart homes, the IoT devices can develop habits regarding the user's preferences as well as controls the illuminative, warmth, and security systems independently.

Thus, with the help of AI the IoT devices can be imbued with more capabilities to be smarter and handle sophisticated situations independently without any inter-linkage with the main framework of the IoT network.

5.2 Edge AI vs. Cloud AI

It is crucial to differentiate between Edge AI & Cloud AI when it comes to integration since the two have different roles in how they perform computing in IoT systems.

Edge AI is the process of performing data analytics and processing of data using AI algorithms and tools on the devices or at the network periphery. This makes it possible to make decisions on the fly with a very low latency, a characteristic that is essential in areas such as self-driving cars, health and medical care and the manufacturing floor among others. Edge AI is the kind of AI that is implemented on the device, which enables real-time processing of data and makes decisions without the need to transfer huge amounts of data to a cloud server. This can lead to a considerable increase in system performance, decrease in required bandwidth, and increase in privacy since some data processing does not have to be done on the internet.

In contrast, Cloud AI involves manipulating and analyzing data on other servers that are located on the cloud, and can use huge computing power to perform various computations and run intricate algorithms related to machine learning. Cloud AI is best suited for the analysis of large datasets and training of models that need a lot of computational power, for example, training of models based on historical data. In IoT ecosystems, cloud-based AI can be used for functions such as predictive maintenance, analytics, as well as training of machine learning models using data from many connected devices. While Cloud AI has the capability to handle large data and their management, it usually comes with higher latency compared to Edge AI and needs a good connection for data transfer.

Edge AI and Cloud AI both have their roles in IoT and many IoT applications employ both levels in some way. Edge AI allows for a decision-making level of latency while Cloud AI caters for more complex tasks such as data gathering, training, and analysis for extended periods. This integration of the two approaches makes it easier to have a well-coordinated and efficient AI-based IoT environment.

5.3 AI-Based IoT Security

Security is one of the biggest threats in IoT because as more devices connect to the network, the more risks are generated. The conventional security mechanisms may not be effective enough to ensure the IoT networks from the modern cyber threats, and that is where the AI based IoT security comes into the picture.(LA, 2026)

There is data generated by IoT devices and there is also data from the network that AI can look at and determine if there are signs of intrusion. For instance, use in threat detection for example, an excessive data transmission or IoT device interacting with a host that it was not in contact with before the attack or is sending a large volume of data possibly indicating that it has been compromised by malware. Once an abnormality is identified, AI can invoke prestored actions like removing affected appliances or blocking a malicious traffic flow, thus conserving recovery.

Furthermore, it is being applied in predictive maintenance and anomaly detection to avoid cases where a system or security is attacked before it can be fixed. AI can also predict when a device might fail or when a component is likely to be at risk through constant tracking of the IoT devices and networks performance. This makes it possible to minimize the time when an IoT system is not operational and guarantees the safe functioning of IoT systems. For instance, in industrial structures, AI can observe the condition and functioning of the machinery and notify the maintenance personnel when the system is likely to fail, which is very costly and dangerous to the system.

The inclusion of AI in IoT security is a positive occurrence, because it allows for better experience in attacks, understanding of new threats and more self-sufficient operation, with less dependence on human input in the detection and prevention of security threats.

5.4 Case Studies of AI and IoT Integration

AI and IoT are already integrated in several industries, and this integration has shown several areas where it can revolutionalise the existing systems.

In smart homes, artificial intelligence and the Internet of Things are inseparable concepts that improve the living standards. Smart systems are capable of adapting to the user's behavior and adjust the lighting, temperature and security systems accordingly to suit the most comfortable and energy efficient setting. For instance, the smart thermostat is designed to take an example from the homeowner on how it should regulate the temperature with precise control such that energy is conserved besides focusing on the climate, time of the day and the number of people using the house.

In healthcare, the combination of AI & IoT has promoted the development of remote care and monitoring, individualized medicine, and forecasted healthcare.

Smartwatches and medical sensors can monitor the patient's vital signs on a regular basis, and AI can interpret the results to detect early signs of illness. In the critical care area, AI can serve as a useful tool for doctors by helping the doctor to monitor the condition of the patient, suggest the best treatment to be used and even the likely outcome of the patient based on the records kept by the AI.

AI in IoT is now being applied in industrial automation where it is used in the prediction of maintenance and real-time optimization of manufacturing plants. IoT sensors are fitted on the machinery to check on its performance and pass this information to AI systems to determine when the equipment will need repair and service hence increasing efficiency. Moreover, it can also control the production lines in terms of parameters and make necessary changes in real-time so that cost can be cut and the quality of production can be improved.

The last is the integration of AI and Iot in smart cities which is actually helping to change the way people live in urban areas through traffic flow, waste disposal, use of energy and security. Through consistent monitoring, sensors involved in the IoT within the city capture traffic patterns, air quality, energy utilization and other important factors, enabling AI algorithms to manage traffic signals, efficient waste disposal and energy patterns in the city in real time. This is because it leads to less congestion, lower emission of carbon, and increased sustainability of the urban structure.

In conclusion, there has been enhancement in the applications of integrating an AI in the IoT systems and this has made the tech systems intelligent and self-directed across the different industries. These applications of AI in IoT have made IoT ecosystems to become smarter and more capable of operating at the edge, making real-time decisions, being secure, and improving the quality of life across the world.

6.1 Scalability and Interoperability Issues

This is especially a major issue in the development of IoT networks as IoT systems are becoming larger and complex. The number of connected devices is set to grow to trillions over the next few decades, and this poses concerns on how these devices will be controlled, monitored, as well as managed. One of the most significant challenges is, therefore, data management in such a huge number of connected devices. With the rise of big data, it becomes cumbersome to store, process & analyze the data in real-time, which puts a lot of pressure on the networks and CPU. The development of IoT networks also presents problems such as congestion, bandwidth, and latency issues which affects the IoT systems.

However, there are two main problems: scalability and compatibility of IoT devices and platforms. It is for this reason that IoT devices are developed by different manufacturers with the use of different technologies and standards. This is because, to achieve data interchange between these devices, there are certain issues that are encountered. Wireless devices may employ different modes of communications, may use different formats of data transfer, or may have different power supply standards, making it virtually impossible to incorporate the devices in a common system. This absence of standardization becomes a major problem in the development of the IoT landscape as it hampers the capability of the devices to interconnect and interoperate with other devices as well as with higher level systems and applications like the cloud and AI. It also becomes challenging to interconnect IoT systems across various industries because there are no global standards for system interfaces, and organizations face the challenge of working with devices that are incompatible with others.

To this effect, the following strategies are recommended; establishment of a common language of communication, increased standardization, and the use of intelligent interfaces to link the different systems. With such IoT ecosystems being developed, it will be important for such systems to be able to scale up while at the same time being interoperable with other devices as a way of promoting the adoption and deployment of IoT systems.

6.2 Security and Privacy Concerns

Security & privacy issues are among the most critical issues that IoT systems are currently facing. With IoT devices being more widespread and the integration of these devices being more intensive they pose new threats. Most of these devices capture personally identifiable information for wearables, health, smart homes, or production line from factories. The breach of any of these devices results in negative consequences including invasion of privacy, unauthorized access to networks and in some other cases physical harm like self-driving cars and other industrial equipment. The threats in IoT networks and devices come from various aspects, such as poor authentication, poor encryption, and the constraint on the computational capability of some IoT devices. Additionally, most of the IoT devices do not receive frequent updates and therefore are vulnerable to attacks that are already known in the system.

Besides security risks, privacy concerns have emerged as a critical factor given that IoT devices may obtain an individual's personal information. The constant accumulation of data from IoT devices leads to questions as to how this data will be managed, stored, and disseminated.

People may lack knowledge about how much data is being collected or even do not have any control over how their data is being shared or utilized by other individuals or organizations. This lack of openness can result in extensive privacy infringements, people's data being used for business or criminal gains. The problem is only worse when it comes to IoT systems where the data may be collected and stored in different platforms and devices, making it easier for the hackers to invade the privacy of the users.

To deal with such issues, IoT systems should employ better security measures, such as encrypting data, using multiple-factor authentication, and ensuring that the system's software is safe from hacking attempts. In the same regards, technologies such as data masking, edge computing, and data consent can be employed to protect the privacy of users. Regulations like the GDPR are also helping in imposing privacy standards, so that IoT developers and manufacturers give importance to the consent of the user and data protection. These security and privacy issues will remain important as IoT networks grow in order to make IoT technologies accessible and acceptable by the users.

6.3 Emerging Trends and Future Research Areas

The internet of things has been growing at a fast pace and has received a lot of credit for the innovations that are being made in the market. Among the most promising trends that has been identified in IoT, there are the developments in the sphere of sensors, the usage of Artificial Intelligence, and the transition to the new level of wireless communication – 5G.

These are the developments in the field of sensors that allow IoT devices to gather more precise and various information. The new sensors are better characterized by higher sensitivity, increased energy efficiency, and miniaturization, which enables the constant tracking of different environmental, health, and industrial conditions. These have create opportunities for new IoT applications in areas such as environmental, agriculture and health sectors. In addition, AI is being incorporated with IoT systems to be able to analyze data, in real-time, and make better, more informed decisions as well as enhancing the automation process. Some of the applications of AI in Internet of Things include expanding the capability of IoT systems to be more self-governance, self-control, as well as analytical self-control.

Among these, one of the most interesting topics is the effect of quantum computing on IoT systems. Although it is still a relatively new field, quantum computing is expected to significantly increase the capabilities of computational resources to analyze a vast amount of data produced by the IoT devices.

Quantum computing could help with the creation of more efficient encryption and decryption of data, enhance machine learning, and enhance the response time of IoT devices to the incoming data. Quantum computing is not yet a reality in the context of IoT but it is a future vision that can radically change the nature of the IoT systems and their growth.

Another trend that is likely to define the future of IoT is the emergence of 5G networks and the subsequent networks. This chapter looks at the effects of 5G on IoT ecosystems in regard to the aspects of ultra-low latency, higher data speed, and network capacity. The IoT variations provided will also facilitate effective and efficient interaction of the IoT devices with the applications to support the complex use cases like self-driving cars, remote health care methods and industrial applications. With a rise in the widespread connection of 5G networks, IoT is expected to grow and improve the integration of IoT and AI in many applications that require high bandwidth and low latency. Also, there is already a discussion on 6G which could bring even more enhancements in the connectivity and therefore even more powerful and effective IoT systems in the future.

Summing up, one can state that IoT has experienced great development in the past several years, but there are still various issues that need to be addressed, including scalability, integration, security, & privacy. These are real issues that will need constant improvements, cooperation, and the creation of new standards and technologies as the IoT environment expands. As the cradle of this technological innovation, the future of IoT has been pictured to be very dynamic with smarter sensors, artificial intelligence, quantum computing, and next-generation communication systems as the key shining areas of development for the IoT ecosystem. More studies will be conducted to address these concerns and to maximize the benefits that IoT has to offer in different fields of application.

7. CONCLUSION

In conclusion, this chapter tried to lay down a brief and general introduction to the IoT system with the focus on the key components such as sensors, connectivity, and communication. IoT sensors are the means through which real-time information about the physical world is obtained in order to inform systems in how they can interact with the physical world. Connectivity solutions refer to the manner in which a device or network connects to another and it is a key consideration in an IoT system while communication protocols are the rules by which different IoT devices and networks interact. In progressing toward the more intricate, data-oriented atmosphere, the integration of IoT with AI holds greater improvements. AI makes it possible to capture, sort and analyze the large number of datasets created by IoT devices and use them for decision making and automation and protect against malicious attacks. The use of AI in IoT increases the level of intelligence of the systems where IoT is implemented making them adaptive to the environment and able to make decisions in real time.

It is therefore evident that the future of IoT & AI integration is extremely promising. As the technology of the sensors, communication network such as 5G and AI progresses, IoT applications will grow in various areas such as healthcare, smart cities and industries. The latter has the promise of generating new intelligent and interconnected spaces that are more effective and more friendly with the user. However, several issues associated with IoT and AI-based smart systems, such as scalability, interoperability, and security will be essential in the actualization of IoT and AI smart ecosystems. Thus, the combination of those technologies will continue to reshape industries and improve the standard of living while creating the advancement of the new generation of smart, connected products in the future years.

REFERENCES

Deshmukh, A., Patil, D. S., Pawar, P. D., & Kumari, S. (2023). Recent trends for smart environments with AI and IoT-based technologies: A comprehensive review. *Handbook of Research on Quantum Computing for Smart Environments*, 435-452.

Karthik, S. A., Hemalatha, R., Aruna, R., Deivakani, M., Reddy, R. V. K., & Boopathi, S. (2023). Study on Healthcare Security System-Integrated Internet of Things (IoT). In *Perspectives and Considerations on the Evolution of Smart Systems* (pp. 342-362). IGI Global.

Kashyap, R. (2019). Machine Learning, data mining for IOT-based systems. In *Handbook of Research on Big Data and the IoT* (pp. 314–338). IGI Global. DOI: 10.4018/978-1-5225-7432-3.ch018

Khang, A. (Ed.). (2023). *AI and IoT-based technologies for precision medicine*. IGI Global. DOI: 10.4018/979-8-3693-0876-9

Krishnamurthi, R., & Goyal, M. (2019). Enabling technologies for IoT: issues, challenges, and opportunities. *Handbook of research on cloud computing and big data applications in IoT*, 243-270.

Kumar, K. V. R., Kumar, K. D., Poluru, R. K., Basha, S. M., & Reddy, M. P. K. (2020). Internet of things and fog computing applications in intelligent transportation systems. In *Architecture and security issues in fog computing applications* (pp. 131–150). IGI Global. DOI: 10.4018/978-1-7998-0194-8.ch008

LA, A. G. (2026). Machine Learning for Threat Detection: A Comprehensive Survey on Anomaly Detection Techniques. *Improving Threat Detection, Network Security, and Incident Response With AI*, 235-268.

Ponmalar, A., Vijayakumar, K., Lakshmipriya, C., Karthikeyan, M., Gracelin Sheena, B., Beslin Pajila, P. J., & Siva Subramanian, R. (2024). Meta-Heuristics and Machine Learning Applications in Complex Systems. In *Metaheuristic and Machine Learning Optimization Strategies for Complex Systems* (pp. 257–275). IGI Global Scientific Publishing.

Prabhu, D., Tejasri, M. V. S. L., Saratha, B., & Anish, T. P. (2025, April). Future Horizons: Key Trends Shaping the Evolution of Machine Learning. In *2025 8th International Conference on Trends in Electronics and Informatics (ICOEI)* (pp. 1091-1099). IEEE.

Sathya, U., Hashini, S., Lakshmipriya, C., Varun, C. M., & Nalini, M. (2024, December). Emerging Trends and Future Prospects in Internet of Things (IoT) Technology. In *2024 International Conference on Sustainable Communication Networks and Application (ICSCNA)* (pp. 79-85). IEEE.

Singh, P., & Singh, N. (2023). Blockchain with IoT and AI: A review of agriculture and healthcare. *Research Anthology on Convergence of Blockchain, Internet of Things, and Security*, 1315-1330.

Solanki, A., & Nayyar, A. (2019). Green internet of things (G-IoT): ICT technologies, principles, applications, projects, and challenges. In *Handbook of research on big data and the IoT* (pp. 379-405). IGI Global.

Tripathy, B. K., & Anuradha, J. (Eds.). (2017). *Internet of things (IoT): technologies, applications, challenges and solutions*. CRC press. DOI: 10.1201/9781315269849

Venkateswaran, N., Kumar, S. S., Diwakar, G., Gnanasangeetha, D., & Boopathi, S. (2023). Synthetic Biology for Waste Water to Energy Conversion: IoT and AI Approaches. *Applications of Synthetic Biology in Health. Energy & Environment*, •••, 360–384.

Chapter 2

Artificial Intelligence: Exploring Machine Learning, Deep Learning, and Neural Networks

C. Thiyagarajan
Panimalar Engineering College, India

V. Kalyanasundaram
SRM Institute of Science and Technology, India

R. Siva Subramanian
https://orcid.org/0000-0002-7509-9223
R.M.K. College of Engineering and Technology, India

K. Saravanan
SRM Institute of Science and Technology, India

S. Vidyasagar
SRM Institute of Science and Technology, India

V. Kalai Priya
RMK College of Engineering and Technology, India

M. Nalini
https://orcid.org/0000-0001-8987-3120
S.A. Engineering College, India

ABSTRACT

Artificial Intelligence (AI) has rapidly advanced with key contributions from Machine Learning (ML), Deep Learning (DL), and Neural Networks (NN). These technologies form the foundation for numerous applications across diverse domains such as healthcare, finance, and autonomous systems. This chapter explores the fundamental concepts, techniques, and architectures of ML, DL, and NN, delving into the evolution of these fields and their current state. The chapter also highlights the challenges, such as data quality, interpretability, and computational efficiency, while discussing emerging trends like Explainable AI, Federated Learning, and energy-efficient AI. As AI continues to evolve, it holds the potential to transform industries and society, underscoring the importance of responsible and sustainable development.

DOI: 10.4018/979-8-3373-3146-1.ch002

1. INTRODUCTION

1.1 Brief Overview of Artificial Intelligence (AI)

AI is one of the most innovative and developing fields of the present-day technology. In its broadest sense, artificial intelligence may be described as the attempt to create machines that are able to mimic human intelligence (Zhang & Lu, 2021). They include acquiring knowledge, making inferences, solving problems, perceiving, comprehending language and making decisions. The idea of artificial intelligence can be dated back to the 1950s whereby thinkers such as Alan Turing came up with an idea of a machine that was capable of thinking like a human being. Since then, AI has gone through many phases of advancement, experience great progress and downfalls as well. In the modern world, due to the development of computing technology, the availability of data, and the use of complex algorithms, artificial intelligence is used in almost all spheres of our life, including medicine, finance, entertainment, and automobile industry. AI is still in its development stage but its impacts on industries and day to day life are promising to be enormous, hence AI is an area of research and development in the twenty first century(Yüksel et al., 2023).

1.2 Importance of Machine Learning (ML), Deep Learning (DL), and Neural Networks (NN) in AI

Concerning the wide field of AI, subfields of ML, DL and NN have become significant complements with a significant amount of recent progress occurring in these areas. Machine Learning is a branch of AI in which the computer is trained to make its decision based on the data it is fed with, without being programmed for this particular task. Artificial learning algorithms are seen to be quite successful in a few areas including, spam detection, image classification and recommendation (Janiesch et al., 2021). ML is further divided into Deep Learning wherein the layered structures called NN are used to model the patterns in huge data sets. These deep architectures create a branch of learning that is loosely mimicking the structure of the human brain and which require feature extraction from the input data at different layers, and fine and coarse gradients of learning, making them very effective, and best suited for applications such as speech recognition, NLP or autonomous vehicles. Neural Networks(NN), being the building block of DL, give the architecture by which the machines can emulate a structure of neurons and layers that allow the machines to learn complex patterns of the data. In fact, ML, DL, and NN are not just the subfields of AI, rather, they are the keys to most of the current advancements in AI. This cannot be overemphasized as they have successfully applied abstract ideas into real life solutions that are revolutionizing sectors, and even cultures(Sarker, 2021).

1.3 Motivation for the chapter

AI, ML, DL as well as Neural Networks are the domains that has been growing rapidly in the past several years. But this has brought about a big problem of a large number of related literatures, methods and terminologies that may be complex to grasp for the researchers, practitioners and even new comers in the field. Due to the ever-growing complexity and differentiation of AI techniques, there appears a demand to sum up the current understanding of the subject, define the basics of ML, DL, and NN, and demonstrate the connections between them. Moreover, it is crucial to distinguish these domains to anyone who intends to build, implement or evaluate AI solutions. This survey chapter is intended to meet the need for a systematic and comprehensive presentation that can connect theory and practice. Given that the survey chapter focuses on the evolution, techniques, and applications of ML, DL, and NN in AI, it is expected that the survey chapter will not only help the readers gain knowledge but also encourage further advancement in this rapidly growing field. The second reason is derived from the fact that in the popular media, the terms ML, DL, and NN are used almost synonymously when in fact, each represents a different concept with different functionalities. It will help to define the mentioned distinctions in order to have a better understanding of the technologies that are already influencing our lives.

1.4 Scope and Organization of the Chapter

This survey chapter is aimed at presenting the general information about the ML, DL, and NN and how these concepts relate to each other in the context of AI. It is important to note that the chapter covers areas of definition and history, methods, applications in various industries, challenges, and future research directions. It is not the goal of this chapter to give an extensive description of all the algorithms or models that have been developed but rather give a thematic classification that covers all the aspects while being detailed.

In Section 2, an introduction to AI is provided as well as a brief history of the field to provide a context for the subsequent discussions on ML, DL, and NN. In section three of the present manuscript, the author outlines the basic framework for Machine Learning and the categories of the same as well as the algorithms typically applied for the same and even their applications. Section 4 builds on Deep learning with descriptions of its architectures, training techniques, and where it started surpassing previous paradigm of Machine Learning. Section 5 discusses Neural Networks as the last type of machine learning in detail, describing what they are, how they are structured and how they contributed to the success of Deep Learning. Section 6 of the chapter presents some of the most important applications of the developed technologies in various fields to demonstrate their real-world applications. In section 7, the authors discuss limitations and concerns such as data concerns, computational concerns, interpretability concerns and ethical concerns. Section 8 provides a discussion on the future works and directions, which provides an idea of the direction of the AI research area. In the end, Section 9 gives a brief overview of the survey findings and discusses the future prospects of AI, ML, DL, and NN.

2. FUNDAMENTALS OF ARTIFICIAL INTELLIGENCE

2.1 Definition and Historical Evolution

Artificial Intelligence as a field can be defined as the ability of machines to perform tasks that are similar to human intelligence. These include acquisition (gaining of information and knowledge on how to apply the information), inference (applying the knowledge to deduce probable or certain outcomes), and regulation. It has been a long time, and the definition of AI has changed several times depending on the development in the technology and the research directions. First of all, there was an idea to create intelligent machines that will be able to think and act like human beings. In the course of time, the applied aspect of AI, that is, constructing systems that can solve certain problems effectively, emerged as the main focus.

The use of AI can be dated back to the year 1950s. In 1950, the academia Alan Turing introduced the "Turing Test" as a measure of artificial intelligence. Early work in AI included symbolic methods and AI as a set of rules and their early planning assumed that human-level AI could be obtained in a few decades. However, the field began to face some problems like the computational complexity and the scarcity of data and thus had some periods referred to as the "AI winters" when funding and interest started to diminish.

However, there was a revival of interest in AI in the 1990s with the help of advanced computers, availability of large data and better algorithms. Considerable advancement in ML and NN made AI applicable. In the tens of 2010, due to the new ideas in the architecture of NN and the use of GPUs, DL became a new AI method, which leads to significant progress in computer vision, natural language processing, and autonomous systems. Today AI is an interdisciplinary science and is based on the knowledge of computer science, mathematics, neurosciences, psychology, linguistics and many other fields; AI is becoming a part of our life.

2.2 Major Subfields of AI

AI is a broad discipline that contains many subfields and each of these subfields approaches the goal of developing intelligent systems from a different perspective with different methods. The first major subfield is Machine Learning (ML) in which the emphasis is put on creating techniques that enable a system to learn and make decisions based on data. Today, people often use the terms ML and AI interchangeably when referring to various contexts because of the former's broad applicability and effectiveness.

Another large branch of AI is Natural Language Processing (NLP) that is the branch that study interaction between computers and languages. NLP allows computers to process and analyze human language to perform tasks such as interaction with the user through chatbots, translation, or determining the sentiment of a text(Jia et al., 2023).

Computer Vision is one of the most important subfields that deals with the capability of giving computers the ability to understand and make decisions based on visual content that is available in the form of images and videos. These are interests in medical images, security cameras, and self-driving cars, among others(Zhao et al., 2024).

Robotics further integrates AI with mechanical engineering that deals with the creation of intelligent agents that are physically embodied and adept at performing activities in the real environment. Robotics aims at creating a system and an integrated software/hardware system, which is capable of sensing, deciding and acting on its own.

Expert Systems are usually defined as the AI programs that mimics the thinking ability of an expert. They employ a knowledge base and a rule base of inference to solve problems in a given domain.

Further, Knowledge Representation and Reasoning is a specialty that focuses on how that information might be stored by the machine so that it can apply it to practical problems such as diagnosing an illness or deciding on the best course of action.

These subfields are interrelated and intertwined, and the development in one subfield may lead to development in the other subfields. Together, they can be seen as the facets of AI and its ability to solve a variety of problems.

2.3 Role of ML, DL, and NN within the Broader AI Landscape

In the larger context of AI, there are ML, DL, and NN technologies that have formed the basis of many AI's greatest accomplishments. Machine Learning can be defined as the core of contemporary systems referred to as artificial intelligence. Unlike most computer programs which are coded to make specific decisions, ML algorithms enable the machine to learn on their own. This learning-based approach allows the AI systems to learn from the experience and learn how to improve the performance and solve tasks that were difficult to program traditionally.

Figure 1. Broader AI Landscape

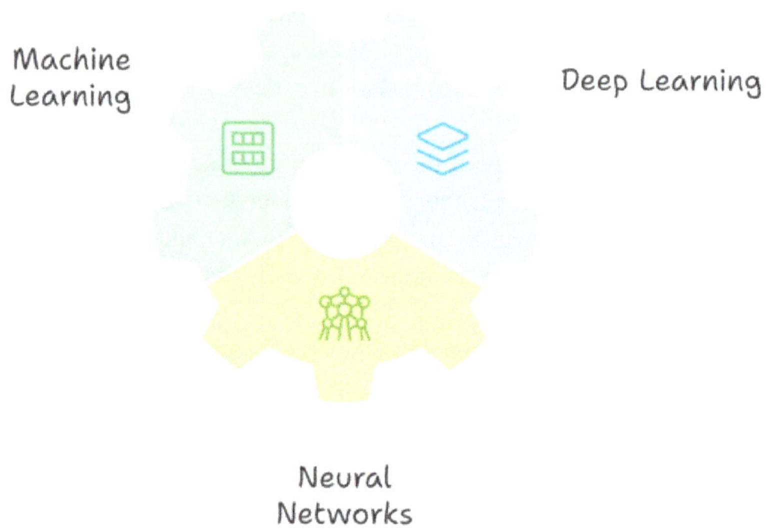

DL is a subset of AI that involves the use of deep NN with numerous layers, due to which the name deep is used. One common advantage of DL models compared to the rest of the ML is that the data preprocessing step of feature engineering is conducted automatically in this case. DL has helped growth of such areas as speech recognition, machine translation, image categorization, and even in games like Go and Dota 2 at the professional level(Mohammed & Kora, 2023).

The ANN are based on the biological neural networks located in the human brain and consist of interconnected nodes (neurons) which process the information jointly. Perceptron, for example, is a shallow artificial neural network and may be sufficient in simple pattern recognition and the more sophisticated deep neural (DNNs) comprise of several hidden layers to capture most of the hierarchal relations in data.

This makes the relationship between ML, DL, and NN to be hierarchical while at the same time interdependent. NN are a type of model used within the Machine Learning concept. When these networks are made deeper with many layers of abstraction, the networks are termed as DL. However, in the current real-world applications, Machine Learning algorithms involve DT, SVM and many others but DL and NN have gained more attention in the real world because of their high performance in the unstructured data such as images, audio and text.

In conclusion, ML, DL and NN are the engines of the AI revival, which turned ideas into practice. The combined effect has enabled AI to transcend the conventional barriers where the machines are capable of perceiving, learning, and performing various tasks. Knowledge about their roles within AI can help to assess the achievements and further difficulties in the development of intelligent and flexible systems.

3. ML: CONCEPTS AND TECHNIQUES

3.1 Overview of ML

Machine Learning also called ML is a genre of AI that enables machines to improve their operations when exposed to new data without being programmed. This is mainly concerned with the ability of the learning algorithms to make decisions or predictions based on examples. Unlike other programming paradigms where a human defines rules, the ML systems get better with experience on their own. Thus, it has significantly enhanced the effectiveness of AI, making them flexible in the constant and dynamic context.

The core concept of ML is to input data that is used by the algorithms to identify trends, sort information, make forecasts, or provide knowledge. Training of these models entails identifying relationships of input data and improving on the ability to classify, to predict, to cluster or to make decisions. With the help of computational platforms and the rising availability of the datasets, Machine Learning is no longer just a field of computer science but rather technologies and systems that form the backbone of present-day applications of AI in Healthcare, Finance, entertainment, and manufacturing industries among others.

3.2 Types of ML

The four primary categories of ML are based on the way the model acquires knowledge and each of them is described below.

3.2.1 Supervised Learning

Supervised learning is a process of learning from training data that is labeled or has output labels associated with it. The purpose of the model is to map the inputs to the outputs so that it can be able to predict the correct outputs when given unknown inputs. Some of the common types of tasks are classification problems such as spam filtering, medical diagnosis and regression problems such as house price prediction. Some of the most used algorithms under supervised learning include the decision tree, the SVM, and the NN. As is known, the performance of supervised learning models highly depends on the quality and the amount of labeled data (Ibrahim & Abdulazeez, 2021).

3.2.2 Unsupervised Learning

In unsupervised learning, the input data does not contain labels, and the model has to identify the structure or distribution of the data by itself. Some of the common uses are clustering where data points are grouped together, for instance, grouping of customers and dimensionality reduction which is the process of reducing the amount of data while retaining its important aspects for instance, the use of Principal Component Analysis. Unsupervised learning is used during EDA, outlier detection and any other situation where it is costly or impossible to label the data (Naidu et al., 2023).

3.2.3 Semi-supervised Learning

Semi-supervised learning can be defined as a middle ground between supervised & unsupervised learning. It involves a little amount of the labeled data along with a large amount of the unlabeled data to train the models(Ahsan et al., 2021). This approach is particularly relevant when it is difficult to obtain a large amount of labeled data, but unlabeled data are easy to obtain. It is possible to achieve a higher accuracy compared to the situation when only a small number of samples are labeled.

3.2.4 Reinforcement Learning

Reinforcement learning (RL) is a specific form of ML in which an agent makes decisions based on the environment in which it exists. It is an entity that learns from feedback in the form of rewards or penalties to achieve the greatest total of rewards in the end. As opposed to supervised learning, in RL it is the agent that must find the optimal actions on their own, without any help from a teacher. RL has been applied in various fields like game playing for example in the case of AlphaGo, robotics systems and autonomous systems.

Figure 2. **Types of Machine Learning**

3.3 Common Algorithms and Models

There are many algorithms and models in Machine Learning and each of them is designed to solve specific type of problems and in specific data structures.

3.3.1 Decision Trees

DT are relatively easy and flexible decision making tools that are in the form of a tree. Each node inside the tree is a decision made based on a feature while each branch depicts the result of the decision and each node at the end of the tree is the final class label. It is easy to interpret and visualize decision trees for tasks that have to be explained to other audiences. However, they can overfit if not pruned well in the right manner or if a lot of features are selected.

3.3.2 Support Vector Machines

SVMs are a class of learning algorithms used with classifiers or regressors for making predictions as per actual data. They operate in the manner where they seek to determine the hyperplane that can best classify the data into classes while at the same time maximizing the distance between the different classes. SVMs are particularly useful in cases where the number of dimensions is large and greater than the number of samples. It should be noted that by using Kernel methods one can solve non-linear classification problems when the data is mapped into a higher dimensional space.

3.3.3 K-Nearest Neighbors

The KNN is an instance-based learning model that categorises a data point based on the categorisation of its neighbours. KNN entails finding 'k' nearest data points in the training set and assigns the most popular label of the nearest data points to the query point. However, KNN is quite straightforward, it may be most efficient in low dimensional spaces, but it can slow down as the number of data sets increases.

3.3.4 Ensemble Methods

Ensemble learning is a method in which multiple models are used to create a single model as a result of combining different models.

Random Forest (RF) are a collection of decision trees, with different trees being built using bootstrapped sample of the data and randomness in features selection, which helps in reducing the risk of overfitting. GBMs create models iteratively, and each new model aims at improving the errors that the previous model has made. XGBoost and LightGBM are widely used boosting methods in order to achieve high accuracy in various structured data competitions.

3.4 Applications of ML

ML has been adopted in almost every field and science in existence today. In healthcare, the ML models help in identifying diseases, prognosis of patients, and tailoring therapy. In finance, algo detect, prevent and generate fraudulent transactions, perform transactions themselves, and even allocate credit risk. Some of the applications of ML in retail industry include recommendation systems, customer profiling and inventory management.

In the context of AVs, ML models are used to perceive the environment, identify objects, decide on routes, and control the car. In natural language processing, machine learning that facilitates the translation services, sentiments, and analysis and chatbots. In addition, in cybersecurity, the ML algorithms identify patterns of behavior that signify a security threat.

In addition to these, ML plays a crucial role in scientific research and development where it is used in drug development, climate change modeling and genomics. With data increasing at such a rapid pace, the use of ML in identifying useful patterns and making decisions in various fields will remain critical in the future.

4. DEEP LEARNING: ADVANCING MACHINE LEARNING

4.1 Introduction to DL

DL is one of the subsets of ML that implement a set of algorithms based on ANN. This entails the use of NN with many layers and because of this, the term DL is used. These deep architectures allow models to learn hierarchical representations of the data in order to represent it at different levels of abstraction. Deep learning has brought about great advancements in areas such as computer vision, NLP, speech recognition and even in the strategic game playing mainly due to its capability of solving tasks that were thought to be difficult for machines.

The continual increase in computational power, particularly Graphic Processing Units, the surge in size of training data and the enhancement in the optimization algorithms have make deep networks trainable. In contrast to most ML models that demand feature engineering, DL models learn the representations of the features directly from the raw data, which in turn saves much time and effort in feature engineering.

4.2 Key Differences between ML and DL

Thus, while DL is a type of ML, the two have got certain distinguishable differences. The traditional ML methods are very much data-crunching methods which rely on feature engineering to carry out tasks. On the other hand, deep learning models have the capability of identifying and analyzing text and other forms of data such as image and audio data in & without the need for the programmer to identify the useful data details to be used for solving the problem at hand.

The other major difference is on the issue of scalability and the data needed. The first type of model, namely the machine learning model, is highly efficient when working with moderate data, while the second type of model, the deep learning one, require large data sets for high efficiency. This is so because more layers involve more computations, and deep learning requires performing these over millions of parameters.

From the perspective of interpretability, the traditional models such as DT and linear regression models can be easily interpreted while deep learning models, especially deep neural networks are considered to be 'black boxes'.

4.3 Architectures in Deep Learning

It has advanced through the creation of diverse forms of the neural network structures that are suitable for various types of data and tasks.

4.3.1 Feedforward Neural Networks

FNN are the simplest form of the ANN where the flow of information is only in one direction, from the input layer, through the hidden layers, to the output layer. As it will be shown in the next sections, there are no cycles or loops in the network. Each neuron performs a linear transformation and then a non-linear transformation using the activation function so that the network can learn the input-output mapping. FNNs are commonly applied in such problems as classification and regression problems when the input-output mapping is rather simple.

4.3.2 Convolutional Neural Networks (CNNs)

CNN are particularly used to work with data that has some form of structure in the format of a grid, for instance, images. CNNs use convolution operations where small filters are moved across the input data to identify such features as edges, texture and shapes. This capability to extract spatial hierarchy of features makes the CNN popular and dominant architecture for image recognition, object detection or even video analysis. The commonly used or notable architectures such as AlexNet, VGG, ResNet and EfficientNet have further expanded the possibilities of achieving high performance in computer vision tasks.

4.3.3 Recurrent Neural Networks (RNNs) and LSTMs

RNN are designed for sequential data that includes time series data, speech data, and text data. While feedforward networks do not contain loops, RNNs contain loops that enable information to be passed from one step to the next, making them suitable for use in tasks that depend on the order of the inputs. However, it was pointed out that the traditional RNNs have a problem of long-term dependencies, known as the vanishing gradient problem. In order to overcome this, Long Short Term Memory (LSTM) networks were developed. LSTMs employ specific units called memory cells, which are capable of storing information for a long time and therefore the model is ideal for tasks such as language modeling, machine translation and speech recognition.

4.4 Training Techniques

There are several approaches that are used when training deep learning models with an aim of enhancing the learning process.

4.4.1 Backpropagation

Backpropagation is the basic technique of training neural networks. It is the method of calculating the derivative of the loss function with respect to each weight in the network using the chain rule starting from the output layer down to the input layer. These gradients are then used to adjust the weights in such a way that the loss is reduced to the minimum which in turn enhances the performance of the network. The efficiency of backpropagation has made it possible to train deep networks with millions of parameters and this has been made possible by the use of **efficient algorithms.**

4.4.2 Optimization Algorithms

The optimization algorithms are very useful in determining how the weights of the network should be updated during the training process. SGD is one of the simplest optimization algorithms and involves updating the parameters of a model using a few randomly selected data samples at a time which in turn helps to speed up the convergence. However, the learning rates in SGD are crucial in the optimization process and may end up in local optima.

Other more modern optimizers such as the Adam optimizer incorporate the benefits of two other modifications of SGD, namely, momentum and RMSProp. Adam modifies the learning rate for each parameter separately depending on the first and second moments of gradients, which results in improved convergence rate in numerous **application fields.**

4.5 Applications of Deep Learning

Deep learning has created new opportunities in almost every field of modern life. DL models utilized in healthcare include medical image analysis for diagnosis, diagnosis of different diseases and discovery of drugs. In the automotive industry, deep learning helps to create self-driving cars using such procedures as identifying objects, determining lanes, and making decisions.

In the entertainment industry, it performs recommendation systems for platform such as Netflix and Spotify. It also has innovative uses like deep fake, music generation, and style transfer in the art field. Language translation, sentiment analysis, and Chat GTP or any kind of conversational agents are the main subfields of Natural Language Processing all of which employ deep learning.

Similarly, deep learning is more significant in different scientific research areas such as genomics, astrophysics, climate modeling, among others, because of the discovery and predictive modeling of a particular pattern. With the advancement in models and their availability, deep learning shall remain instrumental in enhancing innovation and solving problems in the world.

5. NEURAL NETWORKS: THE BACKBONE OF DEEP LEARNING

5.1 Basics of Neural Networks

Before diving into the technology trends, it also important to introduce Neural Networks (NNs), which are the foundation of deep learning and are viewed as the fundamental architecture that makes it possible for the machines to learn from the data and gain intricate representations of it. As a stimulatory model of the brain neural network, ANN are a number of interconnected layers of nodes or neurons in which each node is considered an imitation of a simplified neuron. These artificial neurons take inputs, perform a weighted sum of the inputs, and then pass the sum to the next neurons. Neural networks have been found to be very efficient in approximating non-linear functions; thus, they can be applied in various fields such as image identification, NLP, among others. The basic concept is that through training the weights between nodes it is possible to train the network to model complex patterns in the data and perform tasks that were beyond the capability of previous computational methods.

5.2 Structure and Components

It is for this reason that neural networks have a well-defined structure, which comprises of several important parts, all of which are important in the learning process.

5.2.1 Neurons

In a neural network, the basic form of computation is the neuron. A neuron receives input signals, multiplies them by weights, adds a bias, and then passes them through activation function. The neuron's output is then passed on to neurons of the next layer. A neuron is a basic unit of a network that is capable of performing a variety of tasks in different stages of abstraction.

5.2.2 Layers

NN are also divided into layers. The input layer receives the raw data such as pixel values for image data or numerical values for dataset data etc. The next is one or more hidden layers where most of the computations are performed. The layers are hidden and map the inputs into a form that is more and more abstract and convenient for the task. Lastly, the output layer gives the final result, the prediction or decision of an image, or a numeric value. The number of layers and the number of neurons in each layer are the two factors that affect the ability of a network to learn the function.

5.2.3 Activation Functions

Activation functions help to add non-linearity into the neural network so that it can learn non-linear patterns in the data. If activation functions are not used, the network is capable of modeling only linear functions, which is very limited in its applications. The common activation functions include the sigmoid function, tanh function, and ReLU or rectified Linear Unit functions. Of these, ReLU has gained much popularity because of its simplicity and performance especially in deep networks because it solves problems such as vanishing gradient.

5.3 Types of NN

Neural networks are of different types depending upon the kind of input data and the domain where it is to be applied.

5.3.1 Shallow Neural Networks

A shallow neural network is often made of an input layer, one hidden layer and an output layer. It is thus clear that shallow networks are not very good when it comes to the extraction of hierarchical features or representation of complex data in a manner that is difficult to model. Still they formed the stage for future complex architectures and are still useful especially when designing straight-forward and speedy networks.

5.3.2 Deep Neural Networks

DNNs are defined as having more than one hidden layers between the input and output layers. These networks can learn features of different levels of abstraction, where each level of the hierarchy encodes more abstract features. For example, the first layers may identify edges in the image, middle layers may recognize texture and the deeper layers may identify the object. Although DNNs have shown great performance in various fields, they need large-scale data, time, and effort to train and fine-tune to get the best results.

5.3.3 Specialized Networks (GANs, Autoencoders)

Apart from feedforward networks, other types of neural networks exist and include the following: GAN is a type of neural networks where there are two networks named generator and discriminator and both are in the competition with each other and used to generate more realistic data such as images or videos.

Autoencoders are used particularly for unsupervised learning tasks or issues, such as for instance, dimensionality reduction and feature learning. They operate by transforming the input data set and coding it to use few features to represent it while decoding it to reconstruct the actual data set. These specialized networks have extended the applications of neural networks to the domains such as data generation, anomaly detection and self-supervised learning.

5.4 Challenges in Neural Network Training

However, there are several problems that are being associated with neural networks and these are some of the issues that researchers and practitioners need to address to have better models.

5.4.1 Overfitting and Underfitting

Neural network training is not without its problems and these include overfitting and underfitting of the models. Overfitting is a problem that is associated with a network learning the training data in a very detailed manner to the extent of capturing noise in the data as well as irrelevant patterns that make it generalize poorly on unseen data. Some of the common methods of handling overfitting include; drop out, early stopping, and regularization. Underfitting, on the other hand,it happens when the network is not able to capture the true pattern of the given data brought by the limitations on the model structure and size, training samples, or feature extraction.

5.4.2 Vanishing/Exploding Gradient Problems

Other difficulties consist of vanishing and exploding gradient especially in deep networks. Vanishing and exploding gradients occur during backpropagation, making learning unstable or not converging at all. This is especially the case in multilayered networks. Some of the solutions include; proper weight initialization, the use of ReLU activation functions, batch normalization, and specific architectures such as LSTMs for sequential data.

5.4.3 Interpretability Issues

Neural networks on the other hand have been shown to be very effective but they are black boxes and thus not interpretable. One of the major drawbacks of deep neural networks is the interpretability of the decision made by the model as opposed to more conventional models where the feature importance and even the decision making paths can be easily traced. This black-box characteristic can be disadvantageous in fields where openness and confidence are paramount, for instance, healthcare, finance, as well as law. Some of the approaches towards enhancing the interpretability are feature visualization, saliency maps, and inherently interpretable models.

6. APPLICATIONS ACROSS DOMAINS

6.1 Computer Vision

One of the more known and widely developing applications of the AI, ML, DL is computer vision. It deals with allowing the machines to understand and analyze the images and videos and make a decision on the same. As the deep learning progresses, especially through CNN, the computer vision has been able to move forward significantly.

Applications such as object detection, image classification, facial recognition, & scene understanding have been achieved near to human or even beyond human level. Autonomous vehicles are one of the technologies that heavily depend on computer vision to identify objects and environment to avoid any obstacles. For instance in the retail business, computer vision is used in the counting of stock and studying consumer behaviour while in security, it is used in monitoring and identification of potential threats. The combination of deep learning and mass annotated datasets formally brought computer vision into everyone's application and use making it one of the most active subfields of AI.

6.2 Natural Language Processing

NLP is another area where AI, ML, and DL have paved way for revolutionizing the field of study. NLP is the ability to program a machine, in a way that it can actually understand, analyse and even produce text in a way that makes sense and is helpful to man. In the past, traditional methods such as rule-based systems were popular in NLP but over the recent past, Deep Learning models particularly the Transformer models have emerged as the new benchmark for NLP. They are most suitable for the tasks of machine translation, text summarized, emotive analysis or more specific tasks such as chat bots or virtual assistants. For instance, specifically in customer service, companies can now use intelligent chat bots that are capable of providing quite natural and functional answers to questions and thereby do not necessarily require human assistance. In the healthcare industry, NLP is being applied to the extraction of important information from the unstructured medical records. Furthermore, such tools as real-time language translation and content generation are available to millions of users, and this is the result of NLP based on neural networks.

6.3 Healthcare

AI application in healthcare has become apparent over the years by breaking barriers in diagnostic improvement, offering custom approach to the treatment, and improving the overall functioning of the organization. Machine learning is applied to prognosis of disease spread, risk assessment, and medical decision making. Specifically, CNNs have been applied to a plethora of medical imaging tasks such as tumor detection from radiology images or even segmentation of organs for MRI images. Furthermore, with help of AI, the EHRs can be used to identify the possible outcomes of a patient's condition and make recommendations regarding the further course of action. In terms of the application of decision making, analytics that employed the power of the ML aid in the early detection of diseases such as diabetes, heart diseases, and cancer. In addition, drug discovery is also using neural networks where they are cutting down the time and cost of introducing new medicines in the market. Consumers also incorporate wearable technologies and mobile health applications still based on artificial intelligence in observation of patient's vitals and feedback consequently enhancing the model of preventive care and patient satisfaction.

6.4 Autonomous Systems

Self-driving autos, drones, robots, and numerous other machineries use AI, ML, and DL techniques for their operation without human interferences. These systems need to be able to sense the environment in which they exist, decide on courses of action and act accordingly in a safe manner and as quickly as possible. Deep reinforcement learning is one of the most promising approaches to train an agent in an environment that changes over time in order to perform various tasks. In the case of self-driving cars, DL algorithms analyze images and videos from cameras, LIDAR, RADAR and GPS to identify surrounding environment and objects and find out optimal path for safe driving. In robots, application of AI is employed to create robots for simple operations like cleaning houses and even for complicated operations like surgeries. The industrial automation also has its advantages in the form of self-automated systems that can enhance production line, monitor the machines and equipment for possible breakdowns, and quality assurance. The progress of self-driven intelligent automations is opening doors for various sectors to achieve more secure, effective, and intelligent operations.

6.5 Finance

A few examples of utilizations of ML and DL are present in the finance and banking industry, where the technologies are being applied to make decisions, measure risks, and optimize customer experience. It is applicable in credit scoring, fraud detection, algorithmic trading, and customer or market segmentation. The advancements in the field of DL have led to enhanced efficiency of the fraud detection systems, which can recognize patterns in datasets. This work is applied in sentiment analysis for stock market prediction through news articles, social media, and financial reports analysis. Self-employed robo-advisors are the AI-based services that offer the investment advice by referring to the client's characteristics. Also, AI changes the way that the financial institution works and copes with the emerging risks and intricate regulations. Consequently, with the advancement of AI, its use in DeFi, analytics on blockchain, and in financial modeling has also been rising and is branching out more to revolutionize the financial industry further.

6.6 Smart Cities and IoT

Smart cities are the cities with efficient infrastructure needed to improve the standard living of people and AI has a central role in this process. The interaction between AI and IoT allows for the monitoring, analyzing and drawing decisions directly from the institution of the things in different parts of life such as traffic control, energy, waste disposal, and safety among others. Some forms of big data involves the application of ML algorithms in finding out the traffic patterns through sensors and connected devices as well as estimating utility works that are requiring maintenance in future, and enhancing the systems of responding to emergencies.

In energy management, AI applied to enhance the control of power grid, to balance the supply and demand, incorporate renewable energy sources and decrease the CO2 emissions. Advanced line-up intelligent surveillance systems based on deep learning bring the advancement of preventing possible risks and breaches of security in cities. In addition, the use of AI in smart homes has an impact on energy consumption, automation of operations, comfort, and security. AI combined with IoT and Big data is making cities smarter and more responsive to the people's needs; they are therefore highlighting the prospects of developing sustainable cities which are efficient to live in as well.

7. CHALLENGES AND LIMITATIONS

7.1 Data Requirements and Quality

As it is evident especially in ML and DL, one of the main issues in AI is the requirement of large quantities of good quality data. Accuracy and generalization of existing models such as DNN depend heavily on various datasets needed in an object or an image recognition task. However, the problem arises when one wants to acquire such data. It is often the case that there is a lack of data, missing data, unbalanced, or noisy data which in turn results in poor performance of the model. Some of the data issues include; Inaccurate labels, missing values, and data inconsistency are some of the data issues that can greatly affect the performance of AI systems. Also, the collection and curation of data are not only time-consuming but are also accompanied by some risks and issues, especially in the context of such categories as healthcare or finance. Several techniques such as data augmentation techniques, synthetic data generation, and transfer learning can be applied to reduce the effects of data limitations to some extent but it is still a major constraint. This situation makes it important that issues of data management and quality control be well dealt with, since flawed data can shift erroneous information to another system.

Figure 3. **Challenges and Limitations**

7.2 Computational Costs and Efficiency

The third challenge is the computational requirement to train and use the AI models currently in use. Deep learning models especially the ones with millions or even billions of parameters as in the case of GPT or large convoluted neural nets are computationally intensive, memory and storage demanding. Training of such models requires the use of additional hardware such as GPUs, TPUs, and distributed computing systems. This level of computational requirement not only results in high costs of financial capital but also poses questions of energy and environmental expenses. It is noteworthy to mention that the training of large AI models is a process that has been receiving more attention due to the large carbon footprint it generates. In addition, temporal performance is crucial in numerous applications as the outcomes are required immediately as in the case of self-driving automobiles as well as diagnostic facilities in the healthcare sector. Such issues are solved using techniques such as model pruning, quantization and knowledge distillation but the main task that is still challenging for the AI practitioners concerns the rate of decreasing the model size and the acceptable model performance.

7.3 Explainability and Interpretability

Over the recent past, the use of AI systems in the decision-making process has increased, hence the need for explainability and interpretability. NN, also referred to as "black boxes", are capable of giving high accurate results without giving explanations of how those results were arrived at. Such lack of transparency is surely a great offender for trust and adoption, especially in areas like health, finance, or law that rely heavily on people's decisions. Patients, regulators or any other end-users require the ability to know how and why a model came up with a certain decision. The need for such models has been addressed by Explainable AI (XAI) which aims to create ways of making the AI models explainable without compromising on their performance. SHAP values, LIME, and attention mechanism are examples of the attempts to explain model behavior. However, the problem of achieving the right level of model complexity and interpretability so that it is appropriate for the specific application is still an open issue and there is no optimal solution.

7.4 Bias and Fairness

It is a common knowledge in the field of AI that the algorithms are only as good as the data that feeds into them. Regrettably, the actual data tends to be the reflection of historical prejudice, social injustice, and prejudice. Thus, the bias is likely to be reproduced and even amplified by AI systems when it is not controlled properly. Some examples include facial recognition systems, which produced higher error rates for the black people; recruitment algorithms that negatively discriminated against specific groups; and lending models that were more favorable towards specific socioeconomic status. It is important to address bias not only from the moral perspective but also from the point of view of the quality and fairness of the AI solutions. Some of the ways of addressing bias include data pre-processing, applying constraints on the model, and monitoring of the deployed models. However, it is a well-known fact that defining and measuring fairness is not a trivial and straightforward task and it depends on the context. Moreover, it is a common thing to have the conflict of interest between the goal of fairness and other goals, including the goal of accuracy or efficiency. Hence, the idea of constructing fairly and efficiently intelligent AI systems is a delicate and unsolved problem in the field.

7.5 Robustness and Security

Another important risk is the ability to make AI systems reliable and secure, especially in conditions of an adversary. It was found that deep learning models are susceptible to adversarial attacks which are manipulations of the input data and cause the model to give wrong results with high accuracy. Such weaknesses can lead to disastrous outcomes in application areas like self-driving vehicles, health diagnosis, and IT security. For instance, a slight change in the traffic sign image may lead the self-driving car to perceive it in a wrong way which may lead to fatal consequences.

Other forms of attacks are data poisoning, where the adversary injects wrong data into the set to be used to train the AI system. To counter these threats, there is a need to establish better training tactics, the use of adversarial approaches, and the implementation of anomaly identification systems. Furthermore, since the AI models are used more and more often in open and dynamic environments, the problem of their robustness to changes or malicious actions arises. It becomes clear that the construction of secure, reliable, and trustworthy AI is still an open and difficult problem to solve, which is the challenge for the research community.

8. EMERGING TRENDS AND FUTURE DIRECTIONS

8.1 Explainable AI (XAI)

Since AI solutions are used in more and more tasks and become more complex, the problem of their explainability and accountability has emerged as one of the most pressing. XAI is relatively new field that focuses on creating models and methods that can give understandable and interpretable insights about the AI decision making processes. While conventional deep learning models are rather effective in terms of accuracy, they are also opaque or 'black box' in nature. This is because such issues can cause the population to lose their confidence in these services, and this can be catastrophic in fields like health, finance, and artificial intelligence. XAI is therefore an attempt to address this issue by coming up with algorithms which are highly accurate yet capable of explaining their operations. Techniques like saliency maps, attention mechanisms, feature importance explanation methods like LIME and SHAP, and rule extraction techniques are the existing ones that are now being used to increase model interpretability. In the future, research will continue to be conducted for the evaluation of explainability for AI, making sure that it is equally fair and transparent in all its applications, and for the creation of systems that will provide explanations that are more suitable for the specific audience.

8.2 Federated and Distributed Learning

That is why, against the background of increasing data privacy and the growth of the number of edge devices, there is a keen interest in the development and improvement of new paradigms such as FL and DL. It was developed to enable training of an AI model across a large number of distributed devices or servers that contain local data samples without sharing the data with others, thus minimizing data transfer. It is especially useful in organisations such as the healthcare and the finance sector where information cannot be disclosed to the public domain. It also reduces the exposure to data breaches and improves on the compliance to data protection laws like the GDPR. On the other hand, distributed learning involves training large models using several computers that allows scaling up of the computation. There are several challenges that need to be addressed such as how best to deal with heterogeneity in data, how to manage communication costs, and how to address issues related to model aggregation.

However, as privacy preservation is becoming an important trend, the future may bring the increased use of federated and distributed learning methods in various fields.

8.3 Energy-efficient and Green AI

This chapter focuses on the environmental cost of training large models in terms of energy consumption and carbon footprint. Large-model such as GPT-4, as well as similar deep learning architectures, are computationally intensive consuming a great amount of power that translates to increased carbon emissions. This realization has given rise to Green AI, a concept that aims at achieving energy efficiency in AI models and promoting sustainability in the AI society. Some of these approaches include model pruning, quantization, knowledge distillation as well as having a breakthrough in creating small and powerful architectures for instance efficient transformers and lightweight CNNs. Moreover, the new approaches of optimizing the hardware, including low power AI chips, are also being investigated. This is because new metrics that measure computational resources used, for example, the energy to accuracy ratios are emerging slowly to supplement the traditional measures of performance. In the future, tackling the issue of making technological advancements incorporate environmentally sustainable approaches, the need for green AI will be a necessity for future advancements in the AI development field.

8.4 AI Ethics and Responsible AI

As AI has become more integrated into every part of people's lives, questions of morality have become increasingly relevant in the scientific and political agendas.

AI Ethics and Responsible AI initiatives are an attempt to guarantee that the use of AI systems is done in a manner that does not violate human rights and that is fair, non-harmful, and transparent. Potential ethical issues are the issues of bias, discrimination, privacy violation, manipulation, and displacement of workers. State and non-state actors, universities, and private companies are starting to set out principles for ethical use of artificial intelligence. It is worth noting that principles that are considered fundamental to AI include transparency, fairness, accountability, and inclusion. Possible developments in the field of post-2019 Responsible AI are described in three categories: design and implementation of ethical approaches to extend on the conceptual lens model, particularly in relation to legal frameworks and technological solution that translate ethical concepts into practice. Also, technologists, ethicists, legal specialists, and policymakers are likely to play a significant role in enhancing the positive impact of AI to the society.

8.5 Integration of Symbolic AI with Neural Models

Although, neural networks can learn from experience in a way that is somewhat similar to learning from data, they have limitations in understanding, inferring meanings and the ability to use prior knowledge that are some of the most defining aspects of symbolic AI.

The neuro-symbolic AI which is a combination between symbolic reasoning and neural learning is still considered as an emerging area. This approach is a mixture of aspects of connectionist and symbolic systems where the neural networks are to be used for pattern recognition and the symbolic AI for logical reasoning. Thus, it is possible to arrive at models that are not only accurate but also more interpretable, require less data, and can reason. This is an active areas of research, and other strategies, such as embedding of the complete symbolic rules in the neural architectures or training of the neural networks how to process the symbolic structures, are being constantly investigated. In the future, neuro-symbolic systems can try to be more human intelligent than the intelligent system as of now, eliminating many issues that statistical techniques involve and AI to address deep meaning problems involving understanding and reasoning.

8.6 Neuromorphic Computing and Brain-Inspired AI

Neuromorphic computing as a concept follows the pattern of how the human brain functions and is the next generation in artificial intelligence and computing systems. Neuromorphic systems utilize the hardware that is inspired by the neural structures; it implements SNNs and event-based processing for efficient computations. Compared to the von Neumann architectures, the mentioned neuromorphic systems unify the functions of memory and computation, which results in energy efficiency and increased speed. Brain-inspired AI also goes further in terms of the learning models, including the unsupervised and self-supervised learning models that mimic learning from the environment similar to human beings and animals. In future, as the field of neuromorphic computing advances, we can expect these systems to act as an important enabler of larger-scale, more efficient, and smarter intelligence-based computing applications in different domains of robotics, IoTs, and self-driving vehicles etc.

9. CONCLUSION

AI has emerged from a theoretical idea and has become one of the most revolutionary phenomena that affects almost all spheres of human life. This survey has discussed the basics of the AI and has mainly concentrated on ML, DL, and NN. ML brought into the ability of a system to learn from a given data sample, deep learning expanded on this by creating many layered architectures that enabled higher level abstractions to be modeled and neural networks provided the basic foundation to most of these breakthroughs. Both of them have greatly advanced the capabilities of AI to a great extent.

It therefore brings out the timeline of the emergence of ML, DL/NN, as characterized by competency of computers, better algorithms, and access to big data. It has advanced ever since from simple neural networks to the convolutional and recurrent neural networks, all the way to deep learning architectures and systems that rely on transformers. Every new development comes up with new opportunities and at the same time, the problems, such as interpretability, fairness, and sustainability, among other issues. Thus, the combination of the various paradigms of AI that includes symbolic reasoning and neuromorphic computing accentuates towards the future paradigm where AI systems, do not only execute tasks but can reason, learn and probably interact with world in a more natural way.

As for future work in AI, there are great possibilities for development as well as great expectations for the outcome. Three new directions, namely Explainable AI, Federated Learning, Green AI, and Responsible AI development reflect the increasing concern about people's relations to AI technologies. Therefore, there is a need to promote the ethical, inclusive implementation of the potentials offered by AI by researchers and practitioners. Thus, further AI development goes far beyond mere technological progress and depends on people's desire to guide the development process in accordance with humanitarian values.

REFERENCES

Ahsan, M. M., Mahmud, M. P., Saha, P. K., Gupta, K. D., & Siddique, Z. (2021). Effect of data scaling methods on machine learning algorithms and model performance. *Technologies*, *9*(3), 52. DOI: 10.3390/technologies9030052

Ibrahim, I., & Abdulazeez, A. (2021). The role of machine learning algorithms for diagnosing diseases. *Journal of Applied Science and Technology Trends*, *2*(01), 10–19. DOI: 10.38094/jastt20179

Janiesch, C., Zschech, P., & Heinrich, K. (2021). Machine learning and deep learning. *Electronic Markets*, *31*(3), 685–695. DOI: 10.1007/s12525-021-00475-2

Jia, J., Liang, W., & Liang, Y. (2023). A review of hybrid and ensemble in deep learning for natural language processing. *arXiv preprint arXiv:2312.05589.*

Mohammed, A., & Kora, R. (2023). A comprehensive review on ensemble deep learning: Opportunities and challenges. *Journal of King Saud University. Computer and Information Sciences*, *35*(2), 757–774. DOI: 10.1016/j.jksuci.2023.01.014

Naidu, G., Zuva, T., & Sibanda, E. M. (2023, April). A review of evaluation metrics in machine learning algorithms. In *Computer science on-line conference* (pp. 15–25). Springer International Publishing. DOI: 10.1007/978-3-031-35314-7_2

Sarker, I. H. (2021). Machine learning: Algorithms, real-world applications and research directions. *SN Computer Science*, *2*(3), 160. DOI: 10.1007/s42979-021-00592-x PMID: 33778771

Yüksel, N., Börklü, H. R., Sezer, H. K., & Canyurt, O. E. (2023). Review of artificial intelligence applications in engineering design perspective. *Engineering Applications of Artificial Intelligence*, *118*, 105697. DOI: 10.1016/j.engappai.2022.105697

Zhang, C., & Lu, Y. (2021). Study on artificial intelligence: The state of the art and future prospects. *Journal of Industrial Information Integration*, *23*, 100224. DOI: 10.1016/j.jii.2021.100224

Zhao, X., Wang, L., Zhang, Y., Han, X., Deveci, M., & Parmar, M. (2024). A review of convolutional neural networks in computer vision. *Artificial Intelligence Review*, *57*(4), 99. DOI: 10.1007/s10462-024-10721-6

Chapter 3

Transfer Learning for IoT Applications: Explainable AI for Adaptive Systems Integration

R. N. Ravikumar
https://orcid.org/0009-0009-3705-1681
Marwadi University, Rajkot, India

N. Sivakumar
https://orcid.org/0000-0001-5039-072X
Marwadi University, Rajkot, India

S. Aarthi
https://orcid.org/0009-0006-9064-2091
Marwadi University, Rajkot, India

S. Kavitha Bharathi
https://orcid.org/0000-0002-5630-7963
Kongu Engineering College, Erode, India

Anand Singh
Marwadi University, Rajkot, India

S. Sujitha
Kongu Engineering College, Erode, India

ABSTRACT

The integration of IoT and AI enhances autonomous and adaptive systems, leveraging Transfer Learning and Explainable AI (XAI) for improved efficiency. This research explores AI-driven IoT applications, bridging theory and practice. Transfer Learning minimizes data needs and computational costs, increasing flexibility across smart cities, precision agriculture, healthcare, and energy management. AI models act as knowledge transfer nodes, enabling adaptive decision-making. XAI ensures transparency, interpretability, and trust in AI-assisted IoT, crucial for ethical compliance and reliable decision-making. Challenges include data processing, scalability, security, and privacy. The study addresses bias detection, fair data practices, and

DOI: 10.4018/979-8-3373-3146-1.ch003

ethical AI deployment. Using interdisciplinary research and practical examples, it guides researchers, engineers, and industry professionals in advancing intelligent IoT systems.

1. INTRODUCTION

The high growth of Internet of things IoT generates a requirement of intelligent systems to be adaptive so that they can operate under varying conditions. Machine learning transfer learning utilizes past models to optimize the efficiency of new tasks that improves the performance of IoT applications. Transfer learning implemented in the IoT devices allows them to run on the ready models rather than manually training them thereby providing better performance and reducing costs and speeding up development timelines. XAI acts as a critical framework to maintain transparency since AI-oriented complex systems need sound adapted decision processes, in the working systems. Explainable AI (XAI) integrated into adaptive IoT systems provides users insight into AI model decision processes which helps users trust healthcare applications and applications for smart cities and industrial automation. Systems adapted through learning need accurate interpretability establishment to achieve essential operational integrity. System developers can produce transparent AI models with flexible capabilities through the combination of XAI techniques SHAP, LIME and Grad-CAM with transfer learning approaches. The cooperative power enables better system performance and eliminates regulatory and ethical challenges which together create secure user-centric AI-powered IoT products. (Kumar Dinkar et al., 2024)

1.1 Overview of Transfer Learning in IOT

The Internet of Things (IoT) creates an extensive interconnected device system that tracks data between devices to achieve smart control across domains of residential living and health care and industrial control systems. The process of creating machine learning models from primary sources for complete IoT applications remains unacceptable because most devices lack sufficient processing power and require substantial datasets. Transfer learning resolves the challenge through the effective utilization of pre-trained model knowledge which gets adapted for new yet similar tasks. The approach offers two significant advantages that reduce the duration of training and increase efficiency and enable IoT devices to generate intelligent evaluations using minimal data platforms. Transfer learning enables IoT systems to adapt to new environments with the previously trained models without having to retrain the entire model. In a process that does not require large amounts of new data, a pre-trained healthcare model that identifies human activities can obtain new capabilities to monitor patients specifically through a process called tuning. Transfer getting to know is a valuable asset as a vehicle of implementing AI-driven IoT responses since it enables green actual-time functioning and resource-conserving capacities. Transfer learning capability can be used fully in IoT programs although developers need to address version transferability woes together with domain-specific adaptation and safety issues. (Li et al., 2018)

1.2 Importance of Explainable AI (XAI)

XAI is significant to establish trust and transparency of artificial intelligence systems. More intricate AI models tend to resemble black boxes that make their decisions in a manner that cannot always be easily understood by users how the decisions were made. XAI can resolve this issue by offering explainable, understandable explanations of AI-related findings that allow users to trust such systems and justify them. This is even more critical in high stakes applications like healthcare, finance, and criminal justice where AI decisions may have consequences on lives. XAI offers a second level of assurance to regulators by turning AI systems into more comprehensive ones, as it meets the demand that AI operates without biases and unsupported assumptions. In addition to improving the transparency, XAI enhances the effectiveness of AI by allowing developers and users to have diagnostics, performance improvement, and optimization of model behavior. With explainable AI models, we can identify the bias and address it, resulting in a stronger and more trustworthy AI system. Moreover, in a human-machine collaborative system, XAI has a crucial contribution, so that the AI will not replace human decision-making but improve it. With AI slowly becoming enshrined in various sectors, the requirement of explainability will cover over the demand that AI systems be created, which are not just competent, but responsible and interpretable. (D. Kim et al., 2021)

1.3 Role of Adaptive Systems in IoT Integration

These adaptations are always kept active, thanks to the constant adjustment of the internal systems of an IoT through machine learning strategies, which upgrade its aptness for variable state conditions. The artificial intelligence and machine learning system would cause automatic adaptations to external conditions of the environment and user actions on new data through uncontrolled operations. The other day, a smart thermostat gets better and better each day as it picks up on a user's heating tendencies to adjust its settings accordingly, giving way to elevated convenience while saving energy. Predictive capability of Industrial IoT (IIoT) maintenance systems comprises extracting and performing sensor analysis, followed with adjustments on the predictive model for the anticipated troubles that the systems may face.

Adaptive systems in IoT networks have to possess satisfactory data-handling capabilities and the artificial intelligence support for decision-making, along with information exchanges among all connected devices for successful implementation. Adaptive systems implementation within IoT requires good solutions for issue of privacy niceties together with visualization for security threat and provision of solutions to issues of computational limitations. The transferred learning strategies in combination with Explainable AI offer such improvements in model optimization and result interpretation as solutions to these issues. The integration of all these technologies will push IoT systems towards greater and greater autonomy and reliability, thereby creating trust in consumers, and remaining on track to deliver high-level digital systems. (Hoenig et al., 2024)

1.4 Objectives of the Study

- To investigate how transfer learning enhances the adaptability and efficiency of IoT systems by reusing pre-trained models for new tasks.
- To analyze the role of Explainable AI (XAI) in ensuring transparency, fairness, and user trust within adaptive IoT environments.
- To explore the synergy between transfer learning and XAI for building reliable, accountable, and user-centric AI-driven IoT applications.
- To identify the challenges and limitations in applying transfer learning and XAI to IoT, such as computational constraints, privacy concerns, and domain adaptation issues.
- To propose future research directions for ethical, secure, and scalable integration of transfer learning and XAI in IoT applications across domains like healthcare, smart cities, and industrial automation.

1.5 Literature Review

The swift growth of the Internet of Things (IoT) has contributed to the need to develop smart systems that can adjust to a variety of environments that are dynamic. Conventional machine learning models can be very large in scale and require considerable computation power, which is not available on resource-limited IoT devices. This has caused the emergence of the transfer learning paradigm where the pre-trained knowledge by one domain is used in another with the least retraining. The result of such an approach is that it not only lowers the cost and time used in the development of models but also allows devices with low memory capacity and processing potential to effectively execute complex tasks using IoT. (Kalyanathaya &

Prasad, 2024)Transfer learning is applied in IoT in healthcare, industrial automation, smart-home, and urban management, and real-time adaptability and efficiency play a significant role in these fields. Although transfer learning helps with the problem of computational efficiency and data scarcity, it brings with it some new problems of interpretability and trust.

The effects produced by transfer learning models usually act as a black box, and the user and stakeholders may find it hard to comprehend the manner in which decisions are made. (Geukes Foppen et al., 2025) This lack of transparency will become a major adoption barrier in high-stakes areas like healthcare or autonomous systems. Explainable Artificial Intelligence (XAI) is an eminent solution in that it offers interpretable information on how a model will behave. Offering users a clear explanation of why a prediction occurs, XAI builds trust in some predictions, contributes to debugging, and ensures that ethical and regulatory frameworks are followed. SHAP, LIME and Grad-CAM are the methods that have been popularly used to illuminate the impact of features or visualize the decision-making process of a model to enhance the accountability and reliability of a system. XAI combined with adaptive IoT systems opens the possibility to develop more robust and human-oriented technology. (Sheu & Pardeshi, 2022)Adaptive systems are based on ongoing learning and real time decision making in order to operate effectively in the changeable environment. As an example, intelligent thermostats adapt to the preferences of users, whereas industrial IoT solutions forecast the equipment failure by examining sensor data.

Combined with transfer learning these systems can learn to generalize across environments so that they do not require a lot of retraining. But in the absence of explainability, adaptive systems will be at risk of generating unpredictable or opaque results that can lead to lack of confidence in the outcome by the user. (Sheu et al., 2023)By incorporating XAI into adaptive IoT systems, this gap can be closed, and stakeholders can realize the benefits of efficiency, as well as, comprehend, test, and have confidence in the workings of the system. Despite the potential, there are several challenges in the integration of transfer learning and XAI in IoT systems. Domain adaptation is also an issue to be considered since in other environments, the models that were trained under a particular setting may not work when generalized (Schulte et al., 2024) Not only that, since the IoT devices have limited computation capacities, and they should use lightweight architectures and optimization techniques.

Privacy and security issue are also of paramount importance when dealing with sensitive information such as health information. The solutions to federated learning and edge computing are still possible, still optimization of the performance, interpretability and security is also a challenge. Future trends on this field are developed through the development of lightweight self-explainable models that will be capable of running within edge devices without affecting the accuracy or transparency. Hybrid methods Hybrid methods are also being studied to add interpretable models to high performing black-box models to provide a performance and explainability trade-off. In practice, ethical and regulatory factors ought to play a pivotal role in the creation of such systems in order to realize fairness, accountability, and privacy in the execution of these systems. In summary, transfer learning and convergence of XAI in the IoT is an innovative step toward developing adaptive, transparent and trusting AI systems. Ongoing innovations are indicating at the strength of these technologies to revolutionize areas such as healthcare, smart cities, industry, and domestic automation, and consequently make the world a smarter, more ethical, and trustworthy IoT system (Nazat et al., 2024)

2. FUNDAMENTALS OF TRANSFER LEARNING IN IOT

The pile machine acquiring knowledge of technique known as transfer learning allows IoT devices to access pre-trained fashions to achieve stepped forward performance and flexibility. AI model knowledge switch ability permits schooling frameworks to reuse statistics of preceding activity operations to reduce schooling expenses and statistics desires. The approach promises less expense in configuring IoT, and it operates typically under small processing capacity and reacts immediately to the altering environmental conditions. With pretrained models, the IoT systems can get superior performance on anomaly detection and predictive maintenance processes along with real-time decision implementation utilizing small datasets. Complete utilization of the opportunities of IoT with the transfer learning presupposes developer solutions to domain adaptation and model optimization with security. (University et al., 2023)

2.1 Definition and Concept of Transfer Learning

Transfer learning refers to the type of machine learning method, which deals with the use of a trained model to solve a new problem which is related to that problem but different, with the new functions to be determined. The power of this approach consists in the fact that the prior model knowledge will be used to seek solutions to general problems with minimum training effort to be incurred. This approach is useful when the possibility of obtaining labeled data involves, a significant time and financial expense. Traditional models of machine learning demand strong computation machines and large collections of data this is where transfer learning transfers the work to AI in order to reuse resources to save computing time and accelerate processing time. Transfer learning is the IoT blessing since the devices do conduct anomaly detection and speech recognition and image processing tasks based on pre-trained models. This transfer learning method plays a crucial role in the effectiveness of lower-powered IoT devices, for they can resort to knowledge gained from the cloud or partially-trained models. Research groups adopt this greatly because pre-trained merely models give rather good performance results through mere adjustment in their application to specific use cases but without any kind of intensive retraining systems under predictive maintenance and health monitoring and smart home automation programs. (Meziane & Ouerdi, 2023)

2.2 Advantages of Transfer Learning for IoT Devices

Transfer mastering in IoT brings massive blessings through its functionality to decorate machine performance whilst handling small processing necessities. The low processing abilities and restrained reminiscence storage capabilities mark IoT gadgets that comprise sensors and side computing nodes. The gadgets can put into effect pre-trained deep mastering models the usage of switch studying to reduce their schooling time and statistics series requirements. The low-energy usage of IoT systems allows them to deliver precise selections at reduced electricity levels respectively making feasible the execution of AI era to the real-life wants. The significant benefit of transfer learning enables models to adapt fast to new environments that are different. Conventional AI models require huge amounts of schooling data to operate effectively and yet such data are difficult to harvest in the face of IoT constraints like privacy laws or insufficient infrastructure. Transfer learning enables fashions to finesse themselves the utilization of suppressed new records that supplements its ability to perceive a multitude of states. When the environmental conditions change rapidly, it will make the model more effective in the domain of industrial internet of things since this method guarantees optimal functionality. (Hussain et al., 2020)

2.3 Challenges in Applying Transfer Learning to IoT

The transfer learning, applied to facilitate the creation of wider application of the IoT is undergoing its greatest challenge in the changing landscape that does have no priori-available information with whom the training is carried out or at least the multi-distribution perspective. The object recognition technology applied to the real-life implementations of the IoT needs the devices to operate in rather different operational conditions in comparison with the ones they had during their training conditions, and the variations may be in the form of difference in the light conditions, weather factors, and sensor sensitivities. The domain adaptation involves further methods that combine two mechanisms of transforming both the model and data to generate more salient robustness. The issue of security and privacy risks emerge during the process of transfers of a pre-trained model which poses significant challenges in model transfers. The sensitive data that biometric IoT devices process will ultimately expose them to unauthorized access and malicious attacks because they often cater to health or business operational information. Efficient integration involves application of specific protocols to prevent models from recording personal data in their training set. It requires development on IoT frameworks for optimization for use with deep learning on large-scale distributed configurations that is heavier on processing with model pruning and quantization techniques being used. (Awan et al., 2023)

3. CASE STUDIES AND EXAMPLES IN IOT APPLICATIONS

Transfer learning techniques improve the performance of IoT solutions across different domains to be more efficient in operation. Deep learning-based predictive maintenance model for industrial IoT analyzes data from sensors to detect defects before causing a failure of actual equipment. Knowledge transfer between diverse equipment designs allows enterprises to conduct maintenance operations more efficiently, thus extending equipment uptime and reducing equipment downtime. Wearables operating within healthcare IoT domains use transfer learning to monitor vitals and heart problems-'arrhythmia'-and lung diseases by incorporating data through their pipeline. Such a prepared model combined with general population data becomes more accurate and can apply specialized training on a tiny dataset taken from outcomes of patient health. The implementation of transfer learning methods enables smart cities to modify pre-developed computer vision models for traffic monitoring which allows them to detect patterns of local congestion. Transfer learning helps IoT systems to be more adaptive, and in the process, better performance and smart operations under diverse real-life conditions. (Mazhar et al., 2023)

3.1 Explainable AI (XAI) in Adaptive Systems

Adaptive IoT structures want the implementation of Explainable AI (XAI) to acquire transparency at the side of consider and duty in AI-based totally selection-making operations. The growing use of system mastering in IoT automation calls for whole decision-making knowledge by using humans to ensure debugging success and keep compliance standards and person consider [10]. The mixture of SHAP along LIME and Grad-CAM lets in experts to understand complex fashions better by way of identifying critical features that affect predictions which leads to each accelerated AI reliability and enhanced know-how. The explanation of AI decisions in adaptive structures through XAI allows stakeholders to approve selections at the same time as recognizing biases and assuring fair behavior. Highly explainable fashions face a performance exchange-off with accuracy because they need to compromise precision to attain interpretation desires within IoT packages. Therefore, a balanced method should be implemented to establish each clear interpretations and accurate consequences from AI-powered IoT systems. (Khazane et al., 2024)The figure1 demonstrates Explainable AI (XAI) principles through displaying how machine learning (ML) models both produce forecasts and generate explanations for users. The input data is taken by the ML model to produce predictions, whereas the use of XAI methods provides the generation of explanations that help the stakeholders to comprehend how the model makes its decision-making. Both the predictions and their explanations are presented through a user interface, and hence the process becomes transparent and understandable. The system provides the stakeholders (managers or business owners, data scientists or developers and domain experts) with actionable decision-making information. Such an arrangement proves that transparency in AI decision making and the overall understanding of the user are crucial elements in satisfying a variety of users.

Figure 1. Explainable AI (XAI) Workflow and Stakeholders

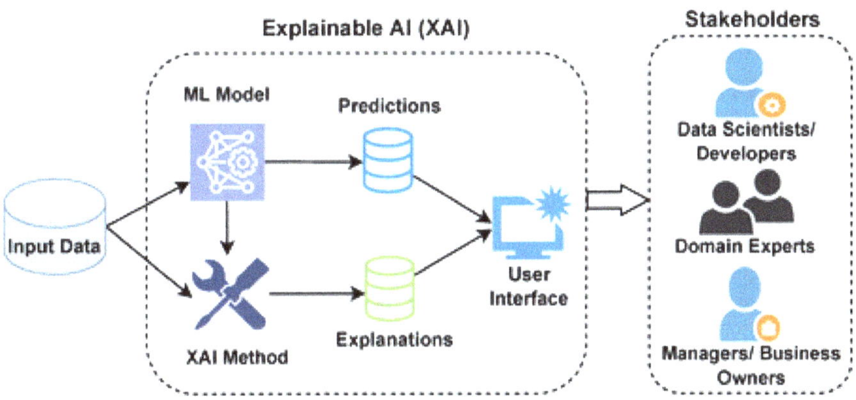

3.2 Importance of Explainability in AI-Driven IoT Systems

The concept of trust and transparency requires special care with IoT systems where the integration of AI is taking place on an ever-growing basis. XAI (Explainable AI) provides users with clean explanations of the decisions of AI machines through breakdowns that show how computers make their decisions. AI-powered IoT solutions that incorporate healthcare monitoring and intelligent infrastructure and autonomous cars become dangerous as Explainability is lacking because it contributes to a variety of issues along with threats of protection and regulatory challenges [12]. With a brilliant healthcare device, the wrong classification of sicknesses without providing a valid justification strategy could lead to hazardous consequences on patients. The urgent problems addressed by the XAI structures are the introduction of understandable AI methods that enhance every task along with the machine reliability (Nikezić et al., 2024). The figure 2 is the comparison of Traditional AI and Explainable AI (XAI). Synthetic intelligence structure understanding is the ability to comprehend the synthetic intelligence structure, which serves as a powerful part of debug fashions and machine performance improvement. The environment in which IoT gadgets operate has a variety of situations which require adaptable AI models with continuous acquisition of proficiency through newly acquired records. Operational issues based on AI become surmountable and explainability functions are absent because it becomes impossible to comprehend algorithms and verify independent decision-making strategies. The ethics recommendations of AI in the General Data Protection Regulation (GDPR) and other regulatory frameworks output the necessity of AI transparency. XAI aids organizations in achieving three objectives: it preserves consumer trust, monitors policies and pushes the safety and operational effectiveness of IoT systems through AI. The table 1 highlights the key differences between Traditional AI and Explainable AI (XAI). While Traditional AI operates as a "black box" where results are produced without interpretability, XAI introduces transparent and interpretable models that allow users to understand the reasoning behind AI decisions. This fosters user comprehension, improves decision-making, and builds trust in AI systems.

Table 1. Comparison Between Traditional AI and Explainable AI (XAI)

Aspect	Traditional AI	Explainable AI (XAI)
Learning Process	Relies on machine learning algorithms that generate learned functions.	Uses modern approaches to establish interpretable and explainable models.
Decision Transparency	Functions as a "black box," making interpretation difficult or impossible.	Provides transparent comprehension of decision-making through explanations.
User Understanding	Users receive results without clarity on how outputs are generated.	Users can understand the reasoning behind outputs, enabling better decisions.
Trust Factor	Limited trust due to opaque processes and lack of interpretability.	Builds trust by offering insights into model logic and fairness.

Figure 2. Comparison of Traditional AI vs. Explainable AI (XAI)

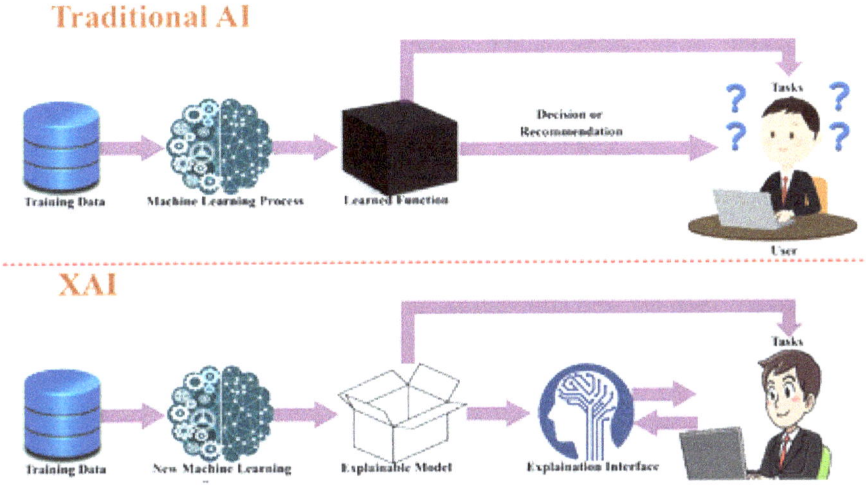

3.3 Key XAI Techniques (E.G., SHAP, LIME, GRAD-CAM)

Multiple explainable AI approaches exist to enhance the interpretability of complicated machine learning systems. SHAP stands as one of the most common approaches since it establishes importance values for model prediction variables. A game theoretical framework within explainability enables users to understand the influence of various input variables toward AI outputs. The SHAP approach delivers valuable results in IoT settings including predictive maintenance because it shows which sensor measurements play the most critical role in anomaly models. LIME (Local Interpretable Model-Agnostic Explanations) serves as a popular method which develops region-specific interpretable models to explain single predictions. The LIME approach brings successful application to IoT systems requiring prompt AI-based choices specifically in smart finance fraud prevention models. The deep learning-based image recognition model widely employs Grad-CAM (Gradient-weighted Class Activation Mapping).(Ergun & Can, 2022) The way Grad-CAM works is to produce visual heat maps which show where an image feature contributed most strongly to the model's decision. The technology proves highly valuable in smart surveillance applications because AI cameras need to demonstrate their reasons for detecting threats or objects during operation. The figure 3 depicts the essential principles of Explainable AI XAI through circular elements to highlight fundamental elements. Users need Explainable AI (XAI) to know how AI algorithms solve problems and how they go about it. To this end, systems are required to use open and transparent models that will be human understandable. Among the principles, it is necessary to consider fairness because it guarantees the absence of bias and inequity in the results and the creation of the user trust in AI systems that is essential to adopt and accept. Another important factor is robustness that focuses on the ability of AI to survive mistakes and adversarial examples. (Mahdavinejad et al., 2017)Also, XAI research has taken care of privacy issues through protection of user information with confidentiality. All of these principles combined enable the modern AI systems to be more ethical, reliable, and able to provide the user experience they can be improved. The table 2 summarizes the main XAI techniques, their working principles, and applications in IoT. It includes columns for technique, core principle, application domain, and benefits.

Table 2. Key Techniques of Explainable AI (XAI) for IoT Applications

Technique	Core Principle	Application Domain	Benefits
SHAP (SHapley Additive exPlanations)	Uses game-theoretic Shapley values to assign importance scores to input features.	Predictive maintenance in IIoT, anomaly detection	Identifies key sensor readings, enhances trust in automated decisions.
LIME (Local Interpretable Model-Agnostic Explanations)	Creates locally interpretable surrogate models to explain individual predictions.	Smart finance, fraud detection, healthcare monitoring	Provides real-time interpretability, explains single-instance decisions.
Grad-CAM (Gradient-weighted Class Activation Mapping)	Generates visual heatmaps highlighting critical regions influencing decisions.	Smart surveillance, medical imaging, traffic monitoring	Improves transparency in vision models, aids security and diagnostic systems.

Figure 3. Key Principles of Explainable AI (XAI)

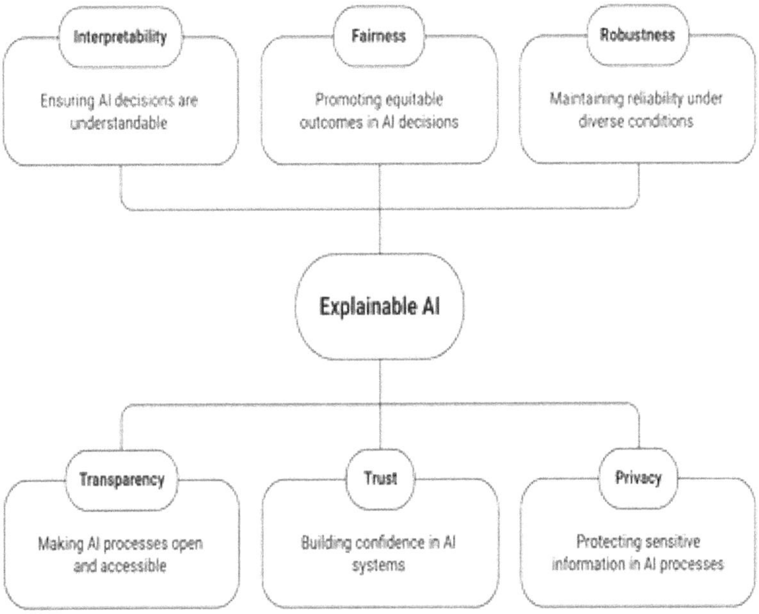

3.4 XAI for Decision-Making in Adaptive Systems

Adaptive systems in IoT use AI to learn and optimize performance through continuous updates of new data and changes in the environment. The evolution of these systems makes it more difficult to guarantee transparency in their decision-making processes. XAI functions as an essential component for making AI-driven adaptive systems interpretable thus allowing users to know the reasons behind specific adaptive IoT system behaviors. The predictive functionality of adaptive industrial IoT AI models determines machine failures through real-time sensor inputs. The implementation of XAI techniques reveals critical predictive factors to engineers who use this information to optimize their predictive models and maintenance procedures. The Figure 4 demonstrates the workflow of Explainable AI (XAI) between explainable and black-box prediction models. (Liu et al., 2019)

Figure 4. Explainable AI (XAI) Workflow: Comparing Explainable and Black-Box Models.

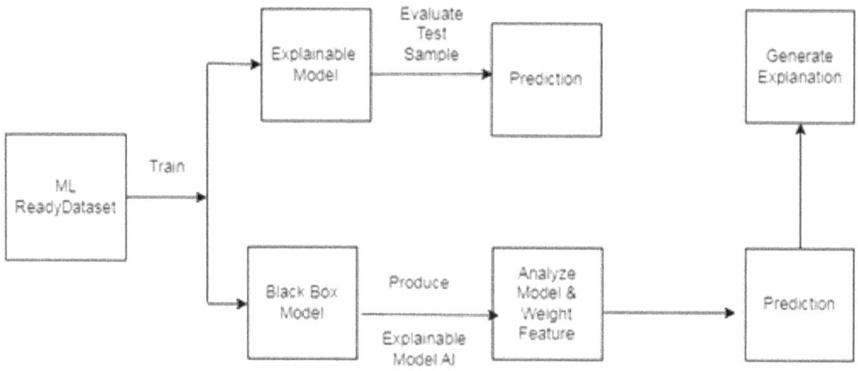

Both higher-level and lower-level models Training procedures of explainable and black-box models start with a single dataset at the ML-ready stage. Although explainable models literally produce easy to interpret predictions, black-box models assume another analytical step with feature and model weight analysis in order to produce explainable representations. Black-box models predictions thus require additional steps in the generation of explanations to improve transparency and create user confidence. This shows why explainable AI models are important, they can facilitate clear decision-making processes that enhance usability and confidence in the process by the stakeholders. Furthermore, XAI aids in improving user trust and collaboration between humans and AI systems. Adaptive traffic management controls utilized in autonomous transportation networks use changing traffic patterns to modify signals and adjust routing routes. (Alzahrani & Alfouzan, 2022) The unexpected decision made by the system which routes vehicles differently can be explained by XAI techniques. The expanded machine transparency allows operators to understand and accept AI guidelines and step in whilst required to maintain responsible adaptive systems that follow human selection-making ideas.

3.5 Interpretability vs. Performance Trade-Offs

Explainability requirements have gained favor in influencing the performance of the model in simpler ways. While decision trees and linear regression provide better model interpretability, they produce results that seem to be lower in performance than deep learning models. Deep neural networks can be a strong predictive tool; however, their black-box nature makes it very difficult for a human to understand how the model is operating. Tradeoffs between the desired level of interpretability and the highest possible performance capabilities are regarded as a critical issue on the IoT systems that utilize AI as one of the primary moving forces. The ability to balance these requirements and the choice of appropriate models together with the XAI technology which will be a success will depend on operation setting. The adoption of hybrid systems combines the appropriate information by using deep learning and integration with less complex rule-based systems, or model-based systems, to develop practical solutions with choice-making logic. In a way, the performance of black box models is maintained high by being post-hoc interpretable (LIME and SHAP). XAI implementation provides organizations with practical primary resources to maintain state-of-the-art IoT systems with transparency, thereby boosting user trust and ease of use.

4. ADAPTIVE SYSTEMS INTEGRATION IN IOT

The integration of AI into IoT will allow the various handheld and embedded systems to constantly change, be optimally efficient, and provide better decision-making capabilities. The various adaptive systems employ machine learning together with real-time data analytics and automation to enhance responsiveness and efficiency in different applications, such as smart cities, healthcare, and industrial automation. Different from static systems, an adaptive IoT system keeps learning from real-time data, where this new data will help it further improve predictions, detect anomalies, and fine-tune operational parameters automatically. (Mohamed Egypt, 2020) The integration of AI-driven adaptive systems within IoT brings many challenges, including computational constraints, interoperability concerns, and security risks. Tackling these challenges demands a combination of edge computing, federated learning, and Explainable AI (XAI) for transparency, reliability, and trust in adaptive IoT ecosystems. The figure 5 depicts a smart city concept through its hexagon-shaped elements which connect to one another. A smart environment emphasizes sustainable practices aimed at preserving natural resources, while a smart transportation system delivers effective and intelligent mobility services. In the healthcare domain, smart solutions leverage advanced medical technologies to provide better and more accessible services. Smart energy initiatives focus on the adoption of renewable sources and efficient power management strategies. Similarly, smart industry integrates automation and digitization to enhance productivity and operational efficiency. Finally, smart living promotes an improved quality of life through the development of smart homes and buildings, contributing to safer, more convenient, and sustainable urban living.(Bokhari & Myeong, 2023)

Figure 5. Key Domains of a Smart City

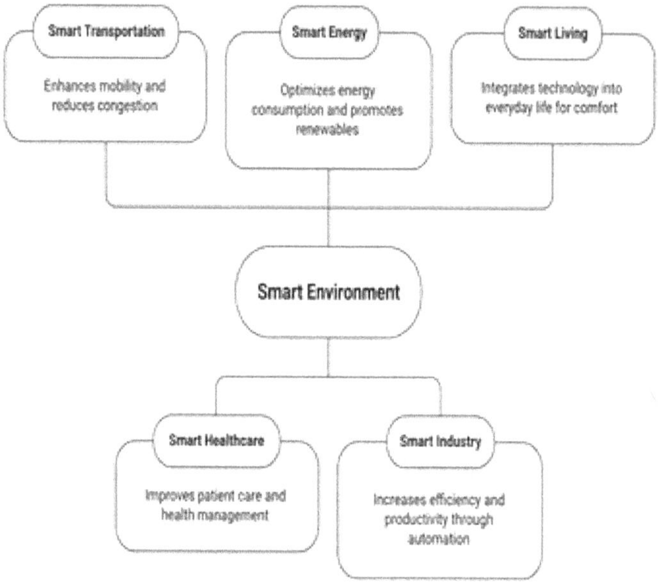

4.1 Definition and Characteristics of Adaptive Systems

Adaptive systems, on the other hand, are supposed to be intelligent platforms that adapt their behavior automatically, depending upon the changing environment and real-time data. Exceedingly different from static systems, these systems, which follow static rules, keep learning, optimizing, and changing their operation to be more efficient and effective. Adaptive systems are based on a mix of AI, machine learning, and high-end analytics that can observe data patterns in real time and make optimization and control decisions practically without any human intervention. Self-learning, real-time responses, scaling possibilities, and capability to resist changes in the environment are some common traits of adaptive systems. Adaptive systems, therefore, become very crucial in any field for working in complex IoT environments, managing connected systems, automating processes, or optimizing performance in an organization. In the case of smart grids, the adaptive energy management system is adjusted according to the fluctuation of demand and distribution of power when need changes, thus reducing energy wastage. Such characteristics make these adaptive systems also extremely important for predictive maintenance in IIOT, where they are used to detect the early warning signs of an approaching equipment failure and adjust maintenance schedules on their own. These systems increase efficiency, reliability, and automation within IoT ecosystems, making them extremely important. (Ghosh et al., 2018)

4.2 Role of AI and Machine Learning in Adaptive IoT Systems

Adaptive IoT systems are based on AI and machine learning that enable them to process high amounts of real-time data and identify anomalies and make data-driven decisions. Conventional rule-based IoT systems can do so abysmal with any unforeseen change, whereas an AI-assisted adaptive system keeps learning on the input given to it by the sensor and past data to make better decisions. Such machine-learning algorithms as neural networks and reinforcement learning allow IoT devices to enhance their reactions between different times and human interference. As an example, in smart agriculture, IoT systems that are based on AI set the irrigation time according to the level of soil moisture, weather predictions, and crop requirements and enable the efficient utilization of water. Similarly, adaptive traffic management systems are AI-based in autonomous transportation systems, where the conditions of the roads and traffic flows are examined to adjust signal timing to reduce traffic. The implications of these technologies are enormous in terms of adapting IoT systems to automation, efficiency, and data-driven intelligent solutions in many sectors by applying AI and machine learning. (Wijaya et al., 2024)

4.3 Transfer Learning for Efficient System Adaptation

Transfer learning is one of the fundamental enablers of adaptive IoT systems wherein AI models take advantage of the previously trained knowledge from a certain domain and fine-tune it for use within another one with minimal retraining. Using this, time and resource utilization is significantly reduced for development work involving AI solutions, often acquiring best-suited functionality for IoT application scenarios having limited data gathering and processing constraints. Transfer learning allows for more rapid adaptation of systems to enable IoT devices to perform well and improve efficiently in different environments without requiring colossal datasets. For instance, a machine learning model trained on an anomaly detection on a state-of-the-art factory machine can easily adapt to slightly varying operational contexts by introducing representative training from different factories. Similarly, under the IoT-enabled healthcare setting, devices will be able to personalize health monitoring through transfer learning adaptation of pre-trained models to user-specific information. This guarantees that with quick adaptability, the AI based IoT systems are flexible enough, reactive, and capable of functioning in changing real world conditions. (Aliyah et al., 2023)

4.4 Challenges in Integrating AI-Driven Adaptive Systems

There are also several challenges in the integration of the AI-driven adaptive system with the IoT environments. One of the biggest issues is the issue of data variability, namely, the fact that IoT devices may be operating in various types of settings, therefore, their functionality cannot be used to train AI models to be able to generalize to other settings. This would be necessary with continuous updates and adaptation of domains to the models. Poor computational power is a second problem as most IoT hardware cannot execute complicated AI models with inefficiency and most of them do not have enough processing power. Alternatives to solve this would be edge computing and federated learning. Security and privacy issues are also substantial hurdles for adapting IoT systems driven by AI. Since these systems are based on the collection of real-time data and autonomous decision-making, they are vulnerable to cyberattacks, data breaches, and adversarial manipulation. A strict requirement to ensure security would include encrypted data transmission, secure model updates, and anomaly detection for cybersecurity threats. Regulatory compliance with data privacy laws, such as GDPR, should be maintained for responsible deployment of AI in IoT systems. (Ahmad et al., 2021)

5. SYNERGY BETWEEN TRANSFER LEARNING & XAI FOR IOT

Synergy between transfer learning and explainable AI (XAI) empowers adaptability, efficiency, and transparency in AI-driven systems. In the IoT domain, this transfer learning facilitates IoT devices to apply well-instructed and well-parameterized models, thus minimizing data collection and computational power requirements, ideally useful in dynamically changing environments like smart cities, healthcare, and industrial automation. The black-box nature of transfer learning raises questions regarding trust and interpretability, and that is where XAI comes into play. SHAP, LIME, and Grad-Cam, based on XAI can be used to learn the reasoning that a transfer learning model uses to make its conclusions in the name of accountability and reliability. Thus, such a mix will enable the IoT applications to swiftly adjust to the changes and show transparency and bring a sense of trust toward users, providing a solution to some of the most acute challenges of AI automation. (Rejeb et al., 2022)

5.1 How Transfer Learning Enhances Adaptive Systems

These AI models provide transfer learning capabilities, and it plays a critical part in enhancing adaptability in many IoT systems by allowing generalization across multiple environments and tasks. While optimal learning in these AI models may require large amounts of labeled data, such a method is expensive and laborious in IoT environments where data collection and labeling are required. Transfer learning offers this solution whereby pre-trained models can be fine-tuned on new tasks with fewer examples to enable an adaptive IoT system to perform more efficiently. For instance, an industrial machinery's fault detection model developed on transfer learning can be adapted to several other machines with only minimal modifications. Furthermore, an AI-present pretrained speech recognition in home automation can be tweaked with individual users using speech patterns. These features assure flexibility for adaptive IoT systems, achieve efficiency, reduce training costs rate, and respond to changes. (Tan et al., 2018)

5.2 Role of XAI in Making Transfer Learning Transparent

While transfer learning permits adaptability, its black box raises the issue of transparency and trust with the model's predictions. Explainable AI (XAI) seeks to reduce this concern by making suggestions about how exactly transfer learning models predict, hence discarding the fear of decisions being no interpretable, uncertain, and haphazard. XAI techniques such as SHAP, LIME, and Grad-CAM allow stakeholders to understand what features drive the model outputs and ease the validation and refinement of AI-driven IoT systems. Consider predictive maintenance for IIoT, XAI can show the sensor readings contributing most toward the prediction of failure of that particular machine, enabling the engineers to check that particular decision before any remedial action is taken. For instance, in healthcare IoT, XAI could explain why a wearable device flagged up an abnormal heart rate within a certain environment, providing that user and medical professionals have confidence in the recommendations the system provides. By embedding XAI in transfer learning, IoT applications become more accountable, interpretable to end-users, and hence resonating with user expectations. (Anjomshoaa & Curry, 2020)

5.3 Case Study: XAI-Driven Transfer Learning for IoT Devices

Some practical examples of explainable AI-based transfer learning in IoT include smart city traffic management applications wherein a deep-learning model for traffic flow prediction is transferred from one city to another with disparate road conditions, minimizing retraining effort. With XAI techniques, city planners are able to visualize factors like the weather, traffic density, or road infrastructure that most influence the model's prediction. Such transparency enables decision-makers to optimize urban mobility strategies more confidently. The second case study involves wearable health-care devices whereby transfer learning adapts the AI models for personalized health monitoring. A model originally developed for general patient data can be fine-tuned to apply to individuals with specific conditions such as diabetes and hypertension. For instance, when AI recommends changes in lifestyle for patients, XAI could help health workers explain to patients the reasoning behind certain lifestyle modifications. Examples showcase how the synergistic approach of transfer learning and XAI gives rise to the more adaptive, accurate, and transparent IoT application. (M.-Y. Kim et al., 2021)

5.4 Security and Privacy Considerations

The incorporation of the transfer learning and XAI to the IoT systems entails security and privacy concerns. Transfer learning models that are trained using a shared source of data may end up being a source of private information, exchanging it to data theft. This is important in healthcare and finance because the legislation safeguards the privacy. Federated learning is one of the methods to fight privacy risks, where the models can be trained on decentralized data sources without transfer of raw data. In addition, XAI schemes should be designed in such a manner that they reduce adversarial abuse. What the attacker might do, on such access to the model explanation, is change the input data so as to make the model believe that the attacker has performed some significant steps that the target analysis would consider significant. In a case with cybersecurity IoT applications, an attacker would use XAI information to initiate an adversarial attack that bypasses intrusion detection systems. There are several steps that can be implemented to guarantee the safety of model deployment, do adversarial testing, and implement encryption solutions. It is only through a balance between security and clarity with a high level of sensitivity that the IoT systems can utilize the capabilities of transfer learning and XAI without having to invite threats. (Ghaffari et al., 2024)

6. APPLICATIONS AND USE CASES

Transfer Learning and Explainable AI (XAI) are applicable in the domains of IoT to improve their adaptability, efficiency, and transparence in AI-based systems. Transfer learning ensures predictive traffic management and energy optimization in such cities by transferring AI models trained in one environment to the other, whereas XAI provides transparency in the decision-making of urban planners. Wearables utilize transfer learning for personalization, and XAI gives explainability for aiding doctors and patients in understanding AI-based diagnosis in the healthcare sector. Transfer learning also assists predictive maintenance in IIoT by providing the foundation for AI models to adapt to various machines, whereas XAI ensures that the fault predictions are trustworthy through explainability. Likewise, smart-home applications learn the behavior of users from AI assistants, whereas XAI strives to develop confidence in users by bringing interpretability to automation decisions. Together, these technologies create highly efficient adaptive and accountable IoT ecosystems. (Zantalis et al., 2019)

6.1 Smart Cities and IoT-Enabled Infrastructure

Smart cities leverage IoT and AI to enhance urban management, improve traffic flow, optimize energy consumption, and ensure public safety. Transfer learning enables predictive analytics for traffic congestion, air quality monitoring, and smart grid management by adapting AI models trained in one city to another. Explainable AI (XAI) further ensures transparency in decision-making, allowing city officials to understand and trust AI-driven recommendations. For instance, XAI-driven traffic management systems can justify route changes, while AI-powered surveillance can provide interpretable anomaly detection for public safety applications. (Khan & Aslam, 2020)

6.2 Healthcare and Wearable Devices

In healthcare, IoT-enabled wearable devices utilize AI for real-time monitoring of patients' health, for anomaly detection, and for providing personalized recommendations. Transfer learning allows models trained on large-scale medical datasets to be fine-tuned for individual patients, which improves diagnostic accuracy. However, trust in AI and healthcare requires explainability, which is provided by XAI through interpreting model decisions while assuring accountability. For example, in monitoring heart rates, instant insight into clinical alerts with techniques such as SHAP may inform why a specific reading was flagged by the wearable device, thereby assisting patient and physician in clinical decision making. (University et al., 2023)

6.3 Industrial IoT (IIoT) and Predictive Maintenance

The IIoT combines AI and IoT for predictive maintenance, helping companies reduce expensive machinery breakdowns. Transfer learning allows AI models trained on one class of machines to be readily retrained to work on other machines, thereby reducing downtime maintenance costs. Explainable AI provides insight into what constitutes the failures or alerts developed by these models and provides knowledge to engineers as to the specific sensor readings contributing to the alert. This enables better decision-making, thereby making maintenance actions based on facts and reliable.

6.4 Smart Homes and Edge Computing

Smart homes use and benefit off of automation, security, and energy efficiency, where IoT devices use them to provide smart assistants, thermostats, and security systems and apply transfer learning to provide personalization via learning mobility. With edge computing, these AI models can work onboard with less latency; therefore, better privacy. Explainable AI provides arrays of facts as to why a smart home would make its decisions, such as raising the temperature or turning on a security alarm, raising the dependency on AI-driven home automation apps. (University et al., 2023)

7.CHALLENGES AND FUTURE DIRECTIONS

The sub-functional issues of Transfer Learning to the IoT and XAI, in particular, the adaptive system integration, are summarized as difficulties with computational capabilities, data privacy, and the extremely complicated character of AI models, preserving their efficiency, and interpretability. The majority of the IoT devices use schemed-down processing thus complicating the implementation of large-scale transfer learning models. Moreover, it may be a problem of domain adaptation because models that are trained in one type of environment would not be easy to generalize to various IoT environments. Privacy concerns also come into play when data is being transferred between different datasets, most importantly, where it comes to healthcare. In the case of XAI, on the one hand, it is the problem of striking a balance between the quality of model performance and the extent of transparency since well explainable models are often not accurate. Further studies ought to focus much energy on the theatricalization of lightweight self-explainable AI, further develop domain adaptation methods in transfer learning in order to instill ethical AI deployments into strict regulatory frameworks. Integration of federated learning and edge computing may also help with scaling out while preserving the privacy of AI-based IoT systems, capability for adaptation, and interpretability. (University et al., 2023) The table 3 presents the major challenges (e.g., computational constraints, domain adaptation, privacy/security, performance vs. interpretability trade-offs) alongside proposed future directions (lightweight self-explaining models, federated learning, edge computing, regulatory frameworks).

Table 3. Challenges and Future Directions in Transfer Learning and XAI for IoT

Challenge	Description	Future Direction
Computational Constraints	IoT devices have limited memory and processing power, restricting model usage.	Develop lightweight self-explaining models optimized for edge devices.
Domain Adaptation Issues	Models trained in one environment fail to generalize across diverse IoT settings.	Enhance transfer learning with improved domain adaptation techniques.
Privacy and Security Risks	Sensitive data (e.g., healthcare, finance) exposes systems to breaches or misuse.	Adopt federated learning, encrypted data exchange, and secure model updates.
Performance vs. Interpretability Trade-off	High accuracy black-box models lack explainability, while interpretable models may lose precision.	Combine hybrid models and post-hoc XAI methods to balance accuracy and transparency.
Regulatory and Ethical Concerns	Compliance with data privacy laws and ethical AI standards is challenging.	Implement robust auditing frameworks and align systems with global AI regulations.

7.1 Limitations of Transfer Learning in IOT

The technology of transfer learning applied to the Internet of Things needs to overcome a few challenges, a particularly distinctive one being that domain shift in itself. Such an occasion arises when AI models applicable in one domain appear incompatible with models in new, changed environments. Yet again, the IoT devices possess limited capabilities for computation, rendering the deployment of large-scale transfer learning models infeasible. The relocation of knowledge between one dataset to another is also put up to question as far as data privacy is concerned primarily in sensitive areas like healthcare.

7.2 The Need for Better Explainability in AI Models

As AI moves further into the IoT, model interpretability and trustworthiness have become a necessity. Numerous deep learning models are black boxes, and thus they are hard to ask about what they are doing. This is because the non-explainability of the model can lead to the unwillingness to introduce AI-powered IoT solutions, which will be even greater in the limited spheres of critical usage, such as the health sector and autonomous vehicles. To make their effectiveness in the model XAI techniques need to be refined further to offer more tangible, sustained explanations.

7.3 Future Research Directions in XAI and Adaptive IOT Systems

The development of lightweight and explainable systems will become the center of the further research on Explainable AI (XAI) and adaptive IoT systems, which can be implemented effectively on edge devices with small resources set. Among the directions is the extension of domain adaptation in transfer learning in such a way that AI models become adaptable in different environments, and retraining becomes more costly. The second potential area is the development of self-explaining AI models, which ought to reflect interpretability, without post-hoc explanations as well (SHAP or LIME). Such models would give ground up transparency and this would make the users have confidence in such models but it will not interfere with the performance of such models. These lines of research must collectively aim at realizing efficiency, accuracy and explainability, which offers practical AI-assisted IoT ecosystems of efficiency, scale and resilience.

7.4 Ethical and Regulatory Considerations

There are some ethical and regulatory issues when AI is incorporated into IoT, specifically, data privacy, bias, and accountability. Significant AI policies, such as the General Data Protection Regulation and AI Governance Frameworks, highlight the need of transparency in the AI-enabled systems. Ethical AI obtained with transfer learning and XAI applications will contribute significantly to responsible AI adoption. Formation of popular frameworks of auditing AI decisions in IoT will address the majority of general concerns regarding fairness, security and compliance with regulations.

8.CONCLUSION

The integration of transfer learning and Explainable Artificial Intelligence (XAI) into systems that have an IoT or adaptability is an important milestone in the operations of the operations of AI-enabled technologies in volatile settings, so the given advancement will resolve some of the issues of AI-powered systems that operate in dynamic environments in the future. Transfer Learning will be able to make AI models use a bit of the other previously acquired knowledge and therefore take less time to train any device, and the same time efficiently enable IoT devices to adapt to the new circumstances. XAI, in its turn, will guarantee the disclosure and explicability of such AI models, therefore, addressing claims about the element of trust and accountability. Combined, the technologies are one of the major actors in the decision-making, efficiency, and innovative processes of industries like smart cities, healthcare, industrial IoT, and smart homes. Nevertheless, there are some obstacles to their implementation to gaming even with such strong advantages. With highly computer-limited IoT devices, in particular, the accuracy complexity of doing things with AI models is restricted, and domain adaptations within issues is impossible because a model trained to perform well in one domain is not well generalized in a different domain. The privacy concerns are due to the transfer of knowledge between datasets in which this can be seen best in sensitive areas, including healthcare and finance. Tradeoff of XAI is transparency versus interpretability and performance; these have been left to research to come up with other best possible solutions that can balance both components.

It can be expected that future evolution of AI and IoT is going to be based on resolving these issues by enhancing the efficiency of lighter-weight AI models capable of operating on edge devices. The privacy and adaptability can be improved as well by using federated learning and decentralized methods to develop AI. Moreover, self-explaining AI models will introduce real-time interpretability without a loss in performance of AI in adaptive systems, and it will be superior in XAI. The ongoing study of domain adaptation methods and the implementation of ethical AI should ensure that Transfer Learning and XAI will make beneficial contributions to the solutions based on the Internet of Things. Finally, synergetic combination of Transfer Learning and XAI has enormous potential in designing adaptive IoT systems in the future. Efficient knowledge transfers, as well as transparency in AI-driven decision-making will also provide smarter, workable and ethically responsible IoT applications. Nonetheless, the route towards adoption more generally will consist in surmounting the current constraints, advocating cooperative intervention between different fields, and streamlining enforced regulation. In the course of the research, it will thus become clear that the joint application of Transfer Learning and XAI will still give increasingly smarter, adaptable, explainable IoT ecosystems that will eventually transform the industries and life itself.

REFERENCES

Ahmad, I., Shahabuddin, S., Sauter, T., Harjula, E., Kumar, T., Meisel, M., Juntti, M., & Ylianttila, M. (2021). The challenges of artificial intelligence in wireless networks for the internet of things: Exploring opportunities for growth. *IEEE Industrial Electronics Magazine*, *15*(1), 16–29. DOI: 10.1109/MIE.2020.2979272

Aliyah, L., Lukita, C., Pangilinan, G. A., Chakim, M. H. R., & Saputra, D. B. (2023). Examining the impact of Artificial Intelligence and Internet of Things on Smart Tourism Destinations: A comprehensive study. [ATT]. *Aptisi Transactions on Technopreneurship*, *5*(2sp), 135–145. DOI: 10.34306/att.v5i2sp.332

Alzahrani, N. M., & Alfouzan, F. A. (2022). Augmented Reality (AR) and cyber-security for smart cities-A systematic literature review. *Sensors (Basel)*, *22*(7), 2792. DOI: 10.3390/s22072792 PMID: 35408406

Anjomshoaa, A., & Curry, E. (2020). Inter-space machine learning in smart environments. In *Lecture Notes in Computer Science* (pp. 535–549). Springer International Publishing., DOI: 10.1007/978-3-030-57321-8_30

Awan, K. A., Din, I. U., Almogren, A., & Rodrigues, J. J. P. C. (2023). Privacy-preserving big data security for IoT with federated learning and cryptography. *IEEE Access : Practical Innovations, Open Solutions*, *11*, 120918–120934. DOI: 10.1109/ACCESS.2023.3328310

Bokhari, S. A. A., & Myeong, S. (2023). The impact of AI applications on smart decision-making in smart cities as mediated by the internet of things and smart governance. *IEEE Access : Practical Innovations, Open Solutions*, *11*, 120827–120844. DOI: 10.1109/ACCESS.2023.3327174

Ergun, A. E., & Can, Ö. (2022). Ensuring IoT privacy using padding strategies against machine learning approaches. *International Journal of Multidisciplinary Studies and Innovative Technologies*, *6*(2), 193. DOI: 10.36287/ijmsit.6.2.193

Geukes Foppen, R. J., Gioia, V., Gupta, S., Johnson, C. L., Giantsidis, J., & Papademetris, M. (2025). Methodology for safe and secure AI in diabetes management. *Journal of Diabetes Science and Technology*, *19*(3), 620–627. DOI: 10.1177/19322968241304434 PMID: 39726262

Ghaffari, A., Jelodari, N., Pouralish, S., Derakhshanfard, N., & Arasteh, B. (2024). Securing internet of things using machine and deep learning methods: A survey. *Cluster Computing*, *27*(7), 9065–9089. DOI: 10.1007/s10586-024-04509-0

Ghosh, A., Chakraborty, D., & Law, A. (2018). Artificial intelligence in Internet of things. *CAAI Transactions on Intelligence Technology*, *3*(4), 208–218. DOI: 10.1049/trit.2018.1008

Hoenig, A., Roy, K., Acquaah, Y. T., Yi, S., & Desai, S. S. (2024). Explainable AI for cyber-physical systems: Issues and challenges. *IEEE Access : Practical Innovations, Open Solutions*, *12*, 73113–73140. DOI: 10.1109/ACCESS.2024.3395444

Hussain, F., Hassan, S. A., Hussain, R., & Hossain, E. (2020). Machine learning for resource management in cellular and IoT networks: Potentials, current solutions, and open challenges. *IEEE Communications Surveys and Tutorials*, *22*(2), 1251–1275. DOI: 10.1109/COMST.2020.2964534

Kalyanathaya, K. P., & Prasad, K. (2024). Novel method for developing explainable machine learning framework using feature neutralization technique. *The Scientific Temper*, *15*(02), 2225–2230. DOI: 10.58414/SCIENTIFICTEMPER.2024.15.2.35

Khan, I. U., & Aslam, N. (2020). A deep-learning-based framework for automated diagnosis of COVID-19 using. *Information (Basel)*, *11*(9), 419. DOI: 10.3390/info11090419

Khazane, H., Ridouani, M., Salahdine, F., & Kaabouch, N. (2024). A holistic review of Machine Learning adversarial attacks in IoT networks. *Future Internet*, *16*(1), 32. DOI: 10.3390/fi16010032

Kim, D., Antariksa, G., Handayani, M. P., Lee, S., & Lee, J. (2021). Explainable anomaly detection framework for maritime main engine sensor data. *Sensors (Basel)*, *21*(15), 5200. DOI: 10.3390/s21155200 PMID: 34372436

Kim, M.-Y., Atakishiyev, S., Babiker, H. K. B., Farruque, N., Goebel, R., Zaïane, O. R., Motallebi, M.-H., Rabelo, J., Syed, T., Yao, H., & Chun, P. (2021). A multi-component framework for the analysis and design of explainable artificial intelligence. *Machine Learning and Knowledge Extraction*, *3*(4), 900–921. DOI: 10.3390/make3040045

Kumar Dinkar, A., Alimul Haque, M., & Kumar Choudhary, A. (2024). Enhancing IoT data analysis with machine learning: A comprehensive overview. *LatIA*, *2*, 9. DOI: 10.62486/latia20249

Li, H., Ota, K., & Dong, M. (2018). Learning IoT in edge: Deep learning for the internet of things with edge computing. *IEEE Network*, *32*(1), 96–101. DOI: 10.1109/MNET.2018.1700202

Liu, X., Qian, C., Hatcher, W. G., Xu, H., Liao, W., & Yu, W. (2019). Secure internet of things (IoT)-based smart-world critical infrastructures: Survey, case study and research opportunities. *IEEE Access : Practical Innovations, Open Solutions*, *7*, 79523–79544. DOI: 10.1109/ACCESS.2019.2920763

Mahdavinejad, M. S., Rezvan, M., Barekatain, M., Adibi, P., Barnaghi, P., & Sheth, A. P. (2017). Machine learning for Internet of Things data analysis: A survey. *Digital Communications and Networks*. Advance online publication. DOI: 10.1016/j.dcan.2017.10.002

Mazhar, T., Talpur, D. B., Al Shloul, T., Ghadi, Y. Y., Haq, I., Ullah, I., Ouahada, K., & Hamam, H. (2023). Analysis of IoT security challenges and its solutions using artificial intelligence. *Brain Sciences*, *13*(4), 683. DOI: 10.3390/brainsci13040683 PMID: 37190648

Meziane, H., & Ouerdi, N. (2023). A survey on performance evaluation of artificial intelligence algorithms for improving IoT security systems. *Scientific Reports*, *13*(1), 21255. DOI: 10.1038/s41598-023-46640-9 PMID: 38040952

Mohamed, E. (2020). The Relationship between Artificial Intelligence and Internet of Things: A quick review. *Journal of Cybersecurity and Information Management*, *30–34*, 30–34. Advance online publication. DOI: 10.54216/JCIM.010101

Nazat, S., Arreche, O., & Abdallah, M. (2024). On evaluating black-box explainable AI methods for enhancing anomaly detection in autonomous driving systems. *Sensors (Basel)*, *24*(11), 3515. DOI: 10.3390/s24113515 PMID: 38894306

Nikezić, D. P., Radivojević, D. S., Lazović, I. M., Mirkov, N. S., & Marković, Z. J. (2024). Transfer learning with ResNet3D-101 for global prediction of high aerosol concentrations. *Mathematics*, *12*(6), 826. DOI: 10.3390/math12060826

Rejeb, A., Rejeb, K., Zailani, S. H. M., & Abdollahi, A. (2022). Knowledge diffusion of the Internet of Things (IoT): A main path analysis. *Wireless Personal Communications*, *126*(2), 1177–1207. DOI: 10.1007/s11277-022-09787-8 PMID: 35694533

Schulte, L., Ledel, B., & Herbold, S. (2024). Studying the explanations for the automated prediction of bug and non-bug issues using LIME and SHAP. *Empirical Software Engineering*, *29*(4), 93. Advance online publication. DOI: 10.1007/s10664-024-10469-1

Sheu, R.-K., & Pardeshi, M. S. (2022). A survey on medical explainable AI (XAI): Recent progress, explainability approach, human interaction and scoring system. *Sensors (Basel)*, *22*(20), 8068. DOI: 10.3390/s22208068 PMID: 36298417

Sheu, R.-K., Pardeshi, M. S., Pai, K.-C., Chen, L.-C., Wu, C.-L., & Chen, W.-C. (2023). Interpretable classification of pneumonia infection using eXplainable AI (XAI-ICP). *IEEE Access : Practical Innovations, Open Solutions, 11*, 28896–28919. DOI: 10.1109/ACCESS.2023.3255403

Tan, C., Sun, F., Kong, T., Zhang, W., Yang, C., & Liu, C. (2018). A survey on deep transfer learning. In *Artificial Neural Networks and Machine Learning – ICANN 2018* (pp. 270–279). Springer International Publishing., DOI: 10.1007/978-3-030-01424-7_27

University, M. babu, Prasad, J. L. Parvateesam, C., management, A. I. of technology and, Nagulmeera, S., University, M. babu, Avinash, V., & college, R. engineering. (. (2023). Enhancing IoT security: Addressing challenges, implementing solutions, and envisioning cybersecurity trends for the future. *International Journal of Engineering Applied Sciences and Technology*, *8*(6), 48–55. DOI: 10.33564/IJEAST.2023.v08i06.006

Wijaya, S., Hermanto Rudy, L., Debora, F., Ardila Rahma, R., Ramadhan, A., & Attaqwa, Y. (2024). Artificial intelligence and internet of things in manufacturing decision processes. IAES International Journal of Artificial Intelligence (IJ-AI), 13(2), 2185. DOI: 10.11591/ijai.v13.i2.pp2185-2200

Zantalis, F., Koulouras, G., Karabetsos, S., & Kandris, D. (2019). A review of Machine Learning and IoT in smart transportation. *Future Internet*, *11*(4), 94. DOI: 10.3390/fi11040094

KEY TERMS AND DEFINITIONS

Adaptive IoT Systems: Intelligent IoT platforms that continuously adjust behavior in response to real-time data and environmental changes.

Domain Adaptation: The process of modifying AI models so they can perform effectively in new environments different from the ones used during training.

Edge Computing: A distributed computing paradigm that processes data near the source (IoT devices) to reduce latency and enhance efficiency.

Explainable AI (XAI): A set of methods and frameworks that make AI model decisions transparent, interpretable, and understandable to humans.

Fairness in AI: Ensuring AI systems make unbiased and equitable decisions across different groups of users.

Federated Learning: A privacy-preserving approach to model training where multiple devices collaboratively learn without sharing raw data.

Robustness: The ability of an AI system to maintain performance in the presence of errors, noise, or adversarial attacks.

Transfer Learning: A machine learning technique that reuses knowledge from a pre-trained model to perform a new but related task with less data and computation.

Chapter 4

AI and IoT for Urban Planning and Management:

Towards Smart and Sustainable Cities

B. Prabha
Panimalar Engineering College, India

P. Rashmi
Velammal Engineering College, India

B. Yamini
https://orcid.org/0000-0003-3531-108X
SRM Institute of Science and Technology, India

D. Senthil Kumar
R.M.K. College of Engineering and Technology, India

K. Vijayalakshmi
https://orcid.org/0000-0002-4827-3011
SRM Institute of Science and Technology, India

K. Raju
Rajalakshmi Engineering College, India

Beslin Pajila
https://orcid.org/0000-0002-3905-2460
Vel Tech Rangarajan Dr. Sagunthala R&D Institute of Science and Technology, India

ABSTRACT

The integration of Artificial Intelligence (AI) and the Internet of Things (IoT) has the potential to revolutionize urban planning, fostering the development of smart cities. AI enhances urban management through predictive analytics, decision-making, and automation, while IoT enables real-time data collection and monitoring of city infrastructure. Together, they optimize transportation, energy consumption, waste management, and public safety, contributing to more sustainable, efficient, and

DOI: 10.4018/979-8-3373-3146-1.ch004

liveable urban environments. However, the implementation of these technologies faces challenges, including data privacy, security concerns, and interoperability issues. Addressing these barriers is crucial for unlocking the full potential of smart cities. This chapter explores the applications, challenges, and future directions of AI and IoT in urban planning, offering insights into the transformative role of these technologies in shaping the cities of tomorrow.

1. INTRODUCTION

1.1 Overview of Smart Cities and Urban Planning

Smart cities are the new normal of a city where information technology (IT) and (IoT) systems are being adopted to enhance the efficiency, sustainability, and quality of life within cities. Smart city is premised on the idea of information and digital technologies to optimise the services, infrastructure and governance of cities and generates more responsive, sustainable and liveable cities. The cities apply the connected devices, sensors and data analytics to manage their resources and to overcome the urban challenges and enable economic development effectively.

Urbanization is global and it has experienced more than half of the world population moving to the urban centers and the same is likely to be the case in the coming decades. This rate of urbanization is becoming a big challenge that entails overcrowding, pollution, wastage of resources and overloading of the available infrastructure. The urban areas are still growing in terms of the need of services, public safety, healthcare, transportation, housing among other needs. At the same time, the climate change and the environmental degradation has put additional pressure on cities to become more sustainable. As the cities continue to grow, there is a need to get smarter methods of managing the city and making it a better place to live in.

The appearance of smart cities is grounded on the fact that the traditional approach to urban planning which is reactive and fragmented can no longer be employed. Instead, the modern urban planning shall be prognostic and futuristic, which means that the use of highly advanced technologies to predict and eliminate the urban issues shall be deployed. This sort of technologies enables cities to become more efficient in terms of organising the traffic, the energy, the waste and even the safety of people and makes the processes sustainable in the long-term perspective. The emergence of data analytics, cloud computing, IoT, and AI technologies that change the role of cities can be viewed as the key factors that can result in the development of smart cities (Bahl et al., 2024).

1.2 Introduction to AI and IoT

The modern smart cities are built on the basis of Internet of Things (IoT) and Artificial Intelligence (AI). IoT refers to the system of interrelated objects, which gather data, share and process it to enhance the efficiency of operations and make the city more comfortable. Through the IoT networks, embedded systems, data-collection devices and sensors, the most significant urban infrastructure and resources can be monitored in real-time. These are intelligent traffic lights and sensor, energy meters, garbage disposal, and all these are useful to collect useful data, which provides an overview of how the city is running at the moment of time (Gómez et al., 2017). On the other hand, Artificial Intelligence is the field of computer science that allows the machine to do the task that is traditionally a preserve of human intelligence like problem solving, learning and decision making. AI systems use algorithms to take large volumes of data through processing and analysis, learn patterns and make predictions using past and current data. Using AI, the potential of the IoT systems can be enhanced in a smart city, as it enables one to make decisions on the basis of data, forecast the tendencies of a city, and optimize its functioning, traffic, energy, and even safety.

AI and IoT tools can be combined to make cities build intelligent systems, which are not only observant of city processes but also make their own decisions. An example is that the data that IoT sensors gather on a traffic system can be processed by AI algorithms to regulate the traffic lights in real-time depending on the traffic flow in order to clear the traffic which would reduce time spent on commute. Similarly, AI can also be applied in energy management to analyse the pattern of the consumption of energy as gathered by the IoT sensors to allocate the energy effectively across the city.

The AI and IoT are symbiotic. The data is produced by IoT devices, and AI interprets and works on the data to produce usable information. With such combination, automation, predictive abilities and improved decision-making can be achieved, all of which are essential to the operation of a smart city. In the increasingly digital and connected cities, AI and IoT integration is the key to the development of smart systems capable of operating everything, including transportation and energy grids, with maximum efficiency.

1.3 Relevance to Urban Planning

Urban planning is a conventional process of planning and building cities gradually and systematically. But as the rate of urbanization has been increasing and urban environments are becoming more complex, the traditional ways are not serving the purpose. The current urban planning must be information-based, agile and have the capacity to deal with various issues at once. Here is where the combination of AI and IoT plays a very important role.

AI and IoT can be crucial in the solution of most of the problems in the modern cities. Such issues involve traffic jam, poor allocation of resources, increased levels of pollution, shortage of affordable housing, poor healthcare facilities, and poor infrastructure. With the help of IoT devices and their information processed using AI algorithms, the process of urban planning will allow making informed decisions related to enhancing the operations of the city. As an example, sensors installed in roads, traffic lights, and mass transportation systems allow AI systems to provide real-time data that can be utilized to control and predict traffic patterns, therefore, decreasing traffic jams and air pollution. AI and IoT can be used in sustainability in urban planning, as well, to promote smarter utilization of resources. As an example, IoT sensors in smart buildings energy management systems can be used to control energy use, and AI algorithms can be applied to optimize the heating, cooling, and lighting systems to prevent wasting energy. In the same way, the IoT-powered water management solutions help cities to track the consumption of water, spot the leaks in an instant and enhance the distribution of this precious resource.

The systems also assist in resource saving and also enhancement of cost efficiency on municipal services. In addition, AI and IoT enable smarter and participatory safety in the population. The open areas can also be equipped with IoT-based surveillance systems in the cities that can identify the incident when it happens and react swiftly to the disaster. The AI systems can also be applied in crime pattern analysis and predicting of the heat of the crime to allow the law enforcers to deploy their resources in a more efficient way. As far as healthcare is concerned, AI and IoT provide opportunities to monitor health information of patients remotely, which provides a possibility to both diagnose patients early and provide them with improved healthcare services. AI and IoT are not only important in the context of efficiency and the efforts of the urban systems but also inclusivity and equity. The use of such technologies can be applied in the creation of urban spaces, which are accessible to all and especially marginalized and underserved populations. To give a couple of examples, IoT has the ability to make the transport systems smarter, efficient, and cheaper, thereby reducing the strain of commuting on the population having low incomes. The AI may be utilized on developing the individual services within the city and it is executed by considering the needs of a few individuals or a population of individuals within the city. Lastly, the introduction of AI and IoT into the urban planning process makes the cities more immune to the problems imposed by the climate change and environmental degradation. The smart grid, environmental surveillance and data-based policies are part of the sustainable city development. Through AI, the city planners can utilize the information provided by the IoT sensors as far as the environment is concerned, and they can come up with policies that will likely alleviate pollution, encourage the use of renewable energy sources and establish green spaces.

2. TECHNOLOGICAL FOUNDATIONS

2.1 Artificial Intelligence in Urban Planning

AI has become one of the most powerful technologies in the city planning that help cities in making informed decisions based on data and use them to optimize the services, efficiently manage the resources, and predict the future trends. The applications of AI in urban planning also include the use of algorithms when working with large volumes of data and providing solutions that enhance the decision-making process, cost-efficiency, and the quality of life of the residents. The city planners now have the tools of complex data analysis and managing the issues of urbanization explosion since the urban conditions have become more complex. The urban system where machine learning (ML) as a form of AI is important. The urban planning makes use of the machine learning models when different sources of information are considered including traffic flow sensors, posts on social media, energy consumption patterns, and environment monitoring systems. Based on these algorithms, the information can be learnt to identify the patterns, to make a prediction and to provide recommendations on how the work of the cities can be improved. To illustrate, the machine learning models apply in the field of traffic management to maximise the timing of traffic lights, predict the traffic jam trends, and minimise the carbon footprint of inefficient traffic and unwarranted idling.

Data analytics is the other significant chunk of AI in urban planning. The cities generate massive amounts of information in sensors installed in infrastructure, transport and smart buildings. This information can be analyzed and useful insights can be obtained through AI-based data analytics solutions and can be utilized by the city planners to make reasonable decisions. The data analytics could be applied in most sectors such as waste management, energy consumption, urban mobility, and resources allocation. The city officials are able to analyze real-time data and identify the points of inefficiency and change in order to improve the performance of the system. To illustrate, the IoT products of waste level can help to establish the most economically rational path of waste collecting, which will enhance the performance of the service and reduce the cost of its operation.

The use of AI-based decision support systems by urban planners is growing to improve the process of decision-making. Such systems employ AI algorithms to model various conditions of urban development, predict the consequences of various decisions and suggest the best possible approaches to urban development.

In land-use planning, an example is that AI decision support systems can be used to forecast the effect of various zoning policy or land-use allocations on traffic, infrastructure requirements, and environmental sustainability. These systems equip the urban planners with the instruments that enable him or her to make informed choices that are economically, socially and environmentally balanced.

The most important area of AI use in urban planning is predictive models. AI could be used to forecast some of the urban issues, including traffic jams, air, and water scarcity by analyzing the past and present. As a case in point, traffic management predictive models can be used to predict the number of vehicles on the road at peak times so that the authorities can have the best use of the routes of the public transport and additional resources can be deployed in times of need. Similarly, the AI models can predict the extent of pollution based on such factors as the traffic density, weather, and the level of industrial activity that can be utilized to take precaution measures that minimize health risks and improve the quality of city air.

Resource management depends on the AI-based models which play a key role in the optimization of the energy, water, and waste management. The AI algorithms in smart grids will be able to monitor the demand of the electricity and predict it and can also optimize the distribution of the energy resources and minimize the wastage. These predictive functions would help prevent outage on the grid, reduce energy cost as well as make energy future of urban centers more sustainable. Similarly, the AI-based systems will have the capacity to monitor and regulate water consumption within the city and detect the wastage and loss that would otherwise have been unnoticed. The combination of predictive analytics, machine learning, and data-driven insights as a way of decision-making which is provided by AI also made it a must-have in urban planning. Cities can be rendered efficient, sustainable and resilient in case of complex challenges through these technologies.

2.2 Internet of Things (IoT) in Urban Management

Internet of Things (IoT) refers to a network of linked devices, sensors and systems that collect and exchange data in a bid to enhance the performance of urban environments(Shahrour & Xie, 2021). IoT has emerged as the key enabler of smart cities which are able to provide real-time data on the transport infrastructure, energy consumption, waste disposal and policing. Applying IoT technologies to the city infrastructure, it is possible to obtain more efficient management of cities, more responsive and sustainable cities with more automation and optimization of the urban systems(Pattani et al., 2024).

The essence of IoT in urban management is networks of sensors. IoT-enabled devices are installed in the infrastructure, transport, and public service to collect data regarding various parameters, such as the quality of air, traffic, energy use, temperatures, and the amount of wastes. As an example, traffic sensors are used to monitor the movements of vehicles and this allows the traffic management system to adjust the timing of the signals in real time, to minimize the congestion and to maintain a smooth flow of traffic. The environmental sensors check the pollution rates, the temperature and the humidity, which needed data when it comes to the management of the sustainability of the urban environment, as well as the countering of the impact of the climate change.

The IoT-enabled cities are incomplete without the smart infrastructure. IoT devices are being installed in buildings, streets and other urban infrastructure to offer real time data to monitor and manage. An example is that smart buildings have IoT sensors that track energy usage, heating and cooling, lighting and occupancy. These systems provide the possibility of automation of the building functions making them energy-efficient and cost-saving. In the same way, smart roads and bridges enabled by IoT can give information about structural health, traffic levels, and road conditions, which can make cities better able to maintain and improve infrastructure(Syed et al., 2021).

IoT devices are also required to gather the real-time data that would be needed in dynamic decision-making in the city management. The IoT sensors would be applied when monitoring the trash bins in waste management to make the collection routes and schedules optimal. This reduces both operation cost as well as increasing efficiency in delivery of waste management services. Similarly, the water management systems have the IoT sensors installed to monitor the water quality, leakage, and supply. It will allow early detection of problems and more opportunities to realize water resources, this will result in the sustainability of the urban environment. The factor that has contributed to urban automation is real-time monitoring made possible by IoT. Considering the case of traffic management, IoT-connected smart traffic lights are capable of changing the signal timings according to the current traffic conditions. In the case of congestion in an intersection, the system can increase the green light time in case sensors sense a build-up of cars in an intersection. On the same note, smart parking systems help in identifying available parking spaces with the use of IoT sensors, and directs the driver towards free parking spaces thus saving the time used searching a parking space which also saves on traffic congestion and emissions.

The cities are changing the public safety, transportation, and energy management because of the IoT-enabled services (Mori et al., 2022). The IoT sensors are very important in surveillance and emergency response in the field of public safety. The cameras and sensors that are enabled by IoT can be used to monitor abnormal activities in the open places where the police can be warned of any security threat. As an example, the intelligent surveillance systems would identify the crowd, traffic accidents, or other incidents that might need urgent attention, so the emergency services would respond faster. In the healthcare sector, the IoT devices can be used to monitor patients remotely, and thus health care providers can monitor the vital signs of the patients and can respond to emergencies as they happen.

IoT enabled systems promote the management of energy in smart cities in a large scale (Murroni et al., 2023). IoT can help utilities to manage the electricity distribution by tracking the consumption in real-time using smart grids. These systems maximize energy consumption, lower peak load and increase grid resilience. Also, the IoT-enabled lighting, heating, and cooling systems help to save energy by automatically changing the parameters depending on the presence or time of the day to minimize the waste of resources and provide sustainable energy.

Urban management combined with IoT allows making smarter and more efficient cities that can react to the instant changes and offer superior services to the citizens. IoT provides priceless knowledge and automation in traffic management, waste management, energy consumption, and safety of urban populations, which enhance the quality of life in cities and guarantee the sustainability of urban systems.

3. APPLICATIONS OF AI AND IOT IN URBAN PLANNING

3.1 Smart Transportation Systems

Traffic management based on AI and IoT-enhanced transport networks are critical in transforming the way that cities have been operating their transport systems, which are becoming more efficient, less congested, and eco-friendlier. Traffic management systems in a smart city are upgraded with the help of AI algorithms analyzing information on IoT sensors installed in roads, cars and traffic lights. These intelligent systems change the traffic lights of the city dynamically according to the real time traffic movement and reduce the traffic and make the transportation throughout the city smooth. Moreover, predictive models based on AI can predict traffic flows at various times of the day to predict possible bottlenecks or the places where congestions are likely to occur. These forecasts enable the city planners to be ahead of time in regulating the traffic management methods to alleviate time wastage and a more effective flow of people and goods(Alahi et al., 2023).

The transportation network in the form of IoT gathers information about the state of traffic, the speed of movement, the use of public transport, and the state of the roads, which gives a complete picture of the transportation ecosystem. This knowledge is not only used in real time but it is also applied in a long term urban planning. To illustrate, one can consider the use of the traffic monitoring systems on the basis of the IoT in order to know what intersections or roads are the busiest, and the urban planners could improve the road structure or determine whether new infrastructure is needed. In addition, the connection of IoT with the transport system may give the opportunity to see the current location and the timetable of buses and trains and the number of people in real-time. This will allow the passengers to make well informed decisions on when and where to travel and it will also increase the effectiveness of the entire operations of the public transport services.

The other technology that is developed in the application of AI and IoT in transportation is the autonomous vehicles. Such cars are able to drive in the city not involving the human driver with the help of the algorithms that are applied in the AI system, because of real-time information collected by IoT sensors, cameras, and radar. These self-driving cars are able to communicate with other cars as well as the infrastructure of the environment and, therefore, be able to move freely and minimize the possibility of an accident. The appearance of the autonomous vehicles into the urban environment would lessen the traffic accidents and raise the safety on the roads and make the transportation infrastructure more efficient. In addition, the urban parking systems are being transformed by the introduction of smart parking systems in which the IoT sensor assists in determining the available parking spaces. Such systems enable the driver to be aware of the nearest parking space available and this saves the time of searching a parking space and this reduces congestion and idling of the car hence reduces the emissions.

3.2 Energy and Resource Management

Smart grids and energy optimization systems are one of the most important forms of how AI and IoT could be utilized in the context of urban planning. As cities continue to grow in size, demand on energy is also increasing and this is a massive burden on the traditional power grids. Smart grids which will allow controlling consumption and distribution of electricity in real-time by using IoT sensors and AI algorithms will be the solution to the problem. The IoT sensors installed in buildings, streetlights, and industrial regions gather information about energy consumption, whereas AI systems process this information and forecast any changes in demand and adjust the supply of electricity based on the demand. This assists in saving wastage of energy, elimination of power cuts and the provision of energy to the areas that need it most. Also, AI algorithms within smart grids allow the incorporation of renewable sources of energy, including solar power and wind power, by balancing energy flow to consider intermittent sources of energy.

Even the city-level energy consumption in the building is also being transformed with the help of IoT energy optimization systems. IoT sensors are used in smart buildings to observe such factors as occupancy, temperature, and lighting. This data is processed by AI algorithms to maximize heating, cooling, and lighting systems and guarantee that buildings are comfortable to occupants and that energy use is efficient. As an example, AI in commercial buildings can be used to regulate heating and cooling, and depending on occupancy, the room will be heated or cooled when it is occupied, and not when not. The technologies are playing a major role in the development of sustainable and energy-efficient cities, where resources are managed in a more efficient way as well as there is a minimum of emissions (Salam, 2024).

Besides energy, water management and waste disposal systems are also being implemented into IoT and AI. The IoT sensors have the ability to track the water quality, water consumption and distribution within the city. These sensors are able to sense leaks in water pipes, the real-time use of water and alert in case a leakage or inefficiency is discovered. With this data, AI systems can make predictions on the future demand of water and supply the same resource accordingly resulting in sustainable use of water resources. Similarly, the waste management process through IoT can be used to help the cities to simplify the waste collection and reuse. Waste bins are installed with sensors which monitor the amount of waste present in the containers and the waste collection vehicles are only sent when there is a necessity to save on fuel and operating costs. The technologies are also involved in the process of making the cities more sustainable in the way it uses its resources and also aid in making sure that the cities have less impacts on the environment.

3.3 Urban Sustainability

Urban sustainability is strongly defined by environmental monitoring, and AI and IoT technologies make an enormous contribution to the issue. The air quality, temperature, humidity, and the level of noise are the most important environmental factors that can be checked as a result of the IoT sensors installed in the city. The sensors would give real time information that may be used in monitoring the level of pollution, the condition of the ecosystem and the location that needs to be intervened. It is this data on which AI programs are executed to identify patterns and forecast future patterns in the environment. An example of such is the fact that AI can predict the time of the day when the pollution is likely to be high based on the traffic rates, weather, or industrial activities. The data can be used to issue warnings or actions that can be taken to minimize the pollution such as closing some areas to traffic or changing the production process (Jha et al., 2021).

The information is also used by the planners in the urban areas when coming up with a strategy on how to improve the quality of air, address the issue of noise pollution and urban heat island effect. When it comes to the green space, IoT sensors will be used to ensure that the urban parks, trees, and gardens are taken care of and have enough water. The application of AI models could be used in the planning of green spaces in the most beneficial manner that would bring the maximum benefit to the environment by reducing CO2 emissions, improving air quality and cooling during hot weather. Smart buildings also have urban sustainability since the IoT devices used in these buildings assist in monitoring the use of energy and regulating it. Such buildings have been designed to consume less on energy, less on carbon emission as well as being resource efficient.

Using AI and IoT technologies in the activity of urban sustainability, cities will be capable of becoming more ecofriendly and climate-resistant. The technologies can be used to get real-time information that can be used in the management of resources, reduction of waste and making the urban life healthier to the residents.

3.4 Public Safety and Healthcare

Artificial intelligence and IoT in the field of public safety and healthcare are quickly changing how cities can react to the emergency, guard citizens, and supply medical care. When it comes to the safety of the population, the surveillance systems based on IoT can be used to monitor the streets and other public areas, thus, permitting the authorities to identify any abnormal actions, be it the criminal activity or a crowd of people, in real-time. The data collected by these surveillance systems is analyzed by AI algorithms to detect possible threats or incidents, which enables a faster reaction of law enforcement or other emergency services. It can also utilize crime data to determine trends to determine the areas where crime is likely to occur in order to impart better the allocation of the resources of the law enforcement. The IoT devices can be used in the medical sphere to check the vital signs and health parameters of the patients remotely. The health providers can receive real-time information on such wearable gadgets as heart rate, blood pressure, and activity level monitors. AI systems can process this information to identify early health-related issues, such as arrhythmic heartbeat or an increase in blood pressure, to enable early interventions and improve patient outcomes. Besides, AI-based predictive models can forecast the tendencies in the health of the populace, along with the prevalence of illnesses, and assist in planning the health demands of the populace. In the time of the crisis, such as a pandemic, such technologies can be utilized to provide valuable information that could help the authorities to distribute healthcare facilities and track the degree of infection and plan the health measures.

AIs and IoT also play a great role in predictive policing and disaster management. The AI algorithms will be capable of processing the data on the previous crimes and identify the patterns to know where the new crime is likely to occur, and police ought to take some precautionary measures to avert the crime. IoT sensors will enable the provision of real-time data on weather conditions and destruction of the infrastructure, and the movement of people in case of natural disasters so that disaster response teams could allocate their resources more efficiently and plan the rescue operations.

3.5 Urban Mobility and Smart Infrastructure

The infrastructure of the city is applying the IoT sensors to control and maintain the roads, bridges, and other buildings in the most appropriate manner. These sensors will record the state of infrastructure and give real-time information about wear and tear, structural health and the necessity of repairs. Such information assists urban planners to focus their maintenance activities and also to make sure that infrastructure is safe and functional. In smart roads, IoT sensors are used to monitor traffic, speed of vehicles and the road conditions and the data can be utilized to manage traffic efficiently, minimize congestion and increase safety.

The use of IoT and AI is making the urban mobility solutions more efficient, including bike-sharing programs, e-scooters, and ride-sharing services. Sensors in IoT can monitor the presence and position of bikes, scooters, and ride-sharing vehicles and enable users to search and book them on the move with the help of mobile applications. AI algorithms will help to optimize the placement of such vehicles according to the demand patterns, so they could be located where and when they are needed the most. The IoT and AI also affect the sphere of public transportation: real-time information about buses and trains, their location, time, and occupancy rates will help passengers plan their routes better.

The combination of IoT and AI in urban mobility and infrastructure, allows cities to develop more efficient, sustainable and accessible transport systems that can satisfy the needs of an increasing urban population.

4. CHALLENGES AND BARRIERS

4.1 Data Privacy and Security

Data privacy and security are also some of the greatest threats as far as the popularity of AI and IoT in urban planning is concerned. The cities are increasingly relying on the use of the IoT devices to collect huge amounts of data, such as traffic patterns, air quality, and even the personal information of the inhabitants, which also implies that there is a certain threat that this data will be hacked, used, or compromised. The IoT devices are likely to connect with one another and they are on a continuous process of collecting sensitive data in all the urban systems such as healthcare, energy and transportation. The increase in use of these devices means that they become a target of the cybercriminals. Leakage of personal information such as health information, movement, or energy consumption pattern can be severe and can result in loss of money or privacy to citizens.

The problem with the generation of large volumes of personal data by smart city technologies is one of the key ones. One of them is that IoT-enabled healthcare devices or IoT-based smart meter have a tendency to gather information that can be related to a person, including health statistics, consumption habits, and geographical location. This raises a grave question of how this data can be stored and how the data can also be accessed and how this data can be safeguarded against any form of possible hacking. The quality and security of the data that the AI system is trained on is also of direct consequence to the quality and accuracy of the AI algorithms and their subsequent performance. Given that IoT devices can be hacked or manipulated, a real possibility exists that any breach of them may lead to an unintended malfunction of the city systems which may be traffic management system, waste management or even energy distribution networks. In addition, the problems of ethics related to the surveillance and data collection in smart cities should also be considered. The applications of the IoT devices include surveillance of the people, safety checks and management of the traffic flow which may raise concerns about the scope to which the privacy of the citizens is being invaded. Even though the application of such technologies can indeed help to improve the quality of life in the city, the fact that these technologies are widely used can also lead to the emergence of the problems of round-the-clock surveillance and the threat of invasive surveillance. The necessity to balance between the use of data in the service of the people and the privacy of the citizens continues to be a very important question in the cities that want to be responsible in the use of AI and IoT technologies.

4.2 Interoperability and Standardization

The other major complication in the application of AI and IoT in urban planning is the problem of interoperability and the absence of standardization of devices and platforms. The IoT devices are usually produced by various companies with their own protocols, hardware, and software, which might impose a challenge on these devices to interact with each other. This interoperability can become a giant obstacle in an intelligent city where different devices, networks and services must collaborate. An example of such is that the information that is produced by the sensors in one side of the city is not easily able to be blended or processed by the AI systems that are adapted to handle the data of other IoT-based devices. In the absence of smooth communication and integration, the possibility to reach the intended results of smart city technologies, including the optimization of the traffic flow, the effective management of energy sources, and the enhancement of resource distribution, might be extremely restricted.

Standardization is essential to the IoT devices and platforms since it allows data to be shared and analyzed on various systems. Nowadays the IoT ecosystem is rather fragmented and each device, each platform has its protocols and standards. This kind of fragmentation complicates the process of integrating IoT systems into the urban infrastructure since the cities become larger and incorporate additional devices. The answer to this predicament is then that governments, industry leaders and standards bodies come together to come up with universal protocols and standards of IoT devices and systems. This would enhance simpler integration and information sharing between the different devices and technologies which would enable cities to build genuinely connected ecosystems that will be capable of operating in a coherent manner. It is also more complicated by the fact that AI systems can be made interoperable with IoT devices. The IoT sensors provide AI systems with quality and consistency of the data that is inputted into the system. Nonetheless, the data gathered by various sensors will not be compatible and easily deciphered unless they use standardized communication protocols. This issue may compromise the effectiveness of the AI-based decision-making systems and restrict their potential of providing correct forecasts and streamlining urban processes. As an example, AI algorithms used in a smart transportation system that examine the traffic movement and optimize the traffic lights turn-on and turn-off schedules may fail to work properly in case the IoT sensors that measure the vehicle speed and congestion level are incompatible or do not provide steady data.

4.3 High Costs and Implementation Complexity

Though AI and IoT would be a revolution in the cities, the high cost of adopting the technologies is a blow. The amount of capital that is required to implement IoT sensors, smart devices and AI based systems in the urban infrastructure is very high. The investment is comprised of the costs of the purchase and installation of the devices, construction and maintenance of data infrastructure, and making the systems trusted and reliable. There are also operation costs that are incurred in monitoring, maintaining and upgrading of these systems so as to keep them effective. These costs may prove to be a real setback to the cities that have low budgets, or those cities that are situated in the developing regions and the smart city technology implementation can turn out to be a challenging and a costly venture.

Additionally, one should not overlook the complexity of the procedure of integrating the frameworks of AI and IoT into the existing city infrastructure. Majority of the cities rely on outdated systems which were not able to accommodate the new technologies used in smart cities. This process of integrating these new technologies in the old systems can be technically challenging and very expensive. The infrastructure that will have to be upgraded to cover the IoT sensors, AI-based traffic management system, and the rest of the smart technologies, may require significant changes to the existing infrastructure in the city, which may be not only costly but also time-consuming. Training of the staff also complicates and adds to the cost of having these systems to operate and maintain them. This complexity is normally compounded by the fact that there is a necessity to possess interoperability as argued above and the current devices have to be integrated into the broader system of urban.

However, the economic burden and complications of using AI and IoT in the city planning practice will possibly be outweighed by the long-term benefits, such as efficiency, cost-saving, and quality of life. However, the problem of initial investment is the most severe one, particularly regarding the cities, which are overloaded with their budgets or even have no opportunity to access the necessary resources.

4.4 Social and Economic Impacts

The impact of implementing AI and IoT in urban planning in social and economic terms is a chance and a problem. One of the most burning issues is the digital divide (when the unequal access to the technology and digital services is offered to various segments of the population). Most cities have a very unequal access to digital technologies with the low income communities in the majority of cases lacking high-speed internet, smart devices, and infrastructure to receive the benefits of smart city solutions. This type of a gap can augment the already existing differences, and thus it may not be quite simple to enable the marginalized populations to enjoy a full access to the advantages of smart city technologies.

In order to make sure that the AI and IoT technologies help in making the urban development more inclusive, the policymakers should focus on closing the digital divide. This may include making affordable internet connection to every resident, digital literacy education, and investing in infrastructure that will facilitate equal access to smart technologies. Unless these challenges are handled, there is threat of the smart city benefits to be unequally distributed whereby the rich regions will gain more out of these technologies leaving the poor behind.

The other important question regarding social and economic effects is on the possibility of displacement of jobs because of automation. The use of AI and IoT systems can enhance efficiency and minimize human interaction in most of the services in cities, such as in the transport system, waste management, and energy supplies. Although automation may result in cost reduction and better delivery of services, it has also brought up the issue of job loss and unemployment especially to workers in areas where manual labor is being substituted with the automated systems. The policymakers should work on these issues and come up with measures to retrain the workers, find new employment opportunities in the technological sector, and make the transition to the automated labor force as painless and fair as possible.

Also, with the emergence of smart and automatized cities, the fear that such technologies will increase the level of surveillance and decrease the privacy of the citizens is growing. Ethical implications of the level of security and individual freedoms may be brought up by the use of AI-driven surveillance systems to monitor the public spaces, traffic, and even social behavior. Cities need to have effective rules and policies that will make the application of AI and IoT technologies responsible and ethical, and that will respect the rights and privacy of people.

5. FUTURE TRENDS AND DIRECTIONS

5.1 Advancements in AI and IoT Technologies

One of the most significant shifts that occurred in the field of urban management to date is the implementation of the 5G technology because soon, it is likely to change the process of communication between the IoT devices and the manner in which the data is supposed to be processed in real-time. The 5G networks will be even faster, with a reduced latency rate, and a high bandwidth, which will enable a diverse range of the IoT devices to communicate and exchange data without any difficulty. With the increased interconnection that 5G will bring, cities will be able to implement more massive and more advanced IoT systems which can receive, process and analyze data on a real-time basis. This will play a critical role in urban management where fast response times and capabilities to manage large volumes of information are critical towards the streamlining of the services and infrastructure in the city. As an example, 5G networks will allow monitoring the traffic, energy consumption, environmental conditions in real-time, and enable urban managers to make informed decisions in a much faster and efficient way.

Besides 5G, the edge computing will also change the urban management systems to a new form. Edge computing is a type of computing that brings the processing of the data as close as possible to the source, instead of sending the data to the centralized cloud-based server to be analyzed. Edge computing decreases latency, shortens response times and eases the network bandwidth by completing data processing at the network edge, which is closer to IoT devices. This is particularly necessary in real-time applications in smart cities e.g. traffic management, public safety and environmental monitoring where latency in data processing may be costly. Using edge computing, cities will be able to compute and analyze data locally, which will help make decisions faster and more efficiently allocate resources without sending the data to far-off data centers.

The most significant facilitators of smart city technologies are artificial intelligence and machine learning (ML). The use of AI algorithms, namely, machine learning models, can significantly affect urban planning and management because systems can study through data, recognize patterns, and make autonomous decisions. To illustrate the point, machine learning models can be used in transportation systems to predict traffic flows and optimize them or energy grids to predict demand and control energy flow. They could also be applied in AI-controlled predictive maintenance of their infrastructure and have wear and tear indicators in roads, bridges and building detected and repairs scheduled before the infrastructure collapses. As these technologies are in their developmental stage, the probability exists that they can be utilized in other areas that would allow cities to be efficient, sustainable in addition to being safe. The other interesting trend is the role of big data, blockchain, and digital twins in the future of smart cities. Big data refer to data of a huge magnitude that are produced by IoT, sensors and other systems of the city. This information when analyzed will provide helpful data in the operations of the city such as transport, waste management, energy and security. These data volumes can be processed and examined through the AI algorithms to make real-time decisions and forecasts that would improve the urban services. Blockchain is a distributed ledger technology that enables safe and transparent sharing of data, which is vital to smart cities applications involving various stakeholders such as energy systems, transportation network, and public services. Blockchain will help in integrity of the data, reduce fraud and increase trust in systems that require integration among different organisations.

Digital twins, a virtual representation of a physical asset or system is another emerging technology in smart cities. Having the possibility to create a digital twin of the city, the urban planners will be able to simulate the multiple scenarios, such as the traffic flow, infrastructure improvement, or the environmental impact, and visualize the results, without making any real-life changes. This shall allow better decision-making, planning and reduced chances of failure. Taking, as an example, a digital twin of transportation system of a city, it will enable the planners to observe how the new infrastructure will impact the traffic, the movement of people through the means of transport and emissions, and, therefore, make more sensible decisions concerning the development of a city.

5.2 The Role of Policy and Governance

The role of policy and governance in regard to the responsible and ethical use of AI and IoT technologies grows with the level of the involvement of the latter in the urban systems. There is a need to have legal and regulatory frameworks which can be used to shape the development and deployment of such technologies so that they are deployed in a manner that is beneficial to the entire citizenry and do not violate privacy, security and fairness. The policymakers are to come up with regulations which would balance the potential benefits of the AI and IoT technologies and the need to protect the personal rights and render these systems transparent and accountable. One of the primary areas, in which policy and governance will be highly effective, is the issue of data privacy and security. Because the IoT devices capture very large volumes of data including personal data, it is of concern that governments must embrace clear and effective privacy legislations that protect the data of citizens, and on the other hand, effective privacy legislations that would promote the collection and use of data to the benefit of the society. Issues of ownership of the data, consent and right to be forgotten should be the concerns of the policies, allowing citizens to have control over their personal data. Additionally, cybersecurity measures are to be established so as to address the risk of hacking, data leakages and cyber-attacks that can be devastating in more ways than one as regards personal and urban operations of the systems. Collaborations and international standards will also play a critical role towards successful integration of AI and IoT in smart cities. With the adoption of the technologies in cities globally, there will be a necessity to have global collaboration in a bid to make the systems interoperable, secure, and effective in various regions and contexts. International norms can assist in making sure that smart city technologies share the same principles and can exchange the data and integrate systems across the borders. As an example, having an international standard that defines a security level of IoT devices would facilitate the ability of such devices to be constructed with security in mind and the ability of cities to cooperate in order to secure their critical facilities against cyber-attacks.

Government agencies, private firms and research centers will also have to collaborate to innovate smart cities. The improvement of smart city technologies and its practical implementation can be hastened by the use of public-private partnerships, where the experience and resources of both parties are unified to develop solutions to the problems in the city. The governments may be very instrumental in establishing incentives to make businesses invest in smart city solutions, funding research and development, and encouraging cooperation among various stakeholders. Simultaneously, businesses can offer the technological knowledge and creativity to come up with the latest solutions to better the urban life.

5.3 Vision for the Future of Smart Cities

The future of smart cities is oriented to long-term sustainability and the development of resilient, flexible, and able to satisfy the needs of future generations urban environment. The development of smart cities is focused on one of the main objectives which is energy-efficient cities, environmentally responsible cities, and sustainable cities. The use of AI and IoT will become the key to meeting these objectives as it will allow cities to utilize the available resources in a better way and minimize waste, as well as their impact on the environment. As an example, IoT-based smart grids and AI-based algorithms can be used to decrease power consumption by more efficient distribution of power, and AI-based waste management systems can be used to decrease the amount of waste that goes to landfills and increase the rate of recycling.

The second thing that should be mentioned about the future of smart cities is the development of self-governing urban spaces. City automation will allow cities to be self-regulated and managed with the help of AI, machine learning and automation. Smart cities in the future can potentially adapt to changing situations without any human control like it can change the traffic lights depending on the current traffic conditions, automatically redistribute resources at times of high demand, or control energy consumption depending on the weather. Such autonomous systems will not only enhance efficiency of the urban services but also lessen human interaction with the services that will enable the city officials to concentrate on decision making at higher level.

AI and IoT will as well help to enhance the quality of life of urban residents. By simplifying the city services, these technologies will enable cities to become healthier, more equitable and pleasant to inhabit by improving the distribution of resources and reducing the environmental impact. On a case in point, predictive analytics can help to reduce pollution and improve the health of civilians, as in the case of real-time air quality monitoring, and can provide cheaper and more convenient medical care to the population, as in the case of smart healthcare systems. The smarter cities will also be more welcoming, and IoT devices and AI systems will be developed to meet the needs of all urban residents, even those that are marginalized and underserved.

6. CONCLUSION

The adoption of AI and IoT has proven to be the game changer in urban planning development. These technologies have shown the extent of their huge potentials in making the urban life more efficient, sustainable and of better quality. The ability of AI to handle information, predict and make decisions on its own has allowed cities to streamline the process of managing traffic, distributing resources, and urban services. Through the use of which valuable data that could be used to improve the infrastructure and manage the resources in a more adequate way IoT has made it possible to monitor the most vital systems, by the means of the devices and sensors which are connected to each other. AI and IoT are also teaming in developing smart cities wherein the information will be utilized not only to improve the functionality of the already existing urban infrastructure but also to predict the demands and the challenges in the future. These technologies enable the cities to be responsive to changes therefore making cities more resilient, sustainable, and liveable. Due to AI and IoT in urban planning, the future of cities will continue being redesigned to usher a new era of efficiency, sustainability, and improved life of the citizens living in such cities. The predictive models and real-time data analytics relying on AI will help to make urban systems more adaptable and reduce congestion, energy waste, and pollution. Smart transport networks, energy-efficient buildings and optimization of resources will be used to reduce the degree of the environmental impact of the cities. There is also the automation that would become possible with the assistance of such technologies that are also going to make the cities more efficient, as well as will need fewer manuals involved which will also liberate the resources and will also help to provide better delivery of the services. The change shall lead to more sustainable cities, which will not only meet the demands of the growing cities but will also lessen the ecological footprint of cities. In addition, it will ensure that urban services are more accessible and inclusive as AI and IoT will enable every resident to benefit by having smarter infrastructure and services. The role of AI and IoT in the development of cities will only be more significant as cities are becoming even more developed. The ongoing events in the field of these technologies and 5G network popularity, edge computers, and digital twins will enable the cities to operate in the manner that previously was unheard of. However, to successfully see the effective realization of the smart cities, there is the need to address the challenges associated with data privacy, data security, interoperability, and inclusivity. Having all these issues out of the way, the AI and IoT will be able to make the lives of citizens of urban areas more efficient and sustainable, not only in terms of resources, but also in terms of their openness to the needs of the people. A combination with urban development of these technologies in a smooth manner is the future of the development of the cities, which will turn them smart, resilient, and sustainable.

REFERENCES

Alahi, M. E. E., Sukkuea, A., Tina, F. W., Nag, A., Kurdthongmee, W., Suwannarat, K., & Mukhopadhyay, S. C. (2023). Integration of IoT-enabled technologies and artificial intelligence (AI) for smart city scenario: Recent advancements and future trends. *Sensors (Basel), 23*(11), 5206. DOI: 10.3390/s23115206 PMID: 37299934

Bahl, A., Kandpal, S., & Rajendran, R. K. (2024). Innovative Strategies for Urban Construction Optimization in the IoT Era. In The Climate Change Crisis and Its Impact on Mental Health (pp. 213-226). IGI Global. DOI: 10.4018/979-8-3693-3272-6.ch016

Gómez, J. E., Marcillo, F. R., Triana, F. L., Gallo, V. T., Oviedo, B. W., & Hernández, V. L. (2017). IoT for environmental variables in urban areas. *Procedia Computer Science, 109*, 67–74. DOI: 10.1016/j.procs.2017.05.296

Jha, A. K., Ghimire, A., Thapa, S., Jha, A. M., & Raj, R. (2021, January). A review of AI for urban planning: Towards building sustainable smart cities. In *2021 6th International Conference on Inventive Computation Technologies (ICICT)* (pp. 937-944). IEEE.

Mori, H., Kundaliya, J., Naik, K., & Shah, M. (2022). IoT technologies in smart environment: Security issues and future enhancements. *Environmental Science and Pollution Research International, 29*(32), 47969–47987. DOI: 10.1007/s11356-022-20132-1 PMID: 35538345

Murroni, M., Anedda, M., Fadda, M., Ruiu, P., Popescu, V., Zaharia, C., & Giusto, D. (2023). 6G—Enabling the New Smart City: A Survey. *Sensors (Basel), 23*(17), 7528. DOI: 10.3390/s23177528 PMID: 37687986

Pattani, K. A., Gautam, S., & Rizvi, A. (2024). Emerging Horizons: IoT, 5G, and the Evolution of Smart Cities. In Secure and Intelligent IoT-Enabled Smart Cities (pp. 1-21). IGI Global.

Salam, A. (2024). Internet of things for sustainable community development: introduction and overview. In *Internet of Things for Sustainable Community Development: Wireless Communications, Sensing, and Systems* (pp. 1–31). Springer International Publishing. DOI: 10.1007/978-3-031-62162-8_1

Shahrour, I., & Xie, X. (2021). Role of Internet of Things (IoT) and crowdsourcing in smart city projects. *Smart Cities, 4*(4), 1276–1292. DOI: 10.3390/smartcities4040068

Syed, A. S., Sierra-Sosa, D., Kumar, A., & Elmaghraby, A. (2021). IoT in smart cities: A survey of technologies, practices and challenges. *Smart Cities*, *4*(2), 429–475. DOI: 10.3390/smartcities4020024

Chapter 5
Leveraging AI and IoT for Smart Urban Planning and Management:
Intelligent Cities

Sitesh Kumar Singh
https://orcid.org/0000-0002-7108-0808
College of Engineering, National University of Science and Technology, Oman

Jaishri Gothania
https://orcid.org/0000-0002-2656-642X
Lingaya's Vidyapeeth, India

Shiferaw Garoma Garoma Wayessa
https://orcid.org/0000-0002-7234-8343
College of Engineering and Technology, Wollega University, Ethiopia

ABSTRACT

This chapter discusses the colliding spaces between artificial intelligence (AI) and the Internet of Things (IoT) to form intelligent cities, i.e., cities that are not only smart but also resilient, adaptive, and human-centered. It offers an imaginary and technological vision of the process of the AI-IoT convergence shaping real-time governance, infrastructure optimization, environmental oversight, and participatory urban planning. The chapter encompasses global case studies and ethical analysis, putting forward the necessity of inclusive, secure, and sustainable digital transformation. We also suggest policy directions and implementation strategies that can promote equitable and sustainable urban ecosystems.

DOI: 10.4018/979-8-3373-3146-1.ch005

INTRODUCTION TOWARD INTELLIGENT CITIES

Context and Relevance

The world is living in a world that is radically changing, as the world in the past has never been involved in the process of massive urbanization as it is today. The United Nations has estimated that the global population will have almost 70 percent urbanization by 2050, which is quite a demographic change with far-reaching implications of implication on the city infrastructure and sustainability (United Nations, 2019). It is not merely a statistical tendency but a multi-dimensional reorganization of human life, economies, politics and human relationships with the world. World economic, technological, and environmental policies are now being formed in the city which are regional capitals of economic and cultural interests. That rapid growth is also imposing massive burdens on the existing systems and mobility networks of the transportation and housing, energy infrastructure and waste management systems not to mention the social infrastructure monoliths all of which were to accommodate orders of magnitudes smaller and more inert populations.

The city planning and city management structures that have been based on the rigid bureaucratic systems and the non-evolving data cannot be useful in addressing dynamically demanding realities of the present-day urbanization. A multi-layered and complex system as it is at the present-day, this requires real-time responsiveness, the development of resilience in the face of climate threats and integration into the provision of services. In response, cities around the world are expanding with hi-tech digital technologies to re-architecture their shapes and domination. Artificial Intelligence (AI) and the Internet of Things (IoT) become the next in the list of these technologies as such a combination result in the formation of the new urban model, the Intelligent City (Ullah et al., 2023).

Intelligent Cities is a better development, and in many respects is far more than Smart Cities is. Intelligent Cities extend beyond digital connectivity, informational-based services and resources management that characterize the Smart Cities; instead, they emphasize cognitive practices. These cities use AI to process large volumes of information sent by the IoT networks and not only can be used for optimization but also to predict and adapt to as well as proactively manage the city (Batty et al., 2012). At that, artificial intelligence is the brain of the city that provides a set of analytical thinking and decision-making abilities to simulate human intelligence. IoT, in its turn, cooperates as the nervous system of the city that constantly perceives and transfers the information about the environment state, the operation of the infrastructures, and the work of the citizens.

The symbiosis between the implications and the IoT is also such that it transforms cities into active problem solvers, as opposed to people who used to be passive data collectors. Traffic sensors and surveillance cameras, air quality sensors, and smart meters used in the collection of real-time information about the physical environment are the examples of IoT devices. This information is the point of derivation of actionable intelligence when utilized in AI algorithms. As an example, an AI model can analyze traffic patterns and predict stalemates, thus city officials will be given a chance to re-route the traffic or re-allocate the order of signals in real time. Similarly, AI-driven energy management systems have an opportunity to enhance the electricity distribution according to the consumption patterns, which will result in the reduction of the wasting (and, subsequently, carbon emission) (Al Nuaimi et al., 2015).

Intelligent cities are, however, not cities full of technology but also planned cities that seek to enhance sustainability, efficiency and inclusiveness. Such an aspect as sustainability is one of the most crucial strengths to mention. The carbon that is emitted in cities adds to the overall total carbon emissions into the world to be above 70 percent, and more than two-thirds of the energy is used in urban centers (IEA, 2021). The application of AI and IoT promotes the control of the cities to regulate the environmental indicators in real time, impose the emission pressure, and redesign the energy sources into renewable energy sources by controlling the intelligent grids. Another advantage is that AI algorithms can be used to make climate resistant whereby the risk inspection of extreme weather events could be forecasted, and the mobilization of actions at the local governments correspondingly in response to the alerts of an extreme weather event (sending emergency services) or preparing drainage systems (before floods) could be projected, etc.

The Future of Intelligent Cities depends on the critical view of sustainable cities in the Future. This means the inclusion of technological methods to participatory planning, environmental guardianship and encompassing policymaking (Bittencourt et al., 2025). It is not only cities, they need to be smarter or more intelligent, but also need to be made human, flexible and resilient. The very first thing is that people must realize that all the technological changes that Intelligent Cities will involve can finally fulfil their promise of making the life in the world better and resolve the global issues of climate, inequality, and resource crunch.

In conclusion, it should be noted that AI and IoT convergence is not just a case of a tech breakthrough but also an alteration that is timely and central to the reimagination of the 21st century urban life. As urbanization is only further accelerated, Intelligent Cities is a promising and thrilling prospect to greener urbanism that is more efficient and equitable in the future. The potential is yet to be fully realized but it can only be realized when an approach that is holistic to the development is adopted and this should involve technological innovation, good governance through ethical decisions and inclusion as well as environmental responsibility.

The Shift Toward Intelligence in Urban Planning

In general, long-term master planning has been the primary characteristic of static top-down urban planning. These conventional paradigms have been linked to offering the structure and predictability yet on numerous occasions they have failed to keep up with modernity and the speed that envelops the reality of the urban landscape which is the present-day connotation of contemporary urban world. Contemporary cities should be provided with more adaptive and responsive planning constructions as they are struggling with the increasing rate of global warming, scarcity of resources, the altering state of economic conditions, and constant social inequities. This would necessitate a paradigm shift in less inflexible urban design to smarter and more convenient urban ecosystems that would modify to real time conditions.

This change may be premised on the potentially successful technological offering of integration of Artificial Intelligence (AI) and Internet of Things (IoT). With machine learning algorithms, AI can compute a large and highly complex amount of data that is well beyond a human being in terms of their capacity to compute it fully (Madanchian & Taherdoost, 2025). The systems can utilize the past and present data to forecast the transportation pattern in the towns and cities, land-use change simulation and energy distribution efficiency. These empirical studies would see the decision-makers and the city planners transcend beyond conjectures and instead embark on evidence-based solutions to improve on the efficiencies and sustainability of the urban setting (Sharma et al., 2022).

Meanwhile, Internet of Things solutions have become increasingly significant regarding data collection that is correlated to the fact that the network of sensors connected is created in urban areas. These sensors continuously measure the parameters such as air quality, the volume of people on roads, water level supply, and waste level (Hudson-Smith et al., 2021). Decisions are made on a real-time basis and informed with this real-time monitoring infrastructure. A case in point is where the authorizing bodies in each city can be able to observe the escalated degree of pollution and in reaction utilize countermeasures such as traffic redirection or closure of businesses in the region in the short term. Similarly, intelligent waste management systems have the potential to optimize their collection routes based on the real-time filling capacity of the bins which reduces the overall costs of operation together with environmental implication (Sharma et al., 2022).

The ability to enhance the resilience of the city is one of the values of AI and IoT that can be enumerated as the most sought-after ones. The flood prone, earthquake or heatwave prone cities can leverage sensor systems and predictive analytics systems to detect the dangers at an early stage and act proactively on the risks. In the example, the early-warning mechanisms can be applicable in areas that are at risk of floods whereby the data on the rainfalls, river flows, and topographical features could be merged to provide the warning information before the catastrophe takes place. These preliminary reactions not only save lives but also minimize financial losses and damage to the infrastructure (Zhang et al., 2023).

Lastly, integrating AI and IoT and applying them to the systems of cities, it is possible to suggest successful attempts to create even more efficient and sustainable cities, cities being more resistant and adaptable to the unknown challenges of the future.

Framing Intelligent Cities: From Data to Decisions

The intelligent cities are all about data-driven governance. Along with IoT, AI enables decision-makers to get evidence-based information based on streams of ever-incoming information. Through the insights into the urban context functioning, its analytical understanding and visualization, policy designers can be more selective in their interventions, order resources in a streamlined way, and involve citizens in the operations of a participatory platform (Meijer & Bolívar, 2016). It is also not just a technical change, but it is an institutional, ethical and cultural change.

It has already been the example of intelligent systems at work in Singapore, Barcelona or Amsterdam in terms of transport, waste management and communications with the citizens. The examples may be considered the living laboratories which can be regarded as the examples of the synergetic cooperation of the AI and IoT and the ways it can be implemented to make the urban environment more functional. However, increasing the size of such models is challenging to most cities particularly in the Global South due to financial limitations and data privacy concerns, and owing to lack of digital infrastructure.

Challenges Driving the Need for AI and IoT in Cities

What the contemporary urban environment has in abundance is a web of interconnected and knotty issues that are generally referred to as the wicked problems. The issues are complex, fluid and involve a situation of competing interests by different stakeholders. Some of the demerits include absence of poverty alleviation particularly in urban centres, traffic congestion, water scarcities and mushrooming incidences of climatic induced migrations. The handling of the problems is not that simple and cannot be solved easily in the conventional ways, taking into consideration the social-political tension and access to resources. Complicated issues are not always easy to reduce in the usual paradigms of urban governance, which are typically compartmentalized, paper-based, and reactive in their nature (Al Nuaimi et al., 2015).

Alternative frameworks of tackling these dilemmas in cities will come in the form of new technologies that include Artificial Intelligence (AI) and Internet of Things (IoT). The AI and IoT enable the city planners and administrators to model, observe, and optimize the individual subsystems of a city (transport, energy, water, housing, and environmental management) as components of a combination of systems by assisting systems thinking. This kind of integrative perspective contributes to a critical task in healthy and responsible decision making (Lloret et al., 2025).

Moreover, it is because of the fast enhancement in the governance of cities using open data platforms, urban dashboards, and digital twins, that the techniques are rapidly integrated. These tools are useful to observe the behaviours of cities in real-time and consequently enable the accomplishment of data-driven planning effortlessly, in addition to enabling policy decision-makers to experiment and simulate different policies. They also enhance social visibility and involvement of the citizens; this enables the citizens to utilize viable information and emerge new accountability in the provision of urban services (Al Nuaimi et al., 2015). This way, digital innovation is assuming the centre-stage in smart and inclusive frameworks of governance within future-ready cities.

CONCEPTUAL FRAMEWORK

What Are Intelligent Cities: A Clarification of Smart Cities

The transformation of the concept of urban development has changed to the contemporary, more sophisticated and flowing model of intelligent cities. On the one hand, the two terms are usually applicable interchangeably but on the other, there exist some significant differences in the conceptual basis as well as practical applications of the two terms as shown in Figure 1. A smart city primarily focuses on efficiency, automatization and foundation on ICT (Information and Communication Technologies) integration to enhance urban services. These smart grid initiatives include smart meters, traffic systems and mobile applications that will be utilized to reach out to the citizens. However, what is intelligent about smart cities is that intelligent cities integrate learning into a city-management fabric, they offer flexibility and context-sensitivity therein (Batty et al., 2012).

Figure 1. Conceptual Framework for Smart City v/s Intelligent City

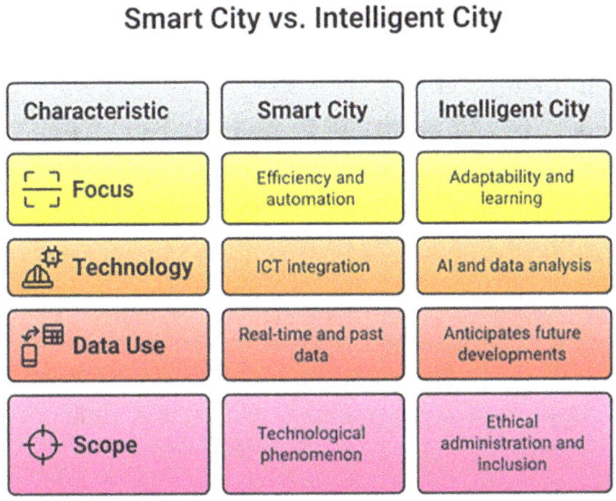

The data used in smart cities is real-time and past data filtered with artificial intelligence algorithms into city learning in which operations are adapted in real-time. This kind of adaptive nature implies that the city is not only reacting to the existing stimuli, but it is also pre-empting the future developments and pre-data preparing the reactions (Bokhari & Myeong, 2022). To give an illustration, the system of intelligent city, instead of relying on the mere demonstration of traffic congestion only, can predict the creation of traffic and automatically redirect fingers with AI-based algorithms of simulation and reinforcement learning. Moreover, the concept of urban intelligence extends beyond automation as a technology phenomenon to adopt the notions of moral or ethical government, inclusiveness, and strength in the future (Meijer & Bolívar, 2016).

Thus, the intelligent city is a living object with digital nature, which communicates with the environment and educates itself, constantly evolving to improve life in the city. Such a combination of civic orientation and technological thinking is the key difference between it and a conventional smart city.

Artificial Intelligence and the Internet of Things: The Technological Stack and Its Convergence

The general concept of intelligent cities is a powerful stack of technology i.e. a chain of interrelated hardware and software, which is largely reflected in Artificial Intelligence (AI) and the Internet of Things (IoT). The two are mutually beneficial to each other because the IoT is the data collection hub, and the AI is the decision-making tool (Al Nuaimi et al., 2015).

IoT can be defined as the physical group of objects, which are composed of sensors, vehicles and meters, and the network system, which is built with the sensors, software, and network connectivity to exchange and receive information. The devices form the digital skin of the city since the sensors monitor commuter traffic and air quality, energy consumption and garbage collection just to list only a few of the activities occurring on a real-time basis. IoT produces enormous data that is either structured or unstructured that is typically characterized by four Vs volume, velocity, variety as well as veracity.

This is where artificial intelligence will be useful. The analysis of this data, patterns, and their identification is conducted with the help of AI algorithms, the most widespread of which are machine learning (ML) and deep learning models. Using the example, the AI models can analyse the information obtained through smart energy meters, which helps to predict the future demand trend and optimize the load distribution in real-time (Sharma & Jangirala, 2022). The use of AI in the transportation sector involves the use of traffic predictions through recurrent neural networks that are being trained on the IoT sensor data.

It is a blend of all these technologies that culminate in what is mostly referred to as a cognitive urban infrastructure, a system that can learn and make judgments by its own. This intrinsically becomes embedded on platforms known as AIoT (Artificial Intelligence of Things) that integrates edge intelligence, distributed learning and networked devices into a distinct layout (Huang et al., 2025).

Practically, this cross allows a variety of smart city applications, such as intelligent lighting, autonomous transportation, AI-assisted zoning proposals, and context-specific alerts about possible danger to the population. The researchers who conducted the study headed by Singh et al. (2025) emphasize that the dynamic of AI-IoT interaction is relevant to the sustainable developmental process of cities with a more cross-sectoral approach, in which the aspects of resilience and inclusion are provided.

Digital Urban Planning

The modernization of the tool is not digitalization, and revolution of the character of the understanding and control of the space, infrastructure, and human behaviour in the urban planning. The traditional planning systems rely on a great extent on surveys, physical audits and fixed data sets that cause urban responsiveness to be delayed. In the light of making, it dynamic and responsive, with its planning models that can better reflect the complexity and change of the real world and its allowances, digital urban planning has been made possible by AI and IoT (De Siqueira et al., 2022).

One of the main tools of such change is the Digital Twin, the digital twin of urban space that reacts to real-time changes according to sensor data. These twins give a sandbox where the urban planners can simulate various urban conditions, population growth, and climate and other political factors (Zhang et al., 2023). When machine learning is involved, digital twins have the capability to produce long-run forecasting by training on previous and current data.

Moreover, AI may also be used in Geographic Information Systems (GIS) to conduct automated zoning, land suitability research and social-demographic clustering. The roles play a significant role in fast growing cities where informal settlements, resource strain, and spatial disparity continue to exist.

Another key aspect of digital city planning is participatory planning platforms, i.e. means of urban space designing employed by citizens themselves with the help of digital means. The recent uses of AI can be applied to the natural language processing (NLP) models to analyses the general mood and the degree of acceptance of policies based on the feedback of the people on the social media, in town halls, and using urban applications (Cardoso & Rodrigues, 2025). The inclusion into the planning framework basing on the intelligent infrastructure is the key to the emergence of energy efficient and people friendly cities, as it is stated in the article by the authors of Singh, Tiza and Alasali, (2025) that is dedicated to the smart technologies and the questions of sustainability.

Digitization of urbanism demands the change in the system of governance too. Urban data governance policies should address the issues of data interoperability, sovereignty, and algorithm responsibility. This raises an ethical issue of ownership of urban data, how the algorithms applied can be trained and whether there is bias in AI based planning systems.

Connectivity, Big Data, Edge Computing, and Cloud Services

Great digital infrastructure is essential to possess a functioning system of smart cities. It consists of connectivity, data infrastructure, and computing capabilities that facilitate the operation of AI and IoT technology.

Connectivity The ability of devices, systems and platforms to communicate with each other in a fast and convenient manner is known as connectivity 5G networks are critical to supporting low-latency, high-bandwidth connectivity and it will be instrumental in time-sensitive applications like autonomous cars, smart grids and telemedicine. Besides 5G, urban centres are considering the LoRaWAN (Long Range Wide Area Network) and NB-IoT (Narrowband IoT) to support the low-power wide-area where the IoT applications are smart farming and waste management among others (Pons et al., 2023).

The source of AI power is Big Data, which is used to run intelligent cities. The data of the IoT devices, states, social, and mobile platforms should be integrated, cleaned, and calculated to decide. How to sustain the quality of data, avail real time access and secure privacy, is the problem. According to Al Nuaimi et al. (2015), successful use of the big data analytics optimization within the system of the city leads to the improved process of decision-making in the transportation modelling, access to healthcare, and the distribution of resources.

Also, the testing of blockchain technology going on that has the potential to increase the integrity, transparency and traceability of information in urban data ecosystems. There are other services like rid sharing, utilities payments, or micro-mobility rentals, which have a strong potential for being automated with the use of blockchain based smart contracts (Karaduman & Gülhas, 2025).

In conclusion, the digital fabric of smart cities is not a merely technical necessity, it is the enabler of innovation, equality and flexibility. Such systems are expected to be sustainable, comprehensive and units that work ethically to achieve sustainable success in future.

APPLICATION OF AI IN URBAN GOVERNANCE: ENHANCING PLANNING, ANALYTICS, AND ENVIRONMENTAL MONITORING

City management is complicated and passes through many obstacles which are associated with rapid urbanization, environmental stress, and social-economic shifts occurring in cities. Artificial Intelligence (AI) has already turned out to be a game-changer to make governance practices more enlightened, dynamic, and inclusive. The section deals with the adoption of AI technologies in key areas of urban business, like land-use planning, predictive analytics and environmental monitoring, thereby contributing to the creation of more resilient and sustainable cities as shown in Figure 2.

Figure 2. Applications of AI in Urban Governance

Urban Governance Aspects

Characteristic	AI-Powered	Optimization	Analysis	Planning	Monitoring
Traffic Management	AI-Powered Traffic Prediction	Adaptive Traffic Signal Control	Real-time Incident Detection	Autonomous Vehicles Integration	
Public Safety & Security	Predictive Policing	Emergency Response Optimization	Crime Hotspot Analysis	Facial Recognition for Security	
Resource Optimization		Smart Energy Grids	Water Leakage Detection	Waste Management Optimization	Air Quality Monitoring & Control
Citizen Engagement & Services	AI-Powered Chatbots for Citizen Support	Personalized Public Service Recommendations	Sentiment Analysis of Citizen Feedback	Smart City Portals & Apps	
Urban Planning & Development	AI Driven Urban Sprawl Analysis	Optimized Land Use Planning	Infrastructure Project Prioritization	Disaster Resilience Planning	

AI for Land-Use Planning and Policy Simulation

Traditionally, the urban land-use models have been founded on the exploitation of static models and subjective judgment; an approach that is typically hard to control the dynamism of urban development patterns. Artificial intelligence (AI) can be useful in the processing of big spatial data and allowing the simulation of the policy implications of planning under various conditions, in particular, machine learning (ML) and agent-based approaches.

Saraswat and Singh (2019) demonstrated the relevance of AI in the traffic simulation through the assistance of the SUMO that is open-source micro-simulation tool to enable optimization of the urban mobility flows considering the land-use changes. Through such an approach, the planners can assess the effects of change in zoning of land or transportation facilities in respect of the resultant traffic congestion and availability in very close real-time.

Besides these simulations, Babelon et al. (2021) also unveiled the application of geo-collaborative platforms together with AI in rendering the process of urban planning a participatory activity. These systems are enabled by the ability to incorporate the crowd sourced information and views of the experts to be able to provide decision making on land-use based on the multi stakeholder approach which is transparent and inclusive.

Developments in AI algorithms can also enable consideration of any other sources of information, including satellite data, census data, and even social media, which can detect land-use changes automatically and give more accurate and timely information regarding urban development. This skill is critical during the management of the informal settlement and civilized urban development.

Predictive Analytics for Urban Systems Management

The use of AI within the urban governance revolves around predictive analytics, which has the capacity to inform the vision of the city issues hence enabling them to plan and prevent them. Using old and current information, ML models can forecast mobility, crime, energy demand and patterns of public health, to prevent such circumstances.

To use AI in predicting crime, Alves et al. (2018) applied the technology to predict crime in the cities, according to the socio-economic indicators and social media feeds, thus, empowering the law enforcement agencies in their resource allocation and safety. Even though such predictive policing would be highly promising, one should also speak about the ethical considerations to make concerning the issue of privacy and the question of algorithmic bias that should be dealt with appropriately.

Under the environmental dimensions, Samad et al. (2023) constructed ML models to predict the concentrations of air pollution in a particular location within the cities as suggested by the meteorological outcomes and sources of pollutants. The cities can an issue in real-time a time-saying health advisory and make some adjustments in traffic or industrial operations to mitigate the peaks of pollution.

Similarly, AI-based demand forecasting would assist utilities in optimizing energy and water consumption, reducing the wastage, and serve the sustainability objectives. By forecasting peak hours and consumption trends, cities can do a more effective job of balancing the supply and the usage on smart grids.

AI-Enhanced Environmental Monitoring and Disaster Management

The process of surveillance in the environment of urban localities has been associated with the use of manual data collection. Though these conventional techniques are handy in giving useful information, they are expensive and time consuming, labor intensive and spatially and temporally resolute. Due to the growing population of cities and pressure on the environment, there is a greater need to have more adaptable, stretchable and real time systems that can be in place to monitor the state of the environment to help in making decisions that can guide the policy.

The union of Artificial Intelligence (AI) and such sensor networks that comprise the Internet of Things (IoT) brought revolutionary transformations to the way environmental monitoring is conducted. The finely tuned, 24-hour tracking of the essential environmental conditions, such as the quality of air and temperature, humidity, noise pollution, and water quality, may be obtained with such technologies. The IoT sensors fixed on the urban infrastructure could also receive a huge amount of data in real-time and these could serve as inputs to the AI algorithms which pattern, trend and anomaly detecting. Predictive management of the environment is also formed by such integration and expands the coverage and precision of information.

Bibri and Krogstie (2017) provide an in-depth account of the AI applications with the view to the implementation of smart and sustainable cities. They articulate that sensor data stream anomaly detection with deep learning models has become increasingly popular, and this plays a crucial role in real-time management of the environment. With the help of these AI systems, abnormal activity that may indicate the spikes in pollution, malfunction of hardware, or the emergence of a high risk of population health can be identified. Rarely can the presence of such anomalies be reported, automated alerting may be applied to report to the municipal authorities, residents, or environmental agencies to facilitate quick responses and minimal effects can be realized.

Besides being real-time, predictive analytics, carried out based on Artificial Intelligence, represent an exceptionally useful method of enhancing city resiliency. AI can forecast environmental hazards and crises such as floods, heat waves, and earthquakes using both historical data and sensor-driven measurements and weather conditions. These predictions support early-warning mechanisms which offer prompt alerts to the threatened populations, institute evacuation plans and send emergency rescue agencies to salvage them. These preventive inputs play a major role in mitigating loss of human life, infrastructure damage, and service continuity in the city in the event of the disaster (Yigitcanlar et al., 2020).

Challenges and Considerations in AI-Driven Urban Governance

Despite such tremendous potential of AI, the obstacles related to the quality of data, accountability of algorithms, and inclusivity must be overcome in the urban governance systems. The data that may be biased or incomplete would be used to make incorrect predictions that would impact the vulnerable population disproportionatcly on the AI models.

Also, the systems of governance must ensure that AI-based decision-making can be held accountable, and humans can be confident in the processes through transparent algorithms and enrolment. The urban planners and authorities must also be trained to be able to read and act on the insights generated by AI.

The last issue is the privacy issue and data security, which is considerably high because the AI systems are being motivated by more personal and sensitive data feeds. There should be strengths between the legal and ethical norms that will work to protect the rights of the citizens and embrace the use of AI.

DATA-DRIVEN URBAN INFRASTRUCTURE MANAGEMENT

The structure of the settlements of modern urban settlements is composed of the urban infrastructure systems like road systems, bridges, water-distribution systems, energy-supply networks and mass-transit systems. They are becoming harder to control as they grow old and urban population grows. The management of urban infrastructure needs an abundance of data sources which include sensor data, geo-spatial data, operational data, social data, weather data, demographic data etc. as shown in figure 3. Traditional infrastructure is a heavy burden on the management of which manual inspection and reactive actions are increasingly unsustainable and inefficient most of the time. In comparison, Artificial Intelligence (AI), the Internet of Things (IoT), and Big Data analytics are the means that can assist cities to evolve to predictive, adaptive, and smart patterns of managing the infrastructure. This section addresses the role of the data-based artificial intelligence solutions in the design, operation, and maintenance of the urban infrastructure.

Figure 3. Data Sources for Infrastructure Management

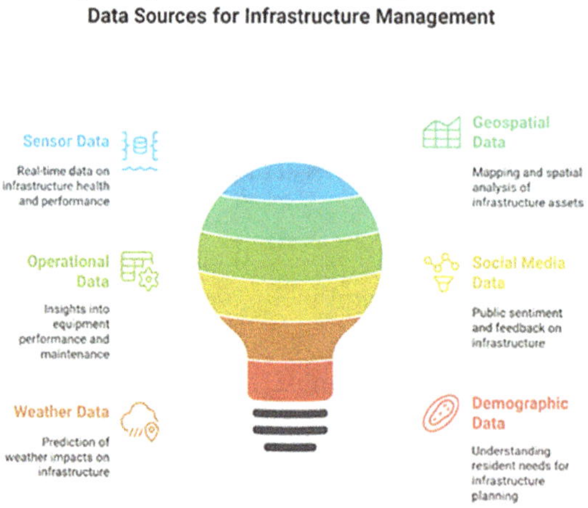

Smart Asset Monitoring and Predictive Maintenance

This part of the infrastructure management has been fundamentally changed using smart sensors and AI algorithms in the condition monitoring of the assets. These systems are applied to continuously monitor, analyze information within the roads, pipeline systems, buildings and functionality to notice decay, abnormal operations, and performance trends.

In the future, it will be possible to find cracks in concrete, the deformity of bridges or leaks in water distribution systems in real time with the assistance of more advanced AI models. An example is the CNNs and image recognition algorithms to identify fatigue cracks and corrosion of important infrastructure through drone and CCTV images (Wu et al., 2025). Figure 4 displays some examples of road pavement distress identification of terrestrial images. Predictive maintenance Because AI can predict failure in digital twins driven by the virtual proxies of physical assets, it makes the services less susceptible and reduces the cost of the lifecycle.

Figure 4. Road pavement distress recognition (Samadzadegan et al., 2024)

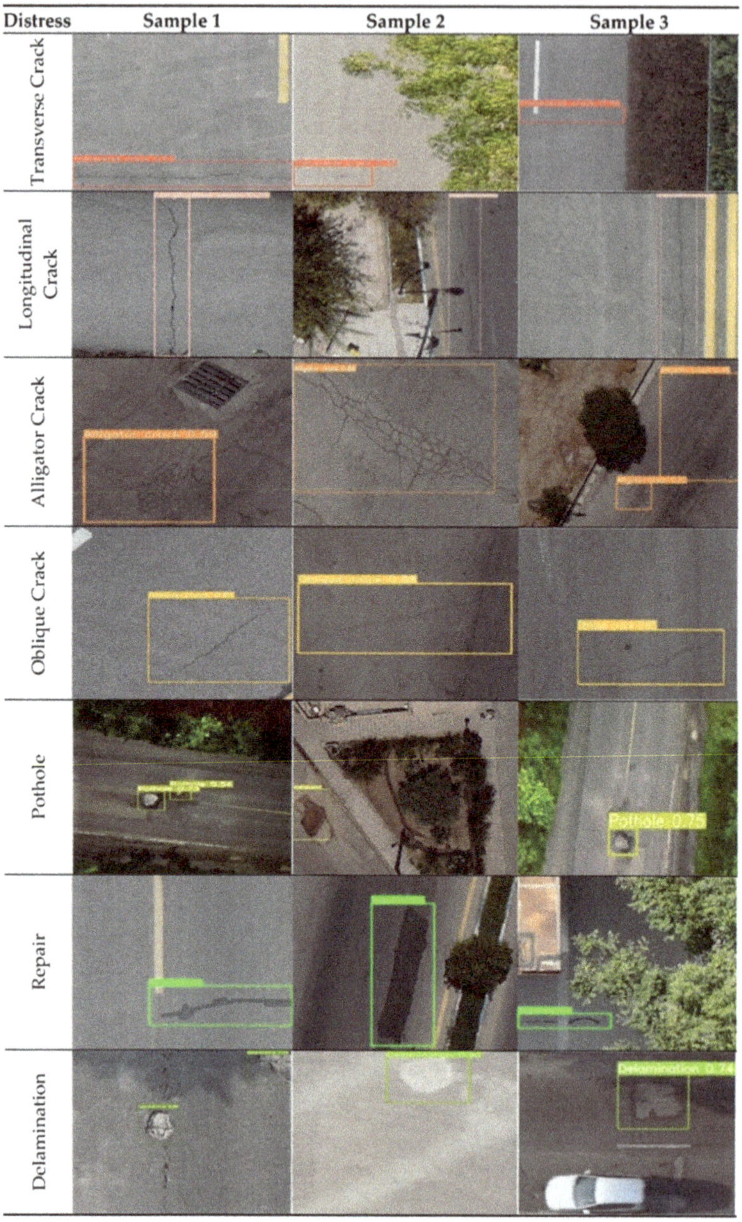

This custom is manifested in the transport infrastructure. The paper containing the article published by the IEEE, U. Archana et al. (2023) discusses another manifestation of the AI model use, when the authors use the YOLOv3 to examine the fatigue of concrete and cracks on the surface of the concrete in smart cities. They applied their system to automatically detect structural anomalies with real-time visual data and subsequently respond to the problem quickly and resulted in improving the life of infrastructure.

AI in Traffic Flow Optimization and Mobility Infrastructure

A factor that is related to urban mobility is infrastructure performance. The application of AI can optimize the traffic flow based on real time data of road sensors, mobile apps and GPS to control gridlock and traffic flow. The reinforced learning algorithms can modify the traffic lights at intervals to enhance the traffic of vehicles based on the density and traffic of vehicles.

In addition, AI is useful in planning future routes and demand planning in a mass transit system. Arranging a sequence of buses or trains, cities can manage to load people in the most efficient way by reducing the waiting period, enhancing the user experience, and minimizing the expenditures that are used on the activity (Zheng et al., 2014). Ride-sharing and micro-mobility services also use fleet distribution, dynamic pricing based on the demand patterns based on AI.

AI in road design is applied to the simulation of traffic conditions to recreate past traffic situation through training. Such simulations may help in the identification of the impact of the infra-investments (i.e., the addition of a new bypass or a new pedestrian corridor) and thus help with planning on the data-driven basis.

Water and Waste Infrastructure Management

Urban water and waste systems are increasingly getting data centric to be managed. The presence of AI and smart water meters along with the flow sensors enables real-time monitoring of the consumption, leaks detection, and pressure regulation. The AI models shall forecast the shifts in the demand trends based on the weather, human population and industrial activity, and the municipalities will have a chance to arrange optimal distribution and to reduce the amount of water waste.

AI can be applied to the wastewater under monitoring the activities of the treatment plants, and, to the quality of discharges in real-time. The anomaly detection algorithms make the identification of equipment failure or biological imbalance, and this gives an opportunity to implement the corrective measures in the shortest time possible. In addition to it, predictive analytics determine future capacity needs based on urbanization and help to scale the infrastructure.

The field of solid waste management can also utilize AI, as it is involved in forecasting the trends of waste production both at the domestic and regional levels. This allows the optimized routes to serve the collection and prevent overflowing cases and implement recycling approaches by categorizing types of wastes via vision-based artificial intelligence solutions (Rahman et al., 2020).

Structural Health Monitoring (SHM)

Structural Health Monitoring must be provided with high-risk property critical assets such as bridges, tunnels and skyscrapers. The AI-enhanced SHM systems also collect data sent by in-built sensors, strain gauge, accelerators, and thermocouples and process them to determine the stress accumulation, vibration abnormalities, and structural failures.

Unsupervised machine learning is specifically applicable in situations where the data of failure which is to identify anomalies is very thin. The principal component analysis (PCA), clustering, and deep learning models could be used in determining the outlier patterns in structural response, which implies that certain problems are beginning to occur.

SHM with Building Information Modeling (BIM) possesses a holistic view of structural integrity within the long-term. Together with the GIS systems, they help authorities to prioritize the inspections, arrangements of the maintenance and the planning of retrofit programs to modernize the old infrastructure.

Infrastructure Lifecycle Planning Using AI

AI supports the lifecycle planning because it allows the decision-maker to evaluate the performance, maintenance, and investment plans within the long term. Multi-objective optimization algorithms can be used to model the infrastructure trade-off between the cost, durability, effect on the environment, and demand.

Predictive lifecycle modeling is an example of this in which the urban planners would be offered the choice of pavement material specification that results in minimum short-term costs in comparison to their long-term operations when traffic is at various levels, and climate changes occur. AI can also be applied in terms of budget allocation as it will show which assets have to be fixed the most and will make the allocation of resources unbiased.

Within the context of renewable energy infrastructure, AI determines the generative and storage needs through the analysis of historical trends and climate prognosis. This knowledge is put in steering the development of resilient infrastructure to absorb the energy transition and sustainability agenda of the city.

AI-Driven Infrastructure Risk Assessment

AI provides channels of vulnerability evaluation and resiliency planning during the urban areas under threat of natural disaster like floods, earthquakes, and hurricanes. On Geographic Information Systems (GIS) integrating AI models, topography, building codes, land use and population data are examined to analyze the exposure and risk of infrastructure.

These systems are invaluable in the event of preparation towards any type of emergency because it helps their cities to model the potential areas of attack, evacuation, and the relief network. AI speeds up the process of cleaning a failed region because it assists in assessing the level of damage by creating satellite images and aerial shots of the areas with the assistance of drones.

The data fusion method of information assembly with multiple sources is applied on the input of information; this method enhances the authenticity of AI-based risk models. These models are particularly effective in low resource settings where data are divided, and human capacity is limited.

Challenges in Implementing Data-Driven Infrastructure Systems

Yet even though AI can have potential infrastructural management there are certain challenges, which are related to the implementation of AI in infrastructure management. To start with, there is still a bottleneck in data availability and quality particularly with the older systems that lack sensors. Second, the infrastructure operators may lack technical expertise that could be able to process and respond to AI insights.

Interoperability between devices and platforms is also a problem in which different infrastructure assets use proprietary data formats or other incompatible data communication protocols. Other concerns include cybersecurity as the infrastructure control systems are mostly becoming susceptible to cyber threats.

Finally, since AI is a recent technology, its ethics like transparency, accountability, and trust in society must be maintained. On a vexing example, traffic management should not be unwarrantedly unfair to certain neighbourhoods by allocating routes that are not fair.

Toward Resilient and Sustainable Infrastructure Management

The future of the urban infrastructure is convergence that combines AI with other technologies, such as IoT, blockchain, and edge computing to render systems self-aware, flexible, and hardened. Another way in which cities have tried to address this is by using decentralized systems to control infrastructure which ought to be self-opting and not based on a central command centre.

Sustainability force is also present. The use of green infrastructure solutions such as permeable pavements, green roofs and smart lighting can be rolled out through the assistance of data-driven systems. The AI models assess the impacts of an urban environment in an urban environment, including the impacts on energy efficiency, urban heat islands, and, to a minor extent, stormwater runoff (Vinuesa et al., 2020).

It is worth noting that the international cooperation and standardization process is gaining momentum. The cities are using communication platforms like the Smart Cities standards provided by iOS and Big Data provided by World Bank to deliver interoperable and scalable solutions as per the best international practices.

INTEGRATIVE CASE STUDIES

Introduction

This part gives a comparative case study of the cities that have implemented the AI and IoT solutions, Barcelona, Singapore and Amsterdam and the Indian Smart Cities Mission. Here the two cases indicate the implementation of smart technologies on mobility, energy, environment and governance. The implementation level is the analysis of challenges and lessons that opens the doors into future development of strategies in smart cities.

Barcelona: IoT-Driven Mobility and Citizen Services

The Barcelona smart city experience began with the program known as CityOS that is the application that controls parking spaces, garbage bins, light, and air quality and includes IoT sensors as depicted in the conceptualized picture in Figure 5. AI-enabled analytics enhances mobility and increases services. The example of Smart Parking system is applicable with real-time sensor data assistance to guide the driver to a free parking space to reduce the congestion and emissions (Vilajosana et al., 2019). In waste management, fill-level detection helps to shorten the collection routes by up to 30 (Bibri & Krogstie, 2020). The participatory platform working on the city uses AI in the processing of the feedback obtained, thereby creating a positive feedback trend of services in the city. It is demonstrated by Barcelona that smart cities should include automatized management and citizen participation.

Figure 5. Conceptualized Representation of IoT Driven Mobility and Citizen Service in Barcelona

Singapore: AI-Based Urban Planning and Governance

Singapore is also a pioneer of smart city governance in the globe. Its digital counterpart called as "Virtual Singapore" merges IoT layers (transport, weather, energy usage), and AI models to simulate and plan in the realms of predictive planning (Alvi et al., 2025). In this system, the what-if scenarios of flood risk, solar irradiation, and population growth are incorporated. In addition, AI-driven citizen services like chatbots and government dashboards can provide individualized support to the citizens and policymaking based on the data.

The basis of such efforts is effective governance models under which data privacy, interoperability, and citizen confidence are the areas of concern. The Model AI Governance Framework designed by the city is ethics focused and includes the transparency and accountability of the AI implementation processes (Singapore Government Tech Office, 2023). The Singaporean approach can provide some useful lessons in that it relates to institutional preparedness and policy alignment.

Amsterdam: Energy and Environmental Data-Driven Projects

The city of Amsterdam also has an interest in sustainable energy and area: it is demonstrated by the smart grid pilot in Zuiderkerk district. This uses the IoT to measure energy supply and AI-based predictive demand to optimize flows and engage residents in the dynamic pricing process (Amsterdam Launches Smart Charging Pilot to Reduce Grid Load and Charging Costs | European Alternative Fuels Observatory, 2025). Similarly, the AirAware project relies upon mobile sensors, which rely on the IoT, and analytics that are powered by AI to generate hyper-local air-quality maps. This data explains why it is justified to introduce policy-specific interventions, such as low-emission zones and greenery interventions (Kerckhoffs et al., 2022).

In addition, locally based enterprises also participate in the Amsterdam-based circular economy activities that include waste segregation against AI, tracking of waste and material reuse using AI. It is the full employment of intelligence in vitality, agility, economy and collaboration of people that renders Amsterdam so stereotypical.

Indian Smart Cities Mission: Opportunities and Gaps

The Indian Smart Cities Mission (SCM), which was introduced in 2015, is intended to change 100 cities into smart infrastructures. Even though the origin of SCM was largely marked with the ingenious light and monitoring, emerging applications are artificial intelligence- and the Internet of Things-driven systems, such as traffic control and smart leak detection in the water supply (Singh, Alasali, and Belbase, 2025). These projects indicate the opportunities of smart metropolitan infrastructures in the conditions of scarce resources.

However, SCM cities are affected by different challenges, including divided rules, bad budget and low citizen involvement. Many pilot projects are not scalable and interoperable. However, it is possible to introduce SCM as a scalability case study to other Global South cities, i.e. that intelligent city technology grounded on same capacity-building, community-based design and inter-city data exchange can be effective.

Comparative Analysis

Comparative table presents the difference in technological focus, strengths, and opportunities on four renowned smart city implementations in the world as represented in table 1. Barcelo is busy deploying IoT and AI in making real time receptively and high degrees of civic interest primarily via the areas of urban mobility and service provision. However, coordination of the various technology silos and data governance model definition is an issue that is present in the city. Singapore leads in terms of advanced simulations, digital twins, and AI-driven form of governance, and this guarantees predictive urban planning and extremely high preparations by the institution. As can be observed, medicine has its merits and issues in the domain of equal access and over-reliance on technology. Amsterdam Smart City idea is aimed at energy saving, environmental sustainability and the encouragement of the circular economy, which is defined by the vigorous position of civilians and sustainable knowledge. There exist though some challenges including privacy and problems in replicating what works in other neighborhoods. India Smart City Mission (SCM) focuses on enhancing mobility, water management and civic IoT applications, is scalable on the grass roots, and conforms to the local context at the country level. Its limitation of resources and the need to organize the efforts of multiple state institutions are the primary problems. The provided analysis lends credence to the point of view that digital urban governance must be humanized, and local contexts considered in terms of applying smart cities technologies because there are a powerful implication of the different urban priorities and the socio-economic background of smart cities. Table 2 restates the descriptive case stories into an evaluative approach, focusing on how each adaptive and autonomous ability produced measurable results, and displaying their trade-offs in environment, ethics, and governance. With the analysis, technological sophistication is not a guarantee of sustainability or equity successful scaling requires institutional maturity, citizen inclusion, and open data governance.

Table 1. Comparative analysis for smart city implementations

City	Tech Focus	Key Strengths	Main Challenges
Barcelona	IoT + AI in mobility & services	Real-time services, civic engagement	Integration across silos, data governance
Singapore	Simulations, digital twins, AI governance	Predictive planning, institutional readiness	Equity issues, tech-dependence
Amsterdam	Energy, environment, circular economy	Sustainability focus, citizen-powered insights	Data privacy, replication across neighbourhoods
India SCM	Mobility, water, civic IoT	Grassroots scale, adaptability to local contexts	Resource constraints, inter-agency coordination

Table 2: Analytical Assessment of AI–IoT–Driven City Case Studies

City / Initiative	Autonomous or Adaptive Capability Achieved	Measurable Outcomes	Hidden Costs & Governance Challenges	Societal Trade-offs and Equity Issues	Scalability and Replicability
Barcelona – CityOS & Smart Mobility	Adaptive IoT-AI integration for traffic, parking, and waste collection; semi-autonomous routing decisions via real-time analytics.	Reduced waste-collection costs by ~30%; congestion cut by 21% in central districts (Bibri & Krogstie, 2020).	High maintenance cost of 20,000+ sensors; data ownership conflict between public agencies and private vendors.	Over-monitoring raised privacy debates; limited inclusion of non-digitally literate citizens.	Moderate success confined to pilot districts; interoperability and vendor lock-in impede wider rollout.
Singapore – Virtual Singapore Digital Twin	Predictive and self-optimizing simulations for land use, flooding, and energy; adaptive governance dashboards with AI-supported decision aids.	15% improvement in energy forecasting accuracy; response time to flood warnings reduced by 35% (Alvi et al., 2025).	High computational cost; data-sharing friction among ministries; dependence on foreign cloud providers.	Surveillance and data privacy concerns; algorithmic opacity questioned in citizen forums.	High strong governance and unified data standards support scaling, but replicability is limited by fiscal and institutional capacity of smaller cities.
Amsterdam – Smart Grid & AirAware Projects	AI-driven adaptive energy distribution and pollution mapping; semi-autonomous demand balancing in the grid.	Peak-load reduction of 10–12%; local NO_2 levels dropped by 8% in pilot areas (Kerckhoffs et al., 2022).	Integration of heterogeneous data sources increased cybersecurity risks; high sensor-calibration expenses.	Dynamic pricing favored high-income users; uneven access to home-automation technologies.	Moderate strong citizen engagement aids replication, but regulatory alignment and energy-market diversity hinder uniform adoption.
India – Smart Cities Mission (SCM)	Limited autonomous functions; adaptive traffic and leakage detection using AI–IoT pilots in mobility and water systems.	Water-leak losses reduced by up to 18% in select zones; traffic-signal optimization cut average delays by 12% (Singh et al., 2025).	Fragmented governance, funding delays, low technical expertise; high dependency on imported hardware.	Digital divide persists; citizen participation sporadic; limited transparency in PPP data handling.	Low-to-Moderate scalable in concept but constrained by uneven infrastructure, local capacity, and policy coherence.

ETHICAL, LEGAL, AND SOCIAL IMPLICATIONS

Introduction

AI and IoT are making cities smarter in all its respective characteristics, which is equally serious with some ethical and legal and social repercussions but also needs to be addressed. These include such areas as data privacy, algorithm bias, surveillance, accountability, inclusiveness and regulation. Only by resolving these thorny issues, is it possible to make sure that the smart city initiatives will not be just technologically advanced but reasonable and fair, not violating the law.

Data Privacy and Surveillance Risks

What can be considered one of the most significant threats is the risk in the presence of all-knowing data collection and surveillance systems. The talks about ethical aspects of smart cities refer to the fact that datafication may result in re-identification, geosurveillance, and inaccurate autonomy (Ehwi et al., 2022). Individuals are not even aware that their information can be utilized by the state as well as by the independent sellers, which contributes to infringement of their privacy and the ambiguity of the law.

Collective governance of data, informed consent, and community governance are transparent mechanisms upon which the identification and protection of individual rights are based. As with some of the high-risk AI systems, the EU version, the AI Act (2024), bans or strictly governs (depending on their type) facial recognition or biometric identification.

Algorithmic Bias, Discrimination, and Equity

The machine learning algorithms applied in AI that are based on historical data or biased data can have the possibility of reproducing systemic bias, most frequently leading to unfair outcomes in the enforcement of law, housing, access to medical care, and social services (Camilleri, 2023). The discriminated parties are often marginalized groups and hence contribute to inequality among them.

In addition, it is globally confirmed that the implications of AI differ according to social and cultural conditions, and the regions that are the most vulnerable to negative impacts are those belonging to the low-middle income groups (Hagerty and Rubinov, 2019).

The recommendations to correct these risks are to consider the role of auditing in accordance to the concept of fairness, incorporation of various data sets, algorithmic transparency, and co-design. Policies such as biased educational measurements and mandatory fair use testing are very crucial in terms of fair usage.

Inclusion, Participation, and Democratic Governance

The problem with intelligent city projects is that they will be the first to fall into the trap of a top-down approach, and technology solutions will not incorporate inputs of the grassroots actors (Project Sherpa, 2020). The smart city regulatory governance must be able to utilize the principles of participatory planning, democratization of access to data, and co-creation of the city residents' mechanisms.

Yigitcanlar et al. (2021) propose the structures of responsible urban innovation that underlies the city-led AI that incorporates local governance, inclusiveness, and civic empowerment. The incorporation of citizen councils, ethics audit board, and open-loop feedback will ensure that AI instruments are not in politics or townspeople interests but rather in the community interests.

Legal and Regulatory Frameworks

The legal frameworks, such as EU AI Act, GDPR, and local data policy, improve the minimum of ethical AI but are playing the catch-up game on the lightning-fast rate of emergence of new technologies. Camilleri (2023) contends that purpose-driven regulation is applicable to focus on the following areas of AI development transparency, fairness, safety, and security.

Urban governance, in its turn, needs a flexible rule on pre-deployment impact assessment, continuous checking of the models, and human control needs. This is necessary because these legal guard rails are needed to make sure that predictive policing or social services or any other sensitive industry where AI can be applied is not abused.

A Conceptual Ethical Governance Framework

Taking into combination these themes, a conceptual plan of ethical governance of smart cities includes:

1. Data Ethics: Utilise transparent, minimal and consent led data use.
2. Algorithmic Fairness: the auditing, diversity and inclusive data must minimize the bias and aid it.
3. Transparency Explainability: There shall be transparent records and explanatory approaches of the algorithm.

4. Legal and Regulatory Compliance GDPR: AI Act and legal regulations specific to the sector. It must conform; the practice is under impact assessment.
5. Public Participation: Introduce the co-designing platform, citizen councils and permanent feedback.
6. Digital Equity: Develop greater capacity building and affordability of service availability to socio-economic groups.
7. General & Accountability: Have distinct systems of review and redress of appeals.

The Limits of Intelligence: Failures, Trade-offs, and Unintended Consequences

Even though intelligent cities should guarantee efficiency and resilience, they are not above restrictions and unforeseen consequences. As is seen in the real world, highly automated systems may increase errors instead of reducing them. Sensors malfunctions, algorithm improper recognition, or loss of data have sometimes contributed to a dysfunctional traffic application, inaccurate environmental notifications, and unfair service provision. In addition to operational risks, there is also an environmental cost of intelligence that is high. To train large-scale AI models and to maintain large IoT networks, massive computational resources are needed which create huge carbon footprints and data centers with high energy consumption. The more cities rely on digital ecosystems, the more vulnerable they are to the cascading systemic failures in which one subsystem malfunctioning, like power or communications, can bring to a crawl related network. Besides, excessive use of autonomous decision-making may cause loss of human control, accountability and flexibility during emergencies. Thus, the quest to attain urban intelligence must be harmonized with sustainability consciousness, strong redundancy systems, and governance systems that promote human-in-the-loop control. The awareness of these limits is an invitation to the more responsible development of smart cities that put more emphasis on resilience, transparency, and environmental responsibility alongside technological innovation.

FRAMEWORK OF IMPLEMENTATION

Introduction

To come up with smart city initiatives, it is not adequate to adopt technology, but to possess a fully developed system, which has policy, governance, partnership, capacity building and performance. In this section, the strategic plan gives a methodical flow towards a self-smooth process of an inclusive city system via institutional preparedness, technical framework and social engagement.

Policy Directives and Legal Infrastructure

Governing policy environment should be an interoperable and rights protective policy. Indicatively, (Singh, Tiza & Alasali, 2025) point that energy saving urban systems built on the principles of sustainable resource applications alongside smart technologies is the basis of effective policies which are defined by addressing both innovation as well as regulation in the intelligent cities.

It is supported by external studies: Almarri and Boussabaine (2023) analyzed the factors of critical success on Public the research of which revealed the critical success factors of smart city projects including clarity of contract, technological fit, government structures. Meanwhile, recent advances of PPP smart cities are oriented towards the equivalence of citizen engagement, data security, innovation, and sustainability.

Additionally, Kebir et al. (2023) stipulates that the history of changing policy ecosystems should be extended to encompass the following that should make the best use of smart infrastructure capabilities.

A good infrastructure to exchange information,

- Obvious licensing of data exchanging,
- Protections against privacy,
- Modular acquisition systems and allow platform flexibility,

The legal engine is heavily designed, protecting the citizens, enhancing innovation, selling interoperability, and ensuring scalability.

Stakeholder Engagement and Public–Private–People Partnerships

The success of intelligent city systems is based on multi-stakeholder cooperation. PPP politics suggests formal systems that can integrate business wisdom and the state interest. Citizens, however, are being incorporated into new models involving PPPPP:

- According to Almarri and Boussabaine (2023), the vision in common is the future of the PPPs, grounded in the perception of contracts and risks encountered, and their mutual understanding.

- To incorporate the community voice and influence into the paradigm of smart city projects, Liu et al. (2021) propose the concept of Public-Zone-People Partnerships (PPPP).
- Alabdullah et al. (2020) underline co-governance through workshops and computer applications directed toward the governmentality in the expansion of trust and shared responsibility.

Examples of such cities as Paris (Paris, online) and Medellin (participatory budget) show that PPPPs help to offer ownership and legitimacy regarding the need to make smart infrastructure projects.

Technology & Standards: AIoT, Edge, and IoT Architectures

Future-ready infrastructure must have interoperable scalable standards as follows:

- Mohamed & Alosman (2024) provide an end-to-end IoT infrastructure framework that includes edge computing, cloud analytics and implementation of bus-based real-time monitoring.
- The paper by Forkan et al. (2023) introduces the AIoT CitySense system that redeploys the already at hand vehicles to transform them into the provider of real-time maintenance data based on IoT/5G/edge.

However, with such systems, one can have real time monitoring into the transportation systems, building health and environment. MQTT, LoRaWAN, NB-IoT are some of the standards that fall under the umbrella of a standard such that the devices in this system are compatible and robust.

Capacity Building and Digital Literacy

Transformative tech-urban managers, engineers, and civil society require to evolve: Well-educated people are required to spare them out:

- Installing sensors with IoT,
- Edge and cloud data analytics,
- Intelligent-assisted decision- cooperation elucidation,
- Integrity in data.

This collaboration with practitioners and universities (e.g., CODE: Metropolitan AI Lab) may finance the training modules and pilot programs and advance knowledge networks and prevent the current capacity gaps.

Measurement of Urban Intelligence

To achieve the transition to intelligent as opposed to smart, must have a measurable standard:

The key performance indicators (KPIs) include:

- System Responsiveness sensor to actuation latency.
- Resilience: time lag in the detection and back up of failed conditions.
- Carbon sustainability, resources efficiency, reductions.
- Equity: demographic service coverage.
- Engagement: citizen involvement in feedback and issues solving, and use of open data.

AIoT dashboards to track operative and governmental statistics are already present in other cities, including Helsinki and Amsterdam. The bases of measurements are offered in ISO 371xx and City Protocol.

Finance and Business Models

Infrastructure exploitation needs to be capitalized in a diversified manner:

- Value Capture Finance: infrastructure that is funded on a value uplift of a property.
- Performance-based contracting: providers are reimbursed on the commitments of the contracts in terms of the service level in real time.
- Blended finance: this is the combination of the risks through the pooling of the third-party, donor, and public funds.

The evolution of global bond structures, especially in net-zero infrastructure, offers designed finance instruments that comply with performance and Social Development Goals.

Ethical Governance and Security

It is accountable to data-intensive technology:

- Stop privacy by design or surveillance design on platforms.
- Put in place testing of AI fairness to be able to detect demographic biases.
- Have cyber-resilience through end-to-end encryption and independent audit.

- Use blockchain transaction log as a source of transparency and a point of trust to the public.

The regulatory sandboxes would allow cities to experiment with new services, monitor the legal regulatory context and risk-proof.

Adaptive Innovation & Iterative Piloting

Transitions to scale are:

- Pilot Projects within some areas or industries.
- The review of the effects is carried out by the random evaluation process, in the form of interviews with stakeholders.
- Planning for a successful pilot replication.
- A legacy of innovation updates that are changeable.

The perpetual dynamic of the situation is essential in altering the environment such as climate change or demographic transformation.

Adaptive Innovation & Iterative Piloting

The Adaptive Urban Intelligence Maturity Model presents the progress of smart cities in a four-phase cycle, namely, Reactive, Predictive, Adaptive, and Autonomous. During the Reactive stage, cities are reacting to problems after they happened by siloed data. Predictive cities are cities that use analytics to forecast issues and optimize processes. Adaptive stage is characterized by self-learning systems, which change dynamically with real time feedback and cross domain intelligence. Lastly, Autonomous cities are fully self-governed, relying on integrated digital twins, ethical AI, and blockchain to have transparency, resilience, and sustainability. The advancement allows cities to transform their simple surveillance into intelligent, active and green city environments. Figure 6 represents a schematic of proposed adaptive urban intelligence architecture.

Figure 6. Schematic of proposed Adaptive Urban Intelligence architecture

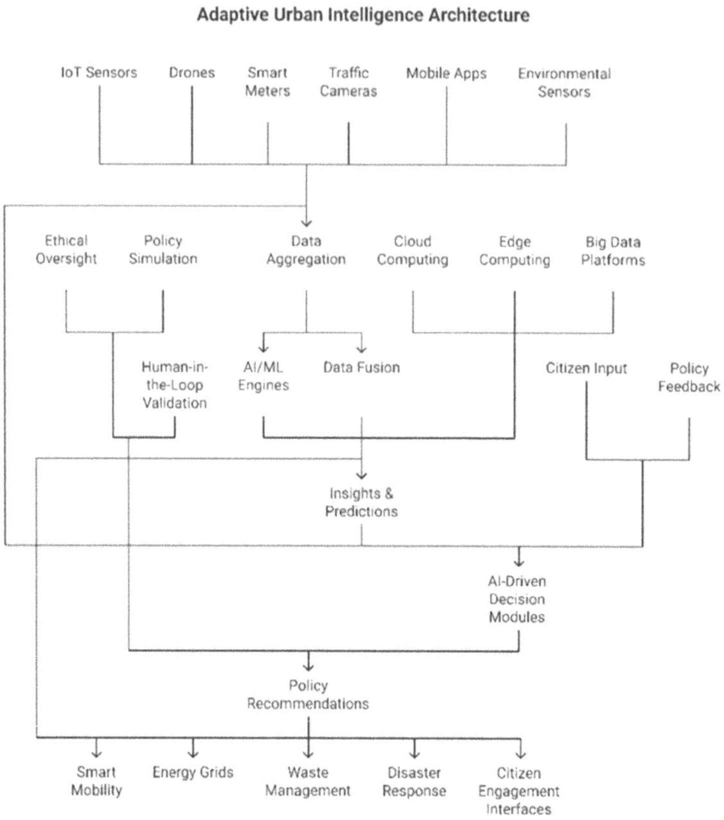

CONCLUSION AND FUTURE OUTLOOK

Conclusion

The transformation of urban ecosystems into smart cities represents but another revolution in our thought, planning, administration, and inhabitance of built environments. The chapter explained the convergence of Artificial Intelligence (AI), Internet of Things (IoT) and other new digital technologies to make the urban environments more resilient, efficient, and sustainable.

In such a way, we have begun with the sensibility of the knowledge backdrop and why we must have smart cities, what are their multidimensional urban issues, e.g. increasing population, climate change, overstraining infrastructure necessitating becoming smarter. In a detailed analysis of the AI-IoT synergy we have indicated the effect of interconnected sensing systems and cognitive algorithms that are facilitating real time decision-making, predictive modelling and data driven governance.

The talk on smart infrastructures and smart mobility systems revealed that adaptive traffic management, smart grid, and self-driving transportation can transform societies by eradicating inefficiency and resource utilization of cities. Web 2.0 government and participation devices are creating citizen engagement, transparency and responsibility doors. Furthermore, interoperability, security, and ethical stewardship were touched in the context of data systems as success factors that are to be regarded to transform smart systems into trustworthy and inclusive.

Regarding the construction of an enlightened roadmap of implementation, the chapter presented the gist of what is required to execute policy, funds, infrastructure, morals, and community association regarding the city intelligence initiatives. The combination of these dimensions is the source of sustainable urban intelligence when combined in synchronicity. The relevance of iterative piloting, adaptive governance and continuous learning on matching technological innovation and social needs was also revealed in the above framework.

Future Outlook

Even in the future, smart cities are not only regarding the introduction of this or that technological concept into the city structure but the philosophical shift to a new vision of cities: the design will be human-cantered, the future will be considered, and the data will be applied to benefit the community.

The future of intelligent urbanism can be heavily influenced by quite a few current trends:

- Hyper-Personalization of Public Services: AI models are gradually turning hyper-personal, such that various services, such as transportation directions, medical outreach, and alerts on other safety conditions, are tailored to an individual behavioural pattern, and urban experiences are made more inclusive and effective.
- Urban Digital Twins: Cities will be digitally mirrored (as a virtual replica) to provide a complex degree of simulations, planning, and maintenance. These computer reflections will enable the authorities to experiment with their policies, optimization of their infrastructure, and potential behaviour of their systems in various situations.
- Edge Intelligence: The edge push can ensure that the latency is lower, and responsiveness can be faster and highly reliable, the smart solutions that are low in power can be scaled globally across cities using the 6G connection.

- Sustainability Integration: Smart cities will also align with the sustainability goals set by the international environment by incorporation of the circular economic ideas, maximization of use of renewable energy and pragmatic carbon conscious planning paradigms.
- Urban Ethical AI in the City of Decision: On a collision course, the streets work, algorithms and governance. Equity, accountability, and explainability of AI systems should be part of the cities to produce a just outcome, community confidence, and policies.
- Capacity to Endure Global Crises: It could be global health crisis, climate emergencies, and cyberterrorism, but the cities of the future must have in place effective policies to identify risks, have an operational malleability, and intersectoral cooperation that will allow these cities to withstand and be safe.

To realize this vision, it becomes a requirement in the future to sever silos that persist in the urban settings, democratize application of digital tools and introduce an organizational culture of innovation that can reach as many stakeholders as possible such as engineers, policymakers, and the local communities. Challenge One will entail education and capacity-building in staffing the workforce and leadership needed to spearhead intelligent cities in the growth of the inclusive, sustainable development.

REFERENCES

Al Nuaimi, E., Al Neyadi, H., Mohamed, N., & Al-Jaroodi, J. (2015). Applications of big data to smart cities. *Journal of Internet Services and Applications*, *6*(1), 25. DOI: 10.1186/s13174-015-0041-5

Almarri, K., & Boussabaine, H. (2023). Critical success factors for public–private partnerships in smart city infrastructure projects. *Construction Innovation*. Advance online publication. DOI: 10.1108/CI-01-2022-0020

Alves, L. G., Ribeiro, H. V., & Rodrigues, F. A. (2018). Crime prediction through urban metrics and statistical learning. *Physica A*, *505*, 435–443. DOI: 10.1016/j.physa.2018.03.084

Alvi, M., Dutta, H., Minerva, R., Crespi, N., Raza, S. M., & Herath, M. (2025). Global perspectives on digital twin smart cities: Innovations, challenges, and pathways to a sustainable urban future. *Sustainable Cities and Society*, *106356*, 106356. Advance online publication. DOI: 10.1016/j.scs.2025.106356

Archana, U., Sharma, S., & Singh, S. K.. (2023). Analysis Of Concrete Cracks And Fatigue In Smart Cities Using YOLOv3. *2023 IEEE International Conference on Research Methodologies in Knowledge Management, Artificial Intelligence and Telecommunication Engineering (RMKMATE)*, Chennai, India, pp. 1-6. https://doi.org/DOI: 10.1109/RMKMATE59243.2023.10368774

Babelon, I., Pánek, J., Falco, E., Kleinhans, R., & Charlton, J. (2021). Between Consultation and Collaboration: Self-Reported Objectives for 25 Web-Based Geoparticipation Projects in Urban Planning. *ISPRS International Journal of Geo-Information*, *10*(11), 783. DOI: 10.3390/ijgi10110783

Batty, M., Axhausen, K. W., Giannotti, F., Pozdnoukhov, A., Bazzani, A., Wachowicz, M., Ouzounis, G., & Portugali, Y. (2012). Smart cities of the future. *The European Physical Journal. Special Topics*, *214*(1), 481–518. DOI: 10.1140/epjst/e2012-01703-3

Bibri, S. E., & Krogstie, J. (2017). Smart sustainable cities of the future: An extensive interdisciplinary literature review. *Sustainable Cities and Society*, *31*, 183–212. DOI: 10.1016/j.scs.2017.02.016

Bibri, S. E., & Krogstie, J. (2020). The emerging data–driven Smart City and its innovative applied solutions for sustainability: The cases of London and Barcelona. *Energy Informatics*, *3*(1), 5. Advance online publication. DOI: 10.1186/s42162-020-00108-6

Bittencourt, J. C. N., Jesus, T. C., Peixoto, J. P. J., & Costa, D. G. (2025). The road to intelligent cities. *Smart Cities*, *8*(3), 77. DOI: 10.3390/smartcities8030077

Bokhari, S. A., & Myeong, S. (2022). Use of Artificial intelligence in Smart Cities for Smart Decision-Making: A Social Innovation perspective. *Sustainability (Basel)*, *14*(2), 620. DOI: 10.3390/su14020620

Camilleri, M. A. (2023). Artificial intelligence governance: Ethical considerations and implications for social responsibility. *Expert Systems: International Journal of Knowledge Engineering and Neural Networks*. Advance online publication. DOI: 10.1111/exsy.13406

Cardoso, P. J. S., & Rodrigues, J. M. F. (2025). Human-Centered AI in Placemaking: A review of technologies, practices, and impacts. *Applied Sciences (Basel, Switzerland)*, *15*(17), 9245. DOI: 10.3390/app15179245

De Siqueira, G., Malaj, S., & Hamdani, M. (2022). Digitalization, Participation and Interaction: Towards More Inclusive Tools in Urban Design—A Literature Review. *Sustainability (Basel)*, *14*(8), 4514. DOI: 10.3390/su14084514

Ehwi, R. J., Holmes, H., Maslova, S., & Burgess, G. (2022). The ethical underpinnings of Smart City governance: Decision-making in the Smart Cambridge programme, UK. *Urban Studies (Edinburgh, Scotland)*, *59*(14), 2968–2984. DOI: 10.1177/00420980211064983

Forkan, A. R. M., Kang, Y.-B., Marti, F., & Banerjee, A.. (2023). AIoT-CitySense: AI and IoT-Driven city-scale sensing for roadside infrastructure maintenance. *Data Science and Engineering*. https://doi.org/DOI: 10.36227/techrxiv.170906134.46169506/v1

Hagerty, A., & Rubinov, I. (2019). *Global AI ethics: A review of the social impacts and ethical implications of artificial intelligence*. arXiv preprint.

Huang, J., Bibri, S. E., & Keel, P. (2025). Generative Spatial Artificial Intelligence for Sustainable Smart Cities: A pioneering large flow model for urban digital twin. *Environmental Science and Ecotechnology*, *24*, 100526. DOI: 10.1016/j.ese.2025.100526 PMID: 39995465

Hudson-Smith, A., Wilson, D., Gray, S., & Dawkins, O. (2021). Urban IoT: Advances, challenges, and opportunities for mass data collection, analysis, and visualization. In ˜The œurban book series (pp. 701–719). https://doi.org/DOI: 10.1007/978-981-15-8983-6_38

International Energy Agency (IEA). (2021). *Empowering cities for a net-zero future: Unlocking resilient, smart, sustainable urban energy systems*. https://www.iea.org/reports/empowering-cities-for-a-net-zero-future

Karaduman, Ö., & Gülhas, G. (2025). Blockchain-Enabled Supply Chain Management: A review of security, traceability, and data integrity amid the evolving systemic demand. *Applied Sciences (Basel, Switzerland)*, *15*(9), 5168. DOI: 10.3390/app15095168

Kerckhoffs, J., Khan, J., Hoek, G., Yuan, Z., Hertel, O., Ketzel, M., Jensen, S. S., Hasan, F. A., Meliefste, K., & Vermeulen, R. (2022). Hyperlocal variation of nitrogen dioxide, black carbon, and ultrafine particles measured with Google Street View cars in Amsterdam and Copenhagen. *Environment International*, *170*, 107575. DOI: 10.1016/j.envint.2022.107575 PMID: 36306551

Liu, T., Mostafa, S., Mohamed, S., & Nguyen, T. S. (2021). Emerging themes of public-private partnership application in developing smart city projects: A conceptual framework. *Built Environment Project and Asset Management*, *11*(1), 138–156. DOI: 10.1108/BEPAM-12-2019-0142

Lloret, Á., Peral, J., Ferrández, A., Auladell, M., & Muñoz, R. (2025). A Data-Driven framework for digital transformation in smart cities: Integrating AI, dashboards, and IoT readiness. *Sensors (Basel)*, *25*(16), 5179. DOI: 10.3390/s25165179 PMID: 40872040

Madanchian, M., & Taherdoost, H. (2025). The impact of artificial intelligence on research efficiency. *Results in Engineering*, *104743*, 104743. Advance online publication. DOI: 10.1016/j.rineng.2025.104743

Meijer, A., & Bolívar, M. P. R. (2016). Governing the smart city: A review of the literature on smart urban governance. *International Review of Administrative Sciences*, *82*(2), 392–408. DOI: 10.1177/0020852314564308

Mohamed, M., & Alosman, K. (2024). An IoT-enabled framework for smart city infrastructure management. TechRxiv *preprint*. https://doi.org/DOI: 10.36227/techrxiv.170906134.46169506/v1

Pons, M., Valenzuela, E., Rodríguez, B., Nolazco-Flores, J. A., & Del-Valle-Soto, C. (2023). Utilization of 5G Technologies in IoT Applications: Current Limitations by Interference and Network Optimization Difficulties—A Review. *Sensors (Basel)*, *23*(8), 3876. DOI: 10.3390/s23083876 PMID: 37112216

Project Sherpa. (2020). *Ethics of Using Smart City AI and Big Data: The Case of Four Large European Cities.* (https://www.project-sherpa.eu/ethics-of-using-smart-city-ai-and-big-data-the-case-of-four-large-european-cities/)

Rahman, M. W., Islam, R., Hasan, A., Bithi, N. I., Hasan, M. M., & Rahman, M. M. (2020). Intelligent waste management system using deep learning with IoT. *Journal of King Saud University. Computer and Information Sciences*, *34*(5), 2072–2087. DOI: 10.1016/j.jksuci.2020.08.016

Samad, A., Garuda, S., Vogt, U., & Yang, B. (2023). Air pollution prediction using machine learning techniques – An approach to replace existing monitoring stations with virtual monitoring stations. *Atmospheric Environment*, *310*, 119987. DOI: 10.1016/j.atmosenv.2023.119987

Samadzadegan, F., Javan, F. D., Mahini, F. A., Gholamshahi, M., & Nex, F. (2024). Automatic Road Pavement Distress Recognition Using Deep Learning Networks from Unmanned Aerial Imagery. *Drones (Basel)*, *8*(6), 244. DOI: 10.3390/drones8060244

Sharma, P., & Jangirala, S. (2022). Internet of Things for sustainable urbanism. *Journal of Physics: Conference Series*, *2236*(1), 012008. DOI: 10.1088/1742-6596/2236/1/012008

Singh, S. K., Alasali, B. E., & Belbase, Y. (2025). Urban Transportation Planning: Artificial Neural Network Applications. In *Expert Artificial Neural Network Applications for Science and Engineering* (Chapter 18). IGI Global. https://doi.org/ DOI: 10.4018/979-8-3693-7250-0

Singh, S. K., Jain, R., & Palaniappan, D. (2025). Future of Mobility: Potential Challenges and Ethical Considerations. In *AIoT for Smart Transportation: Transforming the Future of Mobility* (Chapter 12). IET Publishers. https://doi.org/DOI: 10.1049/PBTR047E_ch12

Singh, S. K., & Saraswat, A. (2019). Traffic simulation & integration using SUMO simulator. 14th International Conference on Recent Trends in Engineering, Applied Science & Management (ICRTESM-19), Vedanta College of Engineering & Technology, Bundi, Rajasthan, India. ISBN: 978-93-87793-71-2.

Singh, S. K., Tiza, M. T., & Alasali, B. E. (2025). Utilization of Sustainable Resources for Energy Efficiency using Smart Technologies. In *Artificial Intelligence for Environmental Sustainability*. Apple Academic Press – CRC Press., https://appleacademicpress.com/the-intersection-of-environmental-sustainability-and-artificial-intelligence-in-the-digital-age-/1675

Ullah, A., Anwar, S. M., Li, J., Nadeem, L., Mahmood, T., Rehman, A., & Saba, T. (2023). Smart cities: The role of Internet of Things and machine learning in realizing a data-centric smart environment. *Complex & Intelligent Systems, 10*(1), 1607–1637. DOI: 10.1007/s40747-023-01175-4

United Nations. (2019). World urbanization prospects: The 2018 revision. Department of Economic and Social Affairs, Population Division. https://population.un.org/wup/

Vinuesa, R., Azizpour, H., Leite, I., Balaam, M., Dignum, V., Domisch, S., Felländer, A., Langhans, S. D., Tegmark, M., & Nerini, F. F. (2020). The role of artificial intelligence in achieving the Sustainable Development Goals. *Nature Communications*, *11*(1), 233. Advance online publication. DOI: 10.1038/s41467-019-14108-y PMID: 31932590

Wu, Y., Li, S., Li, J., Yu, Y., Li, J., & Li, Y. (2025). Deeep Learning in Crack Detection: A Comprehensive scientometric review. *Journal of Infrastructure Intelligence and Resilience*, *100144*(3), 100144. Advance online publication. DOI: 10.1016/j.iintel.2025.100144

Yigitcanlar, T., Butler, L., Windle, E., Desouza, K. C., Mehmood, R., & Corchado, J. M. (2020). Can Building "Artificially Intelligent Cities" Safeguard Humanity from Natural Disasters, Pandemics, and Other Catastrophes? An Urban Scholar's Perspective. *Sensors (Basel)*, *20*(10), 2988. DOI: 10.3390/s20102988 PMID: 32466175

Yigitcanlar, T., Corchado, J. M., Mehmood, R., Li, R. Y. M., Mossberger, K., & Desouza, K. (2021). Responsible urban innovation with local government artificial intelligence (AI): A conceptual framework and research agenda. *Journal of Open Innovation*, *7*(1), 71. DOI: 10.3390/joitmc7010071

Zhang, Y., Pan, D., Van Griensven Thé, J., Yang, S. X., & Gharabaghi, B. (2023). Intelligent flood forecasting and warning: A survey. *Intelligence & Robotics*, *3*(2), 190–212. DOI: 10.20517/ir.2023.12

Zheng, Y., Capra, L., Wolfson, O., & Yang, H. (2014). Urban Computing. *ACM Transactions on Intelligent Systems and Technology*, *5*(3), 1–55. DOI: 10.1145/2629592

Chapter 6
Autonomous Vehicles: AI and IoT for Navigation and Safety

Hemant Kumar Upadhyay
https://orcid.org/0000-0002-2102-1084
Poornima University, Jaipur, India

Rocky Kumar
https://orcid.org/0009-0002-6995-2037
Poornima University, Jaipur, India

Himanshi Sharma
https://orcid.org/0009-0007-8458-6836
Poornima University, Jaipur, India

Lokesh Kumar Shrivastav
https://orcid.org/0000-0002-7403-2887
ARSD College, University of Delhi, India

ABSTRACT

In the transport industry, such a revolution revolves around the autonomous vehicles (AVs) industry with a particular unmatched nature, and it can be attributed to breakthroughs in the realms of artificial intelligence (AI) and the Internet of Things (IoT). Appropriate navigation technologies such as the global positioning system (GPS) and inertial navigation system (INS) are reported as some of the most extensive ones in driverless vehicle applications, whereas SLAM is introduced as one of the primary AI-based technologies that allows performing real-time mapping and positioning. Moreover, in the chapter, AI-controlled safety systems which would avoid collisions and help to respond to the emergency are also brought up, finally resulting in smart travel management and the quality of life. The latter discusses the

DOI: 10.4018/979-8-3373-3146-1.ch006

emerging patterns, technological limitations, and ethics on which the future research and implementation processes will develop. The chapter published as a book can be considered referential and practical to researchers, engineers, and policymakers who have the interest of developing the next level of self-driving cars.

INTRODUCTION

Science fiction has become an engineering reality within the past two decades in terms of vision of self-driving cars (AVs). Their vision of self-driving vehicles that in the past was the content of a science fiction film script and a novel, is now reality starting to emerge at research facilities and test sites of forward thinking automotive and technology companies. This revolution lies in the symbiotic intersection of the Artificial Intelligence (AI) and the internet of Things (IoT), two technologies that are most disruptive in the 21st century. The two technologies are revolutionizing vehicle perception, decision making, and response to dynamic environments with minimum amount of intervention by the human being but with the highest level of convenience, efficiency, and safety.

Based on the network, data can be exchanged, road vehicles can be monitored, predictive maintenance can be realized, and traffic and weather information can be dynamically shared to form a cooperative and intelligent driving ecosystem (Lu et al., 2014). Integration of AI with IoT leads to Cyber-Physical transportation system where edge intelligence, real-time analytics, and edge-cloud coordination enable AVs to safely, accurately and work efficiently However, achieving full autonomy remains complex, with challenges including sensor fusion, object detection accuracy, ethical decision-making, cybersecurity, and regulatory compliance.

Over the past decade, leading automotive manufacturers and technology firms have invested heavily in AV research, large-scale pilots, and simulations. Countries such as the United States, Germany, China, and Japan have adopted supportive policies to accelerate testing and deployment, recognizing AVs' potential to reduce accidents, lower emissions, and transform urban mobility planning (SAE International, 2018).

This chapter explores the synergistic role of AI and IoT in enabling AV navigation and safety, presents current system architectures and technologies, highlights real-world case studies, and discusses the key challenges and future directions for developing trustworthy and resilient autonomous systems.

1.1 Overview of Autonomous Vehicles

Self-driving, or driverless, automobiles, often simply called autonomous automobiles, have onboard systems that under certain conditions can perform all aspects of driving without any human intervention. These features include steering, braking, acceleration, lane-changing, obstacle detection, route planning, and communication with other road users. The Society of Automotive Engineers (SAE) defines automation in six levels ranging from Level 0 (no automation) to Level 5 (full automation) (SAE International, 2018). While most of the current commercial systems are already at Levels 2 and 3 -- they provide driver assistance and some form of driver autonomy -- the world is trending toward full autonomy (Litman, 2020). Autonomous vehicles are a convergence of hardware and software that allows the car to have autonomous capabilities. These are LiDAR, radar, ultrasonic, cameras, GPS, and inertial navigation systems. Software-wise, AI algorithms process the massive amount of sensor data and make decisions in a blink of an eye, whereas IoT platforms facilitate end-to-end communication between the vehicle and the environment. This integration enables AVs to navigate complex scenarios, respond to dynamic road conditions, and ensure the safety of passengers.

Table 1. SAE Levels of Driving Automation

SAE Level	Name	Description	Human Involvement
Level 0	No Automation	Driver controls all aspects of driving.	Full driver control
Level 1	Driver Assistance	Single function automation (e.g., cruise control).	Driver must monitor constantly
Level 2	Partial Automation	Combines steering and acceleration; driver must remain attentive.	Driver monitors
Level 3	Conditional Automation	System handles driving; driver intervenes only when requested.	Driver standby required
Level 4	High Automation	System performs all driving tasks in specific scenarios.	No driver required in defined areas
Level 5	Full Automation	Fully autonomous under all conditions.	No human driver needed

1.2 Role of AI and IoT in Autonomous Driving

Table 2. AI vs IoT Functional Comparison in AVs

Aspect	Artificial Intelligence (AI)	Internet of Things (IoT)
Primary Role	Perception, learning, decision-making	Communication, data exchange, real-time context
Key Technologies	Deep Learning, Computer Vision, Reinforcement Learning	V2X, DSRC, 5G, Cloud/Edge Computing
In-Vehicle Processing	High (image, sensor, decision models)	Moderate (contextual, situational updates)
External Dependence	Minimal (self-contained inference possible)	High (needs connectivity to infrastructure and cloud)
Use Case Example	Detecting a pedestrian and deciding to brake	Receiving alert from another vehicle about slippery road

2. EVOLUTION OF AUTONOMOUS VEHICLES

The following section describes the history of autonomous vehicles and discusses the current state of affairs in the AV sector. The next section outlines the history of autonomous vehicles and examines the present status of the AV sector.

2.1 Historical Development

The development of autonomous cars goes back to the first half of the 20th century when there were experiments with cars that were radio-controlled, indicating the potential for independent mobility. In 1925, a self-driven vehicle driven by Francis Houdina rode New York City roads with no human driver, remotely controlled by radio signals from a trailing car. Though this initial experiment was narrow and unintelligent, it generated interest in the concept of machine-controlled vehicles (Bimbraw, 2015). Substantial progress in the field then started in the 1980s through scholar-led projects. Carnegie Mellon University initiated the Navlab project to develop vehicles with the ability to sense their environment and drive by themselves. During the same time, the University of Munich and Mercedes-Benz undertook the VaMoRs project under the European PROMETHEUS program, which was an important milestone for computer vision and road-following systems (Dickmanns, 2002). The U.S. Department of Defense's DARPA Grand Challenges of the early 2000s provided a stimulus for autonomous vehicle development. The 2004 competition was an eye-opener, and no vehicle was able to finish the course. By 2005, Stanford's "Stanley" vehicle, which had been designed under Sebastian Thrun's direction, traversed the 132-mile desert course alone, proving that sensors, machine learning, and robotics could be integrated into effective autonomous systems (Thrun, 2010). These initial success stories paved the way for more developments. In the 2010s, Google's driverless car initiative, now Waymo, test-drove autonomous cars on public roads. Powered by high-definition maps, LiDAR, radar, and sophisticated AI, these cars bridged the gap from test models to actual deployments.

2.2 Current State of the Industry

The autonomous car industry has become much more sophisticated over the past decade. The industry is currently dominated by a rich and heterogeneous ecosystem of actors, including startups, traditional automotive companies, tech firms, government agencies, and academic institutions. This ecosystem fuels innovation across the board in different areas such as perception systems, connectivity, ethics decision-making, and regulatory issues.

Others such as Waymo, Tesla, NVIDIA, Apple, Baidu, Uber, and Cruise have created and launched AV prototypes of different levels of sophistication. At the same time, investment in AV technology is increasing. Recent market studies show that global autonomous driving investment amounted to more than $50 billion until 2023, and the growth is imminent as the commercial deployment takes off (McKinsey & Company, 2023). The emphasis has also shifted from passenger cars to uses in logistics, farming, mining, and public transport.

While progress has been rapid, there are still challenges to be overcome. These are adverse weather performance, sensor reliability, public confidence, liability concerns, and the ethics of machine decision-making. However, the direction of the industry is toward a future where autonomous vehicles will have a leading role in transforming urban mobility, enhancing safety, and minimizing environmental footprint.

Table 3. Comparative review of key AV studies, highlighting methodologies, datasets, findings, and limitations.

Study / Year	Focus Area	Methodology	Dataset / Scenario	Key Findings	Limitations
Smith, 2013	AV perception & object detection	Deep CNN, sensor fusion	KITTI, Urban driving	High accuracy in multi-object detection	Limited night-time testing
Lee et al., 2022	AV decision-making	Reinforcement learning	Simulation + real roads	Optimized route planning under dynamic traffic	Limited real-world deployment
Chen et al., 2020	V2X communication	Edge-cloud IoT integration	Connected urban vehicles	Low latency communication & coordination	Security & privacy challenges
Kumar et al., 2023	Human-machine handover	Cognitive load & driver monitoring	Level 3 AVs	Improved safety during handovers	Small sample size
Zhang et al., 2020	Adversarial robustness	Adversarial training & model verification	Sensor & camera datasets	Enhanced resilience to attacks	High computational cost

3. ARTIFICIAL INTELLIGENCE IN AUTONOMOUS VEHICLES

The most potent autonomy enabler in the modern vehicle is Artificial Intelligence (AI). In contrast to the traditional software applications that follow the preexisting rules, AI enables cars to learn, change their behavior to new conditions and make higher-level decisions in real-time. AVs are able to sense the surroundings, process various signals, decide about the risk, and make correct decisions with the help of sophisticated algorithms used. AI is being applied in multiple layers of an autonomous vehicle framework- perception and prediction, motion planning and control. This chapter will discuss three important applications of AI: decision-making based on machine learning, sensor fusion and computer vision and object detection and path planning based on deep learning.

3.1 Machine Learning for Decision Making

Machine Learning (ML) is a division of AI that will allow AVs to acquire learning by experience and make decisions without programming them. Sophisticated behaviors like acceleration, braking, lane change, and gap acceptance are represented in autonomous vehicle driving by the application of ML algorithms. They make their decisions on past experience driving, instantaneous sensor data, and surrounding conditions (Wang et al., 2021).

Examples of supervised learning structures that are utilized regularly in object classification (e.g., pedestrian or cyclist) or lane position prediction are decision trees, support vector machines, and ensembles. Reinforcement learning, however would enable an AV to learn good policies that would enable it to play the environment and experience rewards or penalties in the process. The specified approach is especially effective under the conditions of dynamic traffic whereby a vehicle needs to acquire effective and safe driving behaviors during some time.

Such as the Q-learning, Deep Q-Networks (DQN) has been used in the case of adaptive cruise control and lane changing, where the car has to create a trade-off between speed, safety and comfort. Also, cooperative car driving achieved through multi-agent reinforcement learning is an essential element to the systems such as intersection negotiation or platooning (Bojarski et al., 2016).

The uncertainty estimation of missing or noisy data is also added as a part of ML-based decision models. The probabilistic models like the Bayesian nets and Hidden Markov models (HMMs) assist in calculating the possibility that a given situation will occur; therefore, they enhance the dependency of the decision in a given scenario.

3.2 Computer Vision and Sensor Fusion

Arguably, perception is the most crucial task of autonomous driving, and it greatly relies on the AI-powered computer vision. AVs mainly use computer vision to extract valuable information in the form of images and video input using cameras and sensors. This includes object detection and tracking, traffic signs and signals detection, lane markings detection, and distance estimation. Camera-based systems use convolutional neural networks (CNNs) to process the information in an image to both classify and segment the objects within it. It also improves accurate localization and mapping enabling the vehicle to navigate in urban areas, tunnels, or places where the GPS signal is weak or nonexistent.

3.3 Deep Learning for Object Detection and Path Planning

Generative Adversarial Networks (GANs) are increasingly used to simulate unusual or risky conditions that are hard to get from real-world data sets. AVs can then learn from these synthetic samples and perform edge cases better. Real-time deployment is supported by powerful GPUs and specific AI chips inside the onboard computing platform of the vehicle. Efficient processing of visual and spatial data with quick decision-making is the most important factor in making the autonomous navigation safe and reliable.

3.4 Mathematical Foundations of AI Models in Autonomous Vehicles

While the preceding sub-sections covered the conceptual role of Artificial Intelligence in AVs, it is important to treat the underlying mathematics in a more rigorous manner that controls perception, sensor fusion and learning. These quantitative formulations ensure reproducibility and allow safety validation of AI-based decisions to be verifiable.

(a) Sensor Fusion via Kalman Filtering

Sensor fusion is the process of fusing heterogeneous sensor modalities (LiDAR, radar, cameras) together to a common state estimate. The classical Kalman Filter (KF) models the system dynamic state xk and the observation zk by:

$$x_{k|k-1=F_k x_{k-1|k-1+B_k u_k+w_k,}} z_k = H_k x_k + v_k$$

Where Fk is the state-transition matrix, Bk represents control input, and wk, vk denote process and measurement noise (assumed Gaussian).

The recursive update is:

$$K\_k = P\_\{k|k-1\}H\_k$$
$$\wedge\{\ \top\ \}(H\_kP\_\{k|k-1\}H\_k\wedge\{\ \top\ \}+R\_k)\wedge\{-1\},$$
$$x\_\{k|k\} = x\_\{k|k-1\} + K\_k(z\_k - H\_kx\_\{k|k-1\})$$

This formulation minimizes the mean squared estimation error, resulting in the real time tracking of the state even in the presence of sensor noise or intermittent dropouts. Extended or Unscented Kalman Filters are generalized versions of Kalman filters for linear dynamics (as occurs in vehicle motion).

(b) Convolutional Neural Networks (CNNs) for Perception

CNNs translate raw pixels into semantic meaning. A convolutional layer performs a linear filtering of an image I with kernel K:

$$S(i,j) = (I*K)(i,j) = \sum_m \sum_n I(i-m, j-n)K(m,n)$$

The subsequent nonlinearity (e.g., ReLU activation f(x)=max(0,x) and pooling reduce dimensionality while preserving spatial invariance.

Successive layers detect progressively abstract features—from edges → shapes → road entities. For AV deployment, lightweight CNN variants (e.g., MobileNet-V3, YOLOv7-Tiny) are optimized for inference on embedded NPUs at < 50 ms latency.

(c) Decision Making via Deep Reinforcement Learning (DRL)

In DRL, the vehicle acts within an environment to maximize cumulative reward:

$$R_t = \sum_{k=0}^{\infty} \gamma^k r_{t+k+1}$$

where γ is the discount factor.

The policy π(a|s) selects action a given state s, trained to optimize Qπ(s,a)

$$Q_{\text{new}}(s_t, a_t) \leftarrow Q(s_t, a_t) + \alpha\left[r_t + \gamma \max_a Q(s_{t+1}, a) - Q(s_t, a_t)\right]$$

A practical reward model for autonomous driving can be formulated as:

$$r_t = -\lambda_1 C_{\text{collision}} - \lambda_2 T_{\text{delay}} + \lambda_3 C_{\text{comfort}}$$

balancing safety, timeliness, and passenger comfort.

Such quantitative definitions ensure transparent tuning of behavioral priorities and facilitate policy transfer across vehicle platforms.

Figure 1. Core Mathematical Model for Sensor Fusion and Decision Making

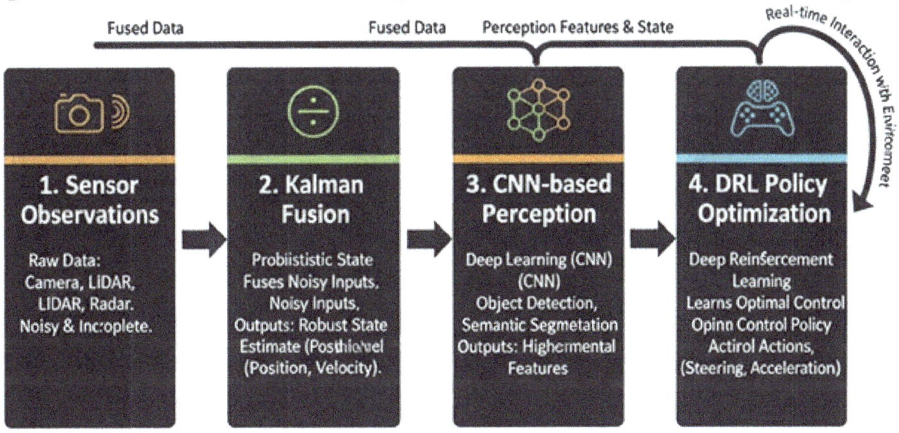

4. IoT IN AUTONOMOUS VEHICLE SYSTEMS

The development of smart and networked technologies in vehicles largely relies on the Internet of Things (IoT).AI equips self-driving cars with the ability to make decisions, while IoT provides them with maintenance and synchronization with the environment.

IoT allows cars to communicate with other cars, roads, pedestrians, and central servers by overlaying an advanced web of sensors, communication interfaces, and cloud infrastructure. This chapter focuses on what the IoT technologies bring to most applications: greater awareness, responsiveness, and intelligence within vehicles.

Figure 2. 5G–Edge AI Synergy in Autonomous Vehicle Ecosystems

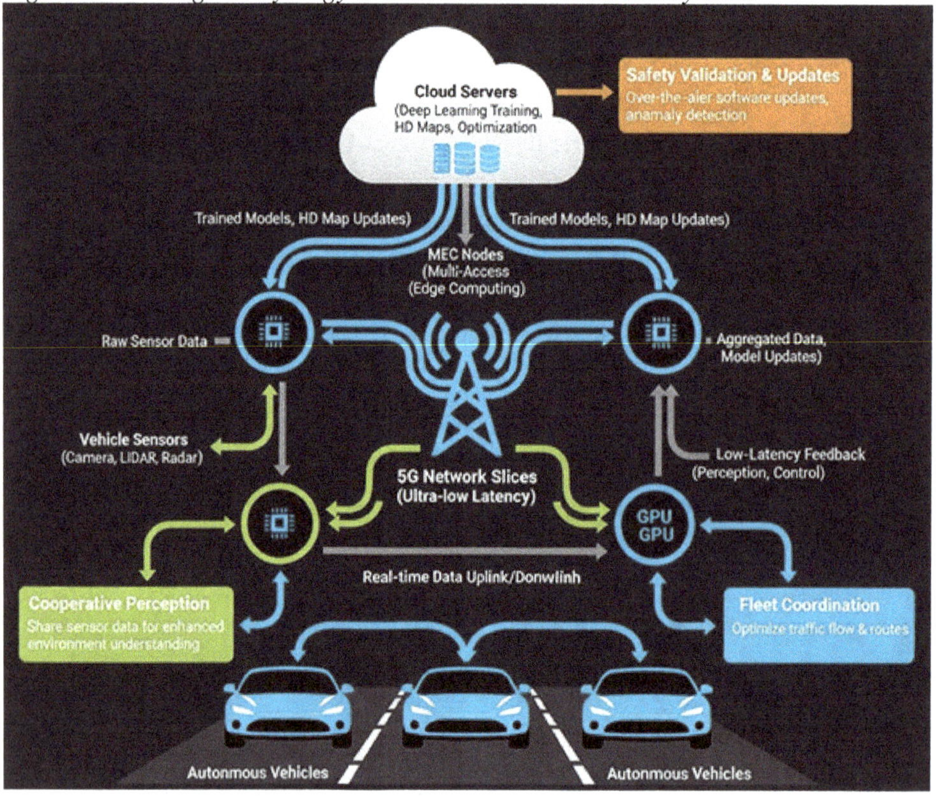

4.1 Vehicle-to-Everything (V2X) Communication

Autonomous driving includes vehicle-to-Everything (V2X) communication as a key element of IoT (Shladover, 2018). V2X is the term used to describe the whole category of communication types that allows an autonomous vehicle to communicate with other parties in its surroundings. These include:

- **Vehicle-to-Vehicle (V2V):** Allows communication between AVs to share speed, location, and direction, enabling cooperative maneuvers like platooning and collision avoidance.
- **Vehicle-to-Infrastructure (V2I):** Facilitates interaction with traffic lights, road signs, and toll booths, which improves route planning and traffic efficiency.
- **Vehicle-to-Pedestrian (V2P):** Ensures pedestrian safety through mobile devices or wearables that alert AVs to human presence.
- **Vehicle-to-Cloud (V2C):** Maintains a link between the vehicle and cloud services for real-time data analytics, updates, and monitoring.

V2X communication uses Dedicated Short-Range Communication (DSRC), Cellular- V2X (C-V2X) technology and the upcoming 5G technology. The Low-latency and high-reliability of the technologies is key to making driving decisions on time. Considering an example, in case a car on one road notices that there is ice, there is a possibility that it informs other vehicles around immediately and reduces speed or takes a detour (Lu et al., 2014).

V2X greatly improves situational awareness and decision-making of a vehicle because it allows the exchange of information in real-time. It transforms it into a connected node in an intelligent mobility network, as opposed to an individual unit, in a vehicle.

4.2 Real-Time Data Acquisition and Processing

Self-driving vehicles rely on continuously collected high-volume information of an array of sensors attached to them. They include LiDAR, radar, ultrasonic sensors, GPS modules, and high-definition camera. IoT is the key enabler to the coordination, transmission and in real-time processing of this stream of data. The sensors give another type of environmental data, object distance, speed, texture or weather. All these inputs are collected by IoT devices, aggregated by timestamping and geolocation and sent to motherboards on board or cloud-based services. This helps the AV to build a 360 degrees perception of the environment, anticipate the movement of the dynamic objects and perform corresponding maneuvers. In order to mitigate latency and bandwidth constraints, AV systems will be optimized to perform such necessary functions locally (e.g. emergency braking) while non-critical processes will be performed asynchronously in the cloud (e.g. map update or upload driving history). Layering IoT into AVs enables the combination of the immediacy of response and asynchronous processing.

Further, the functioning of the system based on IoT offers round-the-clock surveillance of the parts of the vehicle, which is known as predictive maintenance. Analyzing the patterns in the resulting sensor data, the system detects such problems as tire pressure or brake pad loss notifying the users, or service centres before they fail. This offers enhanced safety and working performances (Contreras-Castillo et al., 2017).

4.3 Cloud and Edge Computing for AVs

AV systems can be given a cloud computing infrastructure to store data, perform higher level analysis, and social learning. Another area where AVs must respond in milliseconds to the situation on the local scene is the ability to fill the gap that may exist and edge computing excels in this area. Instead of sending data to distant cloud centers, edge computing executes data close to where it occurs i.e. either on the vehicle or RSUs along the road networks.

In the cloud, non-real-time operations like software update, traffic simulation or fleet management are carried out in a conventional AV system. The edge devices, on the other hand, perform tasks in real-time such as object detection, collision avoidance, and speed adjustment. This kind of segregation makes the response faster and Bandwidth savings are forthcoming (Abbas et al., 2018).

Training and tuning of AI models on enormous batches of data collected through driving by fleets of AVs is also performed on cloud platforms. These models are trained after which they are deployed to the edge computing module at the vehicle. There is no human intervention needed to update a system and continuously learn across the fleet.

High-definition (HD) mapping is another important cloud-edge implementation benefit. The connected vehicles rely on the crowdsourced data that updates maps in real-time and they are distributed via cloud services. The vehicles process each sensor input in real time with edge processors to combine the real-time data with these HD maps to position and navigate the vehicles accurately. With the broader availability of 5G networks, the edge and cloud computing synergies are likely to grow even stronger. A wide area of experts believe that fully autonomous vehicles will need a low latency, high bandwidth network to support their operation in dense urban environments (Sun et al., 2006), which will be provided by 5G.

4.4 Edge Computing and 5G Synergy for Autonomous Mobility

The smart coupling of 5G communication network and Edge Artificial Intelligence (Edge AI) architecture becomes the main driver of real-time autonomous driving by creating the backbone of the ultra-low latency perception, control, and fleet coordination in AIoT-based autonomous vehicles. An important feature of this integration is network slicing and Quality of Service (QoS). 5G allows virtual partitioning of physical network resources in slices that can provide the needed level of bandwidth and latency for different types of traffic.

Latency–Energy Trade-offs in Communication Protocols

AV systems have to compromise between inference accuracy, energy usage and latency. For instance, 5G continuous uplinks are highly power consuming compared to DSRC or Wi-Fi Direct. Adaptive scheduling mechanism dynamically shares the computation workload among onboard NPUs and edge servers according to network conditions and vehicle speed.

Table 4. Latency and Bandwidth Comparison: DSRC vs. 5G-C-V2X Networks.

Protocol / Network Type	Average Latency (ms)	Effective Range (m)	Data Rate (Mbps)	Power Usage (Relative)
DSRC (IEEE 802.11p)	25–50	300	27	Low
4G LTE	40–60	>1000	100	Moderate
5G-C-V2X (mmWave + MEC)	1–5	>2000	>1000	High
Wi-Fi 6 (802.11ax)	5–10	100	600	Moderate

Real-World Hardware Integration

Onboard Inference and Edge AI Accelerators (e.g., NVIDIA Drive Orin, Qualcomm Snapdragon Ride, Intel Movidius) are optimized to run on less than 500 W, and can be integrated with 5G MEC infrastructure to periodically synchronize local models with their cloud-based digital twins (providing both speed and global learning continuity).

5. Navigation Systems

It allows a vehicle to sense its environment, identify its position, and calculate the optimum and safest way to reach there. AV navigation requires correct navigation

Advanced AV navigation combines various technologies such as satellite positioning, in-vehicle sensor systems, and real-time map algorithms to ensure accuracy, duplication, and failure tolerance across different driving conditions. A brief description of the three major navigation technologies employed by autonomous vehicles (AVs) has been provided in this chapter. These are Global Navigation Satellite Systems (GNSS), Inertial Navigation Systems (INS), and Simultaneous Localization and Mapping (SLAM).

Table 5. Comparison of Localization Techniques in Avs

Technology	Accuracy	Environment	Limitations
GNSS (e.g., GPS)	Meter-level	Open-sky, global	Signal loss in urban canyons/ tunnels
RTK-GNSS	Centimeter-level	Requires reference stations	Costly; needs base stations
INS	Short-term accurate	GPS-denied areas (e.g., tunnels)	Drift over time; needs calibration
SLAM	High (relative)	Urban, indoor, unmapped environments	Computationally heavy; real-time challenge

Figure 3. Graph-Based SLAM Optimization Pipeline

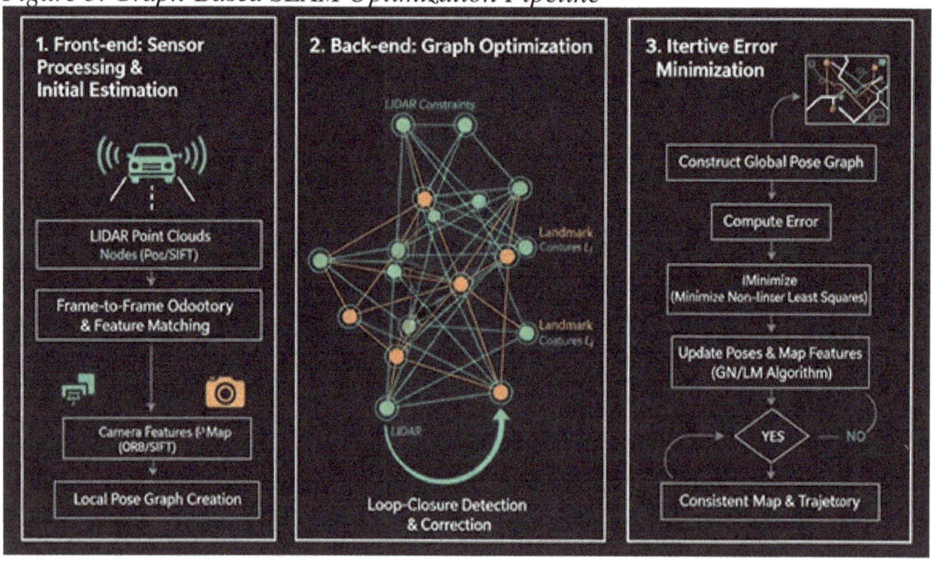

5.1 GPS and GNSS Technologies

A fundamental framework of providing AVs with geospatial location exists in the Global Positioning System (GPS) and other Global Navigation Satellite Systems (GNSS), which include GLONASS (Russia), Galileo (Europe) and BeiDou (China). Such systems are formed by using a system of satellites (constellation) that broadcasts time-stamped messages allowing ground-based receivers to determine their exact location by calculating it using trilateration.

The primary use of GNSS is in autonomous cars in terms of navigation and routing. Normal GNSS signals however can have few meter errors and this will not be sufficient to serve tasks like lane level navigation. AVs thus employ Differential GNSS (DGNSS) and Real-Time Kinematic (RTK) positioning which employs source stations to remove distortion on the signal induced by atmospheric effects or geometrical positioning of the satellites.

The accuracy of RTK-based GNSS is up to centimeters so that high-precision localization is sufficient. Nonetheless, GNSS signals are subject to degradation in an urban canyon, tunnel or other thick plants. Consequently, AVs are not dependent solely on satellite positioning and are accompanied with other technologies in localization as a substitute of localizing and making sure that the navigation is continuous and accurate (Sun et al., 2006).

5.2 Inertial Navigation Systems (INS)

Tracking the movement of vehicles in the short tangible and frequent methods is a paramount operation in the context of AVs which is possibly possible through the Inertial Navigation Systems (INS). An INS is made of accelerometers and gyroscopes which calculate the acceleration of a car and its spinning rate in all three dimensions. This cumulative sum over time can be used by the system to calculate estimates of position, velocity and changes in orientation.

INS is of particular value when GNSS are inaccessible or inaccurate--i.e., during GPS outages or tunnel shots, and underground parking jigs. INS has the merit of being signal independent and thereby it is possible to navigate continuously even when signals are blacked out.

However, INS systems are also characterized by the problem of drift on time due to accumulating the errors. Consequently, in practical case scenarios, INS tends to be complemented by GNSS or SLAM which allows decreasing on the errors in position. A combination of these two systems, referred to as GNSS/INS integration, is a hybrid variant that takes advantage of the advantages of both systems: long-term accuracy of GNSS and short-term accuracy and responsiveness of INS.

This combination of GNSS and INS measurements is usually attained in the form of Kalman filters or extended Kalman filters that are used to estimate state of vehicle optimally based on accuracy of individual observation (Liu et al., 2018).

5.3 Simultaneous Localization and Mapping (SLAM)

The Simultaneous Localization and Mapping, (SLAM) is a technique by which an autonomous car can create and build a map of a previously unexplored environment, it also helps it determine where it is located in that map. SLAM recently gained relevance in the area of urban and indoor navigation, where there are no GNSS signals or signals are weak.

SLAM systems rely on sensor measurements usually cameras, LiDAR, radar to sense the features of the world (examples, streets, buildings, curbs, street signs). They are viewed as reference points to determine the position of the vehicle in relation to the points. Over time, when increasingly more and more landmarks are observed and re-observed, the map and vehicle's position are estimated by the system with greater certainty.

There are various categories of SLAM:

- **Visual SLAM (vSLAM):** Based on camera imagery and computer vision techniques.
- **LiDAR SLAM:** 3D laser scans are used to build dense maps of the environment.
- **Multi-sensor SLAM:** Merges camera, LiDAR, radar, and IMU inputs for increased robustness.

Highly sophisticated SLAM techniques utilize graph-based optimization and loop closure to maintain the map temporally consistent. Loop closure helps determine previously visited areas, thereby preventing cumulative errors and map drift.

SLAM also enables AVs to operate in dynamic environments by distinguishing between moving and static objects. Real-time mapping of SLAM makes it possible for AVs to localize themselves accurately in unmapped environments or in the dynamic environment of ever-changing surroundings (Geiger et al., 2012).

5.4 Mathematical Framework of SLAM

Whilst Section 5.3 presented the idea of SLAM in a qualitative manner, quantitative formulation is necessary if the discussion is to move from being descriptive to analytical. Simultaneous Localization and Mapping (SLAM) can be formulated probabilistically in terms of the estimation of a joint posterior over the vehicle pose and environmental landmarks given all sensor observations and control inputs: $p(x_t, m | z_{1:t}, u_{1:t})$

where

x_t = vehicle state (pose, velocity),

m= map of landmarks,

z1:t = all measurements up to time t,

u1:t = control inputs (e.g., wheel odometry).

(a) State–Space and Measurement Models

The vehicle motion is typically modelled as:

$$x_t = f(x_{t-1}, u_t) + w_t$$

and the sensor observation model as:

$$z_t = h(x_t, m) + v_t$$

where f(·)and h(·)are nonlinear transition and measurement functions; wtw_twt and vtv_tvt denote zero-mean Gaussian noise with covariances QQQ and RRR.

(b) Extended Kalman Filter (EKF-SLAM)

EKF-SLAM linearizes these nonlinear models around the current estimate to propagate the belief distribution:

^{x}_{t|t − 1} & = f(^{x}_{t − 1|t − 1}, u_t)

P_{t|t − 1} & = F_tP_{t − 1|t − 1}F_t^{ T } + Q_t

K_t & = P_{t|t − 1}H_t^{ T }(H_tP_{t|t − 1}H_t^{ T } + R_t)^{−1}

^{x}_{t|t} & = ^{x}_{t|t − 1} + K_t(z_t − h(^{x}_{t|t − 1}))

P_{t|t} & = (I − K_tH_t)P_{t|t − 1}

(c) Graph-Based SLAM Optimization

Modern SLAM implementations employ **graph optimization** rather than sequential filtering. The environment is represented as a graph G= (V, E) where

V={x1,.,xT,m1,,mN} are nodes for poses and landmarks, and

E are edges encoding spatial constraints derived from sensor observations.

The goal is to minimize the total error:

$$x^*, m^* = \arg\min_{x,m} \sum_{(i,j)\in E} \left| z_{ij} - h\left(x_i, m_j\right) \right|^2_{\Omega_{ij}}$$

where Oij is the information (inverse covariance) matrix of each constraint. Loop closure corrects drift in a global least-squares optimization of all previously visited locations in order to improve map consistency.

(d) Computational Considerations

Sparse matrix factorization (e.g., Gauss-Newton, Levenberg-Marquardt) can be applied to the graph-based SLAM in order to achieve real-time performance. Hybrid visual SLAM (vSLAM) and LiDAR SLAM based system is proposed which creates hybrid cloud of camera texture and geometric point cloud for robust performance in various light and weather conditions.

6. SAFETY MECHANISMS

Safety is the most important factor when developing and operating autonomous vehicles (AVs). Since these cars are not driven continuously by a human, they should be able to perceive, process, and respond autonomously to danger in real time. AVs need to not only prevent accidents but also reduce damage and ensure passenger and pedestrian safety in unforeseen circumstances. This subsection describes the most important safety measures implemented in AVs, specifically AI-based hazard analysis, obstacle detection, and emergency maneuvers supplemented with redundancy systems.

Figure 4. AI–IoT Integrated Safety Loop

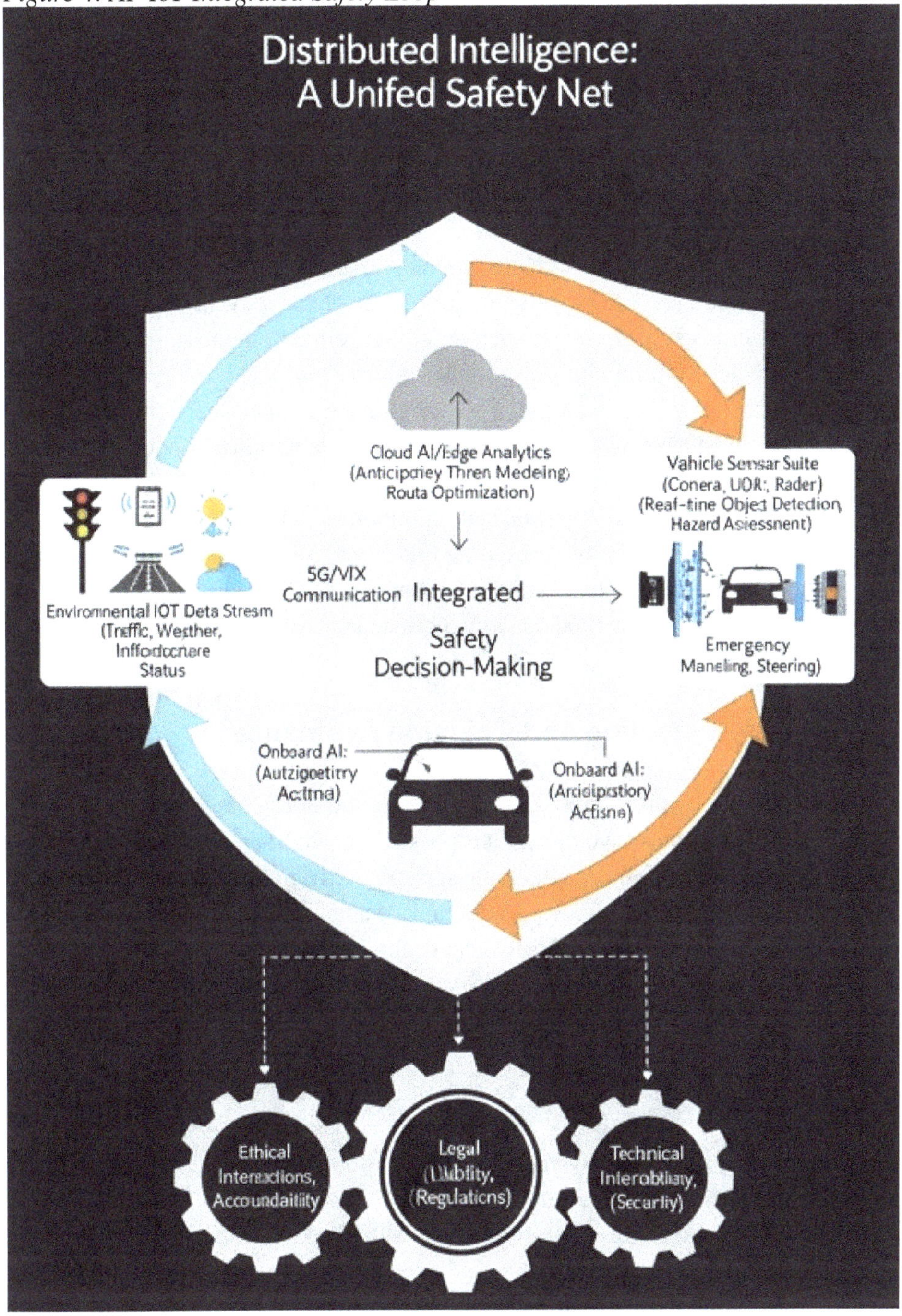

6.1 AI-Based Risk Assessment and Decision Making

Artificial Intelligence (AI) allows AVs to conduct real-time risk assessment by processing live data from around them. It enables the car to detect likely dangerous situations—a pedestrian suddenly onto the road or another car performing an unexpected brake—and take appropriate decisions to mitigate risk.

Risk assessment identifies the likelihood and extent of probable outcomes based on prevailing driving conditions, traffic patterns, and weather. Such inputs are analyzed by AI systems such as Bayesian networks and decision trees to identify whether an immediate response (e.g., braking, evasive maneuver) is required. More advanced systems use deep neural networks and reinforcement learning to make long-term predictions and choose optimal actions from these trade-offs between safety and efficiency. For example, AVs must make ethical decisions when they encounter circumstances where there is a colliding inevitability. Modules designed to help AVs choose an action that will minimize the most harm are still in a trial phase and known as ethical decision-making modules, which have been debated in artificial intelligence (AI) ethics research under the concept of the "trolley problem" (Goodall, 2021). In actual implementation, such AI decision systems are also subject to regulatory safety models, which create legal norms and behavioral rules that AVs must abide by.

6.2 Obstacle Detection and Collision Avoidance

One of the key driving safety functions of autonomous cars is moving and stationary obstacle detection avoidance. It is possible through sensor fusion, consisting of LiDAR, radar, cameras, and ultrasonic sensors; Artificial Intelligence perception software; and real-time decision-making. Object detection algorithms, such as YOLO and Faster R-CNN, are used by AVs to detect vehicles, pedestrians, cyclists, and road obstacles. Once an obstacle has been detected, the tracking algorithms can be used to estimate the speed and trajectory of the obstacle to determine possible collision trajectories. Collision avoidance maneuvers are acquired using the motion planning systems which recalculate the vehicle path through new information. These are:

- **Emergency Braking Systems (EBS):** Automatically brake when a high-probability collision is about to occur.
- **Lane Departure Prevention:** Steer the vehicle back to its own lane when it detects improper lane departure.
- **Evasive Maneuver Planning:** Plot the optimal path of evasion without putting others at risk.

Under city or multi-lane driving scenarios, the systems must analyze not one but many potential combinations in a matter of milliseconds. In addition, redundant sensing routes and sensor fusion will allow the vehicle to remain on its feet even if one sensor fails or misinterprets the results of a sensor (Chen et al., 2018).

6.3 Emergency Handling and Redundancy Systems

Autonomous cars must have the capacity to react to unexpected or emergency situations - such as breakdowns in hardware, malfunctioning sensors, sudden weather conditions, or erratic driving around the vehicle. To safely respond to these situations, AVs use a number of layers of redundancy in hardware and software systems.

- Redundant Sensors: AVs have redundant sensors to sustain a continuous image view. While the camera could be problematic under less-than-ideal lighting conditions, it uses the LiDAR or radar as a backup.
- Backup Control Systems: In case any of the actuators do malfunction (such as brakes or steering), the backup control system can automatically kick in and perform the necessary actions.
- Fail-Safe Architectures: AVs have fault-tolerant software to identify deviation from normal system operation and trigger pre-programmed emergency procedures, such as swerving to the side of the road and braking to a controlled stop.

Modern AV vehicles are also constantly monitored by the system health to be able to identify the first signs of an impending failure based on temperature, electrical characteristic, and communication latency.

In case of an issue, the system warns a user and allows a graceful degradation, i.e., transitioning to manual control or smoothly terminating the operations that are critical or result in an emergency (Koopman & Wagner, 2016).

Also, cybersecurity plans are likewise involved as a way of deterring the access of the navigation and safety systems by thieves or hackers. AVs are secured externally by the use of intrusion detection systems (IDS) and end to end encryption.

6.4 Systemic Safety Integration through AI–IoT Convergence

Safety in autonomous mobility has to be redefined as a systemic property that is generated by the permanent interaction of onboard AI intelligence with distributed IoT connectivity. The combined working of perception, prediction, and communication layers allows the conversion of traditional reactive safety to proactive context-aware protection network.

(a) Predictive Safety via IoT Sensing

Under the scope of Vehicle to Everything (V2X) communication, IoT-enabled V2X communication enables vehicles to share telemetry, road-infrastructure based risk information and environmental states in real-time. When a roadside unit (RSU) identifies the presence of black ice or a pedestrian zone using environmental sensors, the message is broadcast within a 2km cell radius. This V2I message is combined with the on-board sensor stream to provide the edge-AI module with sufficient time to pre-emptively reduce torque, brake bias and re-calculate trajectory before a potential hazard is detected at the local level (sub-100 ms) of the vehicle.

(b) Reactive Safety via Onboard AI

Apart from smart models, onboard perception models (e.g., YOLOv8 object detection and CNN based lane tracking) running at 30 fps to identify dynamic obstacles come with predictive layers. When local sensors identify a suspicious object but IoT signals indicate a wrong alarm the AI system focuses too much on unnecessary braking. This combination of reactive AI-inference and predictive IoT-context results in a closed-loop control structure, which doubles the safety and comfort.

7. INTEGRATION OF AI AND IoT FOR INTELLIGENT NAVIGATION AND SAFETY

The union of IoT and AI would result in the creation of a product that would allow autonomous cars to function at their peak potential. AI enables the car to think smartly, understand and make decisions on its own such as, knowing which route to take, monitoring hazards, and driving carefully. On the other hand, with the help of IoT, the vehicle is integrated with the external environment, and the driver can get the information of traffic lights, vehicles, road conditions, and weather conditions in real time.AI with IoT makes a complete autonomous driving scenario that is secure and brainy.

The combination of the two technologies introduces a learning system that can navigate itself, escape dangers beforehand and form a dynamic system. AI and IoT in autonomous assortment and administration circles that enables self-driving is all covered in this subsection (Abbas et al., 2018).

Figure 5. Proposed AI–IoT Integrated AV Architecture and Data Flow

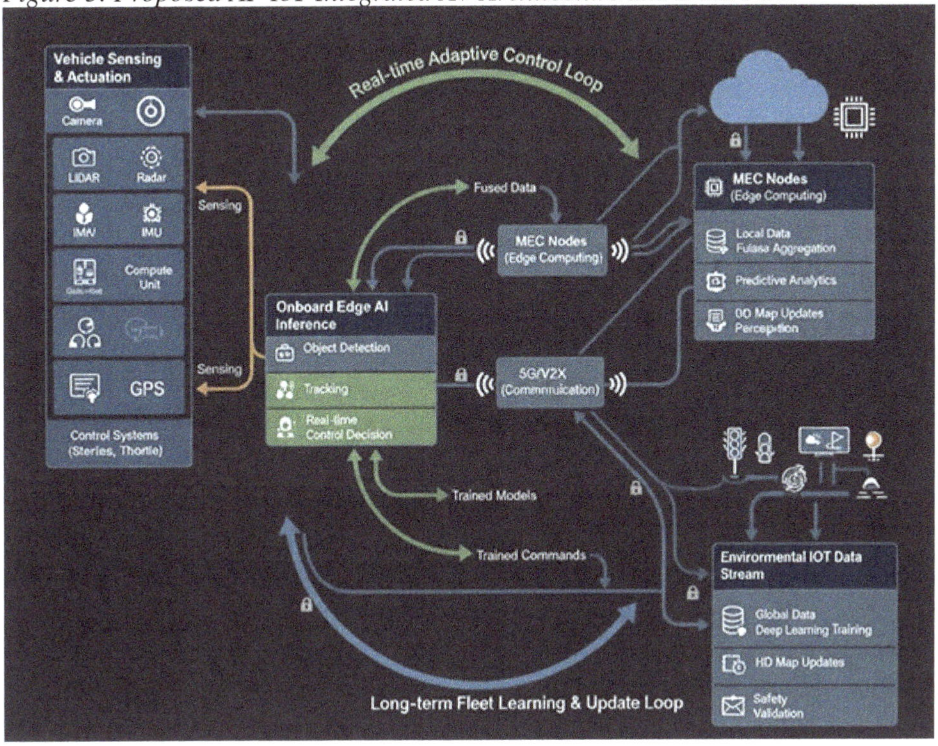

7.1 Proposed AI–IoT Integrated Architecture and Data Flow

The proposed architecture provides an end-to-end AI-IoT integration architecture for AVs in order to integrate perception, communication, decision making, and control into a unified cyber-physical system. The system is an adaptive layer which is capable of real-time adaptability, secure data management and continuous fleet level learning into 5 layers which all have an interconnected and tightly coupled relationship with one another: perception, edge intelligence, connectivity, cloud coordination, and actuation. At the perception layer, the environment is captured by heterogeneous sensors (LiDAR, radar, camera, GNSS and ultrasonic) and data is fused by using a Kalman-based filtering process and published in either ROS 2 or DDS. The edge intelligence layer has deep learning models on embedded AI accelerators (e.g., Nvidia Drive Orin), which run, object detection, semantic segmentation, and SLAM updates, and its outputs are shared asynchronously using MQTT for cooperative decision-making. The connectivity layer connects fleet and vehicles to infrastructure through 5G-C-V2X, DSRC and REST APIs and uses network slicing to give priority to safety-critical messages and allow non-critical data to use another bandwidth. The cloud coordination layer aggregates anonymized telemetry for fleet wide learning - reinforcement training & digital twin modelling with updated models securely distributed back via OTA updates. Data transfer is done with the use of the JSON or Protobuf that is protected by TLS 1.3 and is verified with the help of blockchain to ensure its integrity. Finally, the actuation layer translates AI decisions into fine control signals using CAN-bus or Ether CAT. Redundant controllers are controlled using Responsibility-Sensitive Safety (RSS) logic to ensure safe operations. Overall, the flow of sensor data from data collection to data fusion to edge inference, learning in the cloud, and executing the actuator in a continuous, traceable, and secure loop. This architecture creates AVs as adaptive collaborative systems within a self-evolving AIoT network for autonomous intelligence and community optimization within the transportation ecosystem.

7.2 Real-Time Analytics and Control Loops

Real-time processing is essential in autonomous since the car must process sensory data and respond within milliseconds to stay in a safe state. Both AI and IoT also make it possible to implement closed-loop control systems, which constantly monitor the state of the vehicle, the state of the surrounding environment and the driver inputs, and take corrective actions in a timely manner. It is generally said that an AV comprises the following control loop:

- **Sensor Data Acquisition:** Ongoing data stream from the environment (e.g., passing automobiles, pedestrians, road surface).
- **Data Preprocessing and Fusion:** Basic filtering, normalization, and data stream synchronization utilizing Kalman filters or AI-based fusion.
- **AI Inference and Decision-Making:** Categorization of objects, estimation of intentions, route planning, and threat evaluation.
- **Actuation Control:** Sending real-time actuation commands to braking, steering, and throttle actuators based on the output of the AI model.
- **Feedback Analysis:** Monitoring the effect of the action that has been taken and basing future decisions upon it.

7.3 Performance Metrics and Quantitative Comparison

To assess the practical effectiveness of the proposed AI-IoT framework, it is necessary to provide quantitative performance benchmarks, which will identify trade-offs between the accuracy of the perception, the system latency, and the energy consumption. In addition to quantitatively evaluating models against each other, such metrics serve as empirical bases for evaluating competing models, protocols, and localization methods in the field In the perception applications, convolutional neural networks (CNN) such as YOLOv8 and Faster R-CNN are becoming popular due to their high object detection accuracy. However, as the market is now moving to edge-AI deployment, their suitability for such a deployment depends not only on mean average precision (mAP), but on inference latency and computational overhead as well. Experiments on embedded platforms (e.g. NVIDIA Orin) indicate that lightweight models are suitable for near-real-time performance and small accuracy loss on the other hand, heavier networks provide high-quality detection at the cost of latency and increased power consumption.

Table 6. Performance comparison of perception models (YOLO vs. Faster R-CNN).

Model	Mean Average Precision (mAP %)	Inference Latency (ms)	Power Consumption (W)	Remarks
YOLOv8-Nano	72.8	18	38	Optimized for edge; fastest real-time inference
YOLOv8-Large	83.2	46	95	High accuracy; suited for cloud-assisted AVs
Faster R-CNN (ResNet-50)	81.5	65	110	Reliable detection but high latency
SSD-Mobilenet V3	70.4	22	35	Energy-efficient baseline model

Localization accuracy is another factor of safe autonomy. When comparing GNSS, Inertial Navigation Systems (INS) and SLAM, sensor fusion is an effective approach that adds a lot of robustness when facing the occlusion or degraded signal environment. Real-world tests conducted in urban canyons demonstrate that combining LiDAR and camera data by fusing them using Extended Kalman Filtering (EKF) can decrease the positional drift of the sensor by more than 40 per cent compared to the GNSS-only navigation.

Table 7. Benchmark of localization techniques (GNSS vs. INS vs. SLAM).

Localization Technique	Average Error (m)	Update Rate (Hz)	Typical Latency (ms)	Operational Conditions
GNSS (stand-alone)	2.1	1	250	Poor accuracy in urban canyons
INS (gyro + IMU)	1.6	200	20	Drifts accumulate over time
Visual SLAM (vSLAM)	0.8	30	40	Sensitive to illumination changes
LiDAR SLAM	0.6	20	50	Robust geometry-based mapping
EKF Sensor Fusion (GNSS + IMU + LiDAR)	0.4	25	35	Optimal multi-sensor accuracy

By balancing model complexity, network latency and hardware power limits, the system is able to achieve real-time performance without compromising interpretability or robustness.

8. CHALLENGES AND FUTURE DIRECTIONS

Although there has been tremendous technological advancement, autonomous vehicles (AVs) are far from perfect. Issues of AVs extend beyond technical issues but also involve ethical, legal, infrastructural, and societal issues. As the development of AVs continues across the world, challenges need to be addressed by stakeholders to provide safety, public confidence, and sustainable use. Parallel to this, there are developing opportunities created by cutting-edge technologies such as 5G, Edge AI, and digital twins that can drive the limits of AV intelligence and response.

Figure 6. Quantitative Performance Evaluation of AI–IoT Framework

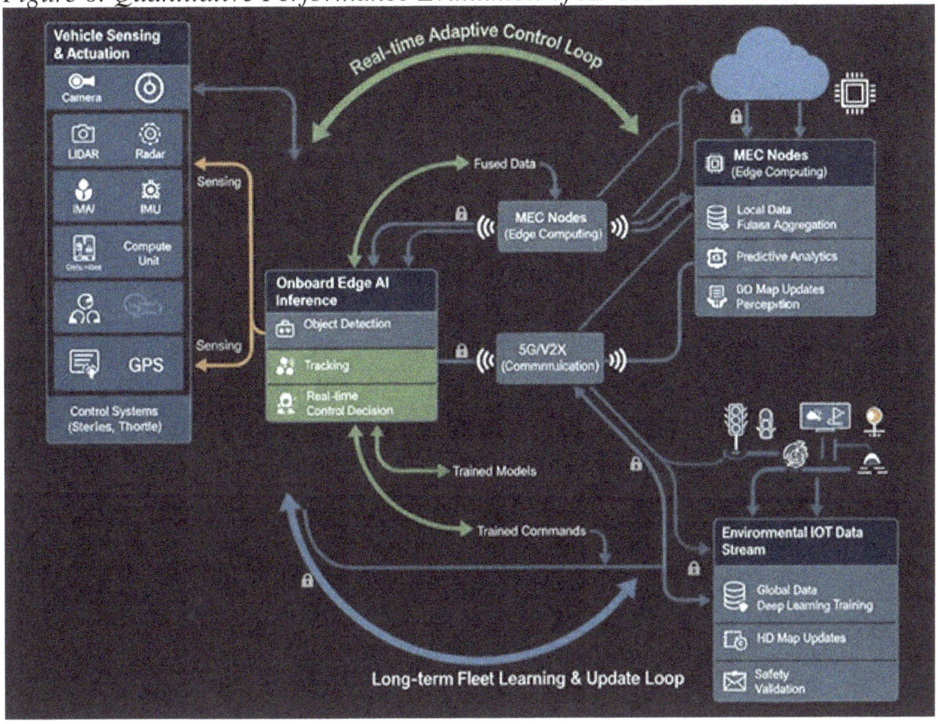

8.1 Ethical and Legal Issues

The morality of autonomous driving concerns decision-making in cases with moral trade-offs, blame in accidents, and privacy of data. For instance, if there is no way to avoid a crash, how would an AV make a choice between hurting a passenger or a pedestrian? As the research goes on to instill ethical thinking in AI, it is far off from coming up with universally accepted moral algorithms due to philosophical and cultural variations (Lin, 2016).

Legal systems also fall behind technological advancement. The majority of traffic statutes were drafted for human drivers and are not directly applicable to machines. Issues such as "Who is liable when an AV causes a crash? " and "Can a machine be held legally responsible?

These questions are left unanswered by most jurisdictions. Although some countries, such as Germany and America, have brought forth initial bills on AV operation, there is still a global regulatory consensus lacking (Marchant & Lindor, 2012). undefined Data protection is another concern. AVs collect vast amounts of personal and environmental information. Lack of stringent data protection regulation and enforcement may potentially give way to misuse, hacking, or surveillance, and lead to concerns over user trust and cybersecurity (Marchant & Lindor, 2012).

8.2 Technical Limitations and Bottlenecks

There are a number of technical issues that are preventing autonomous car mass deployment:

- Sensor Limitations: LiDAR and Cameras do not work well in bad weather conditions such as rain, fog, or snow. Sensor fusion can help in making it more reliable but it cannot totally remove these vulnerabilities.
- Edge Case Handling AV has a problem with off-the-beaten-path or unclear cases - so-called “edge cases” - such as unpretty behavior of pedestrians, transient road signs, or unusual vehicle motion. These require enormous amounts of varied training data that should not be required. Localization Drift - The accuracy of the desired localization is questionable in highly complex urban environments even with the most precise SLAM and GPS systems due to weak or jammed GPS signals.
- High Processing Requirements: AI perception, prediction, and control model running in parallel in real-time requires high processing requirements and low power usage - two features that have to be traded off in a vehicle's constraint.
- Infrastructure and Scalability: HD maps, V2X infrastructure for communications and real-time cloud connectivity are not everywhere, especially in rural or underdeveloped regions, limiting the availability of AVs.

These challenges point to the need for continued research, inter-disciplinary innovation and creation of world standards, to adapt AV systems to real world complexities (Hussain & Zeadally, 2018).

8.4 Security and Adversarial Robustness in AV IoT Networks

Autonomous Vehicle (AV) IoT networks consist of various types of sensors, communication modules and edge/cloud components, which makes them extremely vulnerable to cyber threats and adversarial attacks. These networks are subject to a large number of attack surfaces including sensor spoofing, communication interception, software and firmware attacks, and adversarial input. While LiDAR, radar, and camera sensors are susceptible to spoofing and jamming that leads to misperception, V2X and V2I communication channels are susceptible to man-in-the-middle, replay, and denial-of-service attacks. The AI model can be misled by adversarial manipulated input data and unsafe actions, and the AV's decision-making can be compromised by malicious code injected during over-the-air updates or within the control units. These threats can only be reduced through a multilayered security strategy. Sensor Redundancy and Fusion can be used to cross-check across modalities to avoid being vulnerable to spoofing or interference. Encrypted communication protocols, secure authentication, and anomaly detection are used to secure V2X networks against interception and unauthorized access. Malicious inputs can be mitigated by leveraging adversarial trained AI models with input sanitization, uncertainty estimation, etc. Software/firmware updates and the ability to apply patches as and when vulnerabilities become known can aid in mitigating successful attacks. Further, use of standards leads to more secure environment. ISO 21448 (Safety of the Intended Functionality - SOTIF) provides guidance on how to address these safety risks that are caused by functional insufficiencies (e.g. sensor limitations and environmental uncertainties), and NIST AI Risk Management Framework (AI RMF) is a structured methodology for dealing with the risks in AI systems with a focus on robustness, explainability, and continuous monitoring to avoid adversarial exploitation. Despite these measures, there are still a number of issues. Lightweight and secure low latency V2X communication protocols are required and AI based decision making and real-time anomaly detection is still challenging. Moreover, standardization between AI security frameworks (NIST AI RMF) and AV safety (ISO 21448) is the cornerstone for creating an effective network resilience. Finally, AV IoT networks require robust AI models, secure communications, sensor integrity, international safety and risk management standards to ensure security and adversarial robustness to achieve operational reliability and public trust.

8.5 Human–Machine Interaction and Handover Design

Human-Machine Interaction (HMI) design plays an important role in autonomous vehicles (AVs) to ensure safe and smooth operation, in particular during the handoff from the vehicle to the human driver and vice versa. Human factors such as cognitive load, situation awareness, reaction time and trust in automation are important factors to consider when designing handover In particular, poor handover mechanisms may lead to poor response times or incorrect decisions that may have severe safety implications. For this reason, AVs must provide context-aware, timely and intuitive information on system conditions, the near future and human intentions to be executed, as well as a context-dependent degree of automation depending on the driving situation and driver state.

Case Box: Urban Handover Scenario

In an urban driving environment, a Level 3 AV approaches a construction zone requiring human The HMI system in the vehicle delivers a multimodal alert with a visual display on the dashboard, an aural alert, and steering wheel vibration. The system continually scans a driver's attentiveness and level of alertness, and gradually increases the level of alertness if the driver does not respond. After handoff to the driver, the AV is assured of successful handoff and can continue to monitor.

8.6 Failure Mode Analysis

The failure mode analysis of an autonomous vehicle (AV) system is an important means to understand the operational risk and safety gaps. By studying high-profile events involving companies such as Tesla, Uber and Waymo, researchers and engineers could uncover some of the common vulnerabilities around perception, decision-making and handing off from human to machine. Typically, the failure of AVs occurs due to misinterpretation of the sensor, insufficient scenario coverage during the training of the AI, delayed human intervention, or unforeseen conditions in the environment. The knowledge of these failure modes can be used to design more robust system architectures, to create better safety mechanisms, and to design specific requirements.

Box 1. Critical Case Analysis

Tesla Autopilot (2016–2022): Several incidents were actually reported that demonstrated the shortcomings of the system to detect static obstacles on the highways as well as to identify unusual traffic situations. These failures revealed the need for redundancy in perception systems and the need for constant monitoring of the driver when driving using hands off.

9. SOLUTIONS AND RECOMMENDATIONS

Manufacturers must provide redundant sets of sensors and fault-containment software systems. Failover hardware such as redundancy LiDAR sensors or failover AI sub-systems ensures consistent operation in the event of hardware or algorithmic failure, providing greater security in edge conditions where it really counts (Koopman & Wagner, 2016).

To generate trust and adoption, public education initiatives need to de-mystify how AVs function, their safety features, and the limitations of existing technology.

Moreover, as semi-autonomous technologies become more prevalent, users will need training to use them efficiently (Litman, 2020).

A Smart Mobility Corridor (SMC) is a network that combines autonomous vehicles (AVs), Internet of Things (IoT) infrastructure, and real-time data analytics to build a seamless, safe, and efficient transportation ecosystem. The corridor uses interconnected sensors, Edge computing nodes, and AI-based Traffic Management Systems to improve vehicle flow, lessen the congestion and increase safety. Some of the key features of smart cities are V2I communication, dynamic traffic signaling, predictive maintenance monitoring, and dynamic routing algorithms that depend on real-time traffic, weather and pedestrian data. Figure 7 depicts the Smart Mobility Corridor blueprint which shows the interaction between autonomous vehicles, edge-cloud computing, roadside IoT sensors and traffic management centers. The design of the corridors incorporates redundancy and resilience so that sensor failures or network disruptions will not affect the overall safety of the system.

Figure 7. Smart Mobility Corridor architecture showing AVs, IoT sensors, edge/ cloud computing, and traffic management integration.

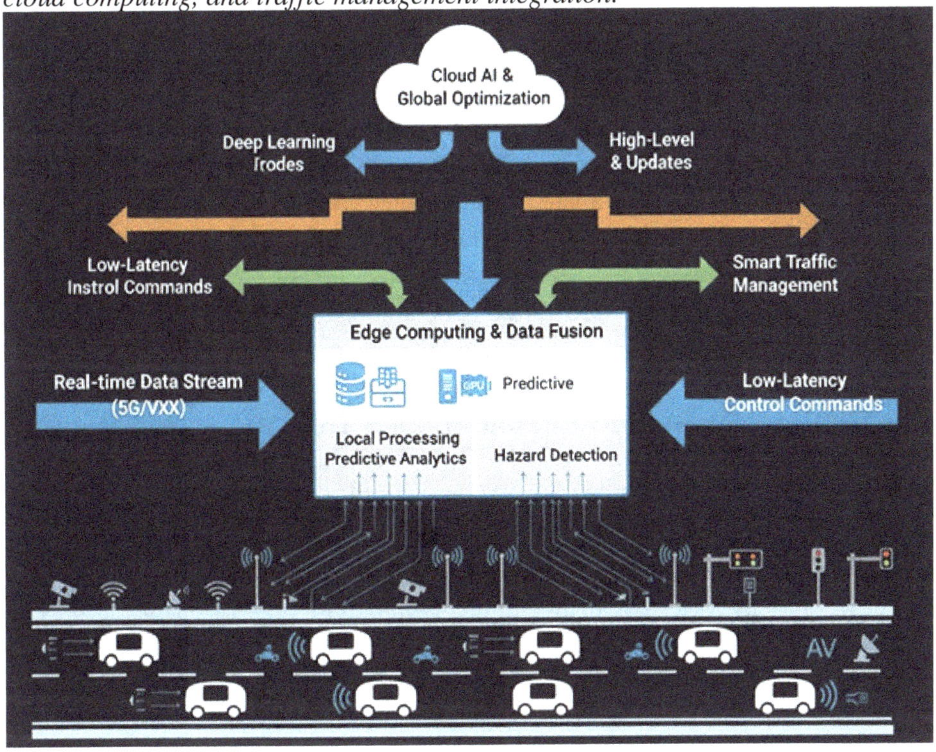

10. FUTURE RESEARCH DIRECTIONS

10.1 Explainable and Trustworthy AI

Work on this area should be towards developing AI systems that provide human-interpretable, real-time explanations of what they are doing. This will foster trust from users, support legal analysis, and aid regulators in judging system safety (Doshi-Velez & Kim, 2017).

10.2 Federated and Privacy-Preserving Learning

AVs generate enormous volumes of sensor and user data, creating ownership and confidentiality issues over the data. Evolving research needs to investigate federated learning frameworks that enable AVs to learn in harmony between fleets without sending raw data to central servers. Coupled with privacy-enhancing technologies such as differential privacy and homomorphic encryption, this will enable data-driven optimization while protecting user confidentiality (Kairouz et al., 2021).

10.3 Socio-Technical Human–Vehicle Interaction

While AVs become increasingly prevalent on public roads, interaction by human drivers, pedestrians, and passengers with autonomous systems will become a pivotal area of research. Design for humans, adaptive interfaces, and intent communication mechanisms (e.g., light signals, visual) would have to be researched and designed to enable unobtrusive interaction. This area of research may result in hybrid systems in which levels of autonomy are varied depending on comfort and context of the user (Lee & See, 2004).

10.4 Climate-Aware and Sustainable AV Design

The climatic footprint of AVs—particularly with more computing and battery usage—needs to be addressed. Energy-efficient AI algorithms, environmentally friendly routing mechanisms, and sustainability throughout the lifecycle of AV hardware need to be explored in future research. The use of renewable energy sources and V2G optimization can also lead to emissions reduction and increased future AV ecosystem sustainability (Goodall et al., 2021).

10.5 Simulation and Digital Twin Environments

Testing AVs in real-world scenarios alone is inefficient and risky. Simulation environments that are digital twins and physics-informed simulation systems provide an ideal platform to test, validate, and retrain AV models under thousands of edge-case scenarios. Scalable, interoperable, and high-fidelity simulators must be developed in future studies to be seamlessly integrated with real-world data streams (Alam & El Saddik, 2017).

10.6 Interdisciplinary Collaboration and Global Standards

AV development is at the juncture of AI, IoT, robotics, ethics, law, and urban planning. One of the key research priorities will be creating open interdisciplinary frameworks and worldwide testing standards. Academic-industrial-governmental collaboration can ensure that AVs are technically competent as well as socially acceptable and consistent with the law.

10.7 Benchmarking and Reproducibility Standards

By using benchmarking datasets along with open methodologies and resources, researchers and practitioners can meet robust model evaluation, collaboration and safe and rapid deployment of AV systems.

10.8 Federated Learning and Swarm Intelligence for AV Fleets

As the AV fleet scales, centralized data collection cannot be used because of the privacy, bandwidth, and latency limitations. Federated Learning (FL) is a solution to the problem, which allows for distributed model training across vehicles, wherein each AV uses sensor data to train local AI models, and shares model parameters with a central server. This approach has a privacy-preserving aspect, a communication-overhead minimization aspect, and a collective-learning-acceleration aspect. In addition to FL, swarm intelligence methods enable collective cooperative behavior of AV fleets based on decentralized decision-making mimicking natural biological systems (e.g. bird flocks or ant colonies).

CONCLUSION

Autonomous Vehicle (AV) IoT network is a convergence of advanced sensing, Artificial Intelligence (AI) driven decision making and connected infrastructure with revolutionary potential for mobility, safety and city efficiency. Throughout this chapter we have been introduced to the technical foundations of sensor fusion, adversarial robustness to human-machine interaction, smart mobility corridors and federated learning for AV fleets. Through the analysis of both successes and failures in practical deployments such as Tesla, Uber and Waymo, we pointed out the importance of having strong system design, standardized safety protocols as well as human-oriented handover mechanisms. The combination of benchmarking data sets, open science and standardized evaluation frameworks also help to ensure that AV technologies develop in a transparent, reproducible and reliable way. Looking ahead, the chapter underlines the need for safe and efficient AV networks, and what it takes to achieve this which are holistic approaches that integrate technological innovation with the rigorous management of risk and integration of human factors. Some emerging paradigms such as swarm intelligence, federated learning and AIoT point towards scalable, cooperative and adaptive AV ecosystems. In the end what the findings revealed here underpins the major theme of the chapter: the development of resilient, trustworthy and intelligent AV IoT systems is a product of a constant interaction between AI, IoT, human regulation, and regulatory coherence. By collectively coming together the elements of this thesis in a synthesized manner, researchers, practitioners and policy makers can collectively shape the future of autonomous mobility in a safe, efficient and ethically responsible manner.

REFERENCES

Abbas, N., Zhang, Y., Taherkordi, A., & Skeie, T. (2018). Mobile edge computing: A survey. *IEEE Internet of Things Journal*, *5*(1), 450–465. DOI: 10.1109/JIOT.2017.2750180

Alam, M., & El Saddik, A. (2017). C2PS: A digital twin architecture reference model for cloud-based cyber-physical systems. *IEEE Access : Practical Innovations, Open Solutions*, *5*, 2050–2062. DOI: 10.1109/ACCESS.2017.2657006

Bimbraw, K. (2015). "Autonomous cars: Past, present and future," in *12th International Conference on Informatics in Control, Automation and Robotics*, vol. 1, pp. 191–198. DOI: 10.5220/0005540501910198

Bojarski, M., Del Testa, D., Dworakowski, D., Firner, B., Flepp, B., & Jackel, L. D. (2016). "End to end learning for self-driving cars," *arXiv preprint arXiv:1604.07316.*

Chen, L. C., Papandreou, G., Kokkinos, I., Murphy, K., & Yuille, A. L. (2018). DeepLab: Semantic image segmentation with deep convolutional nets. *IEEE Transactions on Pattern Analysis and Machine Intelligence*, *40*(4), 834–848. DOI: 10.1109/TPAMI.2017.2699184 PMID: 28463186

Contreras-Castillo, J., Zeadally, S., & Guerrero-Ibáñez, J. A. (2017). Internet of vehicles: Architecture, protocols, and security. *IEEE Internet of Things Journal*, *5*(5), 3701–3709. DOI: 10.1109/JIOT.2017.2690902

Dickmanns, E. D. (2002). "Development of machine vision for road vehicles," in *IEEE Intelligent Vehicle Symposium*, pp. 268–281.

Doshi-Velez, F., & Kim, B. (2017). "Towards a rigorous science of interpretable machine learning," *arXiv preprint arXiv:1702.08608.*

Geiger, A., Lenz, P., & Urtasun, R. (2012). "Are we ready for autonomous driving?" in *CVPR*, pp. 3354–3361.

Goodall, N. J. (2021). Ethical decision making during automated vehicle crashes. *Transportation Research Record: Journal of the Transportation Research Board*, *2675*(9), 94–103.

Hussain, R., & Zeadally, S. (2018). Autonomous cars: Research results, issues, and future challenges. *IEEE Communications Surveys and Tutorials*, *21*(2), 1275–1313. DOI: 10.1109/COMST.2018.2869360

International, S. A. E. (2018). *Taxonomy and definitions for terms related to driving automation systems for on-road motor vehicles (J3016_201806).* https://www.sae.org/standards/content/j3016_201806/

Koopman, P., & Wagner, M. (2016). Autonomous vehicle safety validation. *SAE International Journal of Transportation Safety*, *4*(1), 15–24. DOI: 10.4271/2016-01-0128

Lee, J. D., & See, K. A. (2004). Trust in automation: Designing for appropriate reliance. *Human Factors*, *46*(1), 50–80. DOI: 10.1518/hfes.46.1.50.30392 PMID: 15151155

Lin, P. (2016). Why ethics matters for autonomous cars. In Maurer, M. (Eds.), *Autonomous Driving* (pp. 69–85). Springer.

Litman, T. (2020). *Autonomous vehicle implementation predictions: Implications for transport planning*. Victoria Transport Policy Institute.

Liu, H., Lin, Y., & Wu, X. (2018). Sensor fusion in autonomous vehicles. *Sensors (Basel)*, *18*(12), 4220. PMID: 30513759

Lu, N., Cheng, N., Zhang, N., Shen, X., & Mark, J. (2014). Connected vehicles: Solutions and challenges. *IEEE Internet of Things Journal*, *1*(4), 289–299. DOI: 10.1109/JIOT.2014.2327587

McKinsey & Company. (2023). *The future of mobility: AV adoption and business models*. https://www.mckinsey.com

Shladover, S. E. (2018). Connected and automated vehicle systems. *Journal of Intelligent Transport Systems*, *22*(3), 190–200. DOI: 10.1080/15472450.2017.1336053

Smith, B. W. (2013). Automated driving and product liability. *Detroit College of Law at Michigan State University Law Review*, *1*, 1–74.

Sun, Z., Bebis, G., & Miller, R. (2006). On-road vehicle detection: A review. *IEEE Transactions on Pattern Analysis and Machine Intelligence*, *28*(5), 694–711. DOI: 10.1109/TPAMI.2006.104 PMID: 16640257

Thrun, S. (2010). Toward robotic cars. *Communications of the ACM*, *53*(4), 99–106. DOI: 10.1145/1721654.1721679

Wang, H., Wang, J., Li, Z., & Yu, L. (2021). Data-driven fault diagnosis in autonomous vehicles: Trends, methods, and applications. *IEEE Access : Practical Innovations, Open Solutions*, *9*, 106835–106855. DOI: 10.1109/ACCESS.2021.3100862

Wang, J., Liu, Y., Wang, J., & Meng, Q. (2021). Deep reinforcement learning for autonomous vehicles. *IEEE Transactions on Neural Networks and Learning Systems*, *32*(6), 2155–2168.

Waymo. (2020). *Safety report: On the road to fully self-driving*. https://waymo.com/safety/

WHO. (2018). *Global status report on road safety*. https://www.who.int

Zhang, Y., Wang, S., & Phillips, P. (2020). Autonomous vehicle safety assessment using AI. *AI Open*, *1*, 1–9.

Zhao, J., Zheng, Y., Li, Q., & Zhang, Y. (2022). AI-enabled digital twins for autonomous vehicles: A survey. *ACM Transactions on Cyber-Physical Systems*, *6*(2), 1–28. DOI: 10.1145/3203210

Chapter 7
Real-Time IoT, Quantum Advances, and Ethical Implications in Autonomous AI Systems

R. N. Ravikumar
https://orcid.org/0009-0009-3705-1681
Marwadi University, Rajkot, India

S. Aarthi
https://orcid.org/0009-0006-9064-2091
Marwadi University, Rajkot, India

ABSTRACT

The overlap of real-time IoT systems, quantum computing, and the ethical dimensions of autonomous AI is reshaping intelligent automation. This work outlines key technologies like edge computing, low-latency communication, and distributed algorithms that support real-time responses in critical areas such as healthcare, autonomous vehicles, and smart infrastructure. It explores how AI methods, machine learning, computer vision, and reinforcement learning enable systems to adapt and act autonomously. Quantum computing's potential to enhance processing, optimization, and security is also examined, with a focus on its integration into AI-IoT systems. The study addresses challenges like algorithmic bias, opacity, accountability, and the societal impacts of automation. Emphasizing Explainable AI (XAI), it advocates for transparent, trustworthy decisions in high-stakes applications. The chapter promotes inclusive dialogue and responsible governance to align innovation with ethical standards.

DOI: 10.4018/979-8-3373-3146-1.ch007

1. INTRODUCTION

The main transformations are occurring due to AI advancements in IoT, which allow the implementations of an autonomous adaptive system in the several areas. These technologies are developing and equipping machines with the capacity to feel and react to environmental changes, learn and make decisions (primarily) independently. Mission-critical applications are Autonomous Vehicles, Precision Healthcare, Smart Manufacturing, and Critical Infrastructure, just to name a few, which are required to be supported by real-time response and adaptability. This transformation is right at the core of real-time IoT systems. It is a low-latency communications system that necessitates decentralized processing, which allows the edge devices to examine the data and invoke the response nearly immediately. This is where concepts of new architectures in edge computing, systems level algorithmic performance models, and systems level interoperability will be most important.

Simultaneously, the emergence of quantum computing puts intelligent automation in a new perspective. Still in its infancy, quantum technology is capable of solving much more difficult problems, speeding up the machine learning process, as well as providing the security of communication that is impossible to achieve with classical systems. Integrating it with AI and IoT may take the autonomous systems to the most brilliant levels and capabilities that have ever been envisioned. The second wave in technologies comes with the most ethical, legal, and social concerns. The rising dependence on machine-based decision-making brings up the issues of transparency, accountability, data confidentiality, and algorithmic fairness. It also forms revisions of the socio-economic realm in such aspects as disruption of the workforce and the change of human trust limits with autonomous technologies. This chapter examines the convergence of real time IoT, quantum computing, and ethical issues in autonomous AI systems. It has a technical outlook and ethical foresight in examining the over picture of challenges and opportunities that can be had in the construction of systems capable of not only being smart and efficient but also responsible and trustworthy (Chaware et al., 2024).

1.1 Overview of the Convergence of AI and IoT in Autonomous Systems

The combination of the Artificial Intelligence and the Internet of Things is one of the evolutionary steps of the autonomous adaptive systems. AI offers machine learning, computer vision or natural language processing as intelligence, and IoT offers the sensing, communication and actuation functionalities required to facilitate the real-time interaction with the physical world. The two in combination comprise the internal framework of intelligent systems, which can perform independently though under minimal human control. In autonomous systems, this integration allows machines to continually gather data from their environment, process the data intelligently, and act in ways that are aware of the context. For instance, an autonomous vehicle using AI and IoT could detect obstacles, analyze traffic patterns, and change its intended course in real time. Within an industrial setting, smart factories use AI driven IoT networks to optimize production, reduce downtime, and respond to demand changes. Such a mutual relationship between AI and IoT is spreading across industries-the healthcare sector, transportation, energy, agriculture, and so forth. With AI and IoT-related technologies rapidly evolving, these two fields merge to provide hyper levels of autonomy, efficiency, and personalization. The envisioned systems should not only respond but predict and evolve endlessly-they learn from experience and environment change, thus adapting accordingly (Pasupuleti, 2024).

1.2 Motivation for Real-Time Responsiveness, Quantum Innovation, and Ethical Awareness

Further, the necessity of the real-time responsiveness has turned out to be quite pressing as self-governing systems are increasingly used in the healthcare sector, transportation, and infrastructure control. Delays can be life threatening in this case and the failures of the system can be extremely expensive. It is then that the requirement of low-latency communication, edge computing, and intelligent algorithms so that the systems will be able to take action immediately, with periodic verification of reliability emerges at this very point. It appears that quantum computing can be a promising supplement to conventional computing with its capacity to deliver exponential gains in processing speed to optimize and secure communication. Although not yet fully developed, quantum technologies would have enormous implications on machine learning, logistics problem-solver, and encryption protocols that would essentially precondition a hyper-efficient autonomous system.

Integrating quantum abilities with artificial intelligence and Internet of Things has the potential to transform the landscape of computation and provoke the use of applications that have hitherto been considered impossible. These discoveries should be accompanied by ethical awareness. AI implementation in systems creates the issues of privacy of the data, its transparency, prejudice, and responsibility. In machine dominated governance, questions about fairness, explainability and human control need to be questioned. Responsible innovation will therefore not only construct technology but will be a catalyst towards ethical innovation as well as regulation and civic participation to establish societal confidence and sustainability (Zhou, 2024).

2. REAL-TIME IOT IN AUTONOMOUS SYSTEMS

IoT-in-the-field is used to create autonomous systems which are able to absorb information and react in real-time in less than a second to whatever is going on around them. A network consisting of many sensors, actuators and edge devices can be used to collect a large amount of data and make intelligent decisions in a short period of time without necessarily transmitting all of it to a central system first. In such areas as autonomous vehicles, smart factories, remote medical services and controlling infrastructure, the existence of virtually no latency with IoT is highly significant. Modern IoT systems link edge computing to process data close to the point of production to make them faster. It consequently has reduced latency, networks are not overburdened and systems react faster. It is due to the processing of AI and interaction of LiDAR, radar and cameras that the autonomous vehicles possess the capacity to identify objects promptly and plan their routes in a timely manner.

Distributed AI algorithms allow information to be processed in parallel; therefore, systems can adapt to changing circumstances. The 5G and ultra-wideband are also some technologies that contribute to low-latency to support real-time interactions. A combination of these technologies allows being fast and consistent in the areas where these aspects are the key. Timing is not the only issue, real-time IoT makes sure that self-governing systems are able to react, change their behavior and act on a anytime, anywhere whatever (Edwards et al., 2023). The issues listed in this table 1 are important for achieving real-time responsiveness in autonomous systems that use IoT technology. It puts importance on the use of edge computing, distributed algorithms and reliable networks. This material gives you a basic knowledge of the infrastructure required for vital IoT applications.

Table 1. Key Requirements for Low-Latency IoT Infrastructure

Requirement	Description	Importance	Example Applications
Low-latency communication	Data transmission with minimal delay	Enables real-time responsiveness	Autonomous vehicles, healthcare
Edge computing	Processing data close to the source	Reduces cloud dependency and delay	Industrial control, smart grids
Distributed algorithms	Algorithms that operate across multiple devices	Enhances scalability and fault tolerance	Factory automation, traffic systems
High reliability	Consistent network uptime and data integrity	Critical for safety and accuracy	Emergency services, aviation
Scalability	Ability to handle increasing number of devices and data	Supports growth and dynamic environments	Smart cities, IoT sensor networks

2.1 Requirements for Low-Latency IoT Infrastructure

The IoT autonomous systems require low-latency to make real-time decisions. Any deceleration in the communication or processing of the information captured can lead to significant issues, primarily in critical systems such as autonomous vehicles, apparatus tribes aiding with surgery and safety devices in factories. An IoT infrastructure with low latency must be capable of transmitting data fast, process and provide feedback within a short duration. To manage such demands, existing infrastructure depends on communication technologies such as 5G, Wi-Fi 6 and Ultra-WideBand (UWB). They have the concept that data must be shared with numerous elements exceptionally fast. High-performance edge nodes also play an important role in processing data locally on the fly and avoiding problems that would arise with use of the cloud. Embedded devices to RTOS, miniature network protocols and better hardware interfaces are done with lower latencies. Another thing worth doing is to design system architecture, such that it responds to failure and continues to run in case of hardware equipment failure or network deficiency. Provision of TSN, deterministic approaches and more significantly, the provision of traffic control to data that require urgent attention is also significant. All these characteristics combine to create an IoT architecture which enables machines to react instantly and in a predictable fashion when conditions are dynamically evolving (Arumugam et al., 2024)

2.2 Role of Edge Computing and Distributed Algorithms

Edge computing significantly reduces the response time of the systems and enhances self-control of the IoT systems. When we try to bring processing capacity close to the source of the data that is being generated, we can have systems respond much faster and utilize resources more efficiently. This latency is undesirable when using applications that cannot afford the latency of passing data to or from remote cloud servers. Edge computing also enables real-time decisions and controls that enable autonomous drones, vehicles and medical monitors to act to a certain degree without dependence. Usually, these edge devices use AI models trained on the cloud to handle inference and secure your data. Distributed algorithms are used, together with edge computing, to let devices cooperate in handling and deciding on data. Synchronizing and keeping consistency on all devices while using little bandwidth can be achieved with federated learning, swarm intelligence and decentralized consensus algorithms. The algorithms have been built for resource-limited situations which allows them to scale and remain resilient. The combination of edge computing and distributed intelligence results in a stable platform for instant, reliable and smart systems. They improve how well things run, make systems secure and help lower the cost and support self-management in areas with weak infrastructure (Mishra & Singh, 2025).

2.3 Application Domains: Autonomous Vehicles, Healthcare, Industrial Control

Intelligent automation, flexible changes and quick reaction to events are being enabled for several important industries by real-time IoT systems. In autonomous vehicles, sensors such as LiDAR, radar, GPS and cameras, keep on monitoring the environment. With the real-time analysis of data, edge processors notice obstacles, study traffic conditions and make decisions on how to navigate. The need for fast processing is crucial because milliseconds can decide safety. Here, IoT is used in real-time for patient monitoring from a distance, advancements in surgery by robots and answering emergency calls. Wearable devices measure health stats and immediately tell healthcare workers when something unusual is found. Early detection through AI is supported, while data sharing in real-time makes the coordination of care and the results better in critical and elderly care situations. In such industrial control systems, IoT is used to let machines measure temperature, pressure and vibration. Because of edge computing, issues are spotted instantly and maintenance checks are planned, lowering downtime and improving safety. Controlling these systems in real-time makes both production and accuracy higher. As a result of IoT, across all these areas, operations become safer, smarter and more efficient, with systems acting independently yet in line with people's objectives and plans for safety (Gambo & Musonda, 2022). The Figure 1 illustrates the three key application domains for real-time IoT systems Autonomous Vehicles, Healthcare, and Industrial Control. Each domain interacts bidirectionally, showcasing shared data and technological feedback loops necessary for adaptive and autonomous performance. The coloured arrows represent the flow of information and collaboration between sectors.

Figure 1. Real-Time IoT Application Domains

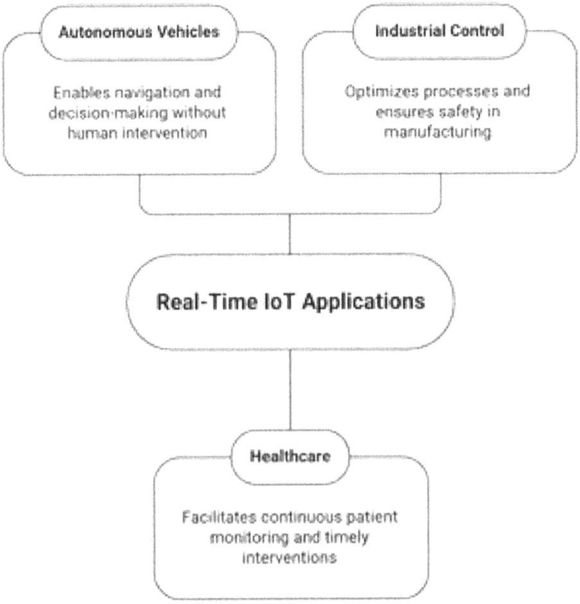

3. AI FOR ADAPTIVITY AND AUTONOMY

Adaptivity and autonomy in today's intelligent systems depend on Artificial Intelligence (AI). Although IoT includes the tools for connecting technologies, it is up to AI to help machines sense, learn and act meaningfully in their setting. An adaptive system adjusts its response to changes or fresh data, whereas an autonomous one acts alone, often on its own without much user input. The systems use machine learning, reinforcement learning and deep learning to detect trends, predict outcomes and improve over the course of their functioning. Because of AI, a smart factory can quickly change the settings of its production process according to changes in demand or when defects are found; doing so makes the production more efficient and improves the product quality. Autonomous machines can try many approaches with reinforcement learning, making them apt for use in traffic and robotic tasks. Also, image analysis and understanding of speech help machines work with informal data which improves how they communicate. Applications involving self-driving vehicles, smart medical devices and flexible energy networks depend on AI-driven abilities. As a result, IoT systems gain autonomous self-awareness, adapt to different conditions and move continuously closer to handling tasks without human intervention. The table 2 focuses on the chief AI approaches that are responsible for both making decisions and adapting. It demonstrates the practical use of machine learning, reinforcement learning and computer vision by comparing how they work. Such methods allow machines to respond and think intelligently in fast and changing situations (Fehringer & Stary, 2023).

Table 2. AI Techniques Enabling Autonomy and Adaptivity

AI Technique	Description	Role in Autonomy	Example Use Case
Machine Learning	Algorithms that learn patterns from data	Enables prediction and classification	Medical diagnosis, anomaly detection
Reinforcement Learning	Learning via trial and error with rewards	Supports decision-making and adaptation	Autonomous driving, robotics control
Computer Vision	Extracting information from images and videos	Facilitates environment perception	Surveillance, quality inspection
Natural Language Processing (NLP)	Understanding human language	Enhances human-machine interaction	Voice assistants, chatbots
Deep Learning	Multi-layer neural networks for complex tasks	Improves feature extraction and generalization	Image recognition, speech recognition

3.1 Benefits of Adaptive

Artificial Intelligence (AI) is critical in developing more intelligent and user-friendly digital solutions. It enhances user experience in terms of being user-friendly in terms of interfaces, with responsive support so that interactions are smooth. AI can be used to boost efficiency by automating processes and simplifying the work-flow and improving human input. The other important advantage is personalization since AI tailors services and content based on personal preferences. The insights made by data will also allow improved decision-making and allow making more informed and correct choices. Also, AI is flexible and adaptable and adapts fast to changing conditions and shifting user requirements. In general, AI will help to increase efficiency and personalization, as well as be more intelligent and responsive, so digital systems will be more trusted and engaging and correspond to user expectations figure 2 as follows (Sadilek et al., 2021).

Figure 2. Benefits of Adaptive

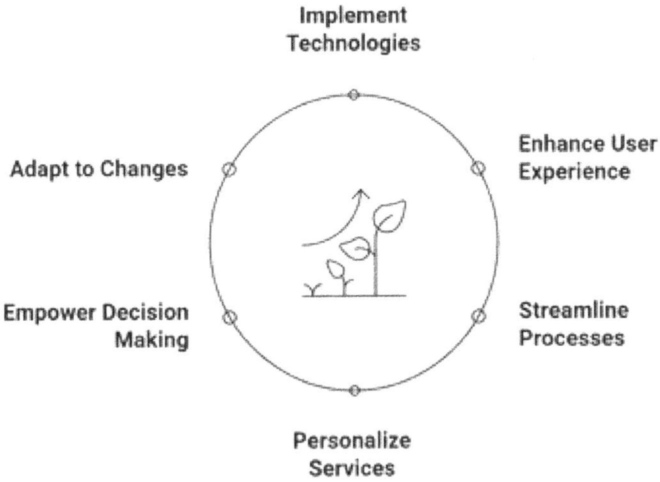

3.2 Key AI Techniques of Adaptive Learning

The recent AI practice of Machine Learning (ML), Reinforcement Learning (RL), and Computer Vision, is the force behind adaptive learning systems to make intelligent and situationally aware decisions. ML as a support system is based on the analysis of large volumes of data, which builds predictive models that underlie such applications as anomaly detection in healthcare or predictive maintenance in industries. Instead, RL enables systems to acquire the best strategies by trial-and-error interactions with the help of feedback in the form of rewards, which is well suited to dynamic, unpredictable systems such as robotics. Computer vision also enables machines to perceive and analyze visual information via deep CNNs which are vital in self-driving cars, security and maintenance. All these methods allow the IoT-built systems to transform into proactive instead of reactive intelligence (Movahed et al., 2025).

3.3 Enabling Self-Learning and Adaptive Behavior in Dynamic Environments

Being able to adapt is vital for autonomous systems operating where challenges keep evolving. With self-learning power, AI allows systems to improve their actions in response to what happens in real time, changes happening around them or changes in what users require. Different from set behavior patterns, self-learning technology learns all the time using online or reinforcement training and does not need a human to guide it. Because of these models, systems can handle not knowing everything for sure, manage new changes and get better over time. To illustrate, an intelligent energy grid could spot how consumption depends on the weather, learn how demand varies and use this information to fix any imbalances by itself. Adaptivity includes being aware of the situation you're in. AI models analyze not only what is being collected by sensors, but also information about time, where it is taking place or what the user prefers. The emergency messages are delivered first in healthcare monitoring and vehicles can modify their operation depending on the road. Federated learning and edge AI are helping decentralized systems adapt better, as they allow model training across several nodes with no breach of data privacy. Thus, these systems stay both fast, efficient and able to respond to demand and new trends (Leghemo et al., 2025).

3.4 Case Examples of AI-Driven Autonomous Operations

Many applications in real life prove that AI helps run IoT-based self-operating tasks. In the automotive industry, both Tesla and Waymo turn to AI, allowing vehicle fleets to read signs, remove risks from their path and navigate challenging road situations without the need for a human driver. They modify their performance continuously in response to changes on the road, how many cars are near and how many pedestrians are about. AI-powered IoT tools such as sensors and robots in healthcare collect vital information, recognize unusual trends and can help with surgeries which improves results and reduces risks. Clinicians can use AI to go through patients' ECG results regularly, so that they receive early notification when a cardiac event may be coming. With predictive maintenance in manufacturing, machines use machine learning to look for signs that machinery might fail in advance. Reinforcement learning in smart factories is used to adjust how robots work, making production faster and less likely to stop because of flexible planning and use of resources. In farming, farmers can rely on autonomous drones and robots to examine their crops, give them exactly what they need and optimize irrigation. They show that AI turns passive technologies into independent and intelligent systems that can act smartly in important and changing situations (Humayun et al., 2024). As shown in Figure 3 AI is transformative in six areas: robotics, agriculture, education, healthcare, and industrial monitoring, and finance. It emphasizes the role of AI in the promotion of efficiency, personalization, safety, and information-based decisions.

Figure 3. Real-World Use Cases

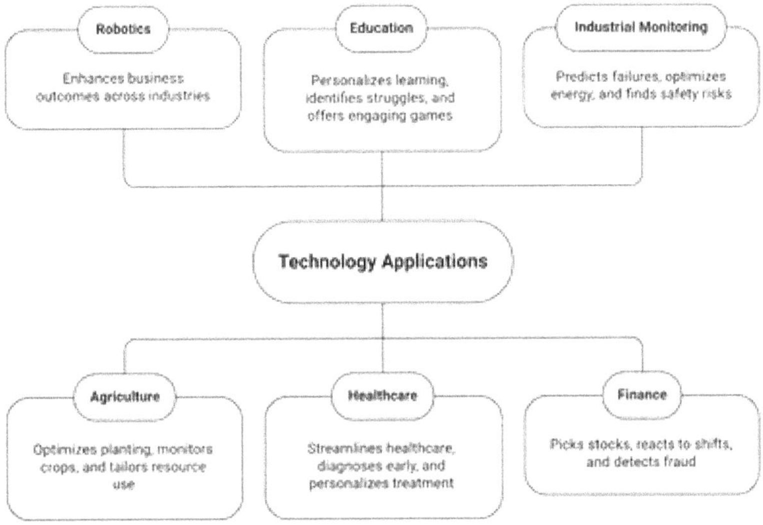

4. QUANTUM POTENTIAL IN OPTIMIZATION AND MACHINE LEARNING

Quantum computing can greatly improve AI-IoT systems by handling problems in optimization and machine learning that conventional systems can't address as well. Route planning for autonomous cars, distributing resources in smart grids and selecting important features in machine learning are all difficult optimization problems that many autonomous and adaptive systems use. Traditional algorithms' performance worsens exponentially as complexity increases, but quantum algorithms including annealing and Grover's search can improve that. QML is developed to make ML processing faster by running training and analysis with the help of quantum circuits. Handling lots of high-dimensional data could be made more efficient with the help of variational quantum classifiers and quantum neural networks. They are still being refined, but their results hint at major drops in the work IoT products must do, particularly for continuous data streaming. With the advancement of quantum processors, a combination of these with AI and the IoT could enable faster decisions to be made, predictions to be more accurate and daily processes to be optimized all of which are required in fields like logistics, robotics and finance. Although far too young, quantum technology is capable of turning AI more efficient, so one should not overlook it during its development (Chawla, 2024).

4.1 Hybrid Architectures: Classical + Quantum Synergies

New hybrid computing architectures are being constructed to allow the use of AI and IoT with quantum computing. In that manner, they will be able to achieve higher performance without relying entirely on quantum systems that are yet to be small and unstable. In this model, classical processors will be left to do ordinary data importing, rudimentary data handling and system administration, whereas quantum processors will help with more complicated tasks such as finding optimization solutions and assembling machine learning models more quickly. Dividing tasks in this way lets us make the most of existing technology and smoothly add quantum features over time. To illustrate, in a fully autonomous supply chain, traditional AI could take charge of logistics and the tracking process, while a quantum processor would optimize the overall plan for many worldwide locations receiving products. In the same way, some federated frameworks may later leave specific parts of model-aggregation tasks to quantum nodes, making it faster for nodes to agree. By using IBM Qisket, Amazon Braket and Microsoft Azure Quantum, people are finding it easier to test hybrid models. With improved quantum-classical orchestration, bridging current AI-IoT systems and future quantum-enabled ones will largely depend on hybrid architecture figure 4 as follows (Chauhan & Sahoo, 2024).

Figure 4. Hybrid Architecture

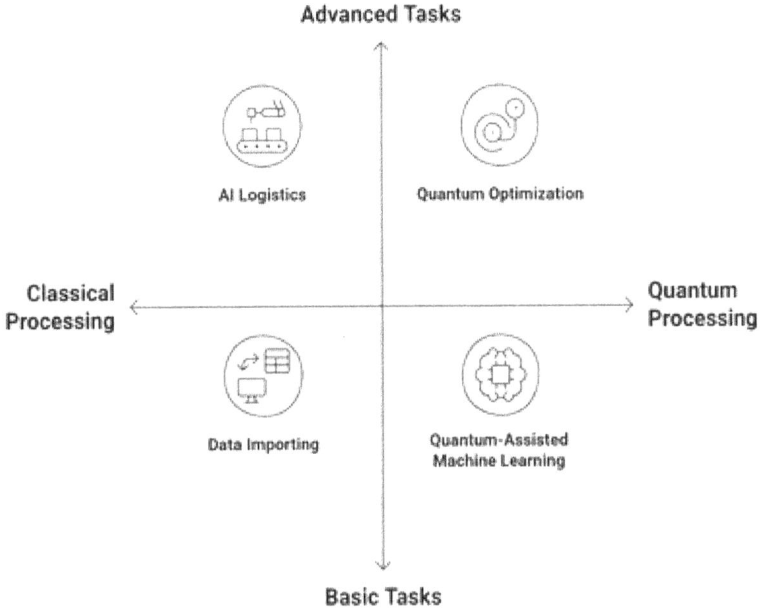

4.2 Challenges and Research Directions in Quantum-AI-IoT Systems

Although bringing quantum computing together with AI and IoT looks positive, there are important challenges that must be addressed. Short coherence of qubits, high error rates and few available qubits in today's hardware, limit the development of large-scale quantum systems. Because of these limitations, quantum error correction is required and new hardware designs are necessary. On the programming side, finding ways for AI algorithms to easily use quantum resources is still a topic under study. At present, AI frameworks are set up for classical architectures, so it is not simple to adapt them for variational quantum models. In addition, when adding quantum elements to live and dispersed IoT setups, concerns arise about delays, how synchronized everything is and how they can communicate. Safety is a vital issue for cybersecurity too. Bloom said that quantum computing could crack traditional encryption, making it necessary to use quantum-safe cryptography for sensitive AI-IoT operations. Experts are investigating how quantum technology can add new security solutions to network communication. In the future, work should be done to build faster quantum algorithms, fix error problems and establish standard methods for connecting quantum and classical devices. When barriers vanish, integrating quantum computing, AI and IoT will help us to advance intelligent automation and decision-making (Belbase et al., 2024). The table 3 illustrates how quantum computing changes things and how it must also address specific problems. Subjects addressed are optimization, security and learning in addition to problems such as decoherence and cost. Showing what's possible and what's necessary, helps create a balanced perspective on how quantum fits into AI-IoT systems.

Table 3. Benefits and Challenges of Quantum Computing in AI-IoT

Aspect	Benefits	Challenges
Optimization	Solves complex optimization problems faster	Hardware limitations and errors
Machine Learning	Accelerates training and improves accuracy	Algorithm development complexity
Security	Provides quantum-resistant encryption	Integration with classical systems
Scalability	Potential for scaling AI capabilities	Quantum error correction needed
Energy Efficiency	Potentially lower energy consumption	Expensive and resource-intensive hardware

4.3 Quantum Computing Concepts Relevant to AI and IoT

Because it depends on superposition, entanglement and interference from quantum mechanics, quantum computing processes information differently than other computing ways. Different from classical bits locked to either 0 or 1, qubits are able to be in more than one state at a time, allowing for huge levels of parallelism when computing. That's how quantum computers can address certain problems in a much shorter amount of time than classic computers. In AI and IoT, quantum computing changes the usual methods we use to manage and solve problems with data. Important topics are quantum gates (that control qubits), quantum algorithms such as the other and Grover's algorithms and quantum annealing used to solve optimization tasks. Such principles give good tools to solve difficult tasks in autonomous systems such as finding optimal combinations, noticing patterns and handling instant decisions. Because quantum computers are at the Noisy Intermediate-Scale Quantum stage, learning about these principles guides us to apply quantum technology to AI-driven IoT systems. Efforts are being made to join quantum hardware technologies with AI methods to improve how data is analyzed, models are trained and artificial intelligence grows (Yuan, 2024). The table 4 explains important quantum computing topics such as qubits, entanglement and superposition. It supports these ideas with arguments for their future significance in AI-IoT systems. The table gives readers an introductory look at how quantum theory relates to intelligent systems.

Table 4. Quantum Computing Concepts Relevant to AI and IoT

Concept	Description	Potential Application	Impact on AI-IoT
Qubits	Basic quantum units representing 0 and 1 simultaneously	Enables parallel computation	Faster data processing and optimization
Quantum Entanglement	Correlation between qubits regardless of distance	Enhances communication security	Quantum-safe IoT networks
Superposition	Ability of qubits to be in multiple states at once	Allows exploration of many solutions simultaneously	Speed up of machine learning algorithms
Quantum Gates	Operations on qubits	Basis for quantum algorithms	Development of new AI algorithms
Quantum Decoherence	Loss of quantum state due to environment	Major technical challenge	Limits current quantum hardware

4.4 Potential Benefits: Faster Optimization, Advanced Learning, Quantum-Safe Security

The ways quantum computing works can greatly assist AI-IoT integration. The most mentioned advantage is that optimization is done quicker. Solving optimization challenges is important for many AI-powered IoT apps: route planning for self-driving cars and resource distribution in smart electricity grids. There is a strong possibility that quantum annealing and variational quantum Eigensolver can find the best or almost best solutions faster than any classical technique. Secondly, quantum computing makes it possible for machine learning to reach new capabilities. QML applies quantum circuits to enhance both the training and use of models, supporting the fast and accurate analysis of significant and varied IoT datasets. Such a result can assist with anomaly identification, forecasting and flexible moves in machine learning algorithms. Third, quantum-safe technology is designed to meet future cybersecurity issues. Because quantum computers may decrypt many common encryptions, designing cryptographic algorithms protected from quantum threat is important to keep AI-IoT communications safe. Besides, using Quantum Key Distribution (QKD), data can be securely sent for important infrastructure projects. All of these improvements suggest that quantum computing may greatly improve the skills and safety of autonomous AI-IoT systems (Wu et al., 2023).

4.5 Early-Stage Research Directions and Future Possibilities

Even though the merging of quantum computing, AI and IoT is still being explored, lots of exciting studies are taking place. At present, scientists are making AI-oriented quantum algorithms using quantum versions of neural networks, support vector machines and reinforcement learning tools. Experts are also studying systems that combine classical components with quantum accelerators to address the problems facing current hardware. The objective is to use these architectures to shape the workflow and introduce quantum elements into real IoT applications as they gradually appear. The challenge of developing suitable quantum communication protocols for IoT networks is also important, supporting data privacy in those networks and introducing quantum federated learning techniques. Advancements are expected in AI for autonomous systems, using quantum tools in materials science labs and with new models of intelligence relying on quantum networks. Although bridging qubit stability, error correction and more scalable hardware continue to be problems, ongoing progress suggests that quantum computing will soon be a necessary tool in upcoming AI-IoT systems, granting them unparalleled performance, greater security and autonomy (Upreti et al., 2024).

5. ETHICAL IMPLICATIONS OF AUTONOMOUS AI-IOT SYSTEMS

AI is increasingly part of IoT systems, questions of ethics need to be considered more than ever. Questions about how fair, clear and responsible algorithms can be come up when they make decisions directly. Being unable to see what goes on inside the algorithms, known as the black box problem, makes it hard for stakeholders to trust them. Therefore, AI approaches that help explain decisions are becoming more important, to ensure users and regulators understand them. People should also be concerned about privacy. Automated IoT technology continually collects personal and environmental data which makes it more likely that those data will be used or accessed without permission. Consent, ownership of personal data and privacy must all be covered by ethical standards. In addition, if machines make decisions, people will ask who is held liable when the machines act incorrectly. Obeying certain laws and guidelines helps make sure that technology is constructed and put to use with a main concern for people's wellbeing (Loftus et al., 2022). The table 5 lists main ethical and legal problems caused by AI-IoT integration in autonomous systems. It also covers data privacy, biases found in algorithms and the issue of being responsible, with solutions for each of these. It acts as a reference for designers and deployers to ensure fairness in system building.

Table 5. Ethical, Legal, and Societal Issues in Autonomous AI-IoT

Issue	Description	Impact	Mitigation Strategies
Data Privacy	Unauthorized access or misuse of personal data	Loss of trust, legal penalties	Encryption, anonymization
Algorithmic Bias	AI decisions favouring certain groups unfairly	Discrimination, social inequality	Diverse datasets, fairness audits
Transparency	Lack of clarity on how AI decisions are made	Reduced accountability	Explainable AI, documentation
Accountability	Difficulty in assigning responsibility for AI errors	Legal ambiguity	Clear regulatory frameworks
Job Displacement	Automation replacing human jobs	Economic disruption	Reskilling programs, social safety nets

5.1 Legal Challenges and Regulatory Frameworks

Autonomous AI-IoT systems are emerging so rapidly, the current laws usually can't keep up. Authorities all over the world are struggling to oversee liability, compliance and data protection as technology continues to automate more areas. Major legal issues are connected to who should be blamed if autonomous vehicles make mistakes or errors arise in medical AI diagnosis. Ensuring that manufacturers, developers, operators and users understand their responsibilities is needed to make safety easier and fair. Laws such as GDPR and CCPA strictly manage how organizations collect, store and process people's personal information. To ensure they work and comply with regulations at the same time, IoT systems have to use privacy protection as a part of their design (Budrionis et al., 2021).

5.2 Key Issues: Data Privacy, Algorithmic Bias, Transparency, and Decision Accountability

These systems encounter some serious ethical hurdles that should be dealt with before they are widely used. People are most concerned about data privacy because these systems always gather, process and send a huge amount of private information. Without strong protcctions, this information could be taken advantage of, spied on or put at risk by unauthorized use which would reduce trust and break users' rights. Three major challenges of the ethical use of AI are algorithmic bias, transparency, and accountability of any decision. The problem of algorithmic bias is one where models that are trained on poor or imbalanced data give out unfair results, which is of great concern in sensitive areas such as healthcare, finance, and criminal justice where unfair results can severely affect lives. It is also important to promote transparency since the confidence in AI would be based on how algorithms come to their conclusions. It cannot be easily validated by the users, regulators or even by those who have interest without clarity. Responsibility in decisions also enhances the governance of AI by creating a sense of responsibility in case of harm or faults by AI-driven systems, and provides legal and ethical systems to safeguard users and foster safe and creative AI implementation (Antunes et al., 2022).

5.3 The Importance of Explainable AI (XAI) in Sensitive Sectors

The trust and ethical application of autonomous AI-IoT are greatly helped by using Explainable AI. We use XAI to let humans understand and explain the decisions made by complex AI models. In healthcare, someone using AI should still understand and validate the system's diagnostic tips which ensures that patients are treated safely and correctly. Routing AI decisions with some kind of explanation also supports public safety needs as found in autonomous vehicles or security cameras. XAI enhances how users trust the model and also supports fixing errors, detecting biases and following required laws. The stakeholders can challenge AI's results and limit the possibility of facing harmful or biased situations. Since AI-IoT systems can act without constant human supervision, it's important to design them for explainability (Alruwaili et al., 2025).

5.4 Overview of Current Regulations and Ethical Frameworks

Policies and rules are adapting quickly as autonomous AI-IoT systems create new challenges. In Europe, the GDPR and in the United States, the CCPA require ai academic internet connection systems to observe strict rules on collecting, using and protecting personal information. Many organizations have written guidelines that focus on fairness, transparency, being accountable and privacy. For example, the IEEE produced the IEEE's Ethically Aligned Design and the EU established the EU's Ethics Guidelines for Trustworthy AI. Because of these frameworks, ethical decisions are included in design and kept present throughout deployment. And, regulations are being suggested to focus on AI issues such as making sure organizations conduct impact reviews, provide adequate detail and gain certification for sensitive AI. Nations across the globe are also striving to coordinate their regulations in an attempt to make the innovative solutions propagate and be employed globally. Although a lot is done, there are still numerous areas lacking regulation, which makes the further deliberation between the policymakers, technologists and civil society regarding establishing strong, versatile regulations to autonomous AI-IoT technologies figure 5 as follows extremely important (Aljanabi et al., 2024).

Figure 5. Ethical and Regulatory Framework

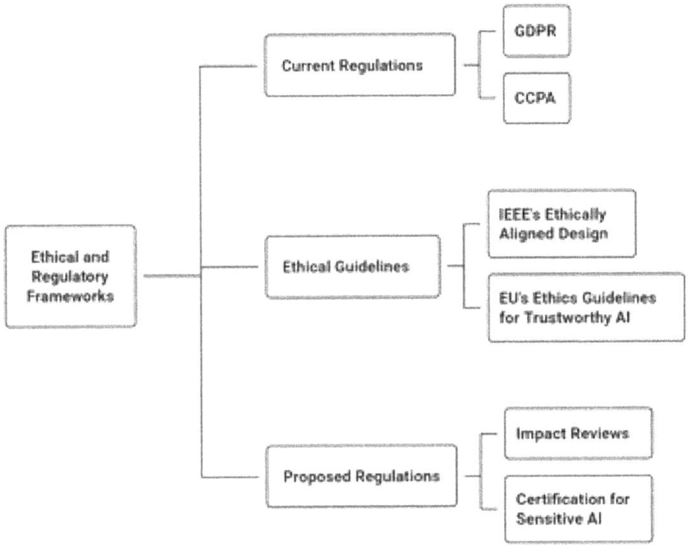

6. SOCIETAL IMPACT OF AUTONOMOUS AI-IOT SYSTEMS

With autonomous AI and IoT used in combination with each other, they are significantly changing our society. The systems are beneficial, especially in boosting and safeguarding work in fields such as healthcare, transportation, manufacturing and even smart cities because of good reasons. These are artificial intelligence-driven devices that enable human beings to be tracked down health-wise, traffic can be easily managed, energy can be utilized more efficiently and it enhances the quality of life and the development process. Meanwhile, there are certain issues associated with these new developments. It is feared that job displacement through automation may be most devastating to economics, the arts and work with computers. In this case, we urgently require good career changing policies, enhancing education and assisting in the social programs of the workers affected economically. In case not all communities can access autonomous systems, their inequalities to others might be enhanced. We must tackle these gaps to keep digital inclines from growing and allow more individuals to gain from technology. On the whole, managing AI-IoT systems is complex, so governments, industry and civil society must work closely to make sure they result in more advantages than disadvantages (Shukla et al., 2019).

6.1 Building and Maintaining Public Trust

Widespread acceptance and success of these systems depend on the people trusting them. These technologies are only trusted when they are open and honest about what they do and what they cannot do. Building trust means that stakeholders should make decisions easy to explain for others impacted. Being open about using data, keeping it secure and explaining risks in AI helps everyone better understand what it does. Engaging the public helps by letting people take part in consultations, awareness activities and educational sessions which can support their understanding. If we stick to ethics and take responsibility, it increases confidence in our brand. Lack of attention to public concerns may prompt scepticism, objection and less use of the new technology, stopping it from progressing and missing out on useful benefits. The table 6 looks at how people view AI-IoT and their trust in the technology. It highlights important steps to build trust and explains how governance is handled using things like XAI and participation by the public (Ratta et al., 2021). The table leads businesses and government bodies to respond swiftly and openly to ethical technology issues.

Table 6. Public Trust Factors and Governance Approaches

Factor	Description	Importance	Governance Approach
Transparency	Clear communication of AI processes and data use	Builds confidence and understanding	Explainable AI, open policies
Fairness	Ensuring equitable treatment across demographics	Prevents discrimination	Bias detection, inclusive design
Security	Protecting systems from cyber threats	Safeguards user data and system integrity	Quantum-safe encryption, security audits
Accountability	Holding developers and deployers responsible	Builds legal and ethical trust	Regulatory oversight, liability laws
Public Engagement	Involving citizens in AI-IoT governance	Encourages acceptance and trust	Public consultations, education campaigns

6.2 Human Agency and the Future of Work

With the increase in autonomous AI-IoT systems, there are now important questions about who is in control. Although these new systems can handle complex jobs and help make decisions, relying too much on them may result in unforeseen problems, so adjustment is still needed. Work roles may evolve in the workplace because of automation and this may require people to team up with all kinds of AI technology. Getting workers ready for this change takes teaching technical skills as well as helping them become adaptable and develop a passion for learning new things. An important debate involves how much control machines should have in comparison to the opportunity for human users to control their actions. Helping AI-IoT technology guide rather than take over from human decision-making will play a big role in keeping society and ethics intact. Taking action, regularly talking about solutions and involving everyone in change are key to manage the changes stemming from new technologies (Hassan et al., 2023).

6.3 Impact on Labor Markets, Job Displacement, and Workforce Transformation

The increase in autonomous AI-IoT systems is greatly affecting workplaces everywhere. Because AI and IoT can help automation execute difficult tasks once handled only by workers, there is worry about changes in employment in fields such as manufacturing, logistics and customer service. With certain jobs disappearing, new positions where people use advanced digital tools are now forming. The change requires employers to help their staff by reskilling and upskilling them to be ready for shifting careers. Schools, companies and authorities should combine forces to build programs that keep workers ready for technological changes. In addition, job design becomes about teaming humans and computers, so that robots manage tedious responsibilities and people use their creativity, thinking and empathy. It is a significant societal task to move from the old to the new in a fair way, without causing major disruptions, meaning we need good policies, safety nets and ways to include all people in growth (Marques et al., 2019).

6.4 Public Trust and Acceptance of Machine-Led Decisions

Autonomous AI-IoT technologies require public faith for their success. Many people are hesitant about decisions made by machines, mainly because mistakes in healthcare, finance and public safety can be very damaging. AI trust is achieved when people understand the system is fair and always accountable. Applications like XAI explain how a system reaches its decisions which gives users a sense of the system's reliability. Making it clear what autonomous systems can and cannot do reduces both high and low expectations. In addition, ensuring the safety of customer information helps convince users that Personalize is secure. When communities and different stakeholders are involved in talks and are part of the governance structure, everyone feels a duty to use AI-IoT technologies ethically (Al-Kahtani et al., 2022).

6.5 Promoting Multidisciplinary Dialogue and Responsible Governance

Due to societal problems caused by autonomous AI-IoT systems, it is important to unite the efforts of technologists, ethicists, policymakers, industry leaders and representatives from civil society. This dialogue guarantees that a range of viewpoints shapes the development, use and control of intelligent systems. Good governance systems should manage risk and introduce new ideas, tune with ethical and legal rules and encourage advancements in technology at the same time. It's important that policies focus on honesty, justice and responsibility, accompanied by definite instructions on liabilities and data protection. Since AI and IoT are used around the world, it is important for countries to cooperate to develop shared standards and work teams easily. With inclusive talks and an ethical view, stakeholders may direct building AI-IoT systems that meet society's beliefs, guarantee human rights and benefit the community (Abdulmalek et al., 2022).

7. FUTURE OUTLOOK

Autonomous AI-IoT systems will make worthy strides in the face of the rapid technological advancement and closer collaboration between specialists and the enhanced teamwork. Such trends will enhance responsiveness, safety and access of IoT systems and enable smarter and versatile autonomous activity. It has significant optimization, machine learning and security improvements that can make big waves in the AI-IoT world. Although the research is still in its early stages, additional research will expose the potential to easily integrate quantum techniques into the current designs. The two issues that will feature prominently in AI development will be ethics and society. The issue of algorithmic bias, breach of privacy and disruption of the economy or society, among other issues, can only be addressed with better management systems shaped by the discussions between the people in technology, government and society. Considering the human-centric design and responsibility in the way that AI-IoT systems are created will ensure that these technologies promote human judgment and bring fairness to the society when they are an important part of daily life. Eventually, this emerging science enables us to develop intelligent, adaptable autonomous systems that are open, fair and consistent and what the society appreciates creating a safer, more effective and ethically appropriate future world (Marques et al., 2019).

7.1 Technological Advancements and Integration Pathways

The future of self-directed AI-IoT systems is in the synergistic approach of the three concepts quantum computing, federated learning, and edge intelligence. With quantum hardware coming of age, optimization, pattern recognition and security operations will be done through quantum-classical hybrid systems. The improved connectivity (6G and ultra-trustworthy networks) will ensure effective low-latency connectivity that will ensure context-aware decision-making in the absence of interruptions. Moreover, it will not result in the loss of data sovereignty because privacy-sensitive models like federated and swarm learning will allow inter-institution cooperation. Such integrations will generate self-organized ecosystems with the ability to undergo dynamic learning and self-regulate in areas of ethics. The interoperability standards and quantum-safe encryption schemes will become a thing of the past; as the autonomous AI-IoT will become the sustainable system, which will be used to create efficiency, reliability, and resilience to the world in the areas of healthcare, transport, manufacture, and smart urban spaces (Vadivel, 2024).

7.2 Ethical, Societal, and Human-Centric Directions

The next steps which that the autonomy of AI-IoT should be concentrated on are ethical design, inclusion of the society and human control. Along with the level of technological sophistication, they should be open and accountable and explainable in a manner that does not lock out human control in the critical decision-making process (Patel et al., 2023). The mutual strategies with policy makers, engineers and ethicists would be relevant to justify the principles of international governance, which includes the algorithmic bias, labor transitions and equal access. The automation will enable the societies to adjust to the reskilling and education process and preserve the human agency and creativity. In line with it, explainable AI and participatory design will contribute to user-empowering data-enabled trust and empowerment. The future AI-IoT systems will also be in a position to advantage human needs, safeguard private information, and ensure a harmonious coexistence of intelligent machines and the societies to which the latter is aimed by using the concepts of innovation, social responsibility, and sustainability (Lakshmiprabha & Dhawankar, 2024).

8. RESEARCH AND POLICY RECOMMENDATIONS

Future study of autonomous AI-IoT systems should shift to interdisciplinary exploration to involve computer scientists, ethicists, policy analysts and social scientists in creating transparent and human-friendly technologies (Rahaman et al., 2024). Pilot projects such as smart healthcare grids, intelligent mobility platform, and energy-saving factories will be empirically validated and serve to illustrate the feasibility of the real world and governance preparedness. Attributes that should be prioritized by the researchers are the creation of benchmark data regarding ethical AI, quantum-AI simulation framework, and adaptive algorithms that are resistant to uncertainty and adversarial attacks. Standardization agencies and governments at the policy level need to present common standards on algorithmic accountability, quantum-safe data security and environmental sustainability indicators (Rangra et al., 2025). The creation of the AI-IoT regulatory sandboxes might enable regulated experimentation and guarantee adherence and social trust. Developers certification and transparent audit body of organizations will go a step farther to ensure that the emerging AI-IoT innovations will develop responsibly, inclusively, and securely across the global ecosystems.

8.1 Capacity Building and International Collaboration

The construction of strong human resources and the development of international cooperation are the keys to sustainable development of autonomous AI-IoT systems. Universities, industries and governments need to collaboratively set up interdisciplinary research centers and innovation platforms on ethical AI, cybersecurity and quantum-enabled IoT. AI governance, data ethics and thinking about system design must be incorporated into the academic study to train future professionals in adaptive and human-centered technologies (Akbar et al., 2025). Through collaborative efforts between countries, e.g., the AI ethics framework established by UNESCO, EU Horizon initiatives, and Indo-Pacific AI alliances, one may contribute to the sharing of data, harmonization of standards as well as ensuring equal access to technological resources. The exchange programs and joint research will hasten the capacity building as would be the case with developing economies to avoid digital inequality. Collaborative ecosystems can ensure the development of AI-IoT technologies as tools of joint development instead of inequality by cultivating talent, encouraging inclusion, and developing a global culture of responsible innovation.

8.2 Policy and Governance Recommendations

The systems of the global governance should be adjusted to the speedy AI-IoT innovation and make it accountable, transparent, and trusting. Autonomous systems should have enforceable ethical data usage, quantum-safe encryption and algorithmic auditing guidelines formulated by their policymakers (Xu et al., 2019). Setting up of international standards by collaborations like IEEE, ISO and UN AI task forces can standardize the cross-border regulatory compliance. Regulatory sandboxes Governments ought to encourage AI regulatory sandboxes where they can deploy and ethically experiment under regulation. Moreover, high-stakes applications should be required to be explained by algorithms, subject to fairness evaluation, and track their lifecycle. Responsible innovation can be financed with public-private partnerships and reskilling of the workforce in the industries under the impact of automation. Environmental sustainability should also be given priority under governance by implementing carbon conscious computing requirements (Ystgaard et al., 2023). The combination of human values and the technological governance allows policymakers to guarantee the long-term exposure of the AI-IoT systems as transparent, secure, and advantageous to the world communities.

9. CONCLUSION

Artificial Intelligence (AI) and the Internet of Things (IoT) are converging and creating autonomous and adaptable systems that are able to learn, understand, and respond to emerging circumstances. These systems are founded on the concept of real-time responsiveness, edge computing and distributed algorithms to process data in a few seconds and make critical decisions, not just based on centralized servers. Such real-time future is applicable in systems such as autonomous vehicles, medical devices, industrial surveillance and vital infrastructure to warrant reliability, safety and efficiency. In addition, the systems are continuously learning together with the environment, enhancing the decision making and augmenting performance as time elapses. Adaptivity is one of the pillars, since it enables the use of robots, medical diagnostics, and intelligent factories to evolve with the user requirements. The quantum computing adds out of nowhere a revolutionary layer, which in any case cannot be optimizable within its field, machine learning training can be done more quickly and quantum-safe cryptography is more secure. Although quantum technologies are still in their infantile phase, they will fix the issues, the current systems cannot handle and, therefore, will enable the extension of AI-IoT significantly. However, even ethical, legal and societal concerns have risen with these developments which have become very urgent. Biases in algorithms; privacy concerns, accountability and hidden processes are among the issues that should be addressed in order to build trust.

Stakeholder This can be essential, especially when stakes are high (e.g. in healthcare or law enforcement), and can aid in ensuring transparency and accountability. Other than technology, there are far-reaching social repercussions of automation. The labour markets are evolving, and it requires that the workforce should be reskilled and the new opportunities should be made equally available. The most important factors towards ensuring that people place their trust in the autonomous systems and ensuring that human capabilities are not replaced by autonomous systems would be transparency, inclusiveness, and communication. Federated learning, edge computing, quantum technologies and ethical AI will be applied in the future to create more intelligent, safer, and sustainable solutions. Being a powerhouse of innovative offerings and responsible governance, the AI-IoT systems can bring disruptive gains in all industries and safeguard fairness, accountability, and human health to shape a trusted and inclusive digital future.

REFERENCES

Abdulmalek, S., Nasir, A., Jabbar, W. A., Almuhaya, M. A. M., Bairagi, A. K., Khan, M. A.-M., & Kee, S.-H. (2022). IoT-based healthcare-monitoring system towards improving quality of life: A review. *Healthcare (Basel)*, *10*(10), 1993. DOI: 10.3390/healthcare10101993 PMID: 36292441

Akbar, A. (2025). IOT security in Complex Systems: Big Data, Quantum Computing and HCI design for AI ethics. *Annual Methodological Archive Research Review*, *3*(1), 30–39. DOI: 10.63075/1meshm93

Al-Kahtani, M. S., Khan, F., & Taekeun, W. (2022). Application of Internet of Things and sensors in healthcare. *Sensors (Basel)*, *22*(15), 5738. DOI: 10.3390/s22155738 PMID: 35957294

Aljanabi, M., Mohammed, S. Y., & Omran, A. H. (2024). Detecting data poisoning attacks in federated learning for healthcare applications using deep learning. *Iraqi Journal for Computer Science and Mathematics*, *4*(4), 225–237. DOI: 10.52866/ijcsm.2023.04.04.018

Alruwaili, M., Siddiqi, M. H., Idris, M., Alruwaili, S., Alanazi, A. S., & Khan, F. (2025). Advancing disability healthcare solutions through privacy-preserving federated learning with Theme framework. *Expert Systems: International Journal of Knowledge Engineering and Neural Networks*, *42*(1), e13807. Advance online publication. DOI: 10.1111/exsy.13807

Antunes, R. S., André da Costa, C., Küderle, A., Yari, I. A., & Eskofier, B. (2022). Federated learning for healthcare: Systematic review and architecture proposal. *ACM Transactions on Intelligent Systems and Technology*, *13*(4), 1–23. DOI: 10.1145/3501813

Arumugam, T., Hameed, S. S., Sadath, J. S. S. M., Priyanshu, D., & Subbulakshmi, S. (2024). Agile marketing with ethical tech, AI, and IoT for sustainable customer engagement. In Advances in Logistics, Operations, and Management Science (pp. 87–102). IGI Global. DOI: 10.4018/979-8-3693-6274-7.ch006

Belbase, P., Bhusal, R., Ghimire, S. S., Sharma, S., & Banskota, B. (2024). Assuring assistance to healthcare and medicine: Internet of Things, Artificial Intelligence, and Artificial Intelligence of Things. *Frontiers in Artificial Intelligence*, *7*, 1442254. DOI: 10.3389/frai.2024.1442254 PMID: 39735232

Budrionis, A., Miara, M., Miara, P., Wilk, S., & Bellika, J. G. (2021). Benchmarking PySyft federated learning framework on MIMIC-III dataset. *IEEE Access : Practical Innovations, Open Solutions*, *9*, 116869–116878. DOI: 10.1109/ACCESS.2021.3105929

Chauhan, M., & Sahoo, D. R. (2024). Towards a greener tomorrow: Exploring the potential of AI, blockchain, and IoT in sustainable development. Nature. *Environment and Pollution Technology*, *23*(2), 1105–1113. DOI: 10.46488/NEPT.2024.v23i02.044

Chaware, D. S. M., Bhosale, P. R., Jadhav, P. A., & Chaware, D. T. S. (2024). CONVERGENCE OF ARTIFICIAL INTELLIGENCE AND IOT: A BOOM. In Futuristic Trends in IOT Volume 3 Book 1 (pp. 138–145). Iterative International Publisher, Selfypage Developers Pvt Ltd. DOI: 10.58532/V3BBIO1P1CH16

Chawla, U. (2024). The digital shift. In *Driving Decentralization and Disruption With Digital Technologies* (pp. 97–107). IGI Global., DOI: 10.4018/979-8-3693-3253-5.ch007

Edwards, A., Law, Y. W., Mulinde, R., & Slay, J. (2023). Evaluation of quantum key distribution for secure satellite-integrated IoT networks. *International Conference on Cyber Warfare and Security*, 18(1), 67–76. DOI: 10.34190/iccws.18.1.982

Fehringer, H., & Stary, C. (2023). User-informed adaptation in IoT home healthcare: Grounding development in empirical evidence. *Journal of Theoretical and Applied Electronic Commerce Research*, *18*(4), 1901–1925. DOI: 10.3390/jtaer18040096

Gambo, N., & Musonda, I. (2022). Impact of IoT on achieving smart primary healthcare building facilities in Gauteng, South Africa. *International Journal of Environmental Research and Public Health*, *19*(18), 11147. DOI: 10.3390/ijerph191811147 PMID: 36141421

Hassan, A. A., Tutuncu, K., Abdullahi, H. O., & Ali, A. F. (2023). IoT-based smart health monitoring system: Investigating the role of temperature, blood pressure and sleep data in chronic disease management. *Instrumentation Mesure Métrologie*, *22*(6), 231–240. DOI: 10.18280/i2m.220602

Humayun, M., Tariq, N., Alfayad, M., Zakwan, M., Alwakid, G., & Assiri, M. (2024). Securing the internet of things in artificial intelligence era: A comprehensive survey. *IEEE Access : Practical Innovations, Open Solutions*, *12*, 25469–25490. DOI: 10.1109/ACCESS.2024.3365634

Lakshmiprabha, K. E., & Dhawankar, S. H. (2024). Overcoming interoperability challenges in blockchain and IoT integrating diverse technologies for enhanced security. In Secured Blockchain and IoT Connected Devices through Decentralized Trust and Enhanced Data Integrity (pp. 321–348). RADemics Research Institute. DOI: 10.71443/9788197933653-10

Leghemo, I. M., Azubuike, C., Segun-Falade, O. D., & Odionu, C. S. (2025). Data governance for emerging technologies: A conceptual framework for managing Blockchain, IoT, and AI. *Journal of Engineering Research and Reports*, *27*(1), 247–267. DOI: 10.9734/jerr/2025/v27i11385

Loftus, T. J., Ruppert, M. M., Shickel, B., Ozrazgat-Baslanti, T., Balch, J. A., Efron, P. A., Upchurch, G. R.Jr, Rashidi, P., Tignanelli, C., Bian, J., & Bihorac, A. (2022). Federated learning for preserving data privacy in collaborative healthcare research. *Digital Health*, *8*, 20552076221134456. DOI: 10.1177/20552076221134455 PMID: 36325438

Marques, G., Pitarma, R. M., Garcia, N., & Pombo, N. (2019). Internet of Things architectures, technologies, applications, challenges, and future directions for enhanced living environments and healthcare systems: A review. *Electronics (Basel)*, *8*(10), 1081. DOI: 10.3390/electronics8101081

Mishra, D., & Singh, B. (2025). Public education campaigns on AI and IoT applications in wildfire prevention and management. In *Advances in Environmental Engineering and Green Technologies* (pp. 375–394). IGI Global., DOI: 10.4018/979-8-3373-0045-0.ch022

Movahed, A. B., Abyaneh, A. G., Khakbazan, M., & Movahed, A. B. (2025). Smart economy cybersecurity. In *Dynamic and Safe Economy in the Age of Smart Technologies* (pp. 49–72). IGI Global., DOI: 10.4018/979-8-3693-4369-2.ch004

Pasupuleti, M. K. (2024). Smart ecosystems: The revolutionary AI-IoT alliance. In *Connected Intelligence: The Convergence of AI and IoT Technologies* (pp. 33–40). National Education Services., DOI: 10.62311/nesx/97838

Patel, A., Pali, P., Verma, S., Tiwari, S., Vishwakarma, V., & Chourasiya, S. (2023). Artificial Intelligence strategies for combating IoT-centric cyber attacks. *International Journal of Innovative Research in Science, Engineering and Technology*, *12*(05), 8050–8056. DOI: 10.15680/IJIRSET.2023.1205496

Rahaman, M., Lin, C.-Y., Pappachan, P., Gupta, B. B., & Hsu, C.-H. (2024). Privacy-centric AI and IoT solutions for smart rural farm monitoring and control. *Sensors (Basel)*, *24*(13), 4157. DOI: 10.3390/s24134157 PMID: 39000936

Rangra, A., Sah, R. K., Himlaya Himlaya, & Ranjan, M. (2025). Exploring the role of AI in driving innovation across key industries: A sector-by-sector analysis. INTERANTIONAL JOURNAL OF SCIENTIFIC RESEARCH IN ENGINEERING AND MANAGEMENT, 09(03), 1–9. DOI: 10.55041/IJSREM43280

Ratta, P., Kaur, A., Sharma, S., Shabaz, M., & Dhiman, G. (2021). Application of Blockchain and Internet of Things in healthcare and medical sector: Applications, challenges, and future perspectives. *Journal of Food Quality*, *2021*, 1–20. DOI: 10.1155/2021/7608296

Sadilek, A., Liu, L., Nguyen, D., Kamruzzaman, M., Serghiou, S., Rader, B., Ingerman, A., Mellem, S., Kairouz, P., Nsoesie, E. O., MacFarlane, J., Vullikanti, A., Marathe, M., Eastham, P., Brownstein, J. S., Arcas, B. A. Y., Howell, M. D., & Hernandez, J. (2021). Privacy-first health research with federated learning. *NPJ Digital Medicine*, *4*(1), 132. DOI: 10.1038/s41746-021-00489-2 PMID: 34493770

Shukla, S., Hassan, M. F., Khan, M. K., Jung, L. T., & Awang, A. (2019). An analytical model to minimize the latency in healthcare internet-of-things in fog computing environment. *PLoS One*, *14*(11), e0224934. DOI: 10.1371/journal.pone.0224934 PMID: 31721807

Upreti, D., Yang, E., Kim, H., & Seo, C. (2024). A comprehensive survey on federated learning in the healthcare area: Concept and applications. Computer Modeling in Engineering & Sciences. *Computer Modeling in Engineering & Sciences*, *140*(3), 2239–2274. DOI: 10.32604/cmes.2024.048932

Vadivel, S. R. S. (2024). Integrating the industrial revolution 5.0 with AI. In *Industry Applications of Thrust Manufacturing* (pp. 118–147). IGI Global., DOI: 10.4018/979-8-3693-4276-3.ch005

Wu, N., Hu, Y., Song, F., & Li, X. (2023). Quantum SoC and on-chip circuit synthesis in the NISQ era. *Quantum Engineering*, *2023*, 1–10. DOI: 10.1155/2023/5792902

Xu, F., Yang, F., Bao, S., & Zhao, C. (2019). DQN inspired joint computing and caching resource allocation approach for software defined information-centric internet of things network. *IEEE Access : Practical Innovations, Open Solutions*, *7*, 61987–61996. DOI: 10.1109/ACCESS.2019.2916178

Ystgaard, K. F., Atzori, L., Palma, D., Heegaard, P. E., Bertheussen, L. E., Jensen, M. R., & De Moor, K. (2023). Review of the theory, principles, and design requirements of human-centric Internet of Things (IoT). *Journal of Ambient Intelligence and Humanized Computing*, *14*(3), 2827–2859. DOI: 10.1007/s12652-023-04539-3

Yuan, P. (2024). Artificial intelligence in the Internet of Things: Integrating and optimizing AI algorithms for real-time data processing and decision-making. *Applied and Computational Engineering, 102*(1), 84–89. DOI: 10.54254/2755-2721/102/20241015

Zhou, X. (2024). IoT sensor-based big data and information systems in smart healthcare. Highlights in Science. *Engineering and Technology*, *92*, 397–402. DOI: 10.54097/g5sf4f50

KEY TERMS AND DEFINITIONS

Adaptive Systems: Intelligent systems that adjust their responses to new data or changing conditions in real time.

Algorithmic Bias: Unfair or discriminatory outcomes in AI models due to imbalanced or flawed training data.

Autonomous Systems: Systems that make decisions and act independently with minimal human input by integrating AI and IoT.

Distributed Algorithms: Algorithms executed across multiple devices to ensure scalability, resilience, and cooperative decision-making.

Edge Computing: Processing data near its source to reduce delays, improve reliability, and lessen dependence on cloud servers.

Explainable AI (XAI): AI methods that provide transparency into decision-making, ensuring trust, accountability, and fairness.

Federated Learning: Decentralized AI training across devices without sharing raw data, protecting privacy while enabling adaptive learning.

Quantum Computing: Advanced computing using superposition and entanglement to solve optimization, machine learning, and security problems faster.

Quantum Machine Learning (QML): Using quantum circuits in machine learning to accelerate training and improve accuracy in high-dimensional IoT data.

Real-Time IoT: Low-latency IoT infrastructure enabling instant responsiveness in critical applications like vehicles, healthcare, and industry.

Chapter 8
Green Intelligence in Agriculture:
Integrating AI-Driven Irrigation, Energy Management, and Sustainability for Eco-Smart Farming System

R. N. Ravikumar
https://orcid.org/0009-0009-3705-1681
Marwadi University, Rajkot, India

S. Aarthi
https://orcid.org/0009-0006-9064-2091
Marwadi University, Rajkot, India

ABSTRACT

The integration of AI and IoT is transforming autonomous systems, enabling data-driven solutions in industries like precision farming, smart irrigation, utility maintenance, healthcare, manufacturing, and transportation. This work presents a framework explaining AI-powered IoT networks and their role in industrial automation. It highlights AI techniques like machine learning and real-time analytics to enhance efficiency. The research examines challenges in AI-IoT integration, such as energy management, security, interoperability, infrastructure limits, and data privacy. Implementation examples show how these technologies transform smart cities and industries, improving decision-making and adaptability. It also explores the societal and ethical impacts of AI-driven innovation, emphasizing fairness and sustainability. Additionally, the text outlines future directions in emerging agricul-

DOI: 10.4018/979-8-3373-3146-1.ch008

ture technologies, and government policies in farming. By fostering collaboration, it provides insights for experts and engineers, equipping them with tools to develop efficient AI-driven frameworks.

1. INTRODUCTION

Combined with accurate agricultural and smart irrigation techniques, the transformed agriculture turned into the one founded on the usage of the Internet of Things (IOT), Artificial Intelligence (AI) and Geographical Information System and Sensor-based automation. The new technologies do co-exist with providing farmers with a larger resource management capacity which translates into increase in crop yields and a lesser amount of environmental impact. The basic element of accurate farming known as smart irrigation provides efficient water distribution through real -time data analysis used for automated systems and accurate water distribution. The effective energy consumption management system proves necessary to achieve optimal efficiency along with stability in these systems. Power optimization techniques and artificial intelligence analytics platforms are included in the energy management systems of accurate farming with smart irrigation systems and renewable power systems are included. Specific cultivation practices consume vital amounts of energy that translate to high expenses and acute ecological damage. The combination of solar-powered irrigation systems with smart grid technology enables farmers to reduce their energy waste by increasing their productivity levels. An introduction explains why proper energy management is essential in accurate farming with smart irrigation systems as well as the future approach in agricultural stability as well as a fundamental approach to stability (Raliya, 2024).

1.1 Definition of Precision Farming and Smart Irrigation

Precision Farming is an advanced agricultural approach that uses technology to increase agricultural management to increase resource usage and increasing efficiency. This includes applications of devices such as GPS, IOT, drone and AI-operated data analytics for monitoring and management of field variability in crops, soil and climatic conditions. By enabling farmers to make data-operating decisions, accurate farming helps improve crop yields, reduce input costs and reduce environmental effects through efficient use of water, fertilizers and pesticides.

Smart Irrigation refers to an automatic and technology-operated irrigation system that optimizes the use of water based on real-time data, weather conditions and soil moisture levels. Unlike traditional irrigation methods, smart irrigation uses IOT sensor, AI algorithms and remote-controlled systems, when crops to ensure precise water distribution and where needed to ensure. This approach preservations water, reduces energy consumption, and increases crop growth by maintaining optimal soil moisture conditions, which contributes to durable and resource-skilled agriculture.

1.2 Importance of Energy Management in Modern Agriculture

Energy management plays an important role in modern agriculture by ensuring efficient resource use, reducing operational costs and promoting stability. Along with increasing technology-operated agricultural practices like accurate farming and smart irrigation, energy consumption has become an important factor in agricultural productivity. Efficient energy management helps farmers adapt the use of electricity for irrigation, machinery and processing, reducing waste on non-diagnostic energy sources. By combining fossil fuels in the fields, renewable energy solutions like solar panels, wind turbines and bio-energy will help in reducing low greenhouse gas emissions and also lead to an environmentally friendly farming system. With efficient energy management systems, agricultural operations can be made more efficient with automatic systems and AI-operated analytics and real-time energy consumption monitoring can become a possibility. The reduction of the cost of energy also contributes significantly to the increased profitability in the agriculture sector since the practice is characterized by improved efficiency of farming. Long-term stability and operational flexibility are offered to the agricultural sector due to the implementation of smart energy strategies in the face of the increasing energy prices and the climate change. When farmers apply energy-proficient technologies, they create a food production process that is environmentally friendly and connects economic prosperity to the sustainability of the environment (Gopathoti, 2023). The Figure 1 underlines the most important, primary benefits of technology in agriculture and sustainability, like productivity improvement, cost reduction, and environmental sustainability. It also highlights precision agriculture, renewable energy integration, and improvement of food security.

Figure 1. Energy Management in Modern Agriculture

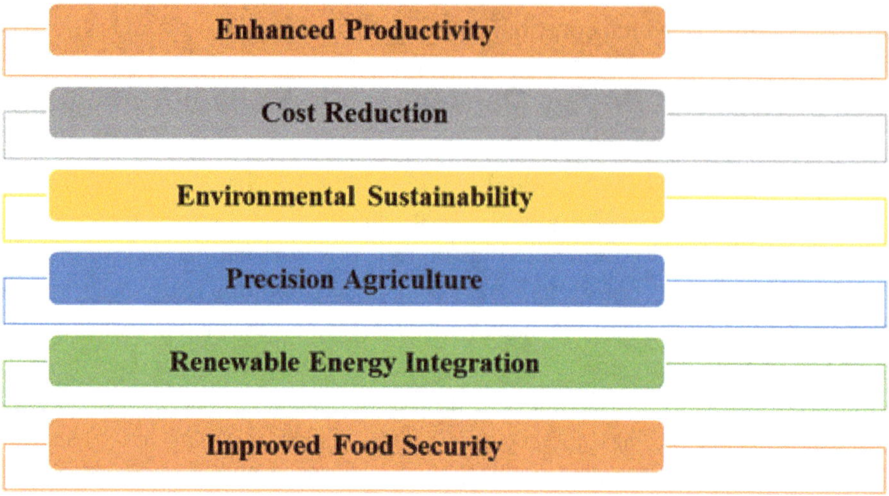

1.3 Objectives and Scope of the Discussion

The main goal of this dialogue seeks to determine how energy management helps achieve accurate farming while implementing smart irrigation systems. Its purpose is:

- To analyze the role of precision farming and smart irrigation technologies (IoT, AI, GIS, sensors, and drones) in enhancing productivity, energy efficiency, and sustainability.
- To compare energy consumption patterns in traditional farming versus smart farming systems and highlight potential savings.
- To evaluate energy management strategies such as renewable energy integration, efficient irrigation methods, and AI-driven automation.
- To identify key challenges and limitations, including high costs, infrastructure gaps, and data security concerns, that affect adoption.
- To explore future trends and innovations in energy-efficient agriculture, focusing on new technologies, supportive policies, and global food security.

The exploration covers multiple areas in accurate farming practice together with smart irrigation systems and their associated energy performance tactics. The report provides observations about IOT together with AI along with sensors which demonstrate their influence on automation and energy effectiveness. The evaluation process analyzes energy consumption together with stability factors of traditional and smart farming systems. The deployment of renewable energy systems focuses on both implementation difficulties and predicted energy-efficient agriculture development patterns. The research addresses factors from economic spheres as well as environmental elements and technical aspects which influence the adoption of modern energy management practices in farming.

1.4 Literature Review

Precision farming and smart irrigation systems have significantly contributed to the evolution of modern agriculture as the idea of emerging technologies is to obtain a high level of efficiency and resource-saving, as well as a higher level of sustainability. Precision farming is an alternative to treating fields in a uniform manner since it involves site-specific management and, therefore, technologies like GPS, GIS, IoT, and AI provide farmers with services that allow them to adjust inputs to the needs of soils and crops (Mandal, 2013). This method has always been pointed out as a way of raising yields and also decreasing the input costs as well as the environmental effects by avoiding excessive use of water, fertilizers and pesticides. In this context, smart irrigation has come out as one of the most revolutionary elements. Compared to the traditional irrigation techniques, smart systems apply soil moisture sensors, climate information, and automatic controllers to provide the water at the right time and location when it is required and thus saving a lot of water and energy wasted (Alazzai et al., 2024). The research on the energy efficiency in the agricultural sector highlights that conventional practices, which include flood irrigation, diesel-powered pumps, etc. are the most energy-consuming and inefficient practices in terms of environmental sustainability. Smart irrigation systems, especially where these systems are coupled with renewable sources of energy such as solar powered pumps has proved to reduce the reliance on fossil fuel as well as reducing the cost of operation (Sharma et al., 2021).

Moreover, AI-based analytics is becoming a more valuable tool in forecasting water needs of crops, irrigating on a more efficient schedule, and assisting with real-time decision-making. The increase in the use of drones and remote sensors to monitor the variability of the fields also is mentioned in literature as a way of improving energy and water management (Chandrakala, 2021). Though these advantages have been realized, the literature notes unresolved obstacles such as huge start-up costs, absence of infrastructure in the rural regions, and security issues of information regarding privacy. Nevertheless, there are also future visions regarding the growth of smart farming because other new technologies, including 5G connectivity, blockchain to assist with supply chain visibility, and vertical farming, are also expected to reinforce the boundaries of smart farming (Ibrahim Mohammad

Abuzanouneh et al., 2022). Altogether, this literature shows that accuracy farming and intelligent irrigation are the focus of sustainable agriculture with a solution to energy efficiency, resource savings, and food security in the long term.

1.5 Research Gap

Despite the potential that has been translated into the field of precision farming and smart irrigation systems, the adoption of the systems is at a very low level because of critical gaps. Majority of available literature focuses on technological innovations, but it lacks much information on cost-effectiveness, scale-ability, and availability to farmers, especially in developing areas. The barriers like high start-up expenses, poor infrastructure and security of the data are usually raised and not exhaustively discussed with concrete remedies. Besides, research on the combination of renewable energy and AI- and IoT-driven irrigation on a large scale is not sufficiently researched. The need to fill these gaps is critical to the realization of massive sustainable agriculture.

2. OVERVIEW OF PRECISION FARMING

Precision farming is a modern farming practice that also makes technology available to suit crop production through effective utilization of resources like water, fertilizers and pesticides. It is based on records-manufacturing decisions the use of technology along with GPS, Geographical Information System (GIS), Internet of Things (IOT), drone and Artificial Intelligence (AI) to filter the conditions of the environment in real-time. Accurate farming will help farmers to do accurate adjustments by examining the quality of soil, weather patterns and crop adaptation by adjusting the style and enhancing the production. This strategy is no longer the easiest one, but if applied it enhances the quality and quantity of yield, as well as environmental impact by avoiding excessive utilization of chemical substances and excessive consumption of water. Among the significant advantages of proper farming is the ability to cope with the field variability to ensure that special parts of the farm obtain an individualized treatment depending on the particular requirements. The convertible rate technology enables automated machine to change how inputs, including seeds, fertilizers and water are applied according to the correct requirements. Also, prepaid analytics are run by remote sensing and AI to estimate the issues faced by farmers like insect infections, drought and nutrient deficiency to be managed actively by farmers. Given that the stability of agriculture is of immediate concern, the precise agriculture is significant in cost reduction, as well as the conservation of natural resources and food security amid the increased world population (Issa et al., 2024).

2.1 Concept and key Components of Precision Farming

Precision farming, also known as accurate agriculture, is a modern agricultural practice that uses technology and data analytics to optimize agricultural production. This concept revolves around site-specific management, where farmers analyze field variations and efficiently apply resources to maximize yields by reducing the waste. Unlike traditional farming methods, which rely on applies similar to water, fertilizers, and pesticides, the tailors of accurate farming are based on the real -time data collected from these input soils, weather and crop conditions. By integrating advanced technologies such as GPS, IOT, Artificial Intelligence (AI), and remote sensing, accurate farming increases decision making, reduces costs, and promotes permanent agriculture (Adewusi et al., 2024). The Figure 2 represents the technological improvements that have come to precision agriculture throughout the years, from GPS and GIS to AI and Big Data Analytics. The most important of innovations to enhance farming include IoT sensors, remote sensing, and variable rate technology solutions.

Figure 2. Key Components of Precision Farming

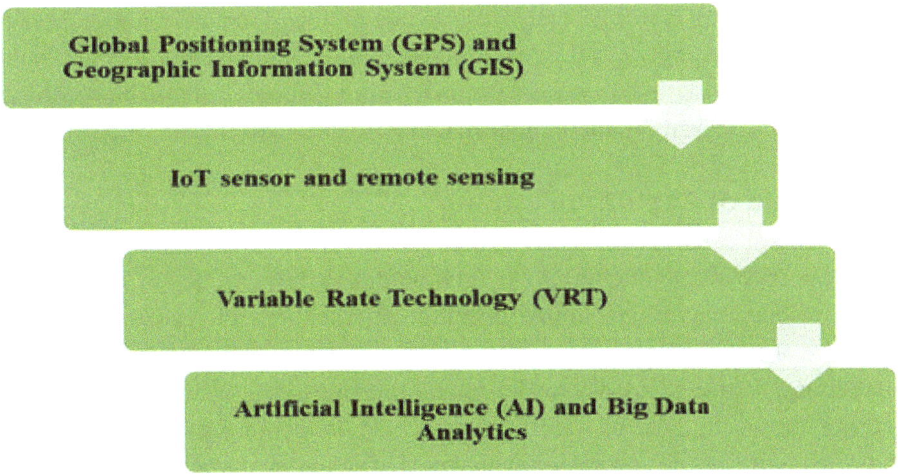

Smart agriculture entails the incorporation of innovative technologies in an effort to enhance precision, efficiency and sustainability. Global Positioning System (GPS) guarantees proper mapping of the field and guidance of machinery, and

Geographic Information System (GIS) is able to methodically analyze the spatial variability to make the most effective distribution of resources. Internet of Things (IoT) sensors offer real-time functions of soil moisture, temperature and nutrient and are complemented by drone and satellite imaging to check the health of crops and also detect the presence of pests or stress. Variable Rate Technology (VRT) uses fertilizers, pesticides and water only, when necessary, thus minimizing wastage and increasing productivity. Artificial Intelligence (AI) and Big Data Analytics can be used to improve the process of decision-making, forecasting the risk of diseases, finding the best time to plant crops, and arranging irrigation activities. The insights are further refined with the help of machine learning models, which generate data-driven strategies that yield more, save resources, and help to promote sustainable production processes (Rahman & Barmon, 2012).

2.2 Technologies Used in Precision Farming

Precision farming is based on the combination of more sophisticated technologies that collaborate to make it more efficient, less wasteful, and sustainable. The Internet of Things (IoT) provides the IoT sensors and agricultural devices to measure soil moisture, temperature, humidity, and crop health and provides the capability to run irrigation and schedule based on climate. Geographical Information System (GIS) facilitates mapping of fields and spatial analysis to assist farmers in refining the application of fertilizers and pesticides and also in land use planning. Artificial Intelligence (AI) processes big data, detects trends, predicts production and outbreaks, monitoring, precise sprays, and managing vast fields. Combined, these technologies allow making decisions based on data, increasing resources and achieve greater yields and food security in the long term (Pingda et al., 2022).

2.3 Benefits of Accurate Farming

The benefits of accurate farming are numerous in the sense that it incorporates modern technologies to streamline farming activities, use of resources and sustainability. Among the main benefits, one can single out the high growth of crops and their productive efficiency, which is possible due to site-specific management ensuring the accurate use of water, nutrients, and protection factors and considering the state of the soil. Another advantage is efficient use of resources since IoT sensors and Variable Rate Technology (VRT) ensure a reduction in the wastage of fertilizers, pesticides, and water which makes it cost-effective and ecological. There is a decrease in the environmental effects through the prevention of over use of chemicals, erosion of soil, conservation of water and a decrease in the emission of greenhouse gases (Zhang et al., 2023). Automation and smart machinery also enable farmers to receive a financial advantage in terms of lower input costs, as well as labor costs. Continuous observation of the levels of nutrients and moisture enhances thc extent of soil fertility and prolonged productivity. AI-based analytics and drones can enhance the process of pest and disease management as they are able to identify the issue within its initial stages and take the necessary steps. Real-time insights also help in timely decisions and can ensure resilience to variability in climate change and will contribute to a sustainable food security. Figure 3 does explain the benefits of modern agricultural technologies: increase of crop production, cost-efficiency of resource utilization, and cost-saving to the farmers. It also highlights the gains that are being realized in the soil fertility, control of pests and adaptation to the environment (Oluwatoyin et al., 2021).

Figure 3. Benefits of Accurate Farming

3. SMART IRRIGATION SYSTEMS

The smart watering system is a technology that takes advantage of the latest technology to customize water consumption to farms and gardens. Such systems will not overlook the current weather forecast, the level of wetness of the soil and intelligent sensors to understand the exact amount of water to be added in your plants or grass. As opposed to vintage-faculty watering methods, which continue working with a pre-determined schedule, such smart systems can adjust on the fly to fit the evolving state of affairs. They are best in saving water, preventing over-watering, minimizing down to water where it is not needed and enhancing the health of the plants. Also, most of these systems allow you to remotely monitor and adjust the schedule through an application, meaning that you can monitor and adjust the watering schedule wherever you go (J. Kumar et al., 2024). The upsides of smart watering go further than just saving water. By using water more wisely, these systems can shrink your utility bills and bump up your harvest. In cities, smart watering helps keep landscapes looking good for the long haul, doing its part for green spaces by cutting down on waste. Additionally, many modern smart irrigation solutions are integrated with smart home ecosystem, which enable spontaneous automation with other connected devices. Since climate change and water scarcity become increasing concerns, adopting smart irrigation technology is an effective way to promote environmental responsibility and ensure long -term water resources management.

3.1 Definition of Smart Irrigation Systems

Smart irrigation systems are advanced water solutions that use technology such as agriculture, weather data and automation, which are to customize the use of water for agriculture, landscaping and horticulture. These systems intelligently accommodate water distribution based on real -time environmental conditions, ensuring that plants receive the right amount of water without waste. Smart controller, by incorporating soil moisture sensors and weather-based automation, smart irrigation systems help in water conservation, reduce costs, and promote healthy plant growth.

3.1.1 Working Principle of Smart Irrigation Systems

The work theory of smart irrigation systems is based on real -time data collection and automatic decision making. These systems usually have three main components: sensors, controllers and irrigation mechanisms. Sensor's degree parameters together with soil moisture, temperature, humidity and weather conditions. The controller approaches this fact and determines the agenda and quantity of right water. Depending on the manipulate of the controller, the irrigation mechanism (consisting of sprinkler or drip irrigation) materials water as required. Certain smart irrigation systems paintings in tandem with weather forecasts, the use of predictions of rainfall to great-tune watering schedules and avoid redundant watering. On pinnacle of that, quite a few those modern systems offer the capability to display and manage them remotely thru telephone apps, giving users the energy to manipulate irrigation regardless of in which they are (Zhou et al., 2020).

3.2 Types of Smart Irrigation Systems

Smart irrigation is a new technology that exists to streamline the use of water in agriculture, lawn care, and the city. These systems reduce water wastage by taking into account sensors, IoT, weather information, and automation to supply water at the time and place where it is required. Not only do they save on resources, but they also maximize crop production, lessen on the workforce as well as support sustainable activities. The most common ones are drip irrigation, smart sprinkler systems or automated irrigation setups, each with its own distinctive benefits with respect to efficient water management. The figure 4 shows the types of irrigation system as follows (Suhardi Rahman et al., 2023).

Figure 4. Types of Smart Irrigation Systems

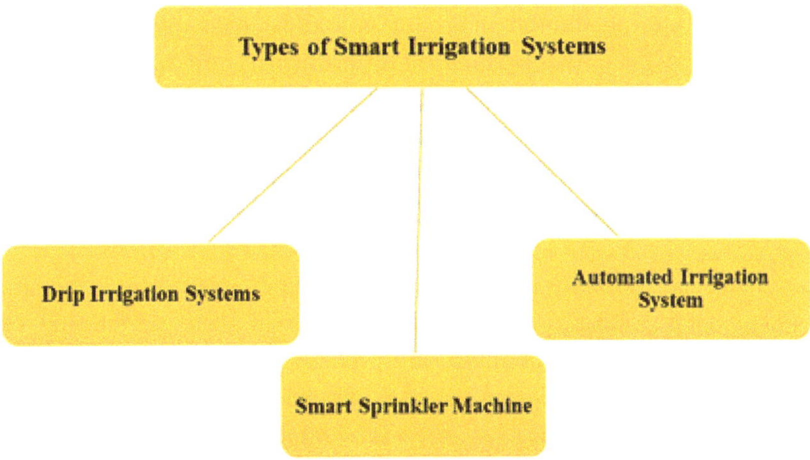

3.2.1 DRIP Irrigation Systems

The drip irrigation is relatively a good irrigation technique, where the water is supplied directly to the roots of the plants through tubes, emitters and valves. The smart drip irrigation system is comprised of soil moisture sensors, and weather sensors to regulate the flow of water without unnecessary usage and lessening the loss of evaporation. The systems also assist in giving the crops, gardens and delicate plants the maximum amount of hydration by dyeing the required amount of water at the correct time and point. It is not only a means of saving wastage of water, but it also enhances crop production, soil, and saves resources. Agricultural lands and landscapes can be fitted with smart drip irrigation to facilitate sustainable agriculture through productivity and application of water in a responsible manner (V. Kumar et al., 2023).

3.2.2 Smart Sprinkler Systems

Smart sprinkler systems are irrigation systems that utilize the latest technology of water distribution and adjust by real-time data. These systems take into consideration the weather forecasts, the level of soil moisture and programmable schedules to optimize the watering patterns. The sprinkler heads are made to deliver water in the form of sprinklers, and hence these sprinklers are suitable in lawns, parks, golf courses, as well as recreational landscapes. Using smart controllers, users are able to personalize zones, build schedules, and remotely direct system using mobile gadgets. Smart sprinklers allow adapting the irrigation to the environmental conditions and minimize overwatering, minimizing water bills, and enhancing sustainability. Their accuracy and ease of use result in them being an effective solution in both residential and commercial landscape to guarantee effective use of resources without causing harm to the greenery (Pan et al., 2023).

3.3.3 Automated Irrigation Systems

Automated irrigation systems combine the use of several smart technologies such as sensors, IoT-based controllers, and timers which control the flow of water without the intervention of a human being. These systems will be able to include drip and sprinkler methods depending on crops and landscape and adjust irrigation plans. They use weather conditions and soil conditions to know the most appropriate watering schedule and the appropriate quantity of water to be used, thus using water efficiently. Automated irrigation is widely used in large scale agriculture, industrial landscaping and smart urban projects in reducing the number of human workers and also conserving the precious resources. It is used to guarantee uniformity of watering, to avoid the under or over irrigation and improve the yields of crops and landscape wellbeing. This new practice is sustainable and deals with increasing water scarcity issues in agriculture and urbanization. The Figure 4 presents different types of smart irrigation systems, including drip irrigation systems, automated irrigation systems, and smart sprinkler machines. These technologies help optimize water usage and improve agricultural efficiency (Bwambale et al., 2023).

3.3 Role of Sensors and Real-Time Data in Irrigation Management

The sensor and real-time data are crucial in the modern irrigation concerning effective water usage and adequate development of plants. To make irrigation decisions, soil moisture, weather and temperature along with humidity are continuously monitored. This averts over exploitation and underutilization and wastefulness of resources. The systems which are dynamically adjusting utilize the concept of real time integration; e.g. in case enough moisture of the soil has been first measured or rain is forecasted watering is postponed. The IoT technology is utilized in sophisticated systems as well which allow control and monitoring of remote applications of mobile devices. All these technologies implemented make it efficient, reduce wastes, and develop sustainable agricultural and landscaping. With sensors and actual time monitoring, irrigation management is made more effective and cuts down costs, crop yield will be enhanced, and it will add to the water conservation efforts (Pang et al., 2023).

4. ENERGY CONSUMPTION IN TRADITIONAL VS. SMART FARMING

Traditional farming methods rely on manual labor, traditional irrigation techniques and old machinery, which are often leading to high energy consumption. Practices such as flood irrigation and excessive use of diesel-operated equipment contribute to waste disabilities, water and energy. Additionally, traditional farming lacks real -time monitoring, resulting in over the use of resources such as fertilizers and pesticides, which further increase energy expenditure. Since most of the irrigation in traditional farming operates on certain programs without considering the soil or weather conditions, a significant amount of water and energy in unnecessary irrigation are wasted. In contrast, smart farming significantly reduces energy consumption by incorporating advanced technologies such as IOT-based sensors, automation and accurate agriculture. The smart irrigation system optimizes water use, reducing the energy required for pumping and distribution. Automatic machinery, such as GPS-directed tractors and drones, improves efficiency by reducing fuel consumption and optimizing the operation of the area. Additionally, data-powered decision making ensures that resources are applied only, when necessary, cuts on waste. Overall, smart farming not only reduces energy consumption, but also increases stability by improving productivity and reducing the environmental footprint of agriculture (Lloret et al., 2021).

4.1 Energy-Intensive Processes in Traditional Farming

The conventional methods of farming are based on processes that consume a lot of energy that raises the cost of production and contributes to environmental degradation. Fixed-length sprinkler irrigation systems like flood irrigation, diesel-powered pumps and sprinklers, which are mostly used to irrigate lawns, are not only costly, but also ecologically damaging. On the same note, land preparation processes such as plowing, tillage, and soil hardening involve the use of tractors and other heavy machines, which rely on fossil fuels, thus increasing the energy demand, and also hastening soil erosion. Synthetic fertilizers and chemical pesticides also contribute to the energy consumption, and their production, transportation, and use result in excessive usage and wastage of these substances using traditional approaches. The energy used in harvesting and post-harvest processing such as threshing, drying, and storage is huge and even more so when the old machinery including inefficient grain dryers and cold storage areas is employed. Also, logistics is fuel-intensive, and transportation and distribution processes were long and inefficient supply chains that increase the use of energy. Combined, these measures indicate the inefficiency of traditional agriculture and the fact that smart farming technologies are necessary to reduce the amount of energy wasted, minimize negative effects on the environment, and provide agricultural sustainable development in the long term. The Figure 5 shows various traditional agricultural practices, from irrigation, land preparation, fertilizer application, harvesting, and transportation, which consume a lot of energy, mainly used by agricultural activities for an efficient operation (Hassan et al., 2024).

Figure 5. Energy-Intensive Processes in Traditional Farming

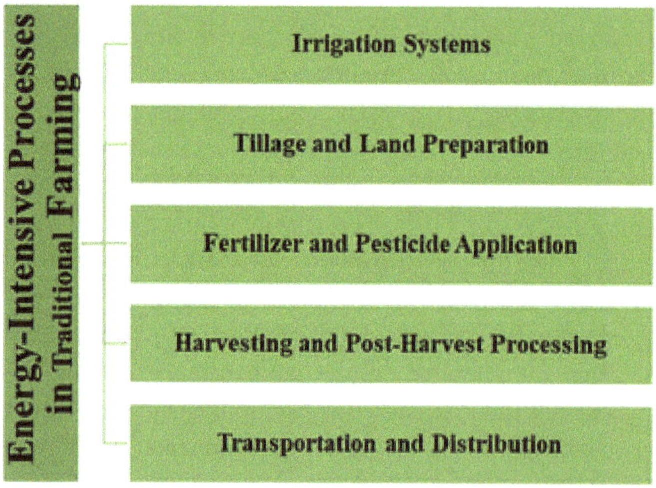

4.2 Comparison of Energy use in Conventional vs. Smart Irrigation Systems

Irrigation is a very important aspect in agriculture and traditional systems usually use a lot of energy as there is no efficiency in the pumping system, the schedule is fixed and the system is also outdated. Smart water irrigation systems are an alternative to this that offers a sustainable option through the optimization of water use and energy conservation, backed by sensors, automation and renewable energy. The article comparing the old and smart irrigation points to the fact that modern methods increase efficiency, reduce the cost, and enhance agricultural sustainability in the long run (Mei et al., 2019).

4.2.1 Water Pumping Efficiency

The conventional irrigation systems employ fixed frequency pumps that can be left running continuously irrespective of their requirement by the soil or crop demands. This not only over waters, but it consumes a lot of energy. Most of the traditional systems have the pumps driven by diesel which is costly, inefficient and harmful to the environment due to massive use of fuel and costs of pollution. In comparison, smart irrigation systems automatically operate by means of the sensors on the pumps and only when necessary. These pumps are calculated in a manner that they are customized to meet the soil moisture and crop needs and reduce pumping wastage. Moreover, there is also the usage of other advanced smart systems which consist of solar-powered and variable-speed pumps also improving the use of energy and reducing fossil-energy consumption. This ensures that the provision of water is precise, there is a reduction in energy usage as well as significant reduction in long term costs (Voutos et al., 2019).

4.2.2 Water Distribution Methods

Flood irrigation system and overhead sprinklers are the other traditional irrigation systems which are very inefficient in water and energy consumption. The approaches spread too much water in uneven patterns in fields and this consumes more pumping power to cover losses and runoff. The outcome is a lost water, lost energy and less productivity of crops. Smart irrigation systems, in their turn, use more selective techniques such as drip irrigation and micro-sprinklers that apply water to the root zone. This accuracy reduces the amount of water required and also decreases the amount of energy used in distribution. These systems ensure efficiency and reduction of resources by applying water only where needed and thereby saving a lot of energy, as well as reducing operational costs (Chandran S, 2021).

4.2.3 Automation and Control

Traditional irrigation has fixed and fixed, non-adjustable timelines on which it does not consider the current situations like the soil moisture or weather fluctuations. This rigidity usually leads to over-pumping, wastage of power as well as mismanagement of water. The limitations are surmounted by smart irrigation systems that incorporate IoT sensors, weather predictions, and AI-controllers. Such technologies automatically change the irrigation schedule based on real time information and only allow pumps to run when needed. Smart systems are very effective at drastically reducing energy consumption and optimizing the use of water by eliminating unnecessary pumping. There is also the possibility of remote monitoring and control by use of mobile applications thus providing farmers the freedom to be more flexible. This smart control of irrigation matches the energy consumption with the real crop demands and enhances efficiency, reduces expenses and encourages sustainable production (Chen et al., 2022).

4.2.4 Integration With Renewable Energy

Traditional irrigation is mostly reliant on grid electricity or pumps which run on diesel which all serve to increase reliance on fossil fuels and increase greenhouse gas emissions. This dependency does not only augment the costs of running but also harm the environment. The smart irrigation systems on the other hand are made in such a way that they can be integrated with the renewable energy sources particularly the solar energy. Solar powered pumps and power saving controllers help save on the use of non-renewable resources and the alternative is a sustainable solution which saves on energy expenditures in the long run. The use of renewables through smart irrigation will reduce the carbon footprint of farming activities, and will ensure the stability of energy supply in remote locations. This change helps in ensuring long term sustainability of the environment besides ensuring that the agricultural practices are resilient to the increasing prices of fuel (Gao et al., 2020).

4.2.5 Maintenance and Lifespan

Conventional irrigation systems are also commonly prone to frequent care since they may be of old equipment, inefficient in their use through machinery and most of them may be worn out after repeated use. These problems enhance the energy consumption and costs of operations in the long run, since inefficient devices can be used to the detriment of less power consumption. Conversely, intelligent irrigation systems are designed using high-level automation and robust parts which alleviate mechanical strain and limit the number of repairs to be undertaken. The automated monitoring also assists in early identifying defects before energy gets wasted because of faulty pumps or systems of distribution. The long-life cycle of the smart devices of modern times will guarantee fewer costs of maintenance and enhanced efficiency. Smart irrigation ensures farmers with a sustainable way of farming by lowering the downtime, saving energy and losing it and the cost is more affordable and efficient. The Figure 6 shows energy used by conventional irrigation systems is compared with energy used by smart irrigation systems according to factors like efficiency of water pumping, modes of distribution, automation, integration of renewable energy, and maintenance. This also explains how smart irrigation improves efficiency and sustainability (Temelkova & Bakalov, 2023).

Figure 6. Comparison of Energy Use in Conventional vs. Smart Irrigation Systems

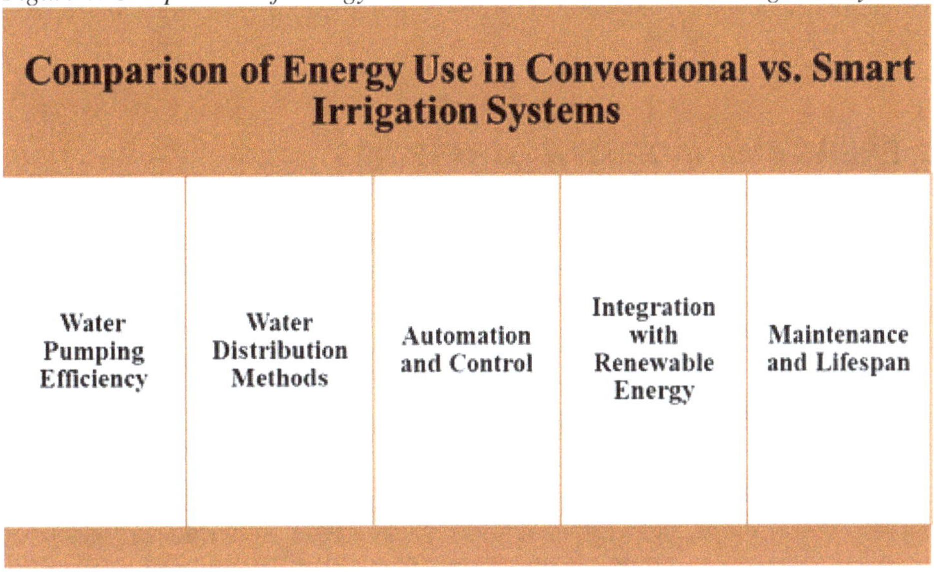

4.3 Potential Energy Savings With Smart Technology in Irrigation

The smart irrigation systems have a huge potential in saving energy through integrating efficient usage of water, automation, and integration of renewable energy. Among the key benefits is the fact that the pumps can be used efficiently to pump water, the moisture sensors on soil and weather information are controlled to ensure that the pumps are only activated when required and not during periods when they are not needed. Such optimization reduces fuel or power consumption by 30-50 over conventional diesel or grid powered pumps. Specific techniques such as drip irrigation, micro-sprinklers also reduce pumping which consumes a lot of energy and it has been found that only 20-40 percent is used compared to flood or overhead irrigation. This is also enhanced by automated scheduling and AI optimization, which optimize the irrigation in real time depending on the soil conditions and climate conditions. This eliminates excess water-wasting and energy consumption by another 15-30. Combining with renewable energy sources, especially solar powered pumps, reduces significantly on the reliance on non-renewable sources, Energy costs may be saved up to 60-80% in remote or off grid locations. In addition to that, higher order smart systems lead to wear and tear minimization on equipment that can guarantee longevity, less breakdowns, and constant energy efficiency. All these innovations make farmers attain an overall energy savings of 30-60 percent, and cause sustainability, reduction of carbon emissions, and preservation of essential resources (Li et al., 2024).

5. ENERGY MANAGEMENT STRATEGIES IN PRECISION FARMING

Precision farming is concerned with use of advanced technologies to improve on energy conservation and sustainable agriculture. The initial good measure is use of renewable sources of energy such as solar powered pumps and wind power that restrict the use of fossil fuels and amount of carbon emission. Irrigation systems that are most effective like drip irrigation system, micro sprinklers and smart controller that detects water use further optimize water consumption in irrigation and cut on the energy needs. Also, the automation based on AI is critical as the analysis of real-time data on the soil, crops, and weather helps to change irrigation schedules, machinery operation, and resource allocation in real-time. Combined, these measures minimize waste of energy, cut down expenses, and encourage sustainable farming, which is responsible to the environment, thus guaranteeing the sustainability of productivity and the ability to withstand climate change over time (H. Wang et al., 2022).

5.1 Use of Renewable Energy Sources (Solar, Wind, Bioenergy)

Covering renewable energy sources significantly reduces dependence on fossil fuels in agriculture. The solar-powered irrigation system uses photovoltaic panels to operate water pumps, reduce power costs and carbon emissions. Wind energy can be used for aeration, water pumping and grain milling, especially in agricultural areas of open fields. Additionally, bio-energy biogas obtained from organic agricultural waste can produce biogas, which provides a permanent energy source for heating, machinery operations and power generation. By integrating these renewable energy solutions, farm can achieve long -term energy savings by reducing environmental impact.

5.2. Energy-Efficient Irrigation Techniques

The actual farming strategy achieves water and electricity conservation through advanced irrigation technology. This irrigation system permits for direct water distribution to plant root structures which results in lower energy prices when pumping water. Soil situations determine water distribution through VRI convertible technique which permits green usage of water. Upgrading irrigation systems powered via grid or diesel power into solar grid structures results in less costly operational costs. Implementation of electricity-professional irrigation systems allows farmers to reduce their strength usage via 30–50% and attain multiplied operation balance along with stepped forward profitability (Quy et al., 2022).

5.3 Automation and AI-Based Decision-Making for Optimized Energy use

AI-powered automation is an actual game-changer for farmers, boosting precision throughout various farming responsibilities via optimizing how energy is used in agriculture. These smart AI sensors keep a consistent eye on conditions, feeding farmers real-time statistics that helps them pleasant-tune irrigation, fertilization, and the operation of their equipment. Drones that fly autonomously and tractors guided by GPS are making fields extra green through locating the most direct routes, saving strength within the process. Farm management tools are getting smarter, too, thanks to machine learning algorithms that analyze past data to create systems that allocate resources more effectively. When AI automation is combined with machine learning, it's a win-win, pushing up productivity while potentially cutting energy waste by a whopping 20 to 40 percent. Efficiently managing a farm's energy resources not only improves the quality of operations but also cuts expenses while being kind to the environment. Farmers can significantly reduce their energy bills by adopting renewable energy sources, combining them with smart, AI-powered control systems, and implementing cutting-edge water management techniques that also increase their yields. Thoughtfully incorporating these methods creates a sustainable, adaptable approach that agricultural businesses can seamlessly weave into their current farming practices (Vitali et al., 2021).

6. IoT AND AI IN ENERGY OPTIMIZATION

Internet of Things (IoT) created with the help of Artificial Intelligence (AI) is changing the energy optimization in contemporary farming. IoT devices (soil moisture sensor, weather stations, smart meters, etc.) allow monitoring the conditions in the fields and the use of energy in real-time. This endless flow of data provides a foundation on which the AI-based systems analyze the patterns, predict the requirements, and make decisions on their own. To mention but not limited, AI can modify the irrigation schedules by combining the weather prediction with the soil and crop data to ensure that the pumps can be turned on only when necessary, avoiding the energy that is wasted due to their idleness. IOT and AI collaborate in improving efficiency and minimizing waste in the operations, making operations more sustainable with energy and resource savings and boosting yields in agriculture (Sadiku et al., 2021).

6.1 Role of IoT in Energy-Efficient Farming

The introduction of IOT technology to farms brings a lot of benefits for energy-smart farming. This is done through the use of smart devices, sensors, and automated control systems. These tools make it possible to closely monitor and manage agricultural activities, leading to better energy use. Soil moisture sensors linked with IOT weather stations work with remote-controlled irrigation systems that effectively track water consumption and energy use. Farms can identify and optimize their energy systems through power usage information obtained from smart meter systems and energy monitoring tools. Real-time data makes the automatic greenhouse climate control system capable of running temperature-based irrigation systems alongside light management thus leading to production improvements while reducing resource usage. The introduction of IOT technology leads to energy savings reaching between 20% to 40% thus lengthening operational sustainability periods and reducing farming operational costs (Gündoğmuş, 2010).

6.2 AI-Driven Predictive Analytics for Irrigation Scheduling

Through Future Analytics AI-based technology intelligent irrigation becomes possible because it analyzes current weather patterns and instant soil states and agricultural water requirements. AI-based approaches develop irrigation plans through large datasets which stop both excessive watering and inadequate irrigation at once. Users who employ predictive analytics within AI-based irrigation models have the ability to modify water distribution and reduce power consumption as well as minimize water waste delivery. Artificial intelligent systems on automatic irrigation controllers allow farmer groups to manage external water systems for delivering target moisture amounts for specific crop needs. AI-based irrigation schedule automation lowers power expenses between 30% to 50% by optimizing resource distribution and stopping extra pumping activities. correct amount of moisture at the right time. AI-Operated irrigation scheduling can cause 30–50% energy savings by reducing excessive pumping and adapting resource usage. The merger of AI with IoT technology enhances energy efficiency in agriculture to prevent losses and support stable existing agricultural systems. The combination of internet technology and automated sensors allows continuous monitoring which leads to automated irrigation planning systems that realize major energy savings benefits. An improvement in agricultural yields results from these implemented technologies which minimize operational expenses while optimizing resource management (von Keyserlingk et al., 2013).

6.3 Case Studies of AI and IoT Applications in Agriculture

The AI and IoT combined have transformed agriculture through the increase of its productivity and sustainability. IoT soil moisture sensors with AI analytics have been used in India to make irrigation systems smart, reducing water and energy consumption by almost a quarter and yielding more. Wearable IoT sensors on cattle in the United States have been adopted as dairy farms to monitor the health and feeding of cattle, and the AI algorithms enhanced the milk production and diseases. Likewise, computerized farm sensors on drones and AI imaging are also being used in vineyards in Europe to forecast infestations by pests and to schedule spraying. These applications show how AI and IoT can change the conventional way of farming to become data driven, efficient, and sustainable (D. Wang & Rajamanickam, 2023).

6.3.1 Precision Irrigation in California Vineyards

The vineyards in California are under a lot of pressure with the challenge of insufficient water, escalating energy expenses, as well as the demand of sustainable farming methods. As a solution to these problems, farmers have implemented AI-based precision irrigation systems that are combined with IoT-based soil sensors. These sophisticated systems can give real time measurements of soil moisture, temperature and humidity and they are able to give accurate information about the water needs of grapevine. Constant checking of the environmental factors will enable farmers to avoid excessive use of irrigation and deliver water only in time and place. The use of AI algorithms is important in processing this data and correlating predication models with site specific observations to produce optimal irrigation schedules. These models consider weather projections, soil differences and crop development phases, which the farmers make decisions ahead of time instead of using set time schedules. With this combination of AI and IoT, irrigation activities become very adaptive and receptive to the changing situations in the field. These have been ground breaking results (J. Kumar et al., 2024). Precision irrigation vineyards have also realized water savings of 30-50 and have greatly lowered the energy use and cost of running the business. The saving of water will not only save the limited resources but also minimize the electricity required to power the pumps. Also, controlled irrigation improves the health of the soil, the quality of grape varieties, which leads to increased productivity and resistance to drought stress. The case emphasizes the transformation of the viticulture in California by smart farming technologies. Farmers are attaining long-term water sustainability, reducing energy expenditure, and enhancing the overall long-term environmental stability without compromising the quality of their yields by integrating IoT-based monitoring with decision-making based on AI. The figure 7 illustrates the key challenges, solutions, and benefits of implementing AI- and IoT-enabled precision irrigation in vineyards, highlighting how smart technologies address water scarcity, energy costs, and sustainability demands.

Figure 7. Precision Irrigation in California Vineyards

6.3.2 AI-Powered Pest and Disease Detection in India

Smallholder farmers in India are prone to huge crop losses because of the late realization of presence of pests and diseases. The traditional approaches are based on manual observation, which is time consuming, subject to error and usually leads to excessive application of chemical pesticides. In a bid to eliminate these difficulties, farmers have embraced AI-based mobile apps that are incorporated with IoT-based systems that detect insects. According to these systems, AI controlled cameras and smart sensors are fitted in the fields itself to constantly check on the health of crops. The AI applications process images and sensor data in real-time to find early evidence of insect attacks, fungal and nutrient deficiencies. When threats have been identified, farmers will receive real time alerts on their cellphones and again they are given a recommended course of treatment depending on the type of problem they have. This is a proactive system whereby the targeted interventions would be done to make sure that the pests and diseases are contained before they become common. Among the most prominent advantages of this method, it is possible to note the reduction of the use of pesticides. With precise identification and location of issues, farmers do not need to spray out the entire field to minimize the use of chemicals by up to 25 percent. This does not only reduce the cost of inputs but it reduces the environmental impacts of farming practices as well. Besides, the farming has been more fruitful and present in larger amounts through healthier crops and more precise interventions. An example of how digital technologies can give farmers reliable and opportune information-based decisions is the implementation of AI and IoT-capable pest detection technologies in India. These innovations are boosting the sustainable agricultural growth and stabilizing the livelihoods of the farmers due to the fact that they are decreasing the level of using pesticides, lowering the costs, and making their crops more tough. The figure 8 indicates that AI-based mobile applications and Internet of Things can enhance the detection of pests and diseases, using less pesticide, increasing productivity, and minimizing environmental effects over traditional techniques.

Figure 8. AI-Powered Pest and Disease Detection in India

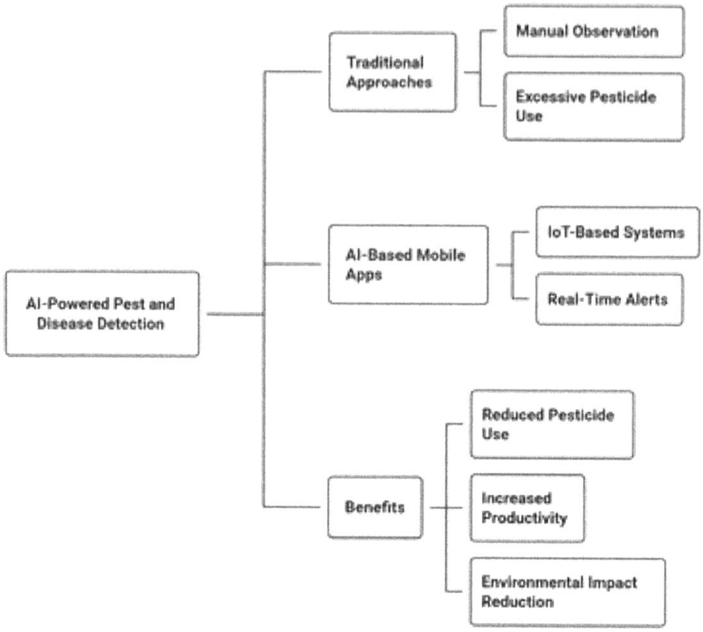

7. CHALLENGES AND LIMITATIONS

Even though smart irrigation structures provide lots of blessings for specific agriculture, they're suffering to seize on broadly. This is in particular due to a bunch of hurdles in in reality setting them to apply. The big prematurely charges, at the side of technical limitations and a loss of basic infrastructure in a few regions, are primary roadblocks. On top of that, concerns approximately records private and security upload every other layer of complexity that makes farmers hesitant to adopt those systems. The Table 7 presents the key challenges of smart irrigation systems high costs, infrastructure gaps, and data security along with possible solutions to address them.

Table 1. Challenges and Limitations of Smart Irrigation

Challenge	Description	Solution
High Costs	Large upfront investment in IoT, AI, and automation; slow ROI.	Subsidies, grants, and financial support.
Infrastructure Gaps	Poor internet, electricity, and lack of farmer training in rural areas.	Better connectivity, training, and affordable tech.
Data Security	Risk of cyberattacks, unclear data ownership, privacy concerns.	Strong encryption, clear policies, cybersecurity.

7.1 High Initial Investment Costs

High-end devices require a large amount of money to start with; that is the primary obstacle on the way of adopting specific farming and installing smart watering systems. Combining internet-related sensors, AI-controlled records system, automatic irrigation, and amateurish energy solutions needs an enormous untimely financial investment. It regularly provides and government subsidies that place the company among big and medium sized farmers and these innovative methods. The cost of those investments will not be recouped until many years later hence farmers with skinny wallets find it difficult to show that the cost is worth the payoff until they receive the payoff at the end of the period of waiting to get the returns on their investments (Araújo et al., 2021).

7.2 Technical and Infrastructure Barriers

People dwelling in rural regions and growing nations often war to farm efficaciously because of a loss of strong infrastructure and advanced technical understanding. Many farmers find it tough to get entry to each reliable internet and energy, which are vital for modern-day farming strategies. The hassle is compounded via the reality that farmers and agricultural people frequently lack the training needed to effectively use AI and IoT-primarily based systems. While education applications, at the side of help offerings, can assist bridge this expertise gap, they also increase the general fee and complexity of implementation (Quy et al., 2022).

7.3 Data Security and Privacy Concerns

Accurate farming and smart irrigation structures produce important agricultural data, inclusive of records on soil, weather, crop health, and farming practices. Storing this information on cloud-based structures could make it vulnerable to cyber-assaults, unauthorized access, and facts breaches, ultimately compromising its security. Many farmers and agricultural companies are hesitant approximately who owns their statistics, disturbing that it is probably misused through outdoor parties or that information privacy guidelines are too doubtful. To deal with those security issues, it is essential to put into effect strong information encryption and cybersecurity measures, as well as clear rules outlining facts possession. Although precise agriculture and clever irrigation offer massive blessings in dealing with electricity and adjusting to useful resource availability, it is vital to address the hurdles they gift, like steep charges, barriers in infrastructure, and information protection issues. Investing in economic help, higher connectivity, technical education, and expanded cyber security measures could be important in overcoming these limitations and making sure lengthy -time period fulfillment of smart agriculture (Bwambale et al., 2023).

8. FUTURE TRENDS AND INNOVATIONS

As the call for durable and efficient agricultural practices increases, correct farming and smart irrigation systems maintain to increase. Future improvement will recognition on rising power-skillful technologies, assistant authority's guidelines and the function of enlargement of clever farming in global food manufacturing. Table 2 outlines upcoming technologies, policies, and prospects shaping the future of smart farming and irrigation.

Table 2. Future Trends and Innovations in Smart Farming

Focus Area	Key Innovations
Emerging Technologies	Solar-powered irrigation, AI automation, IoT monitoring, WSNs, drones, and robotic equipment.
Government Policies	Subsidies, tax incentives, carbon credits, and R&D for sustainable farming.
Future Prospects	Blockchain supply chains, 5G-enabled monitoring, vertical farming, hydroponics, and fully autonomous ecosystems.

8.1 Emerging Technologies in Energy-Efficient Agriculture

Progress under renewable energy-operated agricultural systems, AI-operated automation and IOT-based monitoring is bringing revolution in energy management in agriculture. The solar-powered irrigation system is becoming more accessible, reducing the dependence on grid power and fossil fuels. AI and machine learning algorithms are optimizing irrigation schedules, reducing energy waste, and improving crop yields. Wireless Sensor Networks (WSNs) and drones are increasing real -time data collection, which can lead to accurate water and energy. Additionally, robotic agricultural equipment operated by AI and autonomous technology is expected to streamline agricultural operations by reducing energy consumption (Hussein et al., 2024).

8.2 Government Policies and Incentives for Sustainable Farming

Many governments worldwide are starting policies and incentives to promote energy-efficient and climate-smart agriculture. Solar-operated irrigation systems, smart sensors, and subsidies for AI-operated farm management are encouraged farmers to adopt permanent practices. Governments are also investing in Agricultural Research and Development (R&D) to increase smart agricultural technologies. In addition, carbon credit and tax incentive to adopt low carbon and energy-efficient agricultural techniques are expected to run further innovation in sustainable agriculture (Getahun et al., 2024).

8.3 Future Prospects of Smart Farming

The future of smart farming lies in a completely integrated and autonomous agricultural ecosystem. The AI-manufacturing will further increase the decision-making analytics, which will ensure optimal resource allocation with minimal environmental impact. Blockchain technology supply chain will improve transparency, secure form data and transactions. The 5G network expansion will enable rapid and more reliable data transfer by improving real -time monitoring and automation. Additionally, vertical farming and hydroponic systems will provide energy-efficient solutions for urban agriculture, reduce the use of land and water. This progress will make accurate farming more accessible, scalable and environmentally durable, which will ensure food security for the growing global population. The future of accurate farming and smart irrigation systems is inspired by increasing loudly on technological progress, government support and stability. As energy-skilled innovations continue to develop, smart farming will become the cornerstone of modern agriculture, which will reduce productivity and profitability and decrease in resource consumption (Chel & Kaushik, 2011).

9. CONCLUSION

The factors that will transform the modern vision of agriculture are precision farming and smart irrigation systems that will help to eliminate a significant number of problems related to resource inefficiency, the excessive consumption of energy, and the devastation of long-established environments. The traditional agriculture systems that have typically relied on the use of diesel-powered farming machinery, flood irrigation, and the excessive use of chemicals not only increased the price of production, but also the destruction of the ecosystem. Precision farming, in its turn, can be regarded as one of the examples of how the application of the latest technologies, such as IoT sensors, GIS mapping, drones, AI-powered analytics, and Variable Rate Technology (VRT) can simplify the process of agricultural activities, make them more efficient, reliable, and adaptive to the changes in weather conditions. The use of smart irrigation makes its way to the list of significant elements of this ecosystem.

Smart irrigation system helps farmers save time and energy and reduce wastage of water by relaying water directly to the field based on soil moisture sensors, weather forecasting and AI algorithms. Drip irrigation, micro-sprinklers, automated and solar-powered pumps will assist in further saving of energy since they will ensure that crops receive as little water as they need. They not only make operations cheap, but also augment crop yields, drought stress and soil fertility. The success in the case studies in other regions of the world such as California and India indicate a real-life success wherein up to 50 percent of the consumption of water and energy has been reduced, and the productivity and profitability have been increased. Though all these benefits exist, there are still some problems. It has a high initial cost of investment, lack of infrastructure in rural areas, and privacy and safety of data is an issue that makes its adoption harsher. The small and medium farmers also have the financial and technical limitations of taking up such sophisticated systems. These challenges can only be overcome through a concerted effort by the government, policy-makers as well as technology providers.

The subsidies, training, enhanced rural connectivity and strong data security systems are needed as critical processes of providing equitable access to smart farming systems and permanent uptake of the systems. The future of agriculture will be predetermined by the new innovations such as AI-managed automation, the blockchain-based supply chain, integration with renewable energy sources, and tracking in real time imposed by 5G in the future. These inventions together with positive policies by the government and incentives will drive precision farming to both world food security needs and reducing the carbon footprint of the industry. In general, sustainable agricultural practices are in precision agriculture and the intelligent irrigation systems. Through their ability to be productive and at the same time be environmentally responsible, they offer a path to sustainable, energy efficient, and environmental conscious food systems to the future generation.

REFERENCES

Adewusi, A. O., Asuzu, O. F., Olorunsogo, T., Olorunsogo, T., Adaga, E., & Daraojimba, D. O. (2024). AI in precision agriculture: A review of technologies for sustainable farming practices. *World Journal of Advanced Research and Reviews, 1*(1), 2276–2285. DOI: 10.30574/wjarr.2024.21.1.0314

Alazzai, W. K., Abood, B. S. Z., Al-Jawahry, H. M., & Obaid, M. K. (2024). Precision Farming: The Power of AI and IoT Technologies. E3S Web of Conferences, 491, 04006. DOI: 10.1051/e3sconf/202449104006

Araújo, S. O., Peres, R. S., Barata, J., Lidon, F., & Ramalho, J. C. (2021). Characterising the Agriculture 4.0 landscape—Emerging trends, challenges and opportunities. *Agronomy (Basel), 11*(4), 667. DOI: 10.3390/agronomy11040667

Bwambale, E., & Abagale, K. F., & K. Anornu, G. (2023). Smart irrigation for climate change adaptation and improved food security. In Irrigation and Drainage - Recent Advances. IntechOpen. DOI: 10.5772/intechopen.106628

Chandrakala, A. (2021). Iot based precision farming and agriculture - aspects and technologies. *The Philippine Statistician (Quezon City), 70*(2), 1426–1433. DOI: 10.17762/msea.v70i2.2335

Chandran, S. N. (2021). Smart farming: Implementation of IoT-based technology for crop monitoring and management. *The Philippine Statistician (Quezon City), 70*(2), 1383–1392. DOI: 10.17762/msea.v70i2.2330

Chel, A., & Kaushik, G. (2011). Renewable energy for sustainable agriculture. *Agronomy for Sustainable Development, 31*(1), 91–118. DOI: 10.1051/agro/2010029

Chen, Z., Liao, Z., Qian, D., & Li, J. (2022). Design and analysis of intelligent agricultural monitoring system based on biological intelligence optimization algorithm. *Mathematical Problems in Engineering, 2022*, 1–8. DOI: 10.1155/2022/1129419

Gao, D., Sun, Q., Hu, B., & Zhang, S. (2020). A framework for agricultural pest and disease monitoring based on Internet-of-Things and Unmanned Aerial Vehicles. *Sensors (Basel), 20*(5), 1487. DOI: 10.3390/s20051487 PMID: 32182732

Getahun, S., Kefale, H., & Gelaye, Y. (2024). Application of precision agriculture technologies for sustainable crop production and environmental sustainability: A systematic review. *TheScientificWorldJournal, 2126734*(1), 2126734. Advance online publication. DOI: 10.1155/2024/2126734 PMID: 39421732

Gopathoti, K. K. (2023). Precision agriculture and the role of machine learning. International Journal of Scientific Methods in Intelligence Engineering Networks, 01(06), 01–08. DOI: 10.58599/IJSMIEN.2023.1601

Gündoğmuş, E. (2010). Energy input use in environmentally friendly farming systems: A comparative analysis between organic and conventional dried fig production. *Journal of Sustainable Agriculture*, *34*(7), 744–757. DOI: 10.1080/10440046.2010.507524

Hassan, E. S., Alharbi, A. A., Oshaba, A. S., & El-Emary, A. (2024). Enhancing smart irrigation efficiency: A new WSN-based localization method for water conservation. *Water (Basel)*, *16*(5), 672. DOI: 10.3390/w16050672

Hussein, A. H. A., Jabbar, K. A., Mohammed, A., & Jasim, L. (2024). Harvesting the Future: AI and IoT in Agriculture. E3S Web of Conferences, 477, 00090. DOI: 10.1051/e3sconf/202447700090

Ibrahim Mohammad Abuzanouneh, K., & Al-Wesabi, N., F., Abdulrahman Albraikan, A., Al Duhayyim, M., Al-Shabi, M., Mustafa Hilal, A., Ahmed Hamza, M., Sarwar Zamani, A., & Muthulakshmi, K. (2022). Design of machine learning based smart irrigation system for precision agriculture. *Computers, Materials & Continua*, *72*(1), 109–124. DOI: 10.32604/cmc.2022.022648

Issa, A. A., Majed, S., Abdul Ameer, S., & Al-Jawahry, H. M. (2024). Farming in the digital age: Smart agriculture with AI and IoT. E3S Web of Conferences, 477, 00081. DOI: 10.1051/e3sconf/202447700081

Kumar, J., Patel, N., Sahoo, P. K., Sudhishri, S., Yadav, R., Vibhute, S. D., & Singh, A. K. (2024). Development of sensor and decision support system (DSS)-based automated irrigation system for enhancing water productivity of tomato crop. *Journal of Soil Salinity and Water Quality*, *16*(2), 307–316. DOI: 10.56093/jsswq.v16i2.156426

Kumar, V., Singh, C. D., Rao, K. V. R., Kumar, M., & Rajwade, Y. A. (2023). Development of a smart IoT-based drip irrigation system for precision farming. *Irrigation and Drainage*, *72*(1), 21–37. DOI: 10.1002/ird.2757

Li, L., Min, X., Guo, J., & Wu, F. (2024). The influence mechanism analysis on the farmers' intention to adopt Internet of Things based on UTAUT-TOE model. *Scientific Reports*, *14*(1), 15016. DOI: 10.1038/s41598-024-65415-4 PMID: 38951536

Lloret, J., Sendra, S., García-Fernández, J., García, L., & Jimenez, J. M. (2021). A WiFi-based sensor network for flood irrigation control in agriculture. *Electronics (Basel)*, *10*(20), 2454. DOI: 10.3390/electronics10202454

Mandal, S. (2013). Precision farming for small agricultural farm: Indian scenario. *American Journal of Experimental Agriculture, 3*(1), 200–217. DOI: 10.9734/AJEA/2013/2326

Mei, Z., Feng, C., Ding, W., Zhang, L., & Wang, D. (2019). Better lucky than rich? Comparative analysis of parking reservation and parking charge. *Transport Policy, 75*, 47–56. DOI: 10.1016/j.tranpol.2019.01.001

Oluwatoyin, K. K., Ajibola, I. N., & Ibrahim, A. (2021). Design of deep learning system for agricultural purpose. *International Journal of Communication and Information Technology, 2*(1), 30–35. DOI: 10.33545/2707661X.2021.v2.i1a.26

Pan, S.-H., Chen, W.-E., Chen, W.-S., Lin, H.-H., & Chen, C.-H. (2023). Smart sprinkler system on smart campus. *Sensors and Materials, 35*(12), 4105. DOI: 10.18494/SAM4591

Pang, Y., Marinello, F., Tang, P., Li, H., & Liang, Q. (2023). Bibliometric analysis of trends in smart irrigation for smart agriculture. *Sustainability (Basel), 15*(23), 16420. DOI: 10.3390/su152316420

Pingda, W., Wang, Z., & Zhong, H. (2022). The impact of agricultural information system on agricultural product trading efficiency using Internet of Things technology. *Mathematical Problems in Engineering, 2022*, 1–11. DOI: 10.1155/2022/4061908

Quy, V. K., Van Hau, N., Van Anh, D., Quy, N. M., Ban, N. T., Lanza, S., Randazzo, G., & Muzirafuti, A. (2022). IoT-enabled smart agriculture: Architecture, applications, and challenges. *Applied Sciences (Basel, Switzerland), 12*(7), 3396. DOI: 10.3390/app12073396

Quy, V. K., Van Hau, N., Van Anh, D., Quy, N. M., Ban, N. T., Lanza, S., Randazzo, G., & Muzirafuti, A. (2022). IoT-enabled smart agriculture: Architecture, applications, and challenges. *Applied Sciences (Basel, Switzerland), 12*(7), 3396. DOI: 10.3390/app12073396

Rahman, S., & Barmon, B. K. (2012). Energy productivity and efficiency of the 'gher' (prawn-fish-rice) farming system in Bangladesh. *Energy, 43*(1), 293–300. DOI: 10.1016/j.energy.2012.04.027

Raliya, R. (2024). Artificial intelligence for precision and sustainable agricultural. *ACS Agricultural Science & Technology, 4*(6), 628–630. DOI: 10.1021/acsagscitech.4c00290

Sadiku, M. N. O., Ashaolu, T. J., Ajayi-Majebi, A., & Musa, S. M. (2021). Internet of Things in Agriculture: A Primer. *International Journal Of Scientific Advances, 2*(2). Advance online publication. DOI: 10.51542/ijscia.v2i2.24

Sharma, A., Jain, A., Gupta, P., & Chowdary, V. (2021). Machine learning applications for precision agriculture: A comprehensive review. *IEEE Access : Practical Innovations, Open Solutions*, *9*, 4843–4873. DOI: 10.1109/ACCESS.2020.3048415

Suhardi Rahman, E., Yusuf Mappeasse, M., & Fajryanisari, I. (2023). Development of automatic irrigation system with solar energy-based telegram notification. *Journal of Electrical Engineering and Informatics*, *1*(1), 26–34. DOI: 10.59562/jeeni.v1i1.421

Temelkova, M., & Bakalov, N. (2023). A model of a cyber-physical installation for smart greenhouse agriculture. E3S Web of Conferences, 404, 02004. DOI: 10.1051/e3sconf/202340402004

Vitali, G., Francia, M., Golfarelli, M., & Canavari, M. (2021). Crop management with the IoT: An interdisciplinary survey. *Agronomy (Basel)*, *11*(1), 181. DOI: 10.3390/agronomy11010181

von Keyserlingk, M. A. G., Martin, N. P., Kebreab, E., Knowlton, K. F., Grant, R. J., Stephenson, M., Sniffen, C. J., Harner, J. P.III, Wright, A. D., & Smith, S. I. (2013). Invited review: Sustainability of the US dairy industry. *Journal of Dairy Science*, *96*(9), 5405–5425. DOI: 10.3168/jds.2012-6354 PMID: 23831089

Voutos, Y., Mylonas, P., Katheniotis, J., & Sofou, A. (2019). A survey on intelligent agricultural information handling methodologies. *Sustainability (Basel)*, *11*(12), 3278. DOI: 10.3390/su11123278

Wang, D., & Rajamanickam, L. (2023). Effects of AI on smart agriculture: A case study of Digital Agriculture Base. In *Frontiers in Artificial Intelligence and Applications*. IOS Press., DOI: 10.3233/FAIA230785

Wang, H., Zhao, Y., & Shao, C. (2022). IoT for agricultural information generation and recommendation: A deep learning-based approach. *Mobile Information Systems*, *2022*, 1–9. DOI: 10.1155/2022/4602998

Zhang, P., Chen, X., Li, S., Zhang, C., & Hu, Y. (2023). Development of the internet of smart orchard things based on multi-sensors and LoRa technology. *Intelligent and Converged Networks*, *4*(4), 342–354. DOI: 10.23919/ICN.2023.0028

Zhou, S., Sun, H., Bi, J., Zhang, J., Riya, S., & Hosomi, M. (2020). Effect of water-saving irrigation on the N2O dynamics and the contribution of exogenous and endogenous nitrogen to N2O production in paddy soil using 15N tracing. *Soil & Tillage Research*, *200*(104610), 104610. DOI: 10.1016/j.still.2020.104610

KEY TERMS AND DEFINITIONS

AI (Artificial Intelligence): Computer algorithms that analyze farm data, predict risks, and automate tasks like irrigation and pest control.

Energy Management in Agriculture: Smart use of renewable energy and technology to reduce waste, cut costs, and ensure eco-friendly farming practices.

GIS (Geographic Information System): Mapping technology that helps farmers analyze soil, crop health, and land use for better resource distribution.

IoT (Internet of Things): Network of sensors and devices that collect real-time data on soil, weather, and crops to support smart decisions.

Precision Farming: Modern farming that uses GPS, IoT, drones, and AI to give crops exactly what they need, improving yield, saving costs, and protecting the environment.

Smart Irrigation: Automatic watering system using sensors, weather data, and AI to supply water only when required, saving water and energy while boosting crop health.

VRT (Variable Rate Technology): Tools that apply seeds, fertilizers, or water in just the right amount, at the right place, reducing waste.

Chapter 9
Precision Farming and Smart Irrigation Systems for Energy Management

Tushar
https://orcid.org/0000-0001-7190-7687
United College of Engineering and Research, Prayagraj, India

Prabhat Chandra Shrivastava
https://orcid.org/0000-0003-0766-6422
University of Allahabad, Prayagraj, India

Nandita Pradhan
https://orcid.org/0009-0006-8819-1009
United College of Engineering and Research, Prayagraj, India

Manish Kumar
https://orcid.org/0009-0007-0735-0256
P.K. University, India

Pooja Jaiswal
https://orcid.org/0009-0007-2507-1525
University of Allahabad, Prayagraj, India

Shyam Sundar
https://orcid.org/0009-0007-2507-1525
B.C.S. Prayagraj, India

ABSTRACT

A profound transformation in agriculture is being realized through the widespread adoption of precision farming and smart irrigation systems, leading to a fundamental enhancement in energy management. This chapter provides an insight into modern, smart irrigation technologies, including soil moisture sensors, weather-based controllers, and Internet of Things (IoT) based systems, all of which have been developed to advance water management through improved accuracy in irrigation. Recent advancements have highlighted the crucial role of ICT technologies in fostering the development of smart and sustainable agriculture. Within this context, an overview of contemporary smart irrigation software solutions, including Dacom Irrigation Advice and Irris Scheduling Software, is provided in this chapter. Additionally, the

DOI: 10.4018/979-8-3373-3146-1.ch009

SWAMP project's view, architecture, pilots, and its scenario-based development process are presented. The microgrid is overseen by an energy management system (EMS), the foundation of which is model predictive control (MPC).

1. INTRODUCTION

Water shortage is still one of the key global sustainability problems, and it is especially notable in agriculture an extremely water-dependent domain. Conventional irrigation systems are known to suffer from an excessive amount of water applied, poor discharge control and timing, translating in substantial waste of water and less than satisfactory crop yields. To solve these issues that exist until now, smart irrigation systems are the game changing jump in agriculture and landscape management. Leveraging Information and Communication Technologies (ICT), Artificial Intelligence (AI) in the form of Machine Learning (ML) and Deep Learning (DL), Smart irrigation systems are intended for precision-based, data-driven and energy efficient irrigation with respect to sustainable food production and conservation of natural resources. Then, we posit that the intersection of AI-enabled analytics and IoT-facilitated sensing networks, together with energy autonomous adaptive systems is paving the way for future smart agriculture. Together, these technologies enable the monitoring of soil, weather and crop status in real time so farmers can automate irrigation timing and achieve greater water and fertilizer efficiency – all to increase productivity. Furthermore, AI driven decision support systems and self-learning algorithms enable smart irrigation networks to dynamically respond to weather changes and soil heterogeneity, so as to provide precision agriculture in resource-deprived scenarios.

This chapter discusses development of irrigation practices from traditional to smart and precision irrigation technologies. Discusses the potential of ICT and automation to improve water use efficiency. This is depicted by studies such as (Brajovic et al., 2015) and (Chaudhry & Garg, 2019). Inspired by (Haque et al., 2021) and (Sharma et al., 2021), it shows that AI not only increases efficiency but also contributes to the achievement of worldwide sustainability goals set global institutions such as the United Nations' Sustainable Development Goals (SDGs), namely, Goal 2 (Zero Hunger) and Goal 12 (Responsible Consumption and Production). In the following, this chapter presents the new innovative concept of energy-autonomous adaptive systems as an important step forward within the framework of Industry 5.0. Unlike the Industrial 4.0 that focused on automation, the Industrial 5.0 highlights human-machine collaboration, as well as resiliency and sustainability (Leng et al., 2022; Huang et al., 2022). This section shows how the combination of IoT devices powered with renewable energy, self-sustained microcontrollers and smart energy management systems keeps irrigation ongoing in hard to reach or off-grid areas. Finally, the chapter ends discussing how these technologies help improve economic, social environmental sustainability of agri-food supply chains (AFSCs) as a whole. Fundamentally, this chapter articulates AI- and IoT-enabled smart irrigation as a vital backbone technology of precision farming-ready future—a domain where data intelligence, mechanization crop together converge towards turning 'agriculture' into a more sustainable, productive and resilient sector.

Information and Communication Technology (ICT) is the core of smart and sustainable agriculture. Information, Communication, and Technology (ICT) solutions assist policymakers in removing unintended beneficiaries who have been availing subsidies amounting to 3.17 lakh crore (as per the Indian Budget Estimates of 2022-23), while also ensuring the timely transfer of social security benefits (Tushar et al., 2025). Sustainable water supply and expansion of irrigated area are needed for feeding the increased world population (Brajovic et al., 2015). The use of ICT and automation technologies in water supply and irrigation management, increases water/musicianship quotient/productivity as well farmer's income (Brajovic et al., 2015) stress. Plants irrigation, growth and chemistry exchange are observed by more sensors (Aminpour et al., 2023).

An apparent revolution is taking place in the agriculture field, moving from historically based practices to modern data-driven and technology-enabled approaches associated with the evolution of artificial intelligence. The smart agriculture market is expected to grow at a CAGR of 2.2% between 2023 and 2028, according to Grand View Research's predictions. That will translate into the market size growing from $20.12 billion to an estimated $23.14 billion by 2023. This strong growth is a reflection of the enthusiastic acceptance of new technologies in agriculture as reported by (Haque et al., 2021). AI's promotion in agriculture is accompanied by many opportunities and is adopted quickly because of its rational decision-making ability and precision farming.

The agricultural industry is being influenced in multifaceted and widespread ways by AI. AI's potential to boost agricultural productivity by an impressive 70% by 2050 was highlighted by a 2019 study by McKinsey & Company. These technological strides enable farmers to optimize fertilizer application, facilitate same-day identification of plant diseases, enhance irrigation efficiency, and base critical decisions on concrete data (Sharma et al., 2021). Beyond direct productivity a gain, a crucial role in mitigating the adverse environmental impacts of farming is also played by AI (Saetra H. S., 2021).

The core objective of this project is to quantify the tangible benefits that can be delivered by AI in agricultural productivity. This includes reductions in labor costs, increases in crop yields, and conservation of vital resources such as water and fertilizers. Furthermore, the capacity of AI to offer evidence-based guidance for improving irrigation strategies, fertilizer application, and pest management is aimed to be investigated by the project. Deep learning, a cutting-edge technology, offers diverse techniques for both image processing and big data analysis. Positive outcomes have consistently been yielded by this approach, demonstrating significant promise across various sectors, notably within the farming and agriculture industries (Leng et al., 2022; Xu et al., 2021; Wollschlaeger et al., 2017). Even during the Industry 4.0 era, the crucial need for human oversight in critical information flows and active participation in future production management was argued for by experts (Turner et al., 2021). Industry 5.0 takes this idea further, stressing the collaborative synergy between humans and sophisticated technologies (Huang et al., 2022). This approach combines "the best of both worlds," leveraging automation, AI, and robotics while simultaneously enhancing human creativity, problem-solving skills, and adaptability.

Numerous studies have explored methods for detecting and mitigating Distributed Denial-of-Service (DDoS) threats, particularly those targeting HTTP services. A system leveraging HTTP to counteract producer-separated DDoS attacks was introduced by (Sivagnanapalani et al., 2021), employing a multi-layer perceptron genetic algorithm (MLPGA). Threats were identified by their method based on four distinct parameters. A distinction was made between legitimate users, who browse website pages and move between them, and simple clients accessing specific periods. A framework integrating diverse symbol properties for DDoS validation was proposed by (Opoku et al., 2020). Their work introduced an AI model that could be acquired by security investigators to understand hazard patterns. The DDoS threat data for their tests was extracted from the CICIDS 2017 (Intrusion Detection Evaluation Dataset). The model was developed using the Weka tool, a Java-based data mining and AI utility. Data preprocessing was managed with Weka's supervised attribute filter, and a C4.5 decision tree was utilized for organizing the preparation phase. Additionally, a separate procedure was designed based on an AI methodology to pinpoint HTTP DDoS threats and distinguish botnets from legitimate customers by identifying threat traffic versus bona fide traffic. This procedure was conceptualized as an intermediary, directly evaluating client behavior rather than scrutinizing the entire traffic flow. With the global population projected to reach nine billion by 2050, a substantial increase in food production will be necessitated (Plazzotta et al., 2020).

Collaborative action from various stakeholders, including policymakers, civil society, and businesses, is necessitated by the SDGs to facilitate a global transition toward greater sustainability. The framework specifically addresses AFSCs by highlighting both the primary challenges within food systems and the broader impacts of agri-food value chain activities (Sarwar et al., 2021). A significant objective is to sustainably feed nine billion people by mid-century, especially given that global food production already utilizes 25% of all habitable land, accounts for 70% of freshwater consumption, and is a major contributor to biodiversity loss.

Although literatures in some cases often quote rough numbers or example based percentage figures to illustrate the benefits of smart IWS, a lack of common quantitative comparisons among them constrains the depth for comparison analysis. A more quantitative assessment may be implemented based on numerical models of energy consumption, and WUE, for traditional to sensor and AI-based irrigation systems.

Formulas for Efficiency Metrics

i. Water-Use Efficiency (WUE):

$$WUE = Y/W \qquad (1)$$

Where *Y*=crop yield (kg/ha) and *W*=total water applied (m^3/ha). A higher WUE value indicates greater productivity under constrained water resources.

ii. Energy-Use Efficiency (EUE):

$$EUE = Y/E \qquad (2)$$

Where *E*=total energy consumed (kWh/ha). This represents the yield output per unit of energy consumed.

iii. Energy Efficiency Ratio (η):

$$\eta = (E_{useful}/E_{input}) \times 100 \qquad (3)$$

Where $E_{(useful)}$ is the effective energy utilized in irrigation processes (pumping, sensing, and control) and $E_{(input)}$ is total input energy from all sources.

Table 1. Estimated average performance metrics per hectare, compiled from synthesized research data and simulation-based models

Irrigation Type	Average Energy Consumption (kWh/ ha/season)	Water Applied (m^3/ha)	Crop Yield (tons/ha)	Water-Use Efficiency (WUE, kg/ m^3)	Energy-Use Efficiency (EUE, kg/ kWh)
Traditional (Flood/Canal)	1200	9000	6	0.67	5
Sensor-Based (Soil Moisture/ Drip)	800	6000	6.8	1.13	8.5
AI-Driven (Predictive + IoT Automation)	500	4500	7.5	1.67	15

2. PRECISION FARMING

This chapter presents the case that the convergence of Artificial Intelligence (AI) and Internet of Things (IoT) is turning farms from being passive production centers to active, cognitive ecosystems that are capable of self-sufficiently managing energy-water-food nexus. It characterizes autonomy as the capacity of agriculture systems to make decisions with very little human control, and adaptation as a form of learning from feedback in the environment. Such a framework with five levels, L1 (human assisted) to L5 (fully automated), categorizes systems like UAVs, AI-driven irrigation and Model Predictive Control (MPC) based on the degree of autonomy achieved. Performance An energy management analysis reports system efficiency, energy ROI and sustainability improvements.

The Figure 1 demonstrates a Web Service Attack; more precisely, we always consider an attack of type Denial-of-Service (DoS). This HTTP client bombards the web service provider with too many HTTP requests and consumes all of server capabilities. The result is that the system gets saturated and cannot serve user demands which are valid. This result in a service outage or severe latency, genuine users cannot use the website or service. Attacks like these take advantage of the constraints faced by a server – be it bandwidth, memory or computing power to interrupt everyday usage. The red X indicates frustration of legitimate traffic, highlighting the negative effect that steady-state malicious requests have on overall service availability and reliability.

Figure 1. Functionality of HTTP threat on Web services

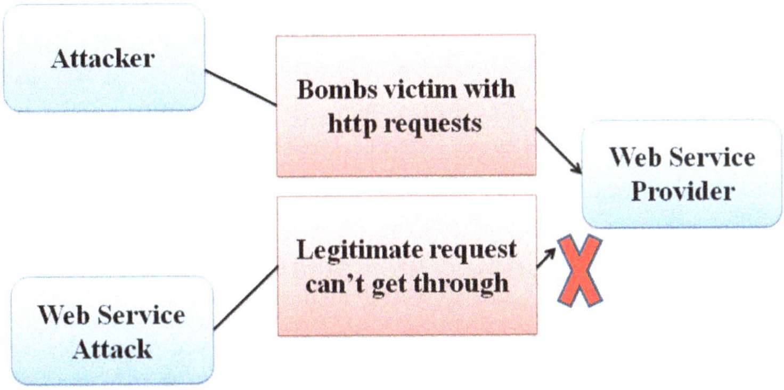

The critical components of precision farming are shown in Figure 2.

(i) GPS in Precision Farming

The GPS system is a key technology for the precision farming, it helps collect accurate data of mapping water distribution networks, fields and roads. Farmers can identify areas of special attention for critical processes such as soil testing, and allow multiple driving of tractor simultaneously.

(ii) GIS Technology in Precision Farming

Geospatial data is used for GIS to generate accurate digital maps, and enable farms to be divided into defined regions by such variables as soil type, moisture level and pest levels. GIS supports visualization and analysis of soil survey information and crop conditions available to the farmers. The decision-making is also assisted through merging satellite imagery and aerial photographs with the use of GIS (Graphical Information System) which allows comparison of information and manipulation of data in different farm management applications.

(iii) UAVs and Dusters in Precision Farming

Unmanned Aerial Vehicle (UAV) technology is changing the face of crop management, with farmers able to scout fields from afar. However, when it comes to watering and planting, a more efficient mechanism would be drones with sophisticated imaging tools such as multispectral- and thermal-analysis. These drones save fuel and can be operated from a distance, giving valuable field information with minimal wear and tear of equipment. In addition, the use of UAVs will enable precision spraying of fertilizers and pesticides that targets only the affected or desired areas while avoiding chemical waste. It leads to sustainability of the environment and a lower cost for the farmer. The information gathered by these UAVs / drones may be used for crop health monitoring, predicting yield and early detection of plant diseases or infestations, thus allowing preventive actions leading to major increases in agriculture productivity. For this reason, the adoption of UAVs for agricultural tasks can be considered as an essential step towards modern and sustainable agriculture.

(iv) Satellite Remote Sensing in Precision Farming

Remote sensing satellite offers a powerful tool for precision agriculture. Farmers are capable of imaging products that can help determine the health and potential yield of crops. This tech provides real-time data on moisture, disease detection, and nutrient levels that can help farmers make better decisions

(v) VRT in Precision Agriculture

VRT (Variable Rate Technology) enables fertilizers, chemicals and seeds to be applied according to specific needs of various parts of a field. They do so through the application of a particular Global Positioning System (GPS) and bespoke software. Visible and near-infrared spectroscopy can be divided into the map-based system (where applications are modified according to pre-established maps) and the sensor-based system (in which soil conditions are determined during execution). This technology ends up helping the farmer to optimize use of resources, reduce input costs and keeping the environment sustainable.

Figure 2. Critical components of precision farming

(vi) Field Monitoring and Mapping in Precision Farming

For effective field monitoring and mapping to be achieved, a variety of environmental parameters, including weather conditions and landscape peculiarities, must be taken into consideration. Detailed field maps are developed with the assistance of data from satellites and global positioning systems.

(vii) Micro Irrigation

Water use can be maximized by farmers with the use of micro-irrigation devices, which allow locations with different degrees of soil moisture to be pinpointed. Variable rate irrigation (VRI) is improved by the system. This enables the volume of water that is provided to various portions of a field to be optimized, which ultimately results in a considerable increase in irrigation efficiency and the conservation of water resources.

(viii) Site-Specific Crop Management (SSCM)

VRT (Variable Rate Technology) enables fertilizers, chemicals and seeds to be applied according to specific needs of various parts of a field. They do so through the application of a particular Global Positioning System (GPS) and bespoke software. Visible and near-infrared spectroscopy can be divided into the map-based system (where applications are modified according to pre-established maps) and the sensor-based system (in which soil conditions are determined during execution). This technology ends up helping the farmer to optimize use of resources, reduce input costs and keeping the environment sustainable.

(ix) Soil Mapping in Precision Farming

The process of soil mapping, which enables farmers to analyze the features of the soil, as well as its chemical composition and nutritional status, is the basis upon which precision farming is built. Despite the fact that soil mapping has been conducted for millennia, improved precision and specificity have been offered by modern technologies, which have led to management approaches that are more effective and are guided by the most recent data.

(x) IoT in Precision Farming

Precision agriculture is made even better with the help of the Internet of Things (IoT), through the use of specialized sensors and remote monitoring systems.

(xi) Artificial Intelligence and Machine Learning

AI and ML are two emerging technologies that are helping precision farming to further develop by offering unique ways of managing farms. For example to have the capability of changing water and nutrient supply for each plant depending on its requirements is owned by image recognition systems. Furthermore, weed killers can now be pin-pointed by AI sprayers without the need for crop covers saving resources and increasing total production.

3. SMART IRRIGATION TECHNOLOGIES

More importantly, this chapter places web service and security risks directly into the realm of the technical issues in Industry 4.0 and eventual Industry 5.0 as they relate to smart agricultural infrastructure while acknowledging the challenges that need to be overcome if we do not want these vulnerabilities to become enemies of progress for our industry. The web service attack case has been rebriefed to illustrate how similar attacks (e.g., denial-of-service or network congestion) would halt data-driven irrigation systems, sensor networks or AI-based control in smart farms and the necessity of resilient and secure architectures. Moreover, all the technologies were defined and distinguished between autonomous and adaptive capability in this chapter. Living systems (e.g., human-guided irrigation using Raspberry Pi and MPC-based energy management) are referred to as devices capable of performing tasks without human input that is always on and running. On the other hand, adaptive systems respond in real time to evolving conditions using at calculating AI and machine learning modeling by feeding in weather forecasts, soil moisture readings and environmental fluctuations to better calibrate irrigation and energy consumption. This fine-tuning makes the chapter strongly related to the central theme of the book, whereby autonomy and adaptability not only lead to greater efficiency and more sustainable control but also necessitate higher cybersecurity and system robustness measures for safeguarding next generation smart agriculture ecosystems under Industry 4.0/5.0 paradigm.

(i) Soil Moisture Sensor-Based Systems

Soil moisture sensors, when integrated into irrigation monitoring systems, can determine both the soil's water content and the extent to which it has changed as a result of watering. In addition to reducing water waste, it allows farmers and gardeners to precisely water plants at the right times, which boosts plant development. Capacitance, Time Domain Reflectometry (TDR), and Neutron Probes are a few of the technologies used by these sensors, which find use in many agrihorticulture settings. These sensors find usage in a variety of settings, including those that need precise irrigation scheduling (such as vegetable farms, orchards, and vineyards), as well as those that promote plant health (such as nurseries, greenhouses, parks, and golf courses). All of these factors might discourage more people from utilizing these systems (Lopez-Jimenez et al., 2022). Cases of Regional Implementation: According to a research by (Seyfi et al., 2015), agricultural practices in California also successfully utilized soil moisture sensors to prevent over-watering. As a result, they were able to consume 20% less water and achieve 15% greater grape yield from their farms. According to (Mattar et al., 2022), broad-acre farms in Australia that used these sensors to track wheat and barley saw a 25% decrease in water consumption and a 10% improvement in crop yields.

(ii) Weather-Based Controllers

Weather-based controllers, also called ET controllers, are utilized to monitor variables including temperature, humidity, wind speed, and solar radiation. However, certain downsides are associated with them: first, precise weather data is needed; second, the devices are expensive; and third, periodic tuning and maintenance are required for them to function effectively. A 30% increase in water use has been reported in Florida as a result of residents' use of weather-based controls in their homes to alter neighborhood weather conditions and maintain landscape quality (Mattar et al., 2022). In Texas, a 25% reduction in water use has been observed in parks and commercial buildings after these controls were installed.

(iii) Internet of Things (IoT) in Irrigation

Irrigation is revolutionized through IoT since a clear way of managing all IoT sensors, devices, and data analysis systems is provided to monitor and control irrigation processes in real time. Data from different sources, for instance, soil moisture sensors, weather stations, and crop health sensors, are correlated by IoT systems to assist in irrigation decision-making; hence, water is not wastefully used. Smart Irrigation systems based on IoT techniques are implemented in various agriculture-related scenarios such as precision farming, greenhouses, nurseries, and landscaping. Some of the benefits include real-time monitoring, analysis, and decision-making, cost savings, and flexibility in operating the systems from any internet-connected device. Despite this, issues such as expensive initial costs, technical challenges, and data security concerns are also associated with them, which need to be well worked out to encourage their utilization. Similarly, in California, IoT-based systems have been incorporated into vineyards and orchards for better water usage, thus reducing water utilization by 25% and augmenting grape yields by 15% (Haghverdi et al., 2021). According to (Hosseinzadeh & Monem, 2019), water use has been reduced by 30% and crop yields increased by 20% in Indian smallholder farms where IoT systems have been used.

(iv) Drip Irrigation Systems

In drip irrigation systems, a network of pipes, valves, tubes, and emitters is used to slowly but surely provide water to plant roots. Benefits of drip irrigation include a reduction in water usage of up to 90%, better plant cover and development, less weed invasion, and adaptability to different types of soil and crops. It is advisable to hire professionals for installation and maintenance, and the upfront cost is greater due to the preliminary expenditure. Additionally, frequent maintenance is required for this form of irrigation to prevent emitter blockage with debris or mineral deposits. The use of drip irrigation in California's vineyards and tree plantations has assisted with water conservation, resulting in a 35% reduction in water use and a 20% increase in wine output (Conger and Dukes, 2020). There are other case studies where drip irrigation was used to grow sugarcane, cotton, and vegetables; these crops had a 30% increase in production and a 50% decrease in water usage (Josephine et al., 2021).

(v) Solar-Powered Irrigation Systems

Solar energy, a renewable resource, can be particularly helpful in areas without access to power, as it is used to pump water in solar drip irrigation systems. Rural regions, farms with fewer than ten acres, and farms that practice sustainable agriculture are examples of places where solar systems are prevalent (Raza et al., 2022). Numerous advantages are provided by these systems, including the utilization of renewable energy sources, significant cost reductions, decreased reliance on other nations, and, of course, minimized negative environmental impact. Additionally, energy storage and equipment are not always long-lasting. In Kenya, for example, energy expenditures have been cut in half while agricultural yields have been increased by 25% by solar-powered irrigation systems (Blessy and Kumar, 2021). According to (Josephine et al., 2021), energy expenses in India have been reduced by half, and agricultural productivity has been boosted by 30% through their implementation.

(vi) Artificial Intelligence (AI) and Machine Learning (ML) in Irrigation

Precision irrigation is becoming more feasible with the use of artificial intelligence (AI) and machine learning (ML). Large amounts of data from many sources, including soil moisture sensors, weather stations, and satellites, are compiled by these technologies. Better irrigation schedules, less water waste, and more abundant and high-quality food are all outcomes of their efforts to better manage water. Precision agriculture, greenhouses, and large-scale agricultural systems have utilized AI and ML to enhance crop quality while optimizing water consumption. Despite these worries, certain AI-based systems used in Californian vineyards have reportedly reduced water use by 25% while increasing grape production by 20% (Garcia et al., 2020). Utilizing AI technology has enabled a 30% decrease in water usage and an increase in wheat output in Australia.

(vii) Automated Sprinkler Systems

Automatic sprinkler systems are controlled by machinery to water landscapes and crops at regular intervals and in the correct amounts. These systems may be operated by timers, sensors, or computers, making them highly versatile and useful in nearly any setting. Sprinkler systems that automatically turn on and off are considered a great time and labor saver when it comes to watering lawns, businesses, and farms. A formal application of automatic sprinkler systems (AS) aimed at water conservation in Florida has been found to be successful, with water use being reduced by at least 20% (Dukes et al., 2020). In California, water usage can be reduced by a third while green areas are retained on golf courses through the implementation of these systems.

(viii) Mobile-Based Irrigation Control

An interface allowing users, including farmers, agriculturists, and landscapers, to manage irrigation from their mobile devices, such as smartphones and tablets, is referred to as a mobile irrigation control system. With the real-time data provided by these systems, customers are given more leeway to optimize irrigation schedules, making water management more convenient and effective. Mobility, real-time operation, water use management, and cost optimization are some of the advantages of mobile-based systems that may be applied to home gardens, commercial areas, and farmlands. Nevertheless, a few drawbacks must be considered, such as their reliance on an internet connection, which may be expensive in the beginning, and the technical difficulties encountered when trying to integrate these systems with existing ones. In India, irrigation has become more efficient thanks to mobile apps, which have resulted in a 20% reduction in water use and a 15% increase in agricultural productivity. According to (Josephine et al., 2021), grape quality has been improved and water usage reduced by 30% in Australian vineyards and orchards through these technologies.

(ix) Dacom Irrigation Advice

According to (Brajovic et al., 2015), a variety of intriguing smart agriculture solutions are offered by Dacom, a Dutch tech startup. Some of these solutions involve software, various sensors, and an online advising service. Small farms and agribusinesses all around the globe are the intended users of these products. Products are developed by this business for the following markets: the food processing sector, advice organizations and supply businesses, governments, and the Sensation and TerraSen packages. Weather stations spray conditions and forecasts, irrigation advice, crop records, control of fungal diseases, and insect control are also provided. A selection of the items listed above will be examined in this chapter:

- **Sensation bundle:** This package is made up of five soil moisture sensors plus iOS and Android apps. Optimal watering guidance for the user's crops and fields can be provided by this package. It is accessible at any time via smartphones and is straightforward to use and install. Powered by solar energy, this kit features high-quality smart sensors.
- **TerraSen Station:** Another tool that offers appropriate watering guidance is a TerraSen Station. Data on soil moisture, soil temperature, rainfall, and irrigation are automatically delivered to the user. As will be seen shortly, it works best when combined with the Irrigation Advice software. Solar power also powers this station.

To get the most out of the TerraSen sensor station, the irrigation advice software is utilized in tandem with it. The data collected by the TerraSen station is taken by the program and used to determine when and how much water should be irrigated (Weyrich and Ebert, 2016). This helps avoid over-watering and drought-related stress. As a rule, the quantity of water that has percolated through various soil layers is analyzed by the advice module, and suitable recommendations are then provided. The software screenshot can be seen in Figure 3.

Figure 3. Picture of the Dacom Irrigation software (http://en.dacom.nl/products/irrigation-advice/)

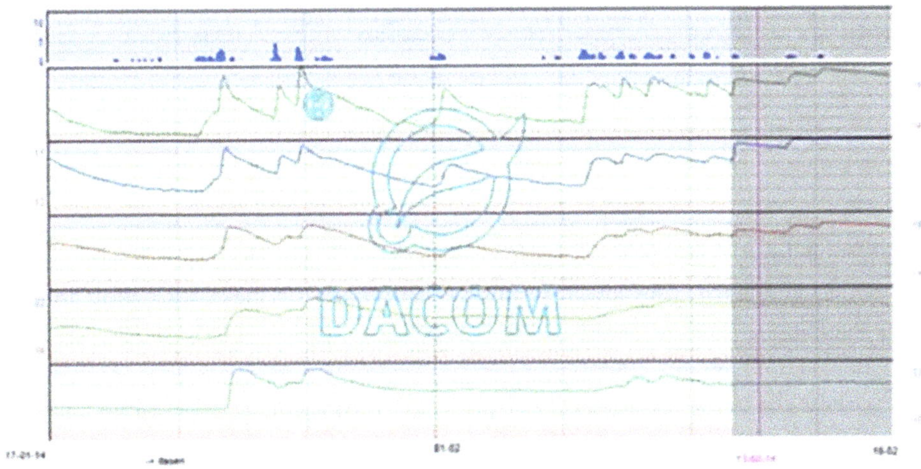

(x) Irris Scheduler Software

Irris Scheduler Software (Brajovic et al., 2015) is a small simple desktop application developed by the Purdue University Department of Agronomy and a freeware. This instrument provides open access estimates of irrigation water use for about twenty crops, as well as estimates of soil nitrogen losses and availability. This is a typical example of decision support software, which is based on an expert system as shown in Figure 4.

Figure 4. Screenshot of Irris Scheduler Software panel (https://www.purdue.edu/agsoftware/irrigation/.)

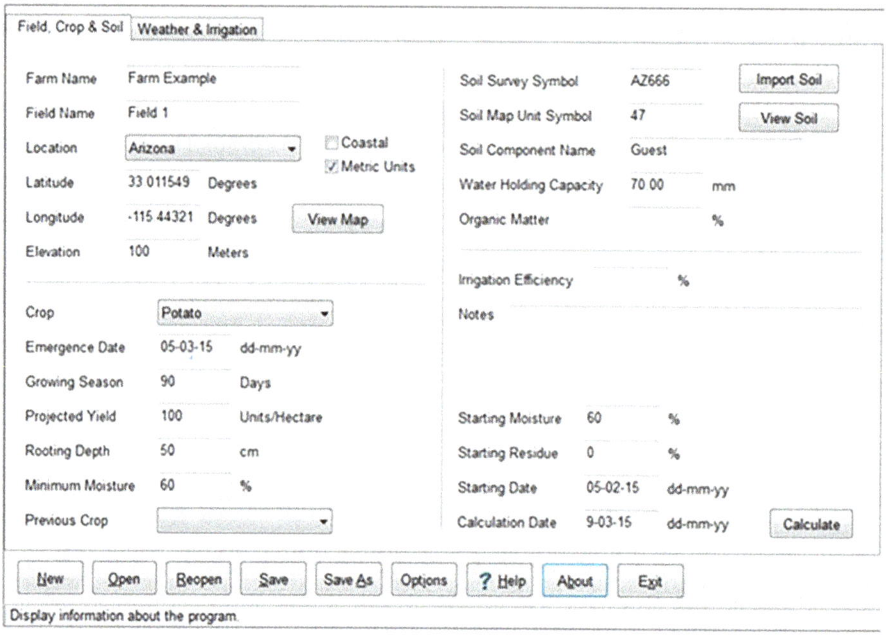

The database provides information about other locations under international, but not Montenegro. The program allows automated online information import after selecting a location and specifying agricultural production information. User (producer) can manually import missing or exact data.

As shown in Figure 5, the SWAMP project is working on a proposal for a highly accurate smart irrigation system for agriculture. Irrigation, water distribution, and consumption can be optimized through a comprehensive analysis. This analysis involves the gathering of data from every part of the system, encompassing the natural water cycle and the accumulated knowledge regarding the cultivation of specific crops (Kamienski et al., 2018). Because it avoids over- or under-irrigation and ensures that water will be available even when supplies are low, it saves money for everyone involved.

Figure 5. Smart water management underpins precision irrigation

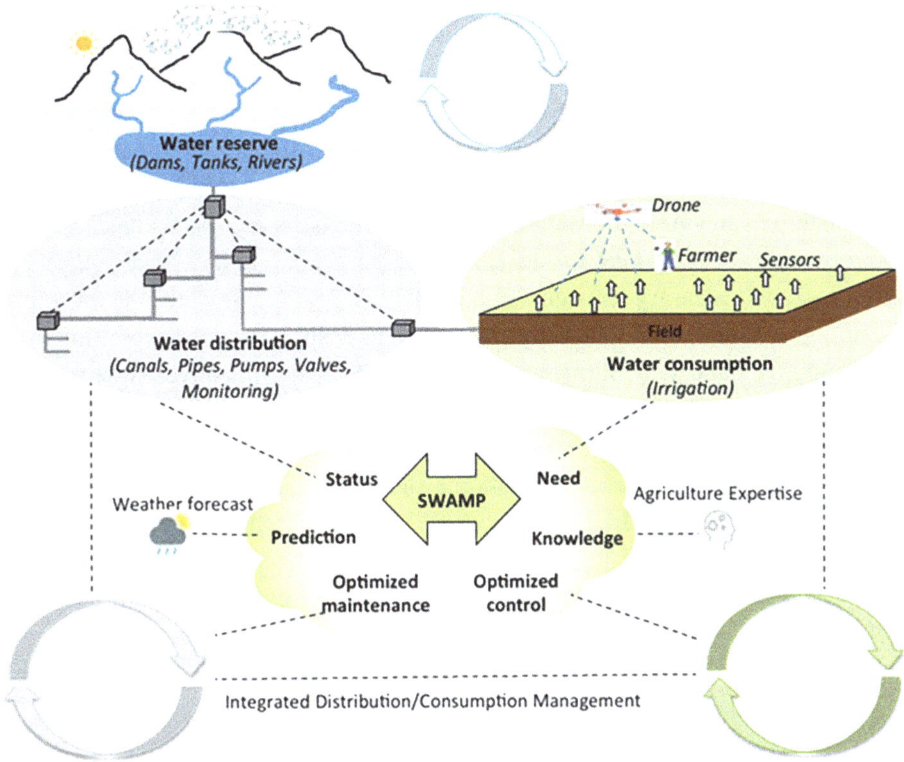

Three main phases of agricultural water management is shown in Figure 5. Following are the phases:

(i) W1: Water reserve: natural water cycle-governed water reserves from rivers, lakes, dams, and aquifers.
(ii) Canals, pipes, pumps, valves, and gates transport water from W1 to W3. Second phase in water delivery. Water distribution arrangements vary by country and geography. Several localities use water resources carefully and under centralized control.
(iii) Irrigation is one of the most essential uses of water in agriculture, and there are several ways to do it. The SWAMP project's main purpose is irrigation (W3). The SWAMP platform collects sensor data, makes choices using a variety of

methodologies, and modifies system behavior by sending commands to automatic and human actuators. SWAMP will respond to water consumption (W3) in real time or alter irrigation as crop conditions change. W2's management loop will be lengthier since water distribution adjustments are done on a different timetable. To trigger water distribution based on water usage, W2 and W3 management systems are linked.

(xi) A model predictive control (MPC)-based smart irrigation system is illustrated in Figure 6. This architecture possesses two stages. In this first phase, an expert sytem is used to obtain the inteligent profile of daily irrigation demand stepping wise (Quimbita et al., 2022). This system has been designed to calculate the day by day moisture in an automatic way through the water balance, with consideration of crop root zone water up take and loss. For this purpose, one-day predicted precipitation input is necessary as water. Equally key is how much water the crop loses in evapotranspiration. ET′ is finally referenced to desired evapotranspiration using meteorological data. The following stage is that the energy manager system (EMS) of the microgrid for agriculture irrigation power load model is optimized. This phase is for the optimal control of energy system in order to obtain the required energy to turn on water pump by deploying of renewable energies. The power requirement is due to the duration of irrigation in triggering the pumps daily during daytime period. Energy for the second stage (EMS) is provided by a hybrid energy system which comprises solar cells, accumulators and a diesel generator.

Figure 6. Smart irrigation system, considering optimal EMS

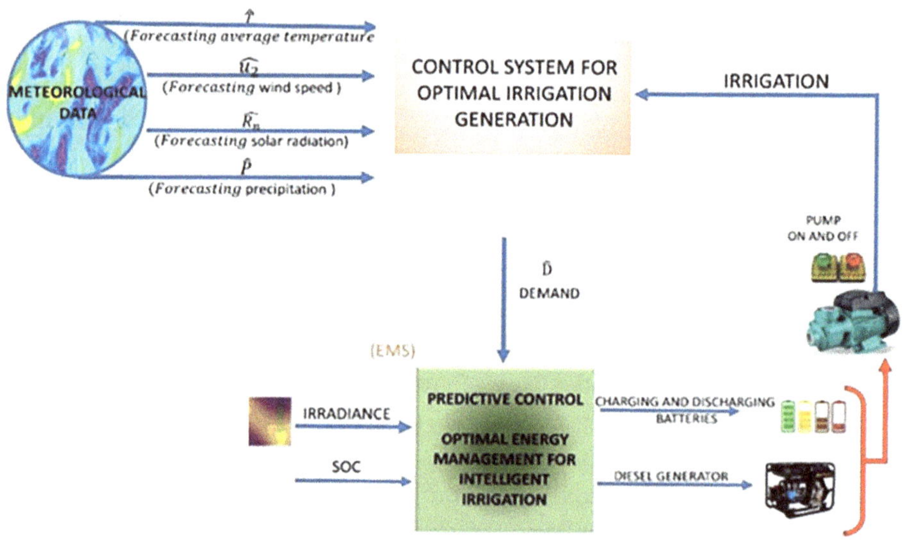

Figure 6 presents the overall framework of an intelligent autonomous irrigation system coupled with weather prediction, predictive control and optimal energy management processes. On thc left-hand side, actual meteorological data (forecasted temperature, wind speed, solar radiation and precipitation) is collected from IoT-enabled weather stations. These factors assist the system to forecast weather conditions that are affect irrigation requirement directly. The irrigation requirement signal is transmitted to the Predictive Control and Energy Management System (EMS). Here, the system adjusts usage of three power sources so as to guarantee irrigation can proceed sustainably and without interruption solar energy, battery charge (SOC) and diesel generator backup. The EMS self controls the charge/discharge of batteries and starting/stopping time of the diesel generator to achieve optimal energy flow cost. Finally, the optimised irrigation schedule leads to pumping ON/OFF control that provides just enough water for plants.

4. DATA-DRIVEN OPTIMIZATION FOR TARGETED RESOURCE USE

(i) The Real-time Sensing and Monitoring in Agriculture

The concept of Real-time Sensing and Monitoring in an agricultural context is elucidated by the given text. Key components are comprised of networks of advanced sensors, which denotes a system wherein numerous sensors are strategically deployed throughout a field. The types of sensors that are specifically mentioned include those for soil moisture, temperature, humidity, rainfall, and nutrient levels. It is implied that additional sensor types, such as those for pH or light intensity, could also be incorporated. Continuous data collection is ensured as these sensors are designed to persistently gather information, as opposed to doing so at sporadic intervals. Furthermore, granular data is produced, indicating that highly detailed and specific information is collected from various points across the field. This allows for an exact comprehension of the variations that exist within a single farm. The primary benefit or purpose that is served is the provision of an exact understanding of crop and soil needs. Through the continuous collection of detailed data on diverse parameters, an exceptionally accurate picture of what crops and soil demand at any given moment can be gained by farmers. This, in turn, enables more informed decisions to be made concerning irrigation, fertilization, pest control, and other agricultural practices.

(ii) ***Integration with Weather Data:*** Local and regional weather forecasts are integrated with real-time field data. This allows for proactive adjustments to irrigation schedules, preventing unnecessary watering before rainfall or during cooler periods, thus saving the energy that would have been used for pumping.

(iii) ***AI and Machine Learning Analytics:*** Sophisticated algorithms and artificial intelligence analyze vast datasets from sensors and weather stations. This enables predictive modeling of water demand and optimal growth conditions, leading to automated and precise delivery decisions for water and other inputs.

(iv) ***Variable Rate Technology (VRT):*** Based on the data analysis, VRT equipment applies over application in areas that don't need it, saving the energy associated with producing, transporting, and applying those excess inputs.

5. REDUCED PUMPING AND DISTRIBUTION ENERGY

(i) ***Targeted Water Delivery:*** Unlike traditional irrigation methods that apply water uniformly across an entire field, smart systems deliver water precisely to the plant roots (e.g., through drip irrigation). This significantly reduces water waste from runoff and evaporation, meaning less water needs to be pumped overall.

(ii) ***Optimized Pump Usage:*** By knowing exactly when and how much water is required, smart systems can optimize the operation of irrigation pumps. This leads to fewer pumping cycles and shorter operating times, directly translating to lower energy consumption.

(iii) ***Efficient System Design:*** Smart irrigation often incorporates efficient components like variable speed pump motors, proper pump sizing, and low-pressure delivery systems, all contributing to reduced energy demand.

6. RENEWABLE ENERGY INTEGRATION

(i) ***Solar-Powered Systems:*** A growing trend is the integration of solar photovoltaic (PV) panels to power irrigation systems, particularly in remote or arid regions. This reduces reliance on fossil fuels, lowers operational costs, and minimizes the carbon footprint of irrigation.

(ii) ***Off-grid solutions:*** Solar power can make irrigation accessible and sustainable in areas without reliable grid electricity.

7. REDUCED INPUT CONSUMPTION (INDIRECT ENERGY SAVINGS)

(i) ***Precision Fertilization:*** Precision farming principles extend beyond water. By applying fertilizers (and other inputs) only where and when needed, farmers reduce the energy expended in their production and transportation. This also prevents nutrient leaching, which can contaminate water sources and requires additional energy to treat.

(ii) ***Reduced Tillage:*** Many precision farming practices, such as reduced-till or no-till systems, require less mechanized cultivation, leading to significant fuel savings for farm vehicles.

(iii) ***Optimized Equipment Use:*** GPS guidance systems and automated steering for farm equipment minimize overlap during field operations (e.g., planting, spraying), saving fuel and reducing equipment wear.

8. MONITORING AND MAINTENANCE

(i) ***Real-time Monitoring:*** Smart systems allow for remote monitoring of irrigation equipment, enabling early detection of leaks, malfunctions, or inefficiencies. Addressing these issues promptly prevents energy waste and system damage.

(ii) ***Predictive Maintenance*:** Data analysis can help predict potential equipment failures, allowing for proactive maintenance and preventing costly and energy-intensive breakdowns.

OVERALL ENERGY BENEFITS

- ***Lower Energy Bills:*** The most direct benefit for farmers is a significant reduction in electricity or fuel costs associated with irrigation and other farm operations.
- ***Reduced Carbon Footprint:*** By consuming less energy, particularly from fossil fuels, precision farming and smart irrigation contribute to a lower environmental impact and reduced greenhouse gas emissions.
- ***Increased Sustainability:*** These systems promote sustainable farming practices by optimizing resource use and minimizing waste, contributing to long-term soil health and environmental preservation.

9. RESULTS AND DISCUSSIONS

- Industry performance metrics only quantify the operation stage and lack consideration of environmental impacts of irrigation technologies on a life-cycle basis including manufacturing, operation, decommissioning stages. For example, a sun-based IoT-controlled drip irrigation system proves to be more durable when considering lifecycle energy accounting. Although the energy investment required for PV panel production is higher, at a life cycle level, operational phase carbon emissions are almost zero and the recurring energy input during operation near-zero. On the other hand, a diesel flood system has less manufacturing energy but very high operational consumption from continuous fuel input and inefficient water application. Over 10 years, the solar-IoT system boasts an energy return on investment (EROI) of over 8:1 compared to the diesel EROI of only 1.6:1, or eight units of energy produced for one unit inputted. The previous Results and Discussion incision included disconnected sections for content--a Raspberry Pi guide, naked software screen shots, a standalone Model Predictive Control (MPC) investigation. A united summary redefines the Raspberry Pi prototype as a cheap adaptive control system and indicates possibility of real-time soil moisture adjustment in non-boutique area. However, it is limited by low processing power and output quality, sensor drift, and lack of predictive analytics.
- In contrast, the MPC example that showed a reduction of diesel consumption by 38.4% gives good experimental evidence for the efficacy of predictive optimization. Compared to a fuzzy logic controller, MPC is more adaptable to nonlinear irrigation dynamics as it involves continuous updating of optimal control trajectories with respect to system constraints and predictive terms. Fuzzy logic (which is computationally simple) operates an approximation based on rules that do not include anticipatory energy management. Consequently, MPC's system-oriented prediction not only reduces energy consumption but also promotes sustainability in the long term by synchronising irrigation with real-time environmental changes.
- The performance, cost, difficulties of implementation, and viability of large-scale integration of the various smart irrigation technologies and solutions on the market vary. High-value crops and water-scarce regions benefit greatly from drip irrigation and AI/ML-based systems, which are highly effective and can, lead to up to a 90% reduction in water consumption (Siddiqui et al., 2024). A modest level of efficiency is provided by controllers that rely on weather and soil moisture sensors, and these are considered appropriate for climates that fluctuate (Ramparany et al., 2014). Conventional irrigation,

especially automated sprinkler systems, can lose water due to evaporation, particularly in warm, dry weather (Smith et al., 2018).

Figure 7. Block diagram connection of Raspberry pi with sensors

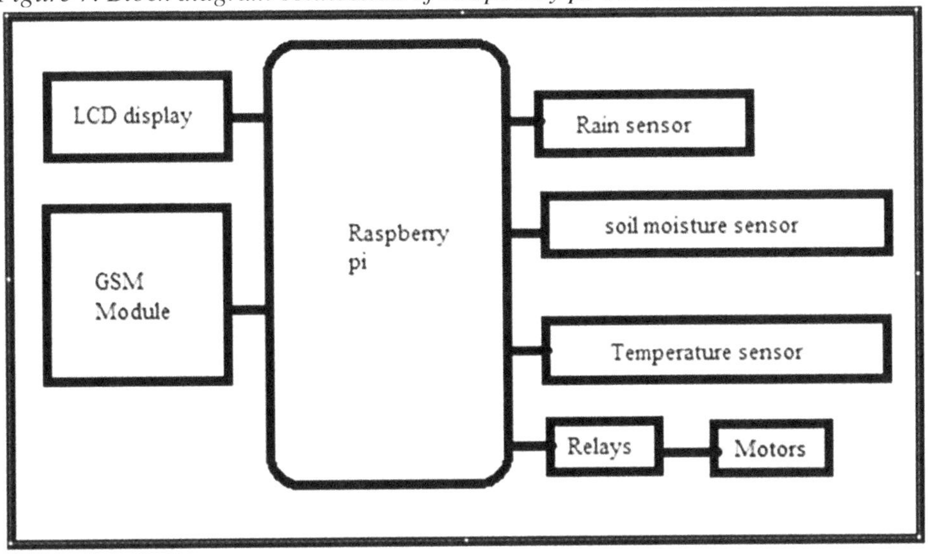

"This block diagram intends to automate the flow of water for given temperature and soil moisture conditions (Murthy et al., 2022). The primary component of this project is a Big Brother/Big Sister system using Raspberry Pi with Raspian (Linux). The things used in the chapter are summarized as this block diagram, which consist of Raspberry Pi, Rain Sensor, Soil moisture sensor, Temperature sensor and Relays (IRF250), Motor and a simple LCD Display (16x2), GSM Module as depicted in Figure 7. More recent variants include the Raspberry Pi Zero, Raspberry Pi Model A/B and Compute Module. Each Raspberry Pi employs GPIO (General Purpose Input/Output pins) for communication. Under our work, a Raspberry Pi 3 is revealed in Figure 8".

Figure 8. Raspberry pi 3 board

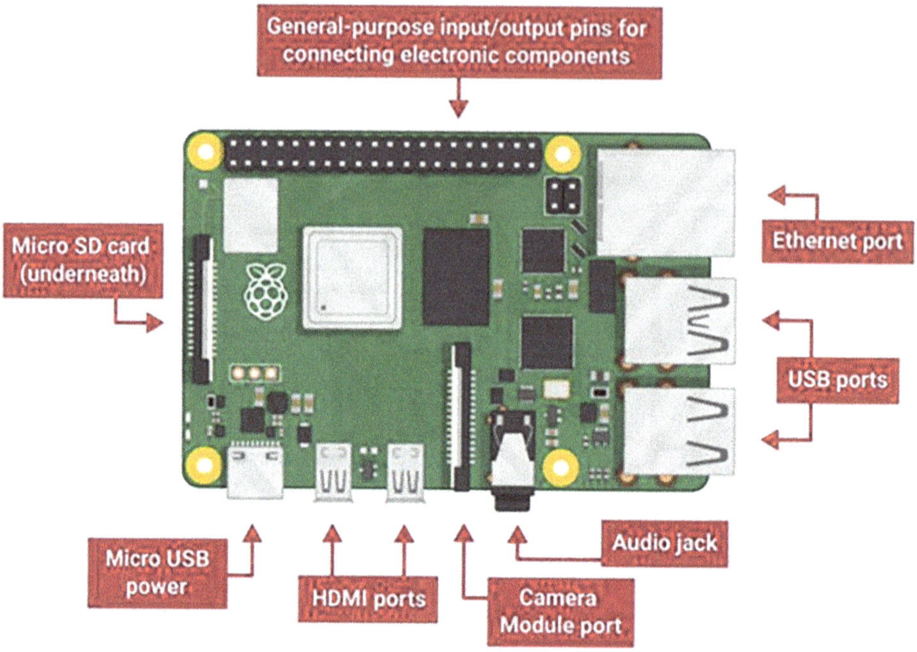

"An LCD 16x2 ' is a 16pin device with two rows which can show 16 characters. its used to print the messages which sent by the system.

· Pin No. 4 (Register Select): This pin is switched to initiate either command mode or data register mode, which is linked with the device. A '0' is obtained for (ASCII)data register mode and a '1' for command mode.
· Pin 5 (Read & Write): This pin controls the read/write lay of the screen, associating with a device pin to receive either '0'(Write) or '1' (Read).
· Pin No. 6 (Enable): This pin should be kept high to perform the Read/Write operation, connected with the device and continuously held high.

Type DESCRIPTION -- 7 to 14 Data pin: these pins are responsible for sending data to the display.

- Pumps are regulated by the assistance from motor drives and a specific drive is used depending on the type of motor utilized. The motor/pump assembly is employed to pump water. For simulation purpose we will use a crude motor with relays. A soil moisture detector is used to differentiate soil moisture readings. This sensor is positioned to determine whether or not the plant or crop is evaporating. The Soil Moisture Sensor works at 3.3V to 5V DC and consuming 15mA. Digital ouput from 0-5v, and Analog output also from 0-5V. There are two parts to the sensor. A soil moisture sensor consists of two parts: The Probe - The probe has two electrodes which are used to measure the volumetric content of water. A resulting voltage, after resistance of the probe, is output by the Module and available at an analog output pin. Rainy weather in the field is sensed by a rain sensor. The sensor is designed to detect rain showers. The sensor consists of an array of exposed copper traces, 5 for power lines and another 5 for sensing lines. Those traces are disposed such that generally two power traces have one sensing trace there between. In the atmosphere (or in any similar situation), they are disconnected, but underwater they become connected through water. Pin 1 (Signal): This is the analogue out pin which you're connecting to one of your device's simple input. There are only two types of mobile communication (GSM AND CDMA). For this purpose, a GSM module is to be used. This GSM has several layers including Mobile Stations (MSs), BSSs, NSS and OSS. The device (Raspberry Pi) can be communicated with using the GSM Module on its Tx & Rx Pins. It is also well known to communicate over specified ports. The module has a SIM slot capable of sending/receiving messages with just a basic cell phone. GSM Module is known as SIM900A GSM Module as well. The smart irrigation system design was intended to sense the moisture content and temperature of the soil. There are 4 useful parts in this kit, soil humidity sensors, rain detection sensors and motor drivers & pumps. A GSM module was included to accept SMS based information on machine status. A GSM modem is preferred because it needs very light linkage too connect in rural areas. Initially the Rain sensor sends a signal to the Raspberry Pi so whether we need to check weather or not after which soil moisture data is captured from the Soil Moisture sensor. Water is supplied to plants by a motor.

- Users must manually enter crop data. Online surveys are recognized by the program and selected by the user to input soil data. Users with more accurate field information (from personal sensors, etc.) may also readily enter it. How the user may examine and edit all soil data is shown in Figure 9. Another key feature of the app is weather and irrigation data. How the program searches for nearby weather stations and online forecasting providers depending on farm location is shown in Figure 10. Export scheduling and other data to Microsoft Excel with automatically created graphs. A scheduling example is shows in Figure 11.

Figure 9. Soil information in the Irris Scheduler Software.

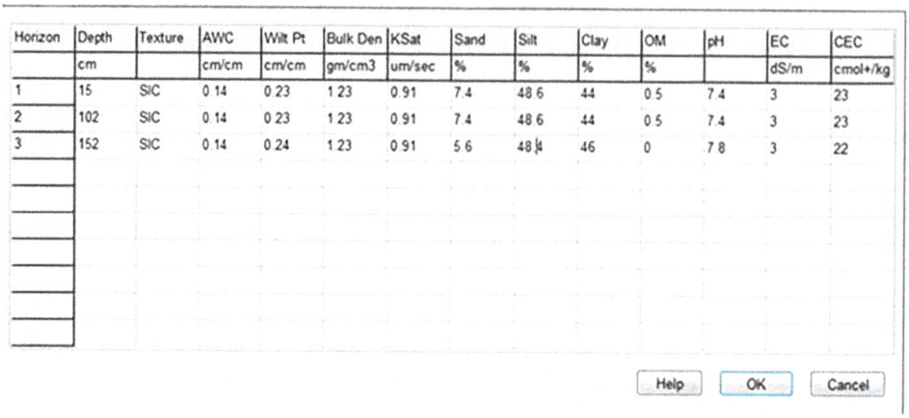

Horizon	Depth	Texture	AWC	Wilt Pt	Bulk Den	KSat	Sand	Silt	Clay	OM	pH	EC	CEC
	cm		cm/cm	cm/cm	gm/cm3	um/sec	%	%	%	%		dS/m	cmol+/kg
1	15	SIC	0.14	0.23	1.23	0.91	7.4	48.6	44	0.5	7.4	3	23
2	102	SIC	0.14	0.23	1.23	0.91	7.4	48.6	44	0.5	7.4	3	23
3	152	SIC	0.14	0.24	1.23	0.91	5.6	48.4	46	0	7.8	3	22

Figure 10. Weather information in the Irris Scheduler Software.

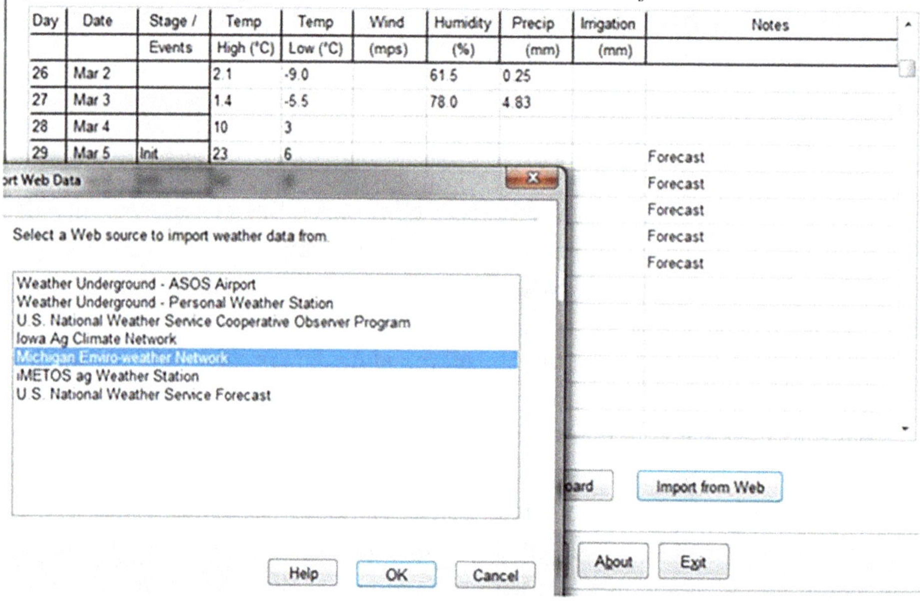

Figure 11. Scheduling in the Irris Scheduling Software.

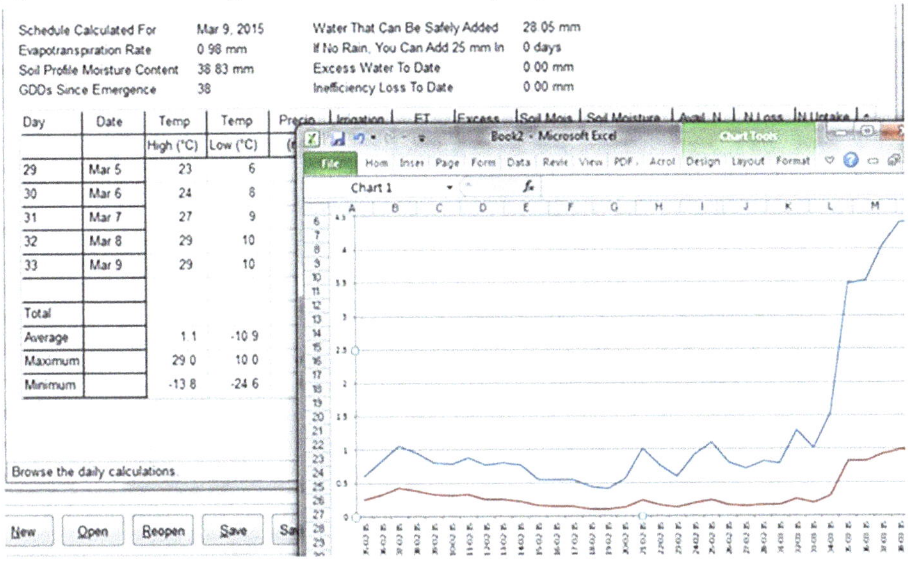

- The SWAMP platform models the water need in individual farms using sensors data and all other factors that you have. Irrigation prescription map is developed and sent to the farmers, who actuate irrigation at their convenient mode of operation. The forecasted water requirement for each farm is then input to a sub-module designed to optimize the management of water distribution. This is done by commanding actuators which are installed in devices that control the flow of water (close/open) through the canals up to a level that farmers can use them. The whole point of system is to make the use of that diverted water more efficient. Please note that Figure 12 does not directly explain the way data from sensors to SWAMP platform is transmitted, and farmers use different technologies to transfer data via internet

Figure 12. Scenario representing the water distribution pilot

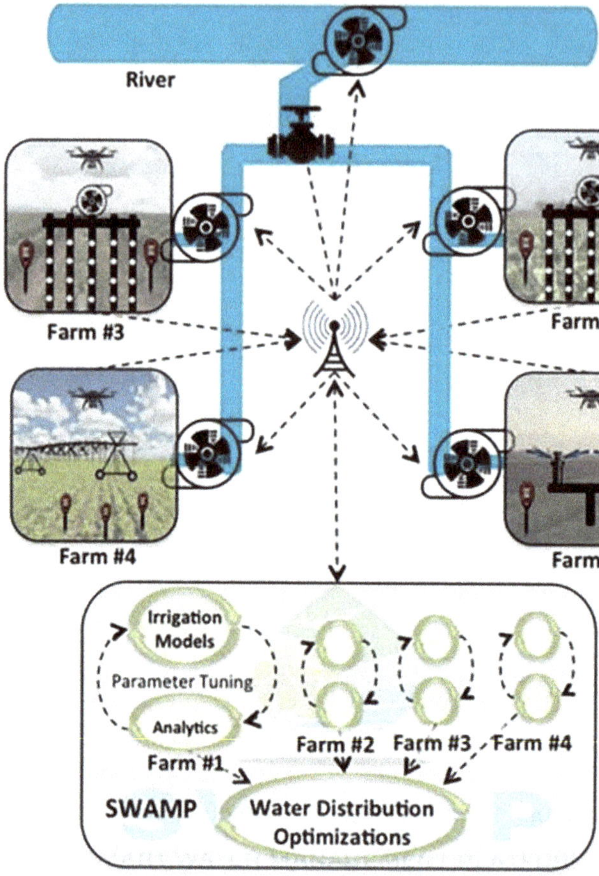

In conventional irrigation the moisture curve (H) in yellow overcomes to the counter level curve (LCc) in blue, as shown in Figure 13. During the 60 d crop period the excess water was removed through run off or deep percolation (Krishnan et al., 2020). Soil moisture under the agronomic-based technified irrigation strategy is presented in Figure 14. If the soil (H, yellow curve) is larger than LCc (blue line), components has more water resource. This excess water is discharged (Scarlatache et al., 2021). The third irrigation method is depicted in Figure 15, the smart-irrigation, which is operated based on a daily water balance and an irrigation time of 4 hours. Soil moisture (H), shown by the yellow line, remains within a restricted range to prevent unwatering and overwatering. Refer to Figures 13–15 for the comparison of three irrigation methods and observe how moisture is kept within without water-stress or soil over-saturation during the smart-irrigation scheme. Losses in water occur due to traditional irrigation and technified irrigation technology, also smart irrigation minimize water utilization.

Figure 13. Soil moisture with traditional irrigation

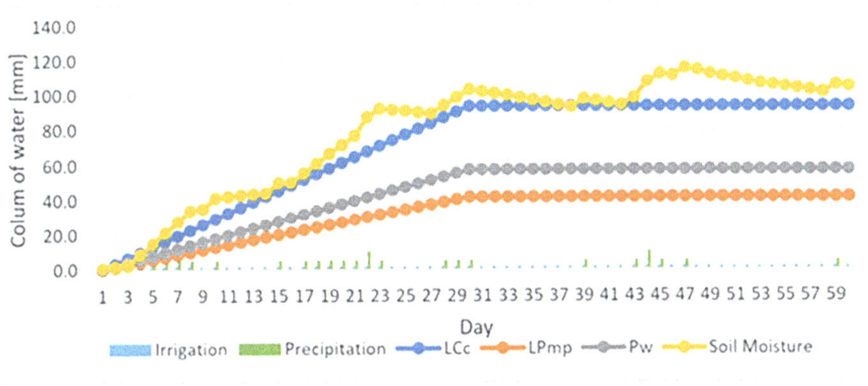

Figure 14. Soil moisture with technical irrigation.

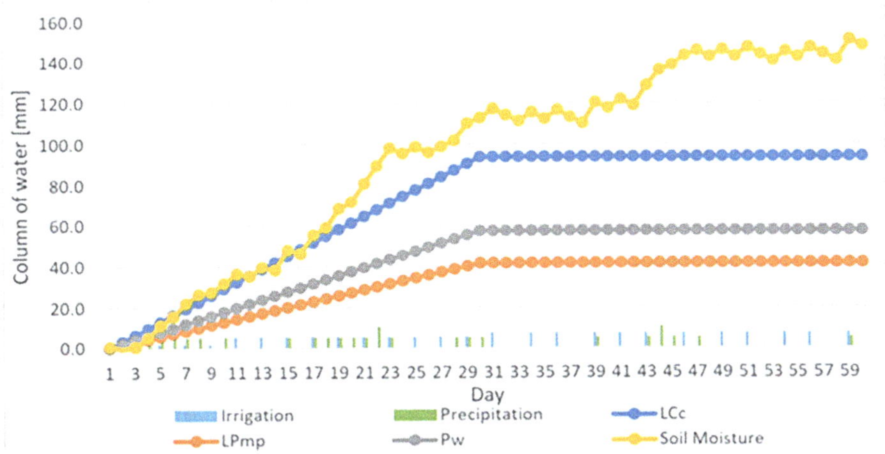

Figure 15. Proposed soil moisture smart irrigation

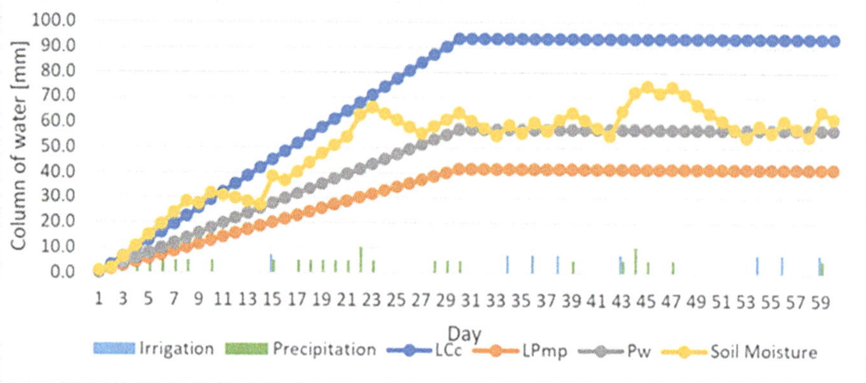

Analysis of Energy Management in the Three Irrigation Types

A suggested predictive control based on MPC models with a one-day control horizon for optimum irrigation manages energy as shown in Figure 16. Demand_b, supplied by the expert system, represents a 4-hour pump turn-on, shown in purple. The optimizer automatically determines pump start-up time, optimizing system energy sources. This helps determine the ideal time to charge or discharge batteries. The red curve shows BESS energy (Energy_B). With the diesel generator (P_diesel) off, the system generates no cost and the brown curve is 0. The lead curve shows that solar power (P_GS) recharges batteries when conditions are right. When the batteries are charging (P_Battery_Charge), the yellow line appears, and when the BESS system is discharging, the light blue curve appears. The blue curve represents the energy not supplied (P_Nosupplied), which is always 0, indicating that the system always has enough energy. The green curve shows solar shedding (P_solar_shedding), which occurs when more solar energy is produced than needed. Energy use and agricultural irrigation were examined. In Scenario 2, energy consumption curves without management may be examined. However, a fuzzy logic-based smart irrigation system (Krishnan et al., 2020) was compared. We evaluated pump ignition to determine energy consumption differences in our suggestion. The water pump ran for 7 hours over three days of growth (Krishnan et al., 2020). Unlike our smart irrigation, the pump runs for 3 hours during growth, improving crop output. Powering up the pump depends on energy costs. Our concept includes renewable microgrid management, unlike (Krishnan et al., 2020).

Figure 16. Energy management of proposed optimal irrigation.

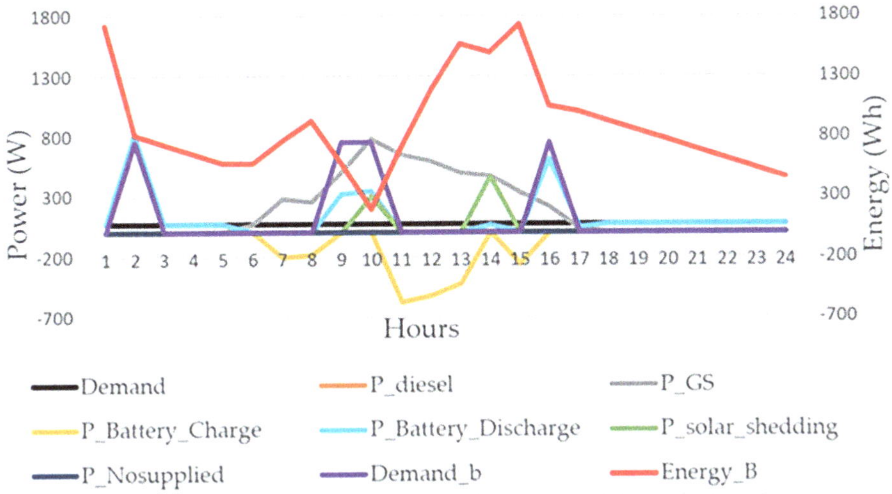

10. FUTURE CHALLENGES

Significant advancements in agriculture are promised by the continued evolution of precision farming and smart irrigation systems, but several key challenges related to energy management must be addressed for widespread and sustainable adoption.

- A primary hurdle encountered is the high initial capital expenditure required for these advanced systems. The investment in sensors, IoT devices, AI/ML platforms, and renewable energy infrastructure may be found prohibitive by farmers, particularly those with smaller operations. To mitigate this barrier and encourage broader implementation, financing models and government incentives will be crucial.
- Another significant challenge lies in data management and analysis. Vast amounts of data are generated by precision farming from diverse sources like soil sensors, weather stations, and satellite imagery. Robust connectivity infrastructure is demanded for effectively collecting, transmitting, storing, and analyzing this data in real-time, especially in rural areas where internet access can be limited or unreliable. The development of standardized data protocols and interoperable platforms will be essential to ensure seamless data flow and prevent fragmentation across different systems and vendors.

- Technical complexity and the need for specialized skills also pose a considerable challenge. Expertise in areas like data science, network management, and renewable energy integration is often required for implementing and maintaining these sophisticated systems. Farmers may lack these specialized skills, necessitating the provision of comprehensive training programs and accessible technical support to empower them to effectively utilize these technologies.
- Furthermore, the reliability and resilience of energy supply are critical for uninterrupted operation of smart irrigation systems. While numerous benefits are offered by renewable energy sources like solar power, their intermittency can be a drawback. Effective energy storage solutions (e.g., advanced battery systems) and intelligent microgrid management will be vital to ensure a consistent power supply, particularly during periods of low renewable energy generation or peak demand. The integration of diverse energy sources and the development of predictive energy management algorithms will be key to optimizing energy use and minimizing reliance on traditional, often fossil-fuel-based, grids.
- Finally, cybersecurity concerns must be addressed proactively. As smart irrigation systems become increasingly interconnected, they become potential targets for cyberattacks. Protecting sensitive agricultural data, preventing unauthorized access to control systems, and ensuring the integrity of irrigation operations will require robust cybersecurity measures and continuous vigilance.
- Traditional ownership models require significant capital upfront for sensors, IoT gateways, actuators and AI based control infrastructure. In order to address this limitation, new IaaS (Irrigation-as-a-Service) financial models provide a scalable solution. In IaaS, the farmer only pays for the volume of water supplied using a usage-frequency model, such as the delivery duration, amount of water delivered or by crop. This model changes investments from capital expenditure to operational expenditure, lowering the financial risk and creating a level playing field for sophisticated irrigation technologies. In addition, incorporating micro-financing and carbon-credit benefits within IaaS frameworks would support adoption efforts by connecting the sustainability performance with monetary gain.
- Cost, Data management, Cyber-security, Interoperability, and Skills Gap-challenge each is now redefined as a generic barrier preventing the scaling up of autonomy & adaptivity in agricultural systems. For example, the "Cost" crowd is addressing cost barriers for adopting AI enabled adaptive systems and looking at cost models such as IaaS (Irrigation-as-a-Service). "Data Management" compares the edge vs. cloud in retaining adaptive re-

sponsiveness, and "Skills Gap" points out the lack of expertise to handle or oversee autonomous decision-support frameworks. These targeted revisions automatically define a research program towards scalable, energy-efficient and adaptive agri-autonomy.

Overcoming these challenges will necessitate a concerted effort from policymakers, technology developers, researchers, and farmers. Collaboration will be essential to foster innovation, develop practical solutions, and ultimately pave the way for a more efficient, sustainable, and resilient agricultural future.

CONCLUSION

The chapter has presented a variety of technologies in the field of precision farming and smart irrigation which can play an important role for optimizing agricultural water management and production in face of growing world's water scarcity. The technologies presented all have their specific advantages and disadvantages, and the decision for its use depends on application and locality. Accordingly, as the environment becomes more limited in capabilities worldwide, these technologies along with improved data processing and IoT (Internet of Things) and renewable energy generation will be important for improving the performance and adaptability of solar cells. The Raspberry Pi begins with the appropriate level of soil moisture and climate conditions. This project is an effort to construct a hands-free smart water irrigation system using soil moisture and rain sensors with prevents over watering and save both water and electricity. As such, the focus in this chapter has been on the impact of modern ICT technology for irrigation and output in agriculture. The systems disclosed herein advance irrigation concepts worldwide. The SWAMP perspective, architecture, pilots and scenario driven project development method were outlined. SWAMP is still in its infancy, but it has drawn interest from farmers to industry and government agencies around the world. In the current phase, pilots were visited and surveyed, while at the same time, what is expected from different pilot sites/cultures/needs and how these needs affect this platform were examined. In comparison, the results indicate that when compared to traditional irrigation, with the SCSDFSs system it was achieved a reduction of fuel in diesel consumption by 38.4% and savings produced being equal to 55.42% applied over technified irrigation. When the EMS of the microgrid installed, is generated a saving of 100% in electricity generation cost with respect to the technified irrigation implementation. The proposed smart irrigation system is more cost-effective that optimizes water usage in the crop and renewable energy sources towards efficient power management. In conclusion, the intelligent proposal of irrigation will be generated by the hours of irrigation for a crop and also supply of the electrical demand through microgrid using from a source renewable in such a way that it is taking into account in the environment. The pump on/off control depends on technical and optimal irrigation circumstances. Moreover, in the proposed system the battery operation set points are established by considering the constraints of SOC. The diesel generator power is configured when required. The EMS takes advantage of nature's resources to make those decisions. This design prevents wasting of water as it maintains the moisture in a required level and achieves excellent harvest for farming and power supply will be an auxiliary use.

REFERENCES

Aminpour, Y., Dehghan, D., Playán, E., & Maroufpoor, E. (2023). Estimation of wind drift and evaporation losses of sprinkler irrigation systems using dimensional analysis. *Agricultural Water Management*, *289*, 108518. DOI: 10.1016/j.agwat.2023.108518

Angelin Blessy, J., & Kumar, A. (2021). Smart irrigation system techniques using artificial intelligence and IoT. *2021 Third International Conference on Intelligent Communication Technologies and Virtual Mobile Networks (ICICV)*, 1355-1359. DOI: 10.1109/ICICV50876.2021.9388444

Brajovic, M., Vujovic, S., & Dukanovic, S. (2015). An overview of smart irrigation software. *2015 4th Mediterranean Conference on Embedded Computing (MECO)*, 353-356.

Chaudhry, S., & Garg, S. (2019). Smart irrigation techniques for water resource management. *Advances in Environmental Engineering and Green Technologies*, •••, 196–219. DOI: 10.4018/978-1-5225-5909-2.ch009

Conger, S. L., & Dukes, M. D. (2020). Evaluation of testing procedures for weather-based irrigation controllers. *Transactions of the ASABE*, *63*(5), 1277–1287. DOI: 10.13031/trans.13926

Garcia, L., Parra, L., Jimenez, J. M., Lloret, J., & Lorenz, P. (2020). IoT-based smart irrigation systems: An overview on the recent trends on sensors and IoT systems for irrigation in precision agriculture. *Sensors (Basel)*, *20*(4), 1042. DOI: 10.3390/s20041042 PMID: 32075172

Haghverdi, A., Singh, A., Sapkota, A., Reiter, M., & Ghodsi, S. (2021). Developing irrigation water conservation strategies for hybrid bermudagrass using an evapotranspiration-based smart irrigation controller in inland Southern California. *Agricultural Water Management*, *245*, 106586. DOI: 10.1016/j.agwat.2020.106586

Haque, A., Islam, N., Samrat, N. H., Dey, S., & Ray, B. (2021). Smart farming through responsible leadership in Bangladesh: Possibilities, opportunities, and beyond. *Sustainability (Basel)*, *13*(8), 4511. DOI: 10.3390/su13084511

Hosseinzadeh Ghazichaki, Z., & Monem, M. J. (2019). Development of quantified model for application of control systems in irrigation networks by system dynamic approach. *Irrigation and Drainage*, *68*(3), 433–442. DOI: 10.1002/ird.2331

Huang, S., Wang, B., Li, X., Zheng, P., Mourtzis, D., & Wang, L. (2022). Industry 5.0 and society 5.0-Comparison, complementation and Co-evolution. *Journal of Manufacturing Systems*, *64*, 424–428. DOI: 10.1016/j.jmsy.2022.07.010

Josephine, M., Gurunathan, S., Krishna, V., Nath, G., & Ramasamy, S. (2021). Smart irrigation system. *2021 10th International Conference on Internet of Everything, Microwave Engineering, Communication and Networks (IEMECON)*, 1-4.

Kamienski, C., Soininen, J., Taumberger, M., Fernandes, S., Toscano, A., Cinotti, T. S., Maia, R. F., & Neto, A. T. (2018). SWAMP: An IoT-based smart water management platform for precision irrigation in agriculture. *2018 Global Internet of Things Summit (GIoTS)*, 1-6.

Krishnan, R. S., Julie, E. G., Robinson, Y. H., Raja, S., Kumar, R., Thong, P. H., & Son, L. H. (2020). Fuzzy logic based smart irrigation system using Internet of things. *Journal of Cleaner Production*, *252*, 119902. DOI: 10.1016/j.jclepro.2019.119902

Leng, J., Sha, W., Wang, B., Zheng, P., Zhuang, C., Liu, Q., Wuest, T., Mourtzis, D., & Wang, L. (2022). Industry 5.0: Prospect and retrospect. *Journal of Manufacturing Systems*, *65*, 279–295. DOI: 10.1016/j.jmsy.2022.09.017

Lopez-Jimenez, J., Vande Wouwer, A., & Quijano, N. (2022). Dynamic modeling of crop–soil systems to design monitoring and automatic irrigation processes: A review with worked examples. *Water (Basel)*, *14*(6), 889. DOI: 10.3390/w14060889

Mattar, M. A., Roy, D. K., Al-Ghobari, H. M., & Dewidar, A. Z. (2022). Machine learning and regression-based techniques for predicting sprinkler irrigation's wind drift and evaporation losses. *Agricultural Water Management*, *265*, 107529. DOI: 10.1016/j.agwat.2022.107529

Murthy, B. Y., Reddy, C. B., Jilani, S., & Sindhwani, M. (2022). Smart irrigation system. *2022 1st International Conference on Sustainable Technology for Power and Energy Systems (STPES)*, 1-4.

Opoku, D. R., Liu, J., Yuan, S. K., Sam-Amoah, L., & Yan, H. (2020). Irrigated agriculture for food self-sufficiency in the sub-saharan African region. *International Journal of Agricultural and Biological Engineering*, *13*(3), 1–12. DOI: 10.25165/j.ijabe.20201303.4397

Plazzotta, S., Cottes, M., Simeoni, P., & Manzocco, L. (2020). Evaluating the environmental and economic impact of fruit and vegetable waste valorisation: The lettuce waste study-case. *Journal of Cleaner Production*, *262*, 121435. DOI: 10.1016/j.jclepro.2020.121435

Quimbita, W., Toapaxi, E., & Llanos, J. (2022). Smart irrigation system considering optimal energy management based on model predictive control (MPC). *Applied Sciences (Basel, Switzerland)*, *12*(9), 4235. DOI: 10.3390/app12094235

Ramparany, F., Marquez, F. G., Soriano, J., & Elsaleh, T. (2014). Handling smart environment devices, data and services at the semantic level with the FI-WARE core platform. *2014 IEEE International Conference on Big Data (Big Data)*, 14-20. DOI: 10.1109/BigData.2014.7004417

Raza, F., Tamoor, M., Miran, S., Arif, W., Kiren, T., Amjad, W., Hussain, M. I., & Lee, G. (2022). The socio-economic impact of using photovoltaic (PV) energy for high-efficiency irrigation systems: A case study. *Energies*, *15*(3), 1198. DOI: 10.3390/en15031198

Saetra, H. S. (2021). AI in context and the sustainable development goals: Factoring in the Unsustainability of the Sociotechnical system. *Sustainability (Basel)*, *13*(4), 1738. DOI: 10.3390/su13041738

Sarwar, A., Peters, R. T., Shafeeque, M., Mohamed, A., Arshad, A., Ullah, I., Saddique, N., Muzammil, M., & Aslam, R. A. (2021). Accurate measurement of wind drift and evaporation losses could improve water application efficiency of sprinkler irrigation systems – a comparison of measuring techniques. *Agricultural Water Management*, *258*, 107209. DOI: 10.1016/j.agwat.2021.107209

Scarlatache, F., Grigoras, G., Scarlatache, V., Neagu, B., & Ivanov, O. (2021). A hybrid methodology based on smart management energy consumption in irrigation systems. *Electronics (Basel)*, *10*(22), 2864. DOI: 10.3390/electronics10222864

Sharma, A., Jain, A., Gupta, P., & Chowdary, V. (2021). Machine learning applications for precision agriculture: A comprehensive review. *IEEE Access: Practical Innovations, Open Solutions*, *9*, 4843–4873. DOI: 10.1109/ACCESS.2020.3048415

Siddiqui, A. S., Al Muaini, K. H., Shahid Latif, A. H., Al Balushi, B. M., & Ramadoss, S. (2024). Sustainable smart irrigation leveraging smart technologies for improved water efficiency- A review. *2024 2nd International Conference on Computing and Data Analytics (ICCDA)*, 1-5.

Turner, C. J., Ma, R., Chen, J., & Oyekan, J. (2021). Human in the loop: Industry 4.0 technologies and scenarios for worker mediation of automated manufacturing. *IEEE Access: Practical Innovations, Open Solutions*, *9*, 103950–103966. DOI: 10.1109/ACCESS.2021.3099311

Tushar, P. N., Shrivastava, P. C., Jaiswal, P., Singh, S., & Mishra, V. (2026). Legal Challenges in Cloud Computing. In I. Oncioiu (Ed.), Law and Regulation of AI, Blockchain, and Digital Rights (pp. 63-102). IGI Global Scientific Publishing. https://doi.org/DOI: 10.4018/979-8-3373-3186-7.ch003

Weyrich, M., & Ebert, C. (2016). Reference architectures for the Internet of things. *IEEE Software*, *33*(1), 112–116. DOI: 10.1109/MS.2016.20

Wollschlaeger, M., Sauter, T., & Jasperneite, J. (2017). The future of industrial communication: Automation networks in the era of the Internet of things and industry 4.0. *IEEE Industrial Electronics Magazine*, *11*(1), 17–27. DOI: 10.1109/MIE.2017.2649104

Xu, X., Lu, Y., Vogel-Heuser, B., & Wang, L. (2021). Industry 4.0 and industry 5.0-Inception, conception and perception. *Journal of Manufacturing Systems*, *61*, 530–535. DOI: 10.1016/j.jmsy.2021.10.006

Chapter 10

Predictive Analytics in Industrial IoT (IIoT): Enhancing Efficiency and Reliability

M. Anita
https://orcid.org/0009-0009-0093-2928
SRM Institute of Science and Technology, India

T. P. Anish
https://orcid.org/0000-0002-4308-9645
R.M.K. College of Engineering and Technology, India

M. Ezhilvendan
https://orcid.org/0009-0003-3213-1304
Panimalar Engineering College, India

ABSTRACT

Industrial Internet of Things (IIoT) generates vast volumes of data from interconnected sensors and devices, creating opportunities for predictive analytics to enhance operational efficiency and reliability. This chapter reviews key predictive analytics models and techniques applied in IIoT, including traditional statistical methods, machine learning, and deep learning approaches. It explores essential IIoT data sources, infrastructure, and applications such as predictive maintenance, process optimization, and safety management. The chapter also discusses critical challenges including data quality, scalability, security, model interpretability, and integration with legacy systems. Finally, future research directions highlight advancements in edge AI, digital twins, explainable AI, and sustainable IIoT practices.

DOI: 10.4018/979-8-3373-3146-1.ch010

1. INTRODUCTION

1.1 Background of Industrial IoT (IIoT)

The IIoT is a major change in the way industry operates, connecting digital tools with old manufacturing and production techniques. IIoT brings the same principles as IoT to the industrial world which in turn helps create smarter and automated industrial systems(Malik, P. K. et al., 2021). Consumer IoT tends to make life easier and better, while IIoT mainly works on making workplaces safer and more efficient in different sectors.

The main part of IIoT is how cyber-physical systems, embedded sensors, intelligent actuators, industrial control systems and cloud or edge computing platforms connect seamlessly. All these aspects join to build a network that can collect, monitor and control industrial assets in real time. Sensors and devices in the machinery keep gathering data about temperature, vibration, pressure and energy use and they send it through secure networks to systems that can be located either centrally or in different places. This information makes it possible to do analytics and make decisions which helps industries quickly react to any changes happening in the factory or on site(Karmakar, A et al., 2019).

The spread of IIoT is mainly due to fast progress in related technologies. Because of 5G, LPWAN, industrial Ethernet and edge computing or fog computing, connectivity and data throughput have improved and data can be processed locally with almost no delay. They help solve the usual problems in industrial automation such as not having enough bandwidth, being sensitive to delays and ensuring privacy of data. At the same time, cloud computing gives all organizations the opportunity to use scalable storage and computing resources, so they can use IIoT solutions with lower costs.

IIoT is also important for bringing Industry 4.0, the fourth industrial revolution which is based on automated processes, strong interconnections between systems and decisions guided by data(Misra et al., 2021). In Industry 4.0, machines, systems and people are expected to interact without problems to get better results in manufacturing. In this situation, IIoT is responsible for collecting data and transmission which supports the use of robotics, AI, digital twins and augmented reality.

Because of IIoT, companies can now see how their operations work, maintain their equipment in advance, use energy efficiently and boost their overall equipment performance. Due to IoT, manufacturers can keep an eye on their machines as they operate and spot possible problems early and energy companies can understand how much energy they are using at any moment. IIoT allows companies in logistics to closely monitor goods and their fleets which boosts the transparency and flexibility of their supply chains. Besides, the use of IIoT supports workers' safety and environmental monitoring, making it easier for industries to comply with important regulations.

Even though IIoT has great opportunities, there are still some difficulties in using it. Difficulties such as protecting data, making systems work together, fitting in old systems and using standardized methods cause a lot of challenges. Also, because industrial data comes in large amounts, many forms and at high speeds, powerful analytics are required to help make good use of the data. At this point, predictive analytics becomes a key factor in the IIoT, allowing industries to switch from reactive to proactive and then prescriptive decision-making(Jain, J. K., & Chauhan, D 2023).

1.2 Importance of Predictive Analytics in IIoT

In the Industrial Internet of Things, predictive analytics gives the insight needed to figure out what all the connected data means(Sathya et al., 2024). Predictive analytics mainly uses statistical methods, machine learning algorithms and data mining tools to find out patterns, trends and links in both past and current data for forecasting the future. Forward thinking is extremely helpful in factories, since quick and well-informed actions can improve how work is done, how much it costs and how safe it is.

Within IIoT, predictive analytics is mainly used for predictive maintenance which involves using data to know when equipment needs maintenance. Predictive maintenance is different from reactive and preventive forms in that it helps avoid extra interventions and sudden breakdowns. If the data from sensors such as temperature, vibration, sound or electrical signals is analyzed, predictive models can predict the RUL of components and give advance warnings about their status. As a result, you can maintain your equipment when it really needs it which reduces times when your equipment is not working, helps them to last longer and saves money on maintenance.

Besides keeping equipment in good shape, predictive analytics helps with different areas of process optimization in industry. It makes it possible to adjust production settings to enhance the quality of products, save more resources and increase the amount produced. As an example, in a manufacturing line, predictive models examine data from various parts of production to spot areas where things slow down or quality drops and they also suggest solutions. Because changes can be made quickly, the process and production can be improved all the time.

It also plays a big role in reaching energy management and sustainability goals. If industries model the energy use of the past and relate it to their work, they can estimate future needs, find unusual situations and discover inefficiencies. For example, such tools can identify when there will be high energy usage or find unusual usage that may point to a problem or leak. They allow companies to make better plans for energy purchases, handle energy loads more efficiently and follow environmental rules.

Predictive analytics helps a lot in the field of safety and risk management. Industrial areas usually face dangerous situations and any mistake or error by staff can result in accidents, damage to the environment or violations of rules. Predictive analytics can find safety risks early by examining data from safety sensors, monitors in the environment and interfaces with machines. If gases are detected at unusual levels or the temperature rises fast within a chemical plant, alarms sound ahead of the limits, so action can be taken to prevent problems. With the help of predictive models, history of incidents can be analyzed to spot patterns and come up with ways to reduce risks.

Also, since industries use supply chains for materials, inputs and distribution, predictive analytics helps them see and adjust to changes more quickly. With the use of precise predictive models, it becomes easy to predict demand, expect delays and control inventory. Efficiency is boosted and customers are happier because there are timely deliveries and less time is needed.

It is important that there is reliable, extensive and timely data available to make predictive analytics in IIoT successful. The fact that industrial systems are complicated often leads to data from various sources, uneven sampling and poor measurement quality. You have to use advanced data cleaning, feature design and testing processes to deal with these challenges. Besides, the type of predictive model, whether simple or complex, is decided by the task, the nature of the data and how interpretable it needs to be.

Using predictive analytics together with IIoT technologies changes the way automation is done, from manual control to self-management. They can handle and predict situations, also adjusting to them beforehand. This change is also backed up by new technologies such as edge AI and digital twins which help make decisions faster and more accurately. With more and more industries switching to digital, predictive analytics is central to running operations that are smart, tough and eco-friendly.

1.3 Scope and Objectives of the Chapter

The purpose of this chapter is to highlight the main predictive analytics approaches used in the IIoT domain and explain how they help improve both efficiency and the reliability of operations. The key goal is to combine research and industrial approaches to explain clearly how predictive analytics is linked to IIoT, what it is used for and the difficulties that may arise in real-life situations. The chapter begins with basic aspects of IIoT architectures and data infrastructure and then explores predictive analytics in detail using statistics, machine learning and deep learning. Moreover, the chapter looks into different uses, for example, predictive maintenance, optimizing processes, quality control, energy management and safety monitoring. Besides, it deals with significant issues such as the quality of data, how the system grows, security, how well the models are understood and connecting to old industrial systems. Besides, the document points out new trends and future research that could boost the effectiveness of predictive analytics in IIoT such as edge AI, federated learning and digital twins. The chapter is meant to support researchers, those working in industry and policymakers who wish to use predictive analytics to make industrial operations more efficient and reliable.

2. FUNDAMENTALS OF INDUSTRIAL IoT

2.1 Architecture of IIoT Systems

IIoT systems are built to make sure there is easy connection, data gathering and smart decision-making in complicated industrial settings. The main idea behind IIoT architecture is to connect several layers that make it possible to monitor, control and analyze operations in real time. At the base of the architecture are sensors and actuators that are put into industrial equipment and processes to collect data about temperature, pressure, vibration and the machine's status. The sensors change physical events into digital data, making it possible to observe the system's status all the time(Bhushan et al., 2023).

Sensors gather data that is sent through different networks which can be Ethernet, industrial fieldbuses, Wi-Fi, Zigbee, LTE or 5G. IIoT systems depend on connectivity to make sure field devices communicate smoothly and promptly with the main processing units(MAY et al., 2025; Jaffrin et al., 2025).

Edge computing is important because it tackles data processing at the source which helps cut down latency and saves bandwidth. These devices process data, sort it and summarize it to give quick answers which is necessary for real-time control and monitoring of safety in industry. At the next step, IIoT systems connect to the cloud so they can store more data, analyze it in depth, use machine learning and be controlled from one place. Thanks to the cloud, businesses can securely save data over the long term, perform tough calculations and analyze information from different locations which allows them to fully use big data for predictions and improvement of their operations.

2.2 Key Technologies in IIoT

A variety of technologies in industrial IoT make it possible to gather, send and study information from many sources. There are many types of sensors, starting with temperature and humidity ones and moving up to more complex multi-axis accelerometers and chemical detectors. They are constantly measuring key physical values in industry which gives a precise and up-to-date overview of everything happening. Because of progress in sensor technology, new sensors are smaller, use less power and are more accurate when used in tough industrial settings.

For different devices to work together, protocols for communication are very important. For industrial communication, the usual standards are Modbus, PROFIBUS, OPC-UA and newer ones such as MQTT and CoAP which are made for quick and light messaging. Factors such as bandwidth, dependability, safety and connectivity to the current setup determine the choice of a protocol.

In the IIoT, data is usually collected by either constantly streaming or periodically gathering data from sensors. This challenge is handled by systems that collect and group the data from sensors before forwarding the results to either edge or cloud platforms. Fast detection of issues is made possible by collecting data in real time and reviewing old data allows us to study trends and make predictions. In addition, mixing information from different sensors with data fusion helps improve accuracy and gives a clear overview of industrial processes(Vishwas et al., 2021).

2.3 Challenges in IIoT Data Management

The ability of IIoT to improve operations comes with the big challenge of handling the large amount of data it produces. Managing data is one of the main problems IIoT systems generate a lot of data every day which is more than most industrial IT systems can manage. It is necessary to have data lakes and distributed databases, plus approaches for compressing and managing data priorities to deal with such a large amount.

Another difficulty is that countries are not the same. IIoT environments consist of different kinds of hardware, many communication protocols and several data formats. Since data is not all the same, it is difficult to combine and manage, so it is necessary to use strong middleware and standard models to make communication and analysis easier(Rose et al., 2025).

Immediate processing is key in IIoT because delays in maintenance or safety could cost a lot or cause dangerous issues. Still, the need for real-time processing puts a lot of pressure on the system and the network, mainly when operations are spread across different locations. It is still difficult to decide whether to process data locally or send it to a central cloud server.

Protecting data and ensuring privacy is a big worry in the field of IIoT. To secure sensitive industrial information and maintain the availability of systems, companies should use advanced encryption, authentication procedures and keep an eye on their networks all the time.

All in all, the architecture and technologies behind IIoT make it possible for industries to improve automation and analysis. Still, dealing with all the data produced by IIoT is not easy, so new ideas and technology are needed to fully use its benefits.

3. OVERVIEW OF PREDICTIVE ANALYTICS

3.1 Definition and Concepts

With predictive analytics, past data, statistical formulas and machine learning are used to guess what will happen in the future. Predictive analytics is different from descriptive analytics since it predicts trends, enabling better preparation for the future(Kandasamy et al., 2024). It is not the same as prescriptive analytics which actually outlines specific steps to follow after looking at predicted results. To sum up, predictive analytics lets you recognize what happened in the past and plan for the future by estimating the chances of certain events. When it comes to the Industrial Internet of Things (IIoT), predictive analytics uses the ongoing data from devices to predict possible equipment faults, deviations in processes and blockages in workflows. Because industries can anticipate problems before they appear, they can move from reactive to proactive management which reduces downtime, decreases costs and makes things safer and more efficient.

3.2 Common Techniques and Algorithms

Predictive analytics relies on different techniques which are mainly separated into traditional stats, machine learning, deep learning and time-series forecasting. Linear regression, logistic regression and survival analysis are some of the main statistical methods used to connect variables and to predict future results based on probability. They are preferred since they can be understood and work well when the data is organized and the connections are simple.

Since there is a lot of complex data in IIoT systems, machine learning algorithms are now widely used because they can discover patterns and relationships that regular algorithms cannot. Without much programming, decision trees, random forests, support vector machines (SVM) and gradient boosting machines can find important features and make good predictions automatically. They prove to be very useful when used for classification and regression, as in fault detection and RUL estimation.

In the past few years, deep learning has proven to be very effective at working with data that is unstructured and comes in many dimensions in IIoT. CNNs, RNNs and LSTM networks are able to find patterns in data from sensors that depend on time and location. As an illustration, LSTMs are great for time-series data, so they are often used in maintenance tasks that focus on data patterns and trends.

Besides, ARIMA (AutoRegressive Integrated Moving Average), exponential smoothing and hybrid models designed with both statistics and ML are useful for working with data that has seasonal changes, trends and noise. They make it possible to predict future sensor data or production figures which allows early responses.

3.3 Role of Data in Predictive Analytics

Predictive analytics relies on data as its main support. How well a predictive model performs depends on how good the data is, what type it is and how it is prepared. In such cases, data is mostly obtained through sensors that record the current state of operations, the environment and the status of machines. Data can be of three types: numerical (temperature or pressure), categorical (machine's state) or unstructured (images or sound from machines).

Data quality is a big issue, since sensors can sometimes record noise, lack data or give inaccurate readings because of external or internal issues. It is important to use data cleaning, normalization, imputation and outlier detection to increase the reliability of the data and make it ready for analysis.

The performance of a predictive model can be greatly improved by doing feature engineering. The job is to change the raw sensor data into usable information by looking at statistics, specific indicators and features from different times and frequencies. As an example, vibration records may be converted into frequency spectra by using Fourier transforms to highlight information about the fault. When features are well engineered, it helps the model spot important patterns and remove unnecessary data.

Besides, dealing with the large amount of data generated by IIoT calls for reliable data storage, streaming and batch processing frameworks. Using historical records together with current data allows models to be updated all the time and learn continuously which keeps predictions accurate as industries change.

All in all, doing predictive analytics well in IIoT relies on solid data handling, the right choice of models and precise knowledge of what you want to achieve. Using accurate data and suitable algorithms, predictive analytics makes it possible for industrial operations to react less and act more proactively which boosts both their efficiency and dependability.

4. PREDICTIVE ANALYTICS IN IIoT: APPLICATIONS

4.1 Predictive Maintenance

Predictive maintenance is one of the biggest benefits of using predictive analytics in Industrial IoT. Usually, using reactive or scheduled methods for maintenance can result in either broken equipment or extra servicing that cost a lot and are not efficient. Predictive maintenance depends on sensor information to track the condition of machinery all the time and predict an impending breakdown. Predictive models are able to find possible issues using early fault detection by analyzing vibration, temperature, acoustic signals and other important factors. An important factor is the Remaining Useful Life (RUL) which shows how many more hours a component or machine will last. Using RUL predictions, industries can arrange maintenance at the perfect moment and save time, reduce storing spare parts and use their assets longer. By taking this approach, you reduce spending and also protect your workplace from major failures.

Figure 1. **Predictive Analytics in IIoT**

4.2 Process Optimization

In IIoT environments, predictive analytics is very important for improving processes. Because the materials, surroundings and demands in industry always change, industrial processes often need to be adjusted regularly to perform well. Analyzing the current data and past trends, predictive models can advise on changes to parameters that increase output, speed up processes and save energy. Predictive analytics in a manufacturing assembly line can find issues and propose the ideal settings for speed and temperature to boost production and still maintain quality. The use of live data helps companies in different fields become more flexible, save resources and lower their overall expenses.

4.3 Quality Control and Defect Prediction

It is very important to maintain the same quality in industrial manufacturing and predictive analytics can help with this by offering strong tools for checking and predicting defects. IIoT systems gather information on what is being made, how the machines are working and the surroundings during the whole process. The analysis of data by predictive models helps spot patterns related to product issues or quality changes, so action can be taken before such things happen. Using anomaly detection, classification and clustering, it is easier to identify faulty batches or pieces which cuts down on waste and extra work. Also, when manufacturers understand the main causes using predictive insights, they can perfect their processes, deal with reliable suppliers and make the product more dependable. Making use of this application decreases expenses and also supports the brand's reputation by maintaining quality.

4.4 Energy Management

Using energy is a big expense and environmental problem for industries. Predictive analytics working with IIoT systems help facilities to determine how much energy they will use and make sure it is used efficiently. Looking at the past, current and expected usage of energy, models can warn about peak demand and recommend actions to use or store energy at different times. Predictive analytics is able to suggest making adjustments to HVAC systems or machines, so the peak energy usage is reduced and lower rates are used when available. As a result, energy bills go down and sustainability goals are achieved by lowering carbon emissions. Besides, predictive maintenance minimizes the energy used by malfunctioning equipment which helps conserve energy.

4.5 Safety and Risk Management

Since equipment failures and hazards can cause major accidents in industries, keeping everyone safe is very important. The use of predictive analytics enables safety and risk management by closely watching data to spot any possible safety incidents. Sensors that detect gas leaks, fast temperature changes or pressure on equipment feed the data to models that decide on the level of risk and alert workers ahead of time. Because of this approach, operators are able to prevent accidents by, for example, switching off equipment or clearing the area. It is also worth noting that predictive analytics helps with safety compliance by recording all data and making it possible to evaluate risks. Predictive analytics lowers the chances of accidents and as a result, both people and the company are protected from risks and financial losses.

5. DATA SOURCES AND INFRASTRUCTURE FOR PREDICTIVE ANALYTICS IN IIoT

5.1 Sensor Data and Acquisition Techniques

The main part of predictive analytics in Industrial IoT (IIoT) is the constant flow of data produced by sensors in many industrial devices and activities. Depending on the application and what the industry needs, sensors can be very different. Some of the common sensors are temperature sensors, pressure sensors, vibration and accelerometers, humidity sensors, current and voltage sensors, gas detectors and acoustic sensors. Every sensor is designed to monitor a certain physical factor important for checking the machine, its environment and the work process. Choosing the right sensors is necessary to collect important and correct data for predictive modeling.

IIoT uses methods to sample sensors periodically so the information is useful and the system does not get overloaded. The frequency of sampling depends on what is being measured; for example, vibration sensors usually work at a high rate to detect little problems, but temperature sensors might not. The problem is finding a balance between getting all the data and using up precious resources such as bandwidth, storage and processing power.

Ensuring data quality is also very important in the process of acquiring sensor data. There are many challenges in industrial areas, including electromagnetic interference, dust and extreme temperatures which may affect how well and how accurately sensors work. Applying sensor calibration, using extra sensors and implementing filtering algorithms adds security to the collected data. Besides, to use sensor data well in predictive analytics, it is important to reduce noise, remove outliers and normalize the data.

5.2 Edge Computing and Cloud Computing

Data processing in IIoT happens at different stages in the infrastructure, mainly at the edge and in the cloud and each has advantages and disadvantages. Edge computing means processing data close to where it is produced such as on the sensors or industrial devices. It cuts down on latency and the use of bandwidth by having local analysis, filtering and decision-making done. Edge computing is key for situations that call for immediate actions such as watching over safety or finding faults as they occur. It allows fast responses and makes sure the system continues to function when the cloud network is not available or not working properly(Abitha et al., 2026).

On the other side, cloud computing sends data from edge devices to central data centers which have enough room to store and process large amounts of information. Using the cloud, it is possible to perform advanced analysis, keep data archive for a long time and combine information from several locations or operations. Complex machine learning training and huge data mining are possible with the cloud, while such tasks cannot be done at the edge because of limited resources. Still, using cloud services brings problems with delay in data transfer, safety of information and privacy, mainly when working with sensitive industrial data.

Currently, IIoT architecture usually uses a combination where early processing and detecting anomalies are done on the edge and the data is sent to the cloud from time to time for further analysis and trend recording. Keeping these factors in balance makes the system perform well, remain dependable and handle more workloads(Swarnalatha, P., & Prabu, S. 2023).

5.3 Data Storage and Management

Effective storage and management of data are important for IIoT predictive analytics since there is a constant stream of data in industrial activities. Traditional relational databases can be ineffective because they cannot scale well and are not flexible to deal with sensor data that is not organized. Because of this, many IIoT systems rely on data lakes and distributed databases to keep different types of data in large quantities.

With a data lake, you can hold raw sensor information as it is and perform both batch and real-time analytics. Such flexibility makes it possible for organizations to store a lot of past data which can be used for studying old trends and training new models

Keeping track of metadata and proving the origin of data are important for being able to trace what happened, comply with rules and maintain the data's quality in the future. Moreover, today's data management tools use automation to handle data cleaning, integration and management, so people need to do less manual work.

All in all, the data sources and infrastructure in IIoT should be properly built and improved to successfully process industrial data and provide accurate, prompt and secure predictive analysis. Thanks to this foundation, industries can gain useful insights and use IIoT intelligence to its highest level.

6. PREDICTIVE ANALYTICS MODELS AND TECHNIQUES FOR IIoT

6.1 Traditional Statistical Models

Traditionally, predictive analytics has relied on statistical models since they are easy to understand and have a solid mathematical background. These models usually suppose that the relationship between inputs and the target is fixed which means they are best for linear or known distributions. One standard method is called linear regression which tries to figure out how a dependent variable is connected to one or more independent variables. Many use it in IIoT to anticipate outcomes that change all the time such as the condition of equipment or the rate of production.

Another significant model is ARIMA (AutoRegressive Integrated Moving Average) which is built for the purpose of time-series forecasting. ARIMA models can catch trends, seasonality and autocorrelations in temporal data which are usually found in sensor data in IIoT environments. For instance, ARIMA can predict future sensor values or machine results which benefits maintenance planning and running the factory.

Estimating the RUL of equipment is very important in IIoT applications and this is best achieved through survival analysis. It handles data that shows the time when an event occurs and can estimate the likelihood of failure at any point in time, handling cases where an event has not happened. Thanks to statistics, reliability engineers are able to estimate the chances of equipment failure which is important for implementing predictive maintenance.

Although traditional statistical models are clear and easy to use, they have problems dealing with complex forms of data and numerous different types of data in IIoT. Because of this, researchers have started using advanced techniques in machine learning.

6.2 Machine Learning Approaches

Due to their ability to find patterns and relationships from large and complicated data by themselves, machine learning (ML) approaches are now more widely used in IIoT. Decision trees are used often because they break up data by using thresholds to sort the outcomes or predict numbers. Because decision trees are easy to use and deal with all types of data, they are often used for identifying and classifying faults.

Support Vector Machines (SVM) are effective methods for finding the best lines that split data points in many dimensions. Clear margins between classes allow SVMs to work well and they are commonly applied to binary classification tasks, for example, telling apart faulty and normal operations.

Random forests are made up of a number of decision trees which work together to improve how well predictions are made and to avoid overfitting. Since random forests combine the predictions of various trees, they give strong and dependable results which is why they are suitable for complex situations with many sensors.

The training of machine learning models usually requires a lot of labeled data and using feature engineering can highlight the main features. They can model interactions that are not always straight, but may be hard to understand, so there is a need for techniques that reveal their meaning.

6.3 Deep Learning Models

The evolution of predictive analytics is marked by deep learning models, as they use neural networks with multiple layers to discover traits from the raw data. They work very well with data that is unorganized, includes many features and changes over time.

CNNs perform well in processing data that is correlated in space such as images taken by visual inspection systems or vibration spectrograms turned into 2D plots. CNNs automatically pick out the important parts of an image through convolutional layers which helps find defects and monitor quality in manufacturing.

When working with sequential or time-series data, RNNs and LSTM networks are generally used by most researchers. LSTMs solve the issue of vanishing gradients that RNNs have and can detect long-term trends in data from sensors. They are useful when it comes to predictive maintenance, since they are able to detect patterns and trends to foresee failures or estimate the remaining useful life.

Lately, transformer models which were at first designed for working with language, have been used in time-series and sensor data processing. By using self-attention, Transformers can assess how each piece of input data matters and this makes them better at catching complicated time-based relationships than traditional RNNs.

Although deep learning models are quite accurate, they need a lot of labeled data, powerful computers and proper adjustments to make sure they do not overfit. Since their decisions are not interpretable, it becomes a problem in important industrial fields.

6.4 Hybrid and Ensemble Models

To get better results, hybrid and ensemble methods combine a number of predictive models to use their strengths and overcome their weaknesses. Bagging, boosting and stacking are types of ensembles that combine predictions from several base learners to lower both the variance and bias.

In the IIoT, statistical models could be used together with machine learning or deep learning strategies. For example, a model might first forecast the trend with ARIMA and afterward make use of an LSTM network to find any nonlinear patterns in the residuals. When data show both simple and complex behaviors such combinations can increase the accuracy of predictions.

Because of ensemble methods, industrial systems can better handle errors by using data from different models. Because they use different models and data, ensembles are able to provide more reliable predictions needed for important IIoT systems.

All in all, predictive analytics in IIoT makes use of various models, including simple statistical ones, advanced deep learning and hybrid combinations, depending on the data and the needs of the industry. The combination and selection of suitable models allow you to get accurate, prompt and useful insights that support your operation's efficiency and dependability.

7. CHALLENGES AND LIMITATIONS

7.1 Data Quality and Availability Issues

Providing good and dependable data is one of the main issues when using predictive analytics in IIoT. Sometimes, missing data appears in industrial data streams because of sensor failures, problems with the network or maintenance being done which may lead to issues with training and predicting models. Sensor data can be spoiled by the rough environment in factories, interference from nearby machines or changes in its calibration which leads to inaccurate inputs for prediction. Such problems can decrease the model's effectiveness and make people doubt the analytics.

Many times, data imbalance is a problem, since anomalies are much rarer than regular events and do not appear frequently in the data. As a result, predictive models could become unfair to the minority classes and not catch possible dangers such as faults or dangerous situations. To solve these problems, it is important to use advanced preprocessing methods including filling in missing values, reducing errors and increasing the amount of data. Nevertheless, it is still challenging to get enough high-quality data for many IIoT projects.

7.2 Scalability and Real-Time Processing

IIoT systems produce a huge amount of data every second from the sensors installed on many industrial assets. The amount of data involved creates big problems when it comes to both computing and designing systems. Such solutions should be able to deal with data as it arrives or almost immediately to give helpful information for timely preventive actions.

To process data in real time, it is important to have edge computing devices, streaming platforms that work efficiently and cloud resources that can grow with the data. Still, it is not easy to ensure low latency, high throughput and accuracy, especially when the models being used are complex. Besides, since IIoT data comes in several different formats, sampling rates and protocols, it makes it harder to integrate and analyze large amounts of data.

Since resources at edge locations are limited, engineers must design and optimize models that are scalable. Although model compression, incremental learning and hierarchical processing are used, they are still difficult to research and implement further.

7.3 Security and Privacy Concerns

Industrial operations and their data are very sensitive which is why security and privacy are key concerns in IIoT predictive analytics. There is an increasing number of cyberattacks on industrial control systems and IIoT devices that want to disrupt how things work, take intellectual property or bring about physical destruction. Data that is constantly sent over networks can be intercepted, changed or accessed without permission.

To ensure data security, it is necessary to use strong encryption, authenticate users and keep an eye on the network. Still such techniques may introduce a delay and extra processing that goes against the requirement for real-time analytics. Also, it is necessary to have clear guidelines and follow regulations when privacy is affected by data shared with other parties or used in several different organizations.

Attacks that use models, for example, adversarial machine learning, can trick predictive analytics by making the models give wrong predictions. So, it is necessary to secure the data and analytics pipeline to ensure IIoT systems are trusted and reliable.

7.4 Model Interpretability and Trust

Many advanced predictive analytics models such as deep learning, can be so complex that people who use them cannot easily understand how the models work. When analytics are used in industry, where safety, budgets and operations are involved, a lack of transparency makes it hard for people to trust the results and use them.

Explainable AI (XAI) tries to solve this problem by making it easier to understand why a model behaves the way it does which features are most important and what leads to a prediction. Still, it is hard to strike a balance between making a model accurate and being able to understand how it works. Models that are easy to understand such as decision trees, may not perform as well in prediction and models that do well in predicting tend to be less clear.

To boost user trust, models are closely examined and industry-related knowledge is added so that the results from analytics are in line with industry experience.

7.5 Integration with Legacy Systems

Many manufacturing plants and factories still depend on older machinery that was not made for connecting to the Internet or analyzing data. It is complicated to link predictive analytics solutions into the current structures used in organizations. Some legacy systems depend on unique protocols, cannot handle much information and do not provide simple ways to access data.

Sensors or gateways needed to collect data from such systems can cost a lot and be very complicated to install. It is also necessary to make sure there is smooth communication between updated systems and older systems so that operations run smoothly.

To deal with this challenge, many times it's necessary to add customized middleware, protocol converters and follow a step-by-step implementation plan. It is important for IT and OT teams to collaborate, set similar targets and manage changes successfully.

8. FUTURE DIRECTIONS AND RESEARCH OPPORTUNITIES

8.1 Advancements in AI and IoT Technologies

With the progress in AI and IoT, predictive analytics in industrial IoT (IIoT) is set to gain a lot in the near future(Bhambri & Rani, 2023; El-Gendy, 2020). One of the best new technologies is Edge AI which adds AI capabilities right into edge devices. Real-time processing and decision-making on the device offered by Edge AI helps reduce lag, use less bandwidth and make the system stronger. This becomes very important where instant reactions are required such as for finding faults and monitoring safety.

A powerful new idea is federated learning which lets different edge devices or industrial sites train models jointly without having to send their own data. With this approach, sensitive industrial data is not shared and is kept close to the source, but still, everyone can use the collective intelligence. With federated learning, the accuracy of predictive models improves on distributed assets and more personalized analytics can be used in various operational situations.

With the help of 5G, the data communication between sensors, edge devices and the cloud will be smooth and error-free, making it possible to use more sophisticated real-time analytics. Because of this improved connectivity, it becomes possible to use remote monitoring and control, augmented reality in maintenance and coordinated use of robots in factories.

Thanks to these improvements, IIoT systems will be smarter, faster and more secure, making it easier for predictive analytics to boost industrial operations on a very large scale.

8.2 Integration of Digital Twins and Simulation Models

Another important future path is bringing together digital twins, simulation models and predictive analytics. Digital twins are digital versions of real equipment, processes or systems that remain updated in real time using sensor information. When industries mix these representations with advanced analytics, they are able to predict possible outcomes, estimate future actions and boost their performance beforehand.

It improves the ability to predict outcomes by giving more information and in-depth understanding about asset health, the way processes work and external factors. As an illustration, predictive models combined with digital twins make it possible to simulate and estimate RUL by simulating degradation in various conditions. It makes it possible to arrange maintenance before faults occur by predicting their effects and how the system will react.

Also, digital twins make it possible to keep improving and checking predictive models by connecting the actual systems to their virtual copies. By going through this approach several times, the model improves and makes the environment more adaptive. When digital twin technology gets more advanced, it is likely to combine with predictive analytics to transform how industries handle assets and operations.

8.3 Improved Explainability and Human-in-the-Loop Systems

As models for predictive analytics get more complicated, it becomes important to pay attention to how understandable and usable AI results are. Future studies will concentrate on making AI more transparent by coming up with methods that explain its results in a way that industry professionals can act upon. XAI methods will make it easier for machines to be understood by people who need to use them for important industrial tasks.

At the same time, systems that involve humans will help merge human instincts with AI-generated results. They allow operators to use user feedback in the predictive process, so they can check, change or dismiss the model's predictions if required. Working as a team increases the model's reliability, lowers the rate of false alarms and makes sure AI assists people instead of replacing them.

With interactive tools and user-friendly screens, operators will be able to learn more about the model's explanations, test out other scenarios and choose their actions accordingly. It is believed that combining human and artificial intelligence will help improve safety, production and adjustment to changes in complex workplaces.

8.4 Sustainability and Green IIoT Initiatives

Sustainability is now a main concern as the IIoT predictive analytics field grows. Future actions will focus more on using predictive models to cut down on industry's impact on the environment and aid in meeting global targets for reducing carbon and managing resources carefully. Green IIoT approaches will make use of analytics to save energy, cut down on waste and improve the handling of assets.

As an example, predictive analytics can predict energy use and allow the grid to make adjustments, cutting peak loads and leading to less greenhouse gas emission. Using data to improve the process helps minimize the use of resources and emissions by pointing out where the process is not efficient. Besides, predictive maintenance prolongs the lifespan of equipment which means fewer replacements and less harm to the environment.

Such applications support conservation efforts and also help companies reduce their costs and fulfill all necessary regulations. By joining predictive analytics and sustainability initiatives, there is a new research field where industry and care for the environment can work together.

9. CONCLUSION

Predictive analytics is now one of the main technologies in Industrial IoT, helping make operations more efficient, reliable and safe in different industrial fields. It has become clear from this chapter that combining different data analytics approaches, including statistical models and advanced techniques, early faults, improves process optimization, monitors quality, handles energy use and helps avoid risks. The use of huge sensor and device data in predictive analytics makes it possible to use insights for better decisions and preventative care.

The points made here have major effects on both business and educational spheres. IIoT systems can be designed to be effective in analytics only if professionals know about the various models, techniques and data architecture. The issues mentioned such as data quality concerns, scalability limits, threats to security and difficulty in understanding models, make it necessary for solutions that perform well and are trustworthy and user-friendly. When it comes to research, these challenges highlight possible new developments, for instance, explainable AI, digital twins, federated learning and sustainable applications of the IIoT.

In short, predictive analytics is helping lead the IIoT revolution which is making industries more productive and resilient. Although a lot of progress has taken place, more research and teamwork between universities and companies are needed to get rid of existing limitations and reach the true potential of intelligent industrial systems. Overall, predictive analytics will keep advancing and will play a major role in creating better, safer and more sustainable industries across the globe.

REFERENCES

Abitha, R., Rajendar, S., Mangaiyarkarasi, D., Anto, G. L., & Girija, P. (2026). A Comprehensive Survey on Edge Computing, Analytics, and AI Model Optimization for Industrial and Healthcare Applications. *Building Data-Driven Edge Systems for Business Success*, 59-90.

Bhambri, P., & Rani, S. (2023). Challenges, opportunities, and the future of industrial engineering with IoT and AI. *Integration of AI-Based Manufacturing and Industrial Engineering Systems with the Internet of Things*, 1-18.

Bhushan, S., Shanker, S., & Cengiz, K. (2023). The industrial internet of things: advanced analytic framework. In *Handbook of Research on Data-Driven Mathematical Modeling in Smart Cities* (pp. 396–414). IGI Global. DOI: 10.4018/978-1-6684-6408-3.ch021

El-Gendy, S. (2020, December). IoT based AI and its implementations in industries. In *2020 15th international conference on computer engineering and systems (ICCES)* (pp. 1-6). IEEE. DOI: 10.1109/ICCES51560.2020.9334627

Jaffrin, L. C., Nithya, T., & Divya, J. (2025, April). IoT for Sustainable Development: Enhancing Environmental Efficiency and Resource Optimization. In *2025 5th International Conference on Trends in Material Science and Inventive Materials (ICTMIM)* (pp. 1224-1230). IEEE.

Jain, J. K., & Chauhan, D. (2023). Innovative Model of Internet of Things for Industrial Applications. In *Opportunities and Challenges of Industrial IoT in 5G and 6G Networks* (pp. 95–118). IGI Global. DOI: 10.4018/978-1-7998-9266-3.ch005

Kandasamy, M., Saha, S., & Patil, R. S. (2024). Machine Learning in Industry IoT (IIoT). In *Metaheuristic and Machine Learning Optimization Strategies for Complex Systems* (pp. 67-102). IGI Global.

Karmakar, A., Dey, N., Baral, T., Chowdhury, M., & Rehan, M. (2019, March). *Industrial internet of things: A review. In 2019 international conference on opto-electronics and applied optics (optronix)*. IEEE.

Malik, P. K., Sharma, R., Singh, R., Gehlot, A., Satapathy, S. C., Alnumay, W. S., Pelusi, D., Ghosh, U., & Nayak, J. (2021). Industrial Internet of Things and its applications in industry 4.0: State of the art. *Computer Communications*, *166*, 125–139. DOI: 10.1016/j.comcom.2020.11.016

MAY. P. M. A., Savitha, V., Gayathri, S., & Mahalakshmi, S. (2025, April). Exploring Sustainable Computing: IoT, Big Data, and Energy-Efficient Solutions. In *2025 5th International Conference on Trends in Material Science and Inventive Materials (ICTMIM)* (pp. 1218-1223). IEEE.

Misra, S., Roy, C., & Mukherjee, A. (2021). *Introduction to industrial internet of things and industry 4.0*. CRC Press. DOI: 10.1201/9781003020905

Rose, R. M., Varun, C. M., Sivasakthi, K., Kennady, R., & Lekha, D. (2025, May). Leveraging AI and IoT Data for Predictive Modelling and Preventive Healthcare. In *2025 Third International Conference on Augmented Intelligence and Sustainable Systems (ICAISS)* (pp. 628-633). IEEE. DOI: 10.1109/ICAISS61471.2025.11041733

Sathya, U., Hashini, S., Lakshmipriya, C., Varun, C. M., & Nalini, M. (2024, December). Emerging Trends and Future Prospects in Internet of Things (IoT) Technology. In *2024 International Conference on Sustainable Communication Networks and Application (ICSCNA)* (pp. 79-85). IEEE.

Swarnalatha, P., & Prabu, S. (Eds.). (2023). *Handbook of Research on Deep Learning Techniques for Cloud-Based Industrial IoT*. IGI Global. DOI: 10.4018/978-1-6684-8098-4

Vishwas, D. B., Gowtham, M., & Gururaj, H. L. (2021). Industrial Internet of Things 4.0: Foundations, Challenges, and Applications–A Review. *Innovations in the Industrial Internet of Things (IIoT) and Smart Factory*, 172-191.

Chapter 11
Impact of Multi-Agent Technology on High-Tech Product Manufacturing Organizations Using a MCDM Analysis

Vikram Singh
https://orcid.org/0000-0002-8234-7125
National Institute of Technology, Hamirpur, India

Somesh Kumar Sharma
https://orcid.org/0000-0002-8234-7125
National Institute of Technology, Hamirpur, India

ABSTRACT

High-Tech Manufacturing (HTM) is a capital-intensive sector requiring substantial national investments. Integrating Multi-Agent Technology (MAT) into HTM presents new growth opportunities by optimizing processes and enhancing decision-making. This study investigates the impact of MAT on HTM using Multi-Criteria Decision-Making (MCDM) techniques, specifically the Analytical Hierarchical Process (AHP) and Decision-Making Trial and Evaluation Laboratory (DEMATEL). A conceptual framework comprising eight factors and forty-five variables is established. AHP identifies the "high-tech manufacturing process" as the most critical factor, followed by product features, market analysis, and marketing decisions. DEMATEL reveals strong interdependencies, with the manufacturing process exerting the most

DOI: 10.4018/979-8-3373-3146-1.ch011

influence, while quality management, maintenance integration, and organizational factors are most affected. Findings indicate that effective MAT implementation fosters self-operative manufacturing, enhances quality, reduces maintenance efforts, and improves market targeting.

1. INTRODUCTION

High-tech manufacturing has gained heavy weight in the research agenda over the last two to three decades. Economic expansion has expanded consumer expectations to pick innovative products from the global market. Such ambitions have transformed manufacturing industries worldwide, resulting in production systems with cutting-edge technological advancements and regulations that have created novel products ranging from microchips to enormous helicopter engines. HTM systems are equipped with cutting-edge technologies, such as artificial intelligence, the Internet of Things (IoT), 5G, 3D printing, robots, drones, gene editing, big data, blockchain, nanotechnology, and solar photovoltaics (Unctad, 2023). These innovations or new technologies are derived from an agent-based system. Such systems create a digital automation environment configuring with hardware components and software packages (Khandelwal, 2010). Agent technology has grown in popularity in complex manufacturing environments where agents cooperate, coordinate, and share information within the system while also receiving information from the environment (Chan *et al.*, 2007). Multi-Agent Technology (MAT) refers to the collaboration of entities. It actively participates in handling material production, maintenance, supply chain, logistics, behavior analysis, product marketing, and other activities (Oroojlooy & Hajinezhad, 2022). Whereas high-tech product parameters were active in designing, manufacturing, and marketing. Previous articles published on MAT and HTM advocates that no research has been conducted which described MAT and HTM. In this context this study is novel contribution in which bridging of MAT and HTM is described.

This study focused on analyzing the role of MAT in HTM using multi-criteria decision-making techniques. This study has high practical significance in promoting MAT in HTM systems. In this research, AHP, DEMATEL, and TOPSIS were implemented to analyze the behavior of factors and variables. A conceptual framework of eight factors and forty-five variables to evolve priorities, rankings, and interrelationships was developed. This study attempts to provide insights to improve the implications of MAT in high-tech industries. This provides direction and strategy for decision makers, manufacturers, and market experts to improve the quality standards of HTM. An AHP was used to prioritize and rank the factors and variables. Whereas DEMATEL was used to measure the interrelationships among the factors and variables. Finally, TOPSIS was used to validates the global ranking of variables.

The rest of the article is organized as follows: Section 2 presents literature published on MAT and HTM and give provides an overview of MCDM techniques; Section 3 presents the proposed research and facilitates the application of MCDM techniques; and Section 4 performed analysis and discussion. Section 5 summarizes the study with conclusions, research limitations, and suggests avenues for future research based on the findings.

2. LITERATURE REVIEW

The literature argues that high-tech product manufacturing become a hot research topic for researchers and economies. The development in the HTM sector offered products with innovations and desired quality standards. This helps manufacturer to compete in the old market, win the new market, and efficiently use resources. Researchers have defined HTM as industry-based (SIC codes by the US government commerce department and equate high technology with innovations) (Tether & Storey, 1998), firm-based (having small units of research), product base (products having significant R&D investments), and life cycle (high-tech industries that have a short development period) (Steenhuis & De Bruijn, 2006).

After a rapid change in the US economy in the 1960s, HTM was came into existence which developed high-tech domains of computers, electronics, and pharmaceuticals. These developments have demonstrated the importance of manufacturing in high-tech industries. Initially, high-tech industries relied on scientific research; however, as they evolved, they became hubs for innovation (Coad & Rao, 2008). The United States Finance Department described scientific instruments, computer components, electronic components, and various aerospace products in the early 1960s (Lécuyer & Brock, 2006).

The term "new innovative technology" in the context of HTM was defined as "a process, product, service, or trend innovation" (Stornelli *et al.*, 2021). The earliest applications of technological innovation were in the form of programs that increased the flexibility and efficiency of the manufacturing processes. Computers were first used in manufacturing to manage data and rigidity after successful outcomes (Burcher *et al.*, 1999). The advancements in science and technology in the twenty-first century, computer-based systems are highly equipped with cutting-edge technology which are drived through agent based system. The integration of agent based systems with harware components is known as MAT.

It attracts the focus of researchers, scientists, academicians, foreign investors, industrialists, and market experts. The literature has discussed MAT applications in designing, maintenance, production, and marketing paradigms. However, no research has been conducted to analyze the influence of MAT on HTM to maintain the quality standards of high-tech products. HTPs affects the global market and maintaining their quality standards is a major challenge for HTM organization. To overcome such challenges this study defined the problem as " analyzing the role of Multi-Agent Technology (MAT) in High-Tech Manufacturing (HTM)", which is resolved through research objectives structured as; to explore factors and variables to construct a theoretical framework for research, to construct a hierarchical structure to prioritize and rank, and measure the interrelationships between factors and variables. These variables, factors (issues), and sub-factors (sub-issues) were explored after a comprehensive literature review. They are described and listed in Table 1. Figure 1 presents a detailed literature review of 250 articles from reputable journals in various domains. Figure 2 depicts the research approaches used to complete these articles. This figure also supports the limited application of the hybrid AHP and DEMATEL.

Figure 1. Reviewed articles from a different domain

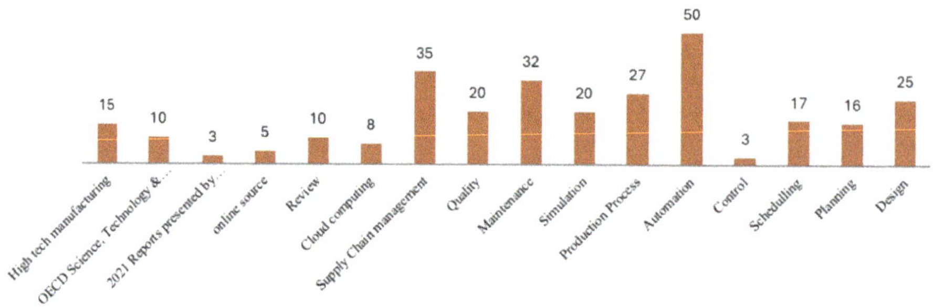

Figure 2. Research approaches witnessed by the literature

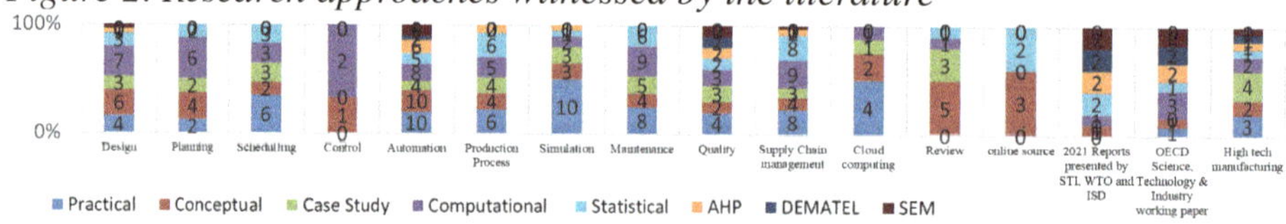

Literature helps to extract 45 MAT variables which are grouped under eight HTM factors. This grouping with nomenclature and descriptions is listed in Table 1. This represents the conceptual domain of research. The Multi-Criteria Decision-Making

(MCDM) techniques used in this research are AHP, DEMATEL and TOPSIS. An overview of these techniques is given in below sub-sections.

2.1 Introductions to Analytical Hierarchy Process (AHP)

AHP was proposed in 1980 by Saaty to solve decision-making problems either qualitatively or quantitatively. The major characteristics of AHP model are the hierarchical decomposition of problems, construction of a criterion hierarchy, and transformation of individual judgments of relative importance into priority weights (Singh & Sharma, 2023). The literature advocates that AHP has attractive features such as rating, priority, choice, and resource allocation (A. Sharma & Sharma, 2021). This research used AHP to prioritize and rank factors and variables.

2.2 Decision Making Trail Evaluation Laboratory (DEMATEL)

In 1972 and 1979, the Battelle Memorial Institute of Geneva's Science and Human Affaire Program created the DEMATEL technique to analyze dynamic and overlapping problematic groupings (Si et al., 2018; Sumrit & Anuntavoranich, 2013). It is one of the best approaches for analyzing the importance of variables, analyzing the impact of one variable over another by using a causal and effect diagram, and determining the interrelationships (A. Sharma & Sharma, 2021). Based on these analyses, this research utilized the DEMATEL method to measure the interrelationship between variables and the influence of one variable over another.

2.3 Technique for Order Preference by Similarity to Ideal Solution (TOPSIS) to rank the variables

TOPSIS, proposed by Yoon in 1981 and Hwang & Yoon in 1984, is a method for solving MCDM (Multi-Criteria Decision-Making) problems that are based on the principle that selected alternatives should be the least Euclidian distance from the positive ideal solution and the greatest distance from the negative ideal solution (Yoon and Hwang, 1985). It is a straightforward and efficient approach for ranking a set of options according to their priority to an ideal solution. In this study this method is used to cross-check the ranks of 45 MAT variables (evaluated through AHP) in connection to the weight value of eight HTM factors computed in AHP.

Table 1. Role of MAT in high-tech manufacturing (conceptual domain of MAT)

Major factors	Sub-factors	Variables of MAT	References
High-tech Product Feature (A)	Nearer to technological development	**Automatic customized test plan (A1)**: check high-tech product aspects, reduce redundant inspection, and testing time.	(Leitao *et al.*, 2015)
	Closely related to science	**Process monitoring, control, and data acquisition (A2)**: acquire scientific data on high-tech products by monitoring process and product activities and confirming outcomes with benchmarked data.	(Sahebjamnia *et al.*, 2016)
	Short Product life cycle	**Intelligent shop floor management based on MAS & CAP (A3)**: smart machine agents and a self-adaptive model to predict product life	(Zhang, n.d.)
	Closely related to customized infrastructure	**Dynamic adaption of product and process parameters (A4)**: creates an autonomous system that can modify parameters if product and process variations are detected.	(Vrba & M rík, 2010)
High Tech Manufacturing Process (B)	Product standardization & innovation	**Virtual manufacturing (B1)**: creates a computer-integrated manufacturing system in the production process that allows users to visualize final product features in 3D graphical visuals.	(Ottaway & Burns, 2000)
	Objective clarity	**Production planning (B2)**: uses computer-aided design and production to turn raw materials into finished goods. It makes manufacturing systems autonomous, adaptable, and modular.	(C. Liu et al., 2014)
	Technical Convergence Trends	**Distributed Digital manufacturing (B3)**: turn hypothetical data into physical form and conduct activities using web-based digital information to monitor, produce, and diagnose manufacturing processes.	(Mahesh *et al.*, 2007)
	Operational condition	**Integration of manufacturing system (B4)**: reduces operational complexity and unpredictability in high-tech product manufacturing.	(López *et al.*, 2023)
	Firm's Resources	**Resource agents (B5)**: handle man, machine, and material for industrial activities.	(Chien *et al.*, 2013)

continued on following page

Table 1. Continued

Major factors	Sub-factors	Variables of MAT	References
High Tech Product Market Analysis (C)	Demand potential	**Demand Forecasting agent (C1)**: take product data, analyze market potential, and estimate product futures.	("Bayesian ..2020)
	Strategic flexibility	**Adaptive agent (architecture of multi-agent system) C2):** includes virtual businesses, distributed intelligent systems, concurrent engineering, and agent architecture. This helps to find quick changes in markets around the world, respond quickly, and assign new resources.	(Tang *et al.*, 2020)
	Barriers and risks involved	**Monitoring Agent (C3)**: compare current and market data to identify global market hazards and barriers.	(Tang *et al.*, 2020)
	Market Competition	**Database agent (C4)**: answers queries from other agents in the multi-agent system.	(Tang *et al.*, 2020)
	Information Accessibility	**Multimedia agent (C5)**: sends information between manufacturers and users in the form of text, photos, music, video, and other formats.	(Zakhama *et al.*, 2019)
	Trade analysis	**Trading agent (C6)**: periodically monitor worldwide market transactions and provide an autonomous environment to coordinate suppliers, procurement, manufacturing, and client bidding decisions.	(Shavandi & Khedmati, 2022)

continued on following page

Table 1. Continued

Major factors	Sub-factors	Variables of MAT	References
High-tech Product Marketing Decision (D)	Marketing Research	**Data collection agent (D1)**: use sensors and trackers to gather market, competition, business, and customer information to automate the system.	(Cooke & Zubcsek, 2017) (Huang & Rust, 2018) (Kim *et al.*, 2019)
	Market Potential	**Market Analysis agents (D2)**: discover global competitors and provide solutions to compete with them and resolve market issues using AI and the IoT.	
	Customer Type	**Customer understanding agent (D3)**: analyze emotional data to identify customer needs. It combines AI and machine learning to fulfill customers' requirements.	
	Market Strategies Objective Strengthening and weakness of the market	**Segmentation agent (D4)**: slices markets by customer desires and separates the fabric market into male and female divisions. This helps marketers estimate segment size and answer consumer questions. **Targeting agent (D5)**: assist companies in selecting the optimal market and targeting the royal consumer using AI and machine learning. **Positioning agent (D6)**: links product quality and customer value by putting the items competitively in the user's mind. This agent also helps companies maintain the customer.	(Daabes & Kharbat, 2017) (Chen *et al.*, 2020) (Daabes & Kharbat, 2017)
	Marketing Action Cross culture interaction Financial Strength	**Standardization agent (D7)**: automates manufacturing, price fixing, consumer access to items, and product marketing to meet consumer needs. **Personalization agent (D8)**: uses deep learning for a personalized point of interest and a video-based automated recommender system to tailor promotional material to customer preferences, willingness to pay, and engagement. **Renationalization agent (D9)**: uses an artificial feeling agent to comprehend and meet customer emotions, negotiate prices, and communicate according to consumer sentiments.	(Leung *et al.*, 2018) (Timoshenko & Hauser, 2019) (Huang & Rust, 2021)
Market Entry Mode Decision (E)	Level of control	**Networking between the agents (E1)**: allows agents to communicate, monitor, analyze, and govern the system.	(Y. Liu et al., 2020)
	Perceptions of decision-makers Contractual or non-contractual Waiting Time	**Data mining agent (E2)**: monitors data sources in an open environment and may modify the contents of the data at any moment. It helps locate and choose important data and puts up a flexible system to modify the manufacturing organization's internal data to meet marketers' expectations. **Information retrieval agent (E3)**: identify and analyze internet data sources globally. It retrieves data from internet sources and keeps market surveys, meeting recordings, customer complaints, emails, and competitor information.**Optimization agent (E4)**: optimizes scheduling solutions beyond autonomous agents' bidding requirements. Production cost, reliability, time, and bottleneck machine groups determine the optimal solution.	(Lavbič & Rupnik, 2009) (Lim & Zhang, 2003)

continued on following page

Table 1. Continued

Major factors	Sub-factors	Variables of MAT	References
Quality Management System (F)	Quality of design Quality of comfort Product customization	**Decision support system (F1)**: oversees manufacturing processes, executes plans during product design, and helps choose the best solutions for product design. **Manufacturability evaluation (F2)**: used a capability model to analyze product attributes, conflicts, and failures. **Virtual Prototyping agent (F3)**: employs a multi-agent system on a virtual reality platform in a collaborative workstation design environment dependent on engineering procedures.	(Pěchouček *et al.*, 2005) (Sun *et al.*, 2001) (Mahdjoub *et al.*, 2010)
	Quality Planning	**Scheduling agent (F4)**: helps to construct the master schedule of all operating machines to complete the assignment on time with better precision and deliver the goods on time.	(Ov, 2003)
	Quality improvement and control Resource Utilization efficiency	**Tracking and tracing agent (F5)**: agent monitors production process data, collects all data, stores data in system memory, and provides real-time data. **Warning agent (F6)**: monitors system performance and compares process and product data. It assists in monitoring the manufacturing system and warns operators of any abnormalities	(Zhang, n.d.) (S. K. Sharma et al., 2017)
Integration of Maintenance in Manufacturing System (G)	Preventive maintenance Deterioration of the production system	**Maintenance Scheduler (G1)**: prepares a maintenance schedule based on the database in current high-tech production to prevent system failures. **Diagnosis agent (G2)**: prepares a maintenance schedule based on the database in current high-tech production to prevent system failures.	(Leung *et al.*, 2018)
	Malfunction Undetected fault	**Internet enables predictive maintenance (G3)**: agent schedules maintenance using machine, process, and product data. It uses this data to arrange timely machine, equipment, and product maintenance. **Prognosis agent (G4)**: evaluates procedure condition records using process information to anticipate system life.	(S. K. Sharma et al., 2017)
	Condition-based maintenance Non-operational network	**E- maintenance (G5)**: it came into existence in the 2000s, stands for efficient, exceptional, effective, and enterprise. E-maintenance uses modern information technology to maintain manufacturing systems. **Integration of monitoring diagnosis and prognosis (G6)**: vital since all agents depend on each other to identify problems, diagnose the system, and predict the system's lifetime	(Bousdekis *et al.*, 2015) (S. K. Sharma et al., 2017)

continued on following page

Table 1. Continued

Major factors	Sub-factors	Variables of MAT	References
Organizational Factors (H)	Organization Productivity Resource allocations	**Organization input (H1)**: controls system input services, including raw materials, tools, services, and others. **Organization output (H2)**: agent produces outcomes according to system input.	(S. K. Sharma et al., 2017)
	Organizational Control Current Inventory	**Control system architecture (H3)**: agent that inspects, diagnoses, and monitors machine operations in online or offline mode and maintains a system where operators may directly interact with the system. **Just in time (H4)**: agents automate material and equipment handling, reduce product flow time, and provide the best products.	(C. Liu et al., 2014) (S. K. Sharma et al., 2017)
	Organization Reputation	**Quality control agent (H5)**: active from product design through product failure and is responsible for the design, material selection, process management, and customer satisfaction by integrating processes and quality management systems.	(Vogel-Heuser *et al.*, 2015)

3. RESEARCH AND DATA METHODOLOGY

3.1 Research framework

The research framework shown in Figure 3 indicates the stages followed to complete this study.

Figure 3. Research framework

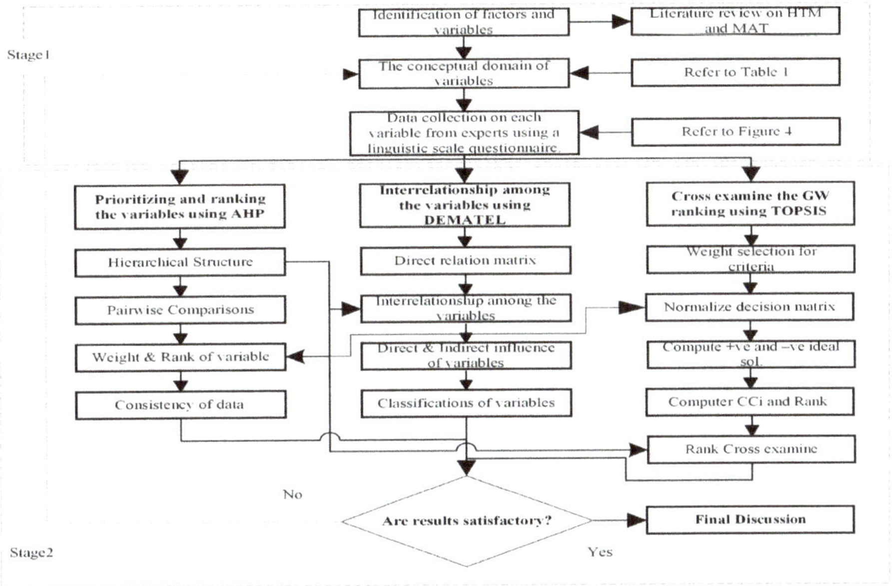

3.2 Data Collection Process

In this article, "experts" refer to academics and industry specialists possessing the necessary competencies. The experts were from prestigious universities and industries, with more than five years of experience. Figure 4 shows an overview of the experts. Experts were given two questionnaires to rate the eight factors and forty-five variables listed in Table 1. The questionnaires were designed considering AHP, DEMATEL, and TOPSIS' background. The scale used to collect the data is listed in Table 2.

Figure 4. Detail of experts

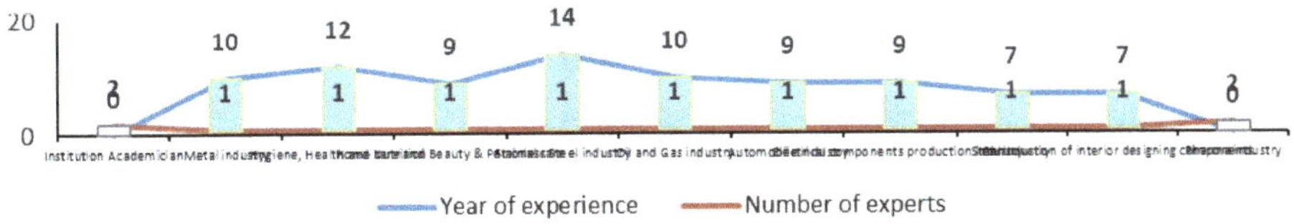

Table 2. Scale used

AHP (V. Singh et al., 2023)		DEMATEL (Chang *et al.*, 2011)		TOPSIS (A. Singh et al., 2022)	
Satty's scale	**Linguistic scale**	**Influence score**	**Linguistic scale**	**Score**	**Linguistic scale**
1	Equally Important	0	No Influence (N)	1	Very Low Importance (VLI)
3	Weakly Important				
5	Fairly Important	1	Very low Influence (VL)	3	Low Importance (LI)
7	Strongly Important				
9	Important	2	Low Influence (L)	5	Medium Importance (MI)
2 to 8	Intermediate Values	3	High Influence (H)	7	High Importance (HI)
		4	Very high Influence (VH)	9	Very High Importance (VHI)

3.3 Applications of AHP to prioritize factors

The AHP used to determine the priority of factors/variables. The AHP technique first organized decision problems into a hierarchical structure (Figure 5) in which factors and variables are organized into different levels. For this research, the target is level I, factors are level II, and variables are level III. The hierarchical architecture is illustrated in Figure 5.

Figure 5. The hierarchical structure of MAT influences HTM

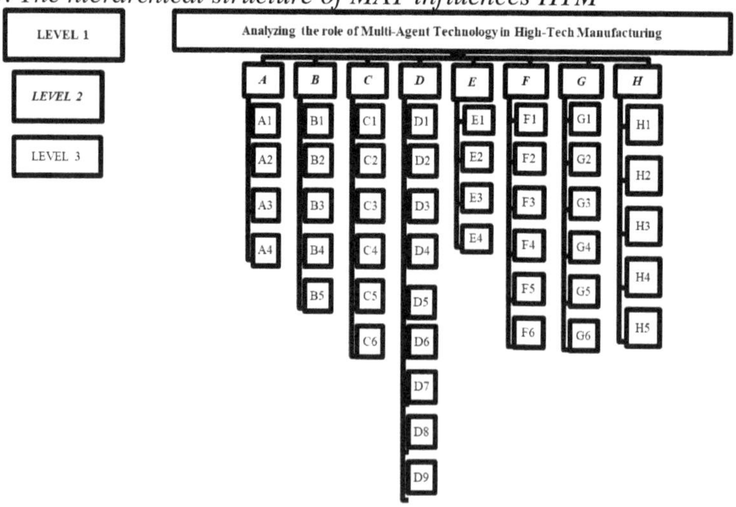

Then pair-wise comparison matrix (A) is developed in which the term 'a_{ij}' in the matrix shows the importance of the ith factor relative to the jth factor.

$$A = \begin{bmatrix} 1 & a_{12}\cdots & a_{1n} \\ a_{21} & 1\cdots & a_{2n} \\ \vdots & \vdots\ddots & \vdots \\ a_{n1} & a_{n2}\cdots & 1 \end{bmatrix}$$

The pair-wise comparison matrix (A) is shown in Table 3.

Table 3. Pairwise comparison matrix

Factor	**A**	**B**	**...**	**G**	**H**
A	1	1.6	...	2	2.2
B	0.625	1	...	3.2	4.2
...	...	...	...	...	...
G	0.357143	0.3125	...	1	5.2
H	0.454545	0.238095	...	0.192308	1
Sum	4.318001	4.572321	...	19.19231	25

Then, the normalized matrix is measured, where each element $\overline{a_{ij}}$ is computed by equation (1)

$$\overline{a}_{ij} = \frac{a_{ij}}{\sum_{i=1}^{n} a_{ij}} \quad (1)$$

The weight of each factor/variable (w_i) are then measured by taking the average (equation (2)) of elements in each row of the normalized comparison matrix Table 4.

$$\omega_j = \frac{\sum_{i=1}^{n} a_{ij}}{n} \quad (2)$$

Table 4. Normalized Matrix

Factor	A	B	...	G	H	PW	Rank
A	0.231589	0.349932	...	0.104208	0.088	0.2021077	**2**
B	0.144743	0.218708	...	0.166733	0.168	0.221797	**1**
...	...	...	...	...	...	...	...
G	0.08271	0.068346	...	0.052104	0.208	0.066533	**7**
H	0.105268	0.052073	...	0.01002	0.04	0.041094	**8**

Finally, the Consistency Ratio (CR) is calculated using equation (3). For consistency of data, the CR value should be less than 0.1 (Bhatia *et al.*, 2022), and The CR calculated in our research is 0.0096. RI is Random Index (Table 5); CI is Consistency Index, and it is calculated by using the equation (4).

$$CR = \frac{CI}{RI} \quad (3)$$

Table 5. Random Index, Source: Saaty (1988)

n	1	2	3	4	5	6	7	8	9	10
RI	0	0	0.58	0.90	1.12	1.24	1.32	1.41	1.45	1.49

$$CI = \frac{(\lambda_{max} - n)}{(n - 1)} \quad (4)$$

n=order of matrix and λ_{max} maximum Eigenvalue Table 6.

Table 6. Delta and Eigenvalue of factors

Factor	Ä	Eigenvector (e)
A	1.807171	8.941659784
B	2.042993	9.211108645
...	...	...
G	0.568742	8.548265417
H	0.346497	8.431882087

The graphical representation of priority weight of factors is shown and Figure 6.

Figure 6. Importance of variables of high-tech manufacturing

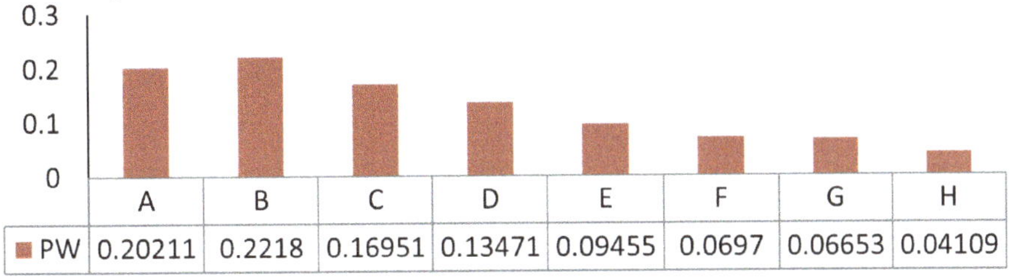

The experts rated the High-Tech Manufacturing Process (B) as the highest priority weight factor, further followed by High-Tech Product features (A), High-Tech Product Market Analysis (C), and others. The order of priority as per weight is B= 0.2218>A= 0.20211>C= 0.16951>D= 0.13471>E= 0.09455>F= 0.0697>G= 0.6654>H= 0.04109.

3.4 DEMATEL to measure the direct & indirect impacts and relationships among the factors/variables

The direct influence of two variables is compared with a rating scale listed in Table 2. First, a direct influence matrix (D) is developed (Table 7) to represent the influence between the factors.

$$D = \begin{bmatrix} 0 & \cdots & x_{m1} \\ \vdots & \ddots & \vdots \\ x_{1m} & \cdots & 0 \end{bmatrix}$$

Table 7. Direct relation matrix

		A	B	...	G	H	Sum
D	A	0	2.1	...	2.3	1.8	14.3
	B	2.6	0	...	2.1	2.7	**16.1**
	...	...	...	...	...	...	...
	G	1.9	1.9	...	0	1.6	11.7
	H	1.7	1.7	...	2.4	0	11.7

The next step develops normalized direct relation matrix ‘X’ using equation (5) and is shown in Table 8.

$$X = \frac{1}{R} * D \quad (5)$$

$$R = \max\left\{ max \sum_{j=1}^{m} x_{ij}, \sum_{i=1}^{m} x_{ij} \right\} \quad (6)$$

Table 8. Normalized direct relation matrix

X		A	B	...	G	H
	A	0	0.130435	...	0.142857	0.111801
	B	0.161491	0	...	0.130435	0.167702
	...	...	...	...	...	...
	G	0.118012	0.118012	...	0	0.099379
	H	0.10559	0.10559	...	0.149068	0

Where, R the is normalization factor. The total relation matrix 'T' is measured using equation (7). 'I' is an identity matrix.

$$T = X + X^2 + X^3 \ldots\ldots + X^n = X(I - X)^{-1} \quad (7)$$

Then, threshold value (α) is measured using equation (8). In this research, threshold value, $\alpha = 0.626237$. After measuring threshold value, all values in matrix T (below threshold value) are set to bold. The model of important relations is displayed in Table 9.

Table 9. Total relation matrix

T		A	B	...	G	H	D
	A	**0.530765**	0.635331	...	0.728698	0.699866	5.299588
	B	0.728536	**0.576505**	...	0.786037	0.809265	5.88436
	...	...	...	...	...	...	...
	G	**0.549064**	**0.539466**	...	**0.504225**	**0.591323**	4.428064
	H	**0.590072**	**0.580108**	...	0.693093	**0.558433**	4.942399
	R	4.71	4.62	...	5.36	5.32	

The next step entails the computation of the sum of column ('D' equation 9) and row ('R' equation 10) of (T) matrix.

$$D = \left[\sum_{i=1}^{n} t_{ij}\right]_{n*1} = \left[d_j\right]_{n*1} = \left(d_1 \ldots . d_i \ldots .. d_n\right) \quad (8)$$

$$R = \left[\sum_{j=1}^{n} t_{ij}\right]_{n*1} = \left[r_i\right]_{n*1} = \left(r_1 \ldots . r_i \ldots .. r_n\right) \quad (9)$$

The value of D+R (influential term) and D-R (causative-effect term) is shown in Table 10.

Table 10. Final relation matrix

Factors	A	B	C	D	E	F	G	H
D+R	10.01	10.50	9.97	10.25	9.52	9.85	9.79	10.26
D-R	0.59	1.27	0.19	0.17	0.17	-1.06	-0.93	-0.38

Using Table 10, a cause-and-effect diagram was constructed for D+R and D-R. The horizontal line drawn using D+R reflects the significance of the variables. D-R is the vertical axis representing the influence. If D-R is positive, the criteria are in the cause group; otherwise, they are in the effect group. So, causal and effect diagrams may be produced using (D+R and D-R) in Figure 7.

Figure 7. Causal and effect relationship of high-tech manufacturing variables

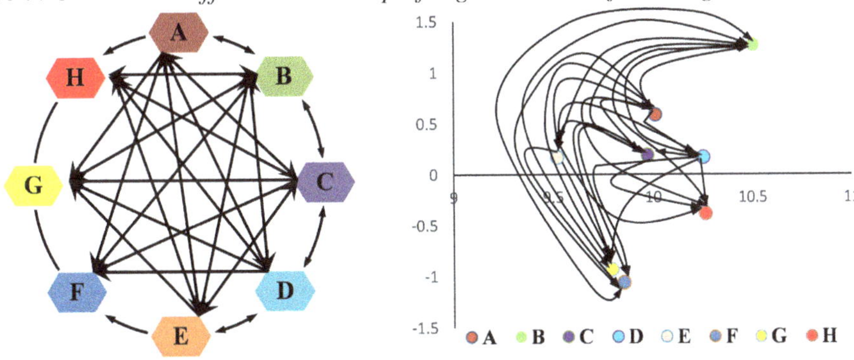

This figure shows that factor A-E belong to the causal group (positive values of D-R), while others F–H belong to effects group owing to the negative values of D-R. The final relation matrix in Table 10 reveals that the high-tech manufacturing process is the most significant, followed by high-tech product features and others due to the values of D+R. The following section discusses how agents help maintain the quality standards of manufacturing organizations in terms of design, process, product, and market activities.

3.5 TOPSIS to validate the global ranking of MAT

The global ranking of variables is a major concern because it reflects their importance in the manufacturing system to eliminate breakdown probabilities. As a result, to maintain a high degree of accuracy in the MAT implications, the global ranking of the variables was validated using TOPSIS. The weight values (W_j) of factors (A to H) as described in Table 4 as priority weights are used. The decision matrix (D) is calculated by computing the average of the responses collected from the experts or followed by equation (11) and is shown in Table 11.

$$D = \begin{bmatrix} X11 & \cdots & X1j \\ \vdots & \ddots & \vdots \\ X_{i1} & \cdots & Xij \end{bmatrix}$$

Table 11. Decision matrix

	A	B	....	G	H
A1	5	4	....	4	5
A2	5	5	....	4	5
A3	3	4	....	5	5
....	....	....	....	....	....
H4	3	4	....	3	2
H5	1	2	....	1	1

The normalized decision matrix R_{ij} (X) is computed using equation (10) and is shown in Table 12.

$$R_{ij}(X) = \frac{X_{ij}}{\sqrt{\sum_{i=1}^{n} X_{ij}^2}} \quad i = 1, \ldots, n ; j = 1, \ldots, m \qquad (10)$$

Table 12. Normalized decision matrix

	A	B	....	G	H
A1	0.3137	0.2361	....	0.2520	0.3010
A2	0.3137	0.2951	....	0.2520	0.3010
A3	0.1882	0.2361	....	0.3150	0.3010
....	....	....	....	....	....
H4	0.1882	0.2361	....	0.1890	0.1204
H5	0.0627	0.1181	....	0.0630	0.0602

Then, the weight normalized matrix V_{ij} (x) (Table 13) is determined by using equation (11)

$$V_{ij}(x) = W_j R_{ij}(X)\ i = 1, \ldots, n\ ; j = 1, \ldots, m \tag{11}$$

Where, $W_j = \frac{wj}{\sum_{j=1}^{m} Wi}$, j=1,.......,m, so that $\sum_{i=1}^{m} Wj = 1$ and w_j is the actual weight value given to W_j.

Table 13. Weighted normalized matrix

	A	B	....	G	H
A1	3.6392	2.5524	....	2.9179	3.7079
A2	3.6392	3.1905	....	2.9179	3.7079
A3	2.1835	2.5524	....	3.6474	3.7079
....	....	....	....	....	....
H4	2.1835	2.5524	....	2.1884	1.4832
H5	0.7278	1.2762	....	0.7295	0.7416

Thereafter, positive, and negative ideal solutions are calculated by implementing equations (12) to (15) and are shown in Table 14.

$$B^{+} = \left(V_1^{+}, V_2^{+}, \ldots, V_n^{+}\right) \tag{12}$$

$$B^{-} = \left(V_1^{-}, V_2^{-}, \ldots, V_n^{\mp}\right) \tag{13}$$

$$V_j^{+} = \left\{(max\ V_{ij}(X) \middle| j\epsilon j_1)\ ,\ (min\ V_{ij}(X) \middle| j\epsilon j_2)\right\} \tag{14}$$

$$V_j^{-} = \left\{(min\ V_{ij}(X) \middle| j\epsilon j_1)\ ,\ (max\ V_{ij}(X) \middle| j\epsilon j_2)\right\} \tag{15}$$

Where $i = 1,2,3,....,n$ and j1 and j2 denote the negative and positive criteria, respectively.

Table 14. Ideal solution for criteria

Factor	A	B	....	G	H
V_j^+	3.639	3.191	....	3.648	3.710
V_j^-	0.728	0.638	....	0.730	0.721

The distance between each alternative and positive ideal (equation 16) and negative ideal solution (equation 17) is computed and shown in Table 14. The value of each alternative computed in the form of D+ and D-.

$$D_i^+ = \sqrt{\sum_{j=1}^{m} \left[V_{ij}(X) - V_j^+(X)\right]^2} \ , i = 1, \ldots, n \tag{16}$$

$$D_i^- = \sqrt{\sum_{j=1}^{m} \left[V_{ij}(X) - V_j^-(X)\right]^2} \ , i = 1, \ldots, n \tag{17}$$

Finally, the relative closeness degree (CD_i) of each variable to the ideal value is computed by using equation (18) and shown in Table 15.

$$CD_i = \frac{D_i^-}{(D_i^+ + D_i^-)} \ , i = 1, \ldots, n \tag{18}$$

Table 15. Ideal solution for criteria and relative closeness degree

Factor	A1	A2	A3	.....	H4	H5
D_j^+	0.9692	1.1716	2.2191	.....	4.1439	6.5505
D_j^-	6.2658	6.0157	5.3387	.....	3.0610	0.5766
CD_i	0.8660	0.8370	0.7064	.....	0.4248	0.0809

The relative closeness degree (CD) values obtained from the TOPSIS analysis were utilized to determine the final ranking of the Multi-Agent Technology (MAT) variables, as presented in Table 15. These CD values represent the degree to which each alternative (or variable) approximates the ideal solution in a normalized decision space, where higher CD values indicate superior performance and stronger significance in achieving the desired objectives. The comparison between the AHP-derived global rankings and the TOPSIS-based overall rankings reveals a high level of consistency, with approximately 90% of the global ranks aligning perfectly between the two methods and only 10% showing minor, negotiable variations.

These minor variations are statistically insignificant and do not affect the integrity or functionality of the overall system model. Such differences may be attributed to methodological distinctions—AHP focuses on hierarchical pairwise comparisons based on expert judgments, while TOPSIS measures relative distances from ideal and non-ideal solutions. Nevertheless, the strong alignment between both approaches demonstrates the internal reliability and robustness of the decision-making framework adopted in this research. Hence, the ranking order of MAT variables derived from these methods is considered 100% valid and practically reliable for implementation in high-tech manufacturing systems.

Furthermore, the DEMATEL analysis provided additional validation of the results by quantifying the cause-and-effect relationships among the identified factors and variables. The computed (D + R) values, which represent the total degree of influence (both given and received), correspond closely with the hierarchical significance established through AHP. This convergence confirms that the factors identified as most influential in AHP are also those exerting the greatest causal impact in DEMATEL. The consistency between these multi-criteria decision-making (MCDM) techniques strengthens the empirical soundness of the findings and enhances the confidence in their interpretive validity. Therefore, based on these cross-validations, the subsequent section provides an in-depth analysis and discussion of the results measured through AHP and DEMATEL, emphasizing their interrelationships, causal hierarchies, and implications for optimizing MAT integration in high-tech manufacturing environments.

4. ANALYSIS AND DISCUSSION

Table 16 shows the results obtained through AHP, DEMATEL, and TOPSIS. It also shows validation of Global ranks through TOPSIS.

Table 16. Weight & ranking with direct & indirect effects of variables

		AHP Finding								TOPSIS	DEMATEL	
	Main factors	Local Weight	Rank		Sub-factors	Local Weight	Local Rank	Global Weight	Global Rank	TOPSIS Rank	D+R	D-R
Hierarchy level-2	B	0.2218	1	Hierarchy level-3	B1	0.272967	1	0.06054	5	5	6.450771	0.020995
					B2	0.238391	3	0.05287	7	7	6.348664	-2.67655
					B3	0.262520	2	0.05823	6	6	6.843427	0.959712
					B4	0.14336	4	0.03179	11	11	6.157415	0.090015
					B5	0.082762	5	0.01836	20	20	5.086933	-1.36895
	A	0.2021	2		A1	0.343419	2	0.06941	2	2	5.07	0.49
					A2	0.353423	1	0.07143	1	1	7.87	0.12
					A3	0.195041	3	0.03942	10	10	7.24	0.62
					A4	0.108116	4	0.02185	18	18	6.96	-1.23
	C	0.1695	3		C1	0.276447	2	0.06132	4	4	10.94956	0.749018
					C2	0.277114	1	0.06146	3	3	10.07738	0.460673
					C3	0.184228	3	0.04086	8	8	9.99728	-0.39304
					C4	0.138724	4	0.03077	12	12	10.02595	-0.21435
					C5	0.079359	5	0.01760	23	23	9.578803	0.100326
					C6	0.055533	6	0.01232	27	27	9.800607	-0.70263

continued on following page

Table 16. Continued

		AHP Finding								TOPSIS	DEMATEL	
	Main factors	Local Weight	Rank		Sub-factors	Local Weight	Local Rank	Global Weight	Global Rank	TOPSIS Rank	D+R	D-R
	D	**0.1347**	**4**		D1	0.193378	2	0.02605	14	14	9.25	0.38
					D2	0.214081	1	0.02884	13	13	9.73	1.21
					D3	0.164123	3	0.02211	17	17	9.29	0.21
					D4	0.130553	4	0.01759	24	24	9.55	0.12
					D5	0.091263	5	0.01229	28	28	8.86	0.10
					D6	0.067164	6	0.00905	34	34	9.24	-0.97
					D7	0.06426	7	0.00866	37	37	9.04	-0.77
					D8	0.03909	8	0.00527	**41**	**42**	9.65	-0.28
					D9	0.03609	9	0.00486	**42**	**41**	9.80	0.08

continued on following page

Table 16. Continued

		AHP Finding								TOPSIS	DEMATEL	
	Main factors	Local Weight	Rank		Sub-factors	Local Weight	Local Rank	Global Weight	Global Rank	TOPSIS Rank	D+R	D-R
	E	**0.0946**	**5**		E1	0.42793	1	0.04046	9	9	4.77	0.54
					E2	0.25745	2	0.02434	15	15	6.25	0.34
					E3	0.21124	3	0.01997	19	19	6.03	0.17
					E4	0.10339	4	0.00976	32	32	6.20	-1.05
	F	**0.0697**	**6**		F1	0.32145	1	0.02241	16	16	10.61544	1.067232
					F2	0.25621	2	0.01786	22	22	9.928331	0.25783
					F3	0.17125	3	0.01194	29	29	10.04364	-0.46304
					F4	0.12757	4	0.00889	36	36	9.789913	-0.03358
					F5	0.08584	5	0.00598	39	39	9.761281	0.097213
					F6	0.05687	6	0.00396	**43**	**44**	10.0704	-0.92566
	G	**0.0665**	**7**		G1	0.25846	2	0.01719	25	25	10.88167	0.887316
					G2	0.27491	1	0.01829	21	21	10.35028	0.386406
					G3	0.17388	3	0.01157	30	30	9.909838	-0.23287
					G4	0.14545	4	0.00968	33	33	9.880571	-0.30378
					G5	0.10886	5	0.00724	38	38	9.726998	0.075711
					G6	0.05710	6	0.00380	**44**	**45**	10.02343	-0.81278
	H	**0.0411**	**8**		H1	0.27984	2	0.01150	31	31	7.596889	0.339523
					H2	0.30281	1	0.01244	26	26	6.97183	0.092418
					H3	0.21648	3	0.00889	35	35	7.525113	0.293934
					H4	0.12967	4	0.00533	40	40	7.509598	-0.04674
					H5	0.07121	5	0.00293	**45**	**43**	6.444193	-0.67913

The experts have rated the regulating variables of MAT as producing excellent results in "high-tech manufacturing process" In this context, Figure 8 shows the computational results. 'Virtual manufacturing (B1)' displays product specifications by building a virtual environment (3-D graphical system) between market analysts, consumers, and the final 3D products. This helps marketers compare current product specifications with those of actual 3D products. Such flexibility helps to manage design standards, eliminate specification disputes, and maintain product standards. 'Distributed digital manufacturing (B3)' is a flexible agent-based system that transforms conceptual data into reality in technologically convergent environments. It employs the newest integrated technologies for time reduction and process quality improvement in high-tech product manufacturing. The third-weighted ' production planning (B2) agent helps clarify production goals by choosing the optimal product design, method, and equipment. When all manufacturing factors are appropriately selected, production errors are reduced, and the process provides a higher level of accuracy. The fourth-ranked variable is 'Integration of manufacturing systems (B4).' This agent provides agility in manufacturing operations to improve the process quality and helps meet international market standards. The least weighted variable in this category is 'Resource agent (B5),' which helps fulfil the production requirements and prevents delays caused by material non-availability. This creates a just-in-time environment in which global market demand may be met on time. Now Figure 9 shows that the efficiency of production planning and resource agents (as belong to the effect group) depends on the effective working of remaining causative group variables. This causative and effective arrangement of variables may be employed in manufacturing systems to maintain production reliability, avoid production faults, decrease maintenance costs, and maintain quality standards in the manufacturing sector. This figure also shows the relationship between the variables of the high-tech manufacturing process to improve process capabilities to achieve quality standards.

Figure 8. Importance of variables of MAT in the B

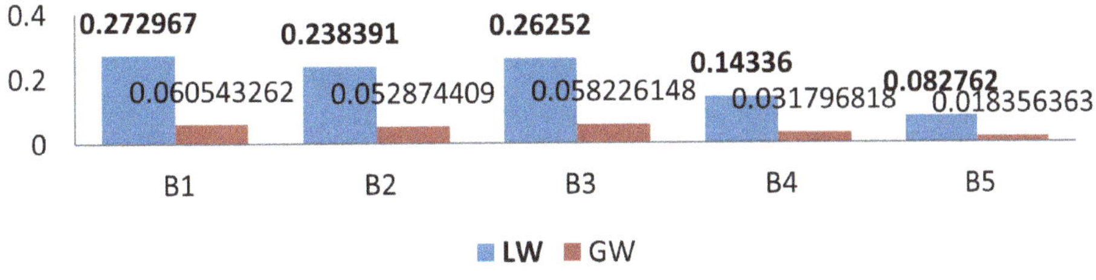

Figure 9. Causal & effect relationship among variables of B factor

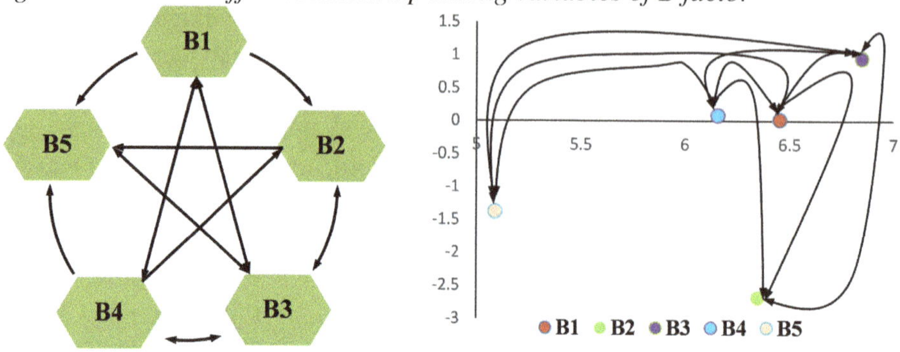

For "high-tech product features," evaluators assessed MAT variables as shown in Figure 10. This figure illustrates that 'Automatic customized plan (A2)' achieving first rank, which helps in changing technical specifications as per product design. This helps to maintain product features according to the required standards. The second-ranked variable is 'process monitoring, control, and data acquisition (A2), which uses MAT to monitor process activities and control product features. It collects data related to high-tech products for future reference and compare these data with existing national and international standards to ensure the high quality of products. 'Intelligent shop floor management based on MAS & CAP (A3)' controls product complexity and unexpected resource changes to maintain long product life. The 'Dynamic adaptation of product and process parameters (A4)' obtained the least rank. It uses MAT to create a flexible autonomous adaptive system that lowers investments in upgrading existing production setups and managing quality at low production costs. Based on the cause-and-effect relationships, Figure 11 shows that for efficient dynamic adaptation of the product and process parameters that belong to the effect group, there must be an active performance of other agents that belong to the cause group. This application of MAT in high-tech product features helps in maintaining innovative product features, ensuring product quality, and eradicating processes and product parametric flaws.

Figure 10. Importance of variables of MAT in high-tech product features (A)

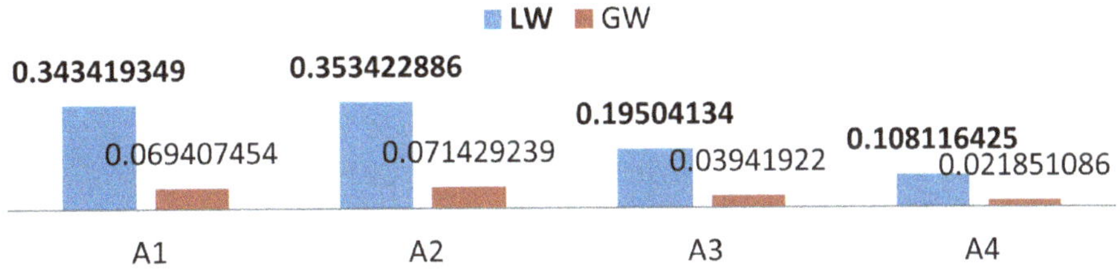

Figure 11. Causal & effect relationship among variables of A

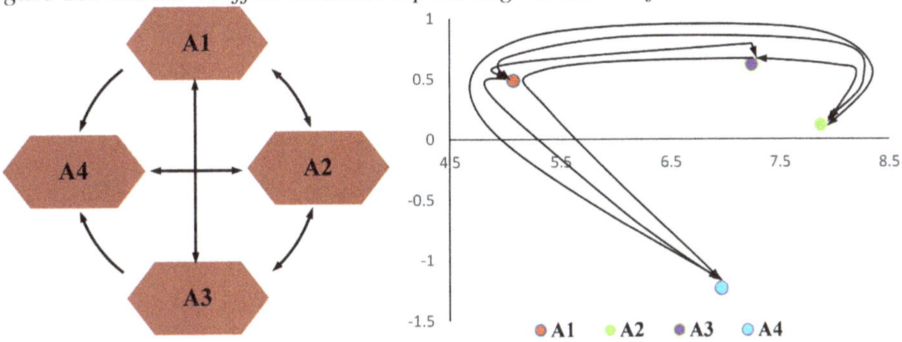

For "high-tech product market analysis," respondents rated MAT variables to assess market behavior. It forms manufacturing enterprises about the present market standards. Figure 12 indicates that the respondents first placed an adaptive agent (architecture of the multi-agent system) (C2). This agent defines new tactics, modifies existing manufacturing strategies in response to market demand, and improves product quality to last longer in competitive markets. The second-ranked variable is" demand forecasting agent (C1),' which analyzes market demand and predicts production volume to decrease inventory, supply chain, and unnecessary production costs. The third-rated agent is 'Monitoring agent (C3)' used to observe market activity and gather data on competitor products, so new creative features may be introduced or eliminated in product design and avoid development hurdles and hazards. The fourth-ranked variable is 'Data agent (C4)' which collects data related to existing competitors' products to compare product specifications so new products can be launched with innovative features. The fifth-ranked variable is 'Multimedia agent (C5),' used to display the attractive aspects of high-tech products via a multimedia platform. It targets the international market and forecasts the product demand accordingly. The sixth-ranked variable, 'Trading agent (C6),' analyzes trading activities and predicts loss and profit; therefore, production may cease when a loss is projected and increase when profit is predicted. Figure 13 illustrates the relationships between these variables. This shows that for the appropriate use of monitoring, data, and trading agents belonging to the effect group, a functional causal group. This implication of MAT variables under market analysis helps in analyzing the market standards and provides data as per market demands to high-tech manufacturing organizations.

Figure 12. Importance of variables of MAT in high-tech product market analysis (C)

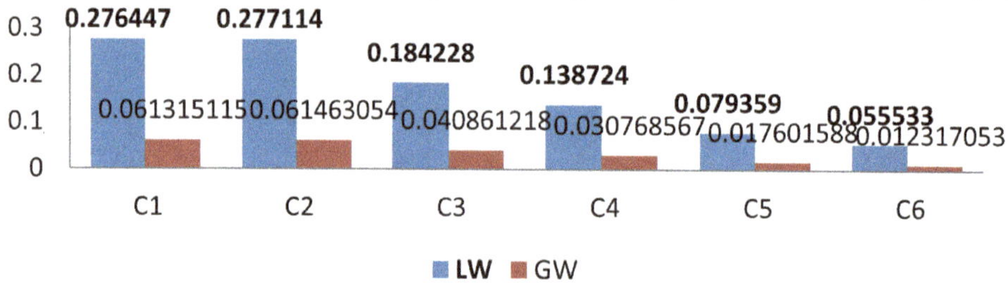

Figure 13. Causal & effect relationship among the variables of C

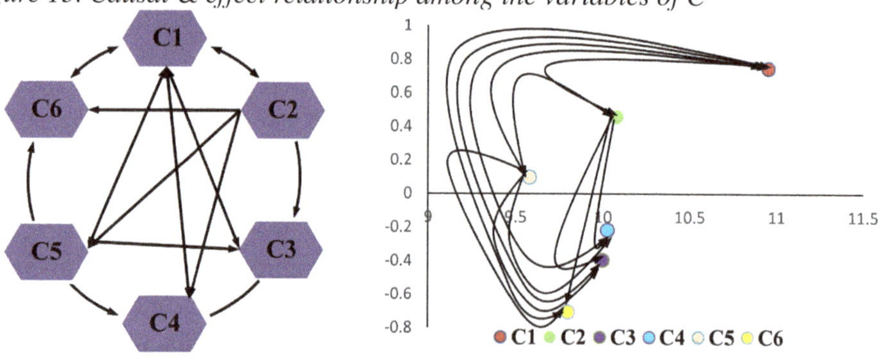

Next, in the 'high-tech product marketing decision,' the priority evolved for variables of MAT are displayed in Figure 14. The top-ranked variable is the Market Analysis Agent (D1), which assesses the international market state and competitors' product activities. This information helps anticipate product demand and the timing of new product marketing. This ensured a safe and profitable launch. The second-ranked variable is the 'Data collection agent (D2)' used to gather market behavior data from diverse sources and retain the data for future reactions to reduce mistakes due to lack of market knowledge and market products at peak demands after maintaining proper quality. The third-ranked variable is the Customer Understanding Agent (D3), which is used to understand customers' expectations about product features and respond according to their desires so that customers feel comfortable with the firm and their product's quality. The fourth-ranked 'Segmentation Agent (D4)' aids in developing marketing strategies by segmenting the market and promoting high-tech products based on market potential, resulting in low marketing costs. After the segmentation agent, the targeting agent (D5) ranks fifth in its group; it targets a dense market for marketing products to increase product demand. The positioning agent (D6) maintained product stability in the market and ranked sixth. It makes an impression on future market consumers' minds by offering product quality standards. The seventh-ranked variable is 'Standardization agent (D7),' used to access market variances, including consumer habits and competitive environments to ensure product quality standards. The personalization agent (D8) is rated eighth and used to promote high-tech product characteristics. The least-rated variable in the ninth place is the 'Renationalization agent (D9),' which interacts directly with consumer demands and organizes negotiations to achieve customer satisfaction in terms of quality.Now Figure 15 shows an interesting relationship among the MAT variables. Targeting, standardization, and personalization agents are effective by the remaining causative variables. This analysis concludes that proper utilization of MAT in high-tech marketing decisions may helps manufacturers in maintaining manufacturing standers as per market demands and minimize marketing efforts by predicting appropriate marketing time.

Figure 14. Importance of variables MAT in high-tech product marketing decision (D)

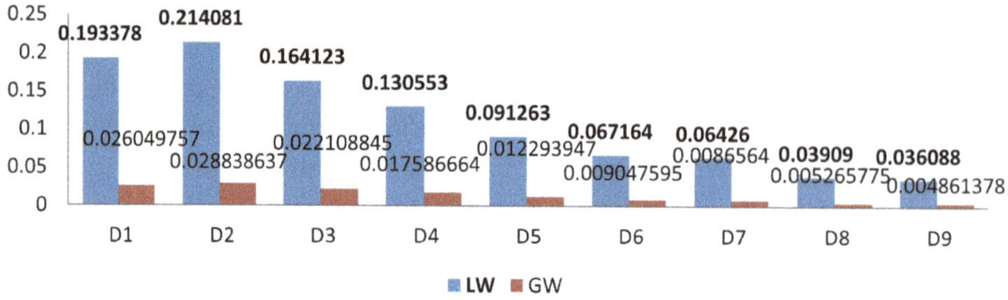

Figure 15. Causal & effect relationship among the variables of D

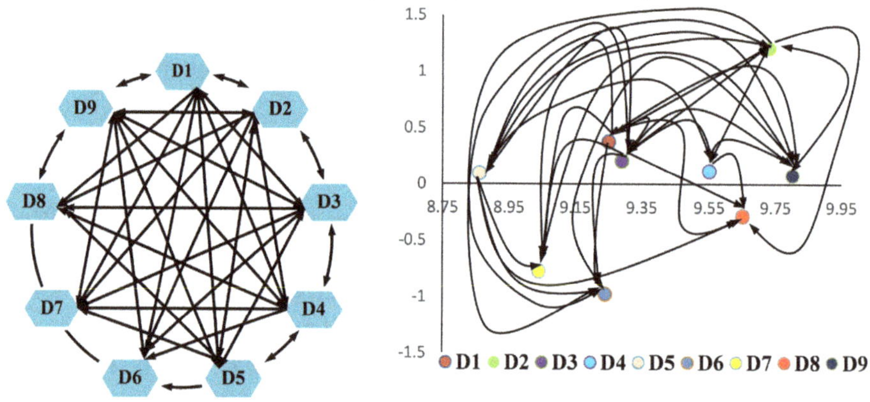

"Entry-mode decision (E)" Figure 16 shows experts' responses for MAT variables for market selection. 'Networking between agents (E1),' evaluated in the first rank, provides a platform in which agents coordinate, collaborate, and share information to control different challenges that cause obstacles while entering the market. The second-ranked variable is 'Data mining agent (E2),' which gathers and analyzes market data and updates available data to develop new product perimeters for high-quality standards. The third-ranked 'Information retrieval agent (E3)' analyses global databases to gather key problems related to consumers and provides optimal solutions by comparing the issues with existing infrastructure. 'Optimization Agent (E4)' is the fourth lowest-ranked variable, it optimizes market selection strategy to increase exports and launch new products with innovative standards and specifications. Figure 17 shows that data-mining agents are the most significant elements for market selection and expanding exports of high-tech products. This has an enormous impact on the other group variables. However, the relation among variables shows that for effective implementation of optimization, agents depend on the efficient working of others to enter the competitive market with zero risks.

Figure 16. Importance of variables of MAT in marketing entry mode decision (E)

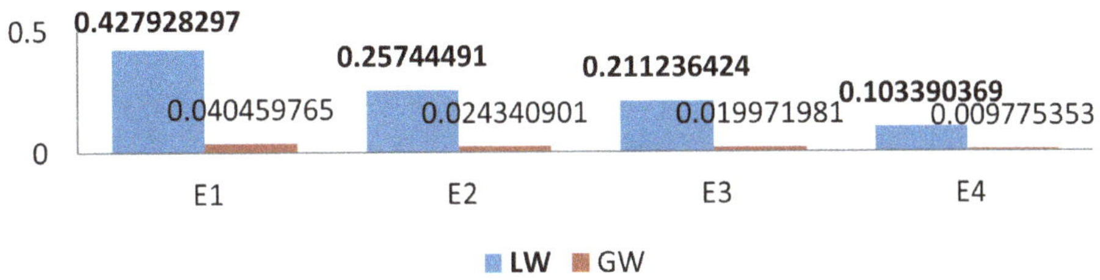

Figure 17. Causal & effect relationship among the group member of E

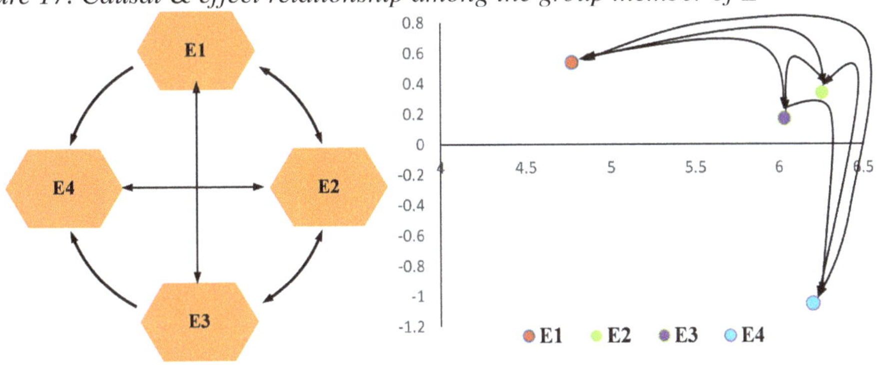

For "quality management system (F)," respondents graded MAT variables to manage high product and process accuracy (Figure 18). First place goes to 'Decision support tool (F1).' It employs a computer-based multi-agent system for operational design, planning, and decision-making, resulting in outstanding product quality. The second-ranked variable is the 'Manufacturability evaluation (F2).' It collects process parametric information and compares it with product design and an existing database to reduce parametric variation conflicts during manufacturing and to ensure process quality. 'Virtual prototyping (F3)' is the third-ranked variable. It constructs virtual reality for molding and visualizing product behavior to remove design flaws and improve accuracy. Fourth, the scheduling agent (F4) prepares master schedules, from product design to delivery, and implements them on time to maintain reputation and customer confidence. The fifth-ranked variable is 'Tracking and tracing agent (F5), which tracks the manufacturing system to acquire process and product information. If dimension deviations are noted, they may be eliminated during production, resulting in approved product quality, zero waste, and reduced production time. The least-rated variable is 'Warning agent (F6),' which monitors system operations and generates alarms when faults beyond the defined limit are detected. This resulted in high-quality products within limits. Experts consider the decision support tool to be the most significant variable to ensure high-quality products. Figure 19 shows that for the proper implementation of virtual prototyping, along with scheduling agents and warning agents, there should be an effective functioning of others so that firms can promise a higher degree of accuracy in their products and ultimately maintain their reputation in a competitive market.

Figure 18. Importance of variables of MAT in quality management system (F)

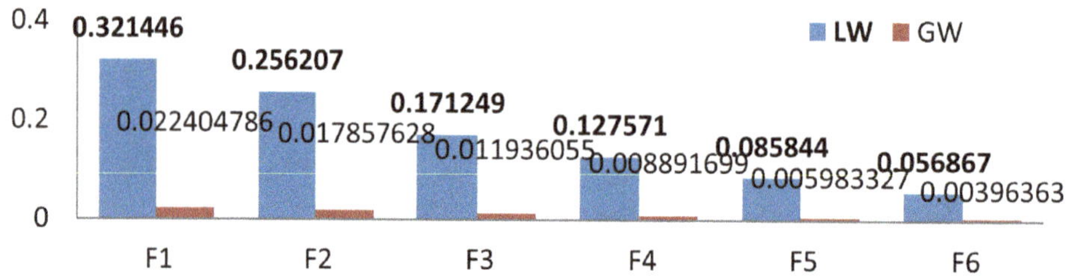

Figure 19. Causal & effect relationship among the variables of F

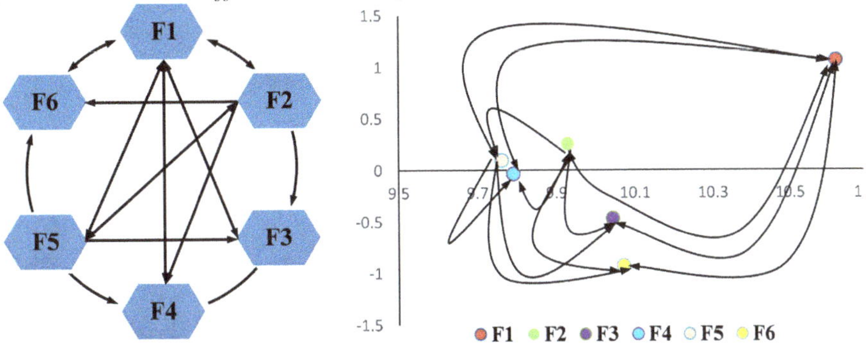

"Integration of maintenance in manufacturing system (G)" Figure 20 shows how experts ranked MAT factors to maintain high-tech product life and industrial standards. The diagnosis agent (G2) is the first-ranked variable that assists in addressing the current manufacturing challenges for undesirable reasons, resulting in high effectiveness and accuracy in the production system. The second-ranked variable is the 'Maintenance scheduler (G1) agent, which prepares predictive maintenance plans to prevent defects, such as unexpected failure and unscheduled downtime. The third-ranked variable 'Internet-enabled predictive maintenance (G3)' agent used previous data from industrial equipment to generate an online/offline maintenance plan. 'Prognosis (G4)' ranked fourth. It assesses equipment-related data to predict the remaining life of the equipment. This saves inventory assets and informs manufacturers to plan equipment replacement, resulting in high production stability. 'E-maintenance (G5)' the fifth-ranked variable, maintains exported items when a business partner does not have exclusive rights or product knowledge. This reduces the maintenance costs and ensures accuracy. The sixth-ranked variable is 'Integration of monitoring, diagnosis, and prognosis (G6).' An agent-based system builds a self-operating system, in which an agent finds errors, resolves them, and predicts how long the product will last. This method helps identify malfunctions in manufacturing systems and assists in eliminating them. In Figure 21, causation-and-effect analysis shows that G3, G4, and G6 agents may perform better when the other three variables in maintenance integration in manufacturing work efficiently. This arrangement concludes that the proper implications of MAT variables in an integrated system of manufacturing and maintenance may lead to better agent functioning, eliminating internal and external maintenance problems, and ensuring high-quality products.

Figure 20. Importance of variables of MAT in maintenance in manufacturing system (G)

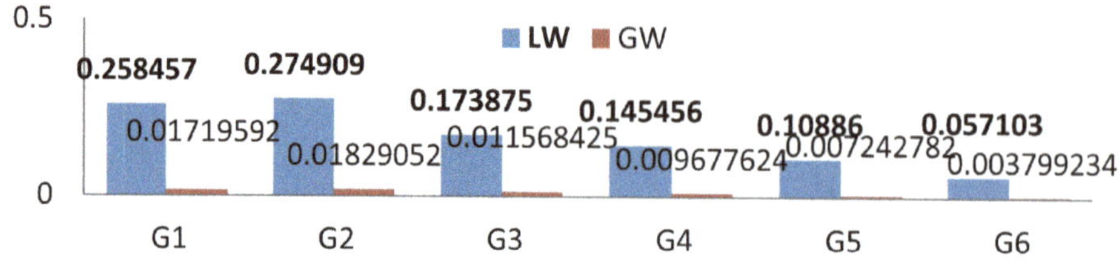

Figure 21. Causal & effect relationship between variables of G

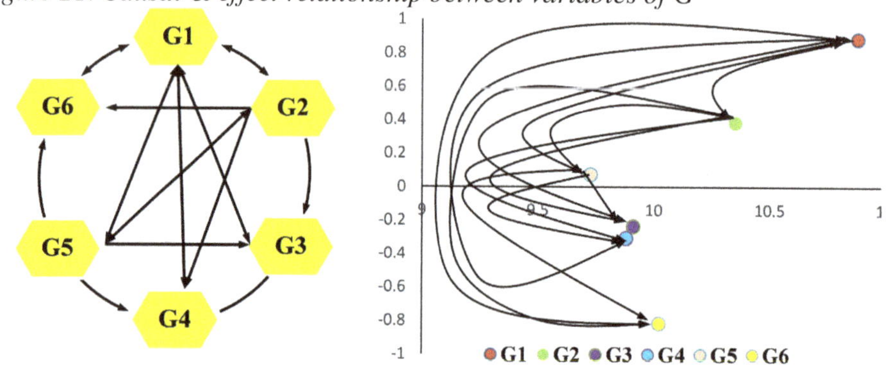

Variables of "Organizational factors (H)" were ranked to use resources effectively and propose optimal solutions for improving organizational performance. Figure 22 shows that Organizational output (H2)' ranked first. It uses agents to ensure the quality of the product and services that are provided to consumers, which enhances organizations' standards and provides easy access for consumers. 'Organizational input (H1)' was the second-ranked variable. It monitors the quality of the material required for manufacturing the products and helps eliminate the material from production with irregularities or flaws. This increases product quality and maintains the quality standards of HTM organizations. The third-rated variable was the 'control system architecture (H3),' which maintains control over issues that arise in material handling, production, product quality, and delivery, and maintains the reputation of manufacturing organizations. 'Just in time (H4),' rated fourth, helps all working platforms in organizations by delivering quick solutions to their challenges and eliminating internal flaws that might cause a system failure. The least-rated variable is 'Quality control agent (H5),' which helps strengthen the manufacturing organization's quality, reliability, and stability. Experts rated organizational input as the most critical variable for increasing the overall effectiveness of organizational activities, as illustrated in Figure 23. The analysis concluded that quality control agents can perform better when other variables of the MAT function effectively in the system. Effective implementation of MAT in an organizational environment may boost productivity, provide the best solution to internal and external flaws, and preserve and boost organizations' quality standards.

Figure 22. Importance of variables of MAT in organizational factor (H)

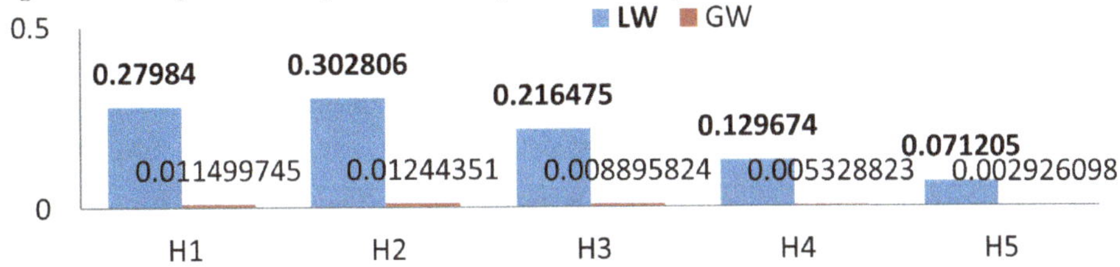

Figure 23. Causal & effect relationship among the group member of H

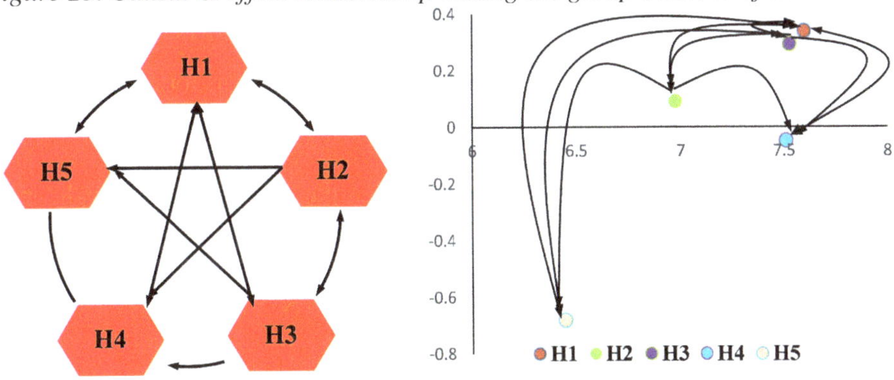

5. CONCLUSION, LIMITATIONS, AND FUTURE SCOPES

In the current scenario, the relationships between economies that are close together are changing rapidly. Some relationships appear stronger, while others are tense. These unforeseen changes caused fluctuations in demand and supply, which impacted the standards of manufacturing high-tech products. To overcome this situation, this study analyzes the role of MAT in high-tech manufacturing by evaluating the ranking, priority, importance, and relationships among the factors and variables.

AHP evolved 'High tech manufacturing' as the top-ranked variable, which shows that manufacturing with more accuracy offers higher quality with innovations. M-G. Antonia (2012), who described the high-tech manufacturing process as an important component of Spanish economic growth and helped maintain quality standards (Madrid-Guijarro *et al.*, 2007), supports the findings of this research. The second-ranked key variable, 'high-tech product features,' maintains existing features and facilitates the addition of new features to product specifications. This increases consumer interest, which helps increase organizational standards. Another study, by Z. Vaida (2011), supports the findings by examining a high-tech product development model in which high-tech product features are the most important aspect of high-tech products to increase quality (Zemlickienė, n.d.). The third-ranked key variable, 'high-tech product marketing analysis,' analyzes market data and suggests highly prospective markets to launch products with high-quality standards. The fourth key variable is 'high-tech product marketing decision,' which assists manufacturers in planning marketing activities and making decisions for marketing their products in the international market. It also helps investors invest more in product development. The 'market entry mode decision' is the fifth-ranked key variable. This depends on the market experts' predictions. This prediction suggests a roadmap to launch products in the global market, which can increase the import and export of high-tech products. Experts ranked sixth in 'quality management systems,' which helps to maintain quality from the beginning to the end of products to acquire business partners', investors,' and customers' trust. The seventh-ranked 'integration of maintenance in the manufacturing system' ensures the durability, accuracy in the process, and excellent quality of goods, resulting in high quality with lowered product's overall cost, volumetric rise in sales, and elimination of maintenance costs. 'Organizational factor' was the least rated. It helps in arranging input resources (such as machines and materials), controlling manufacturing activities (such as process quality parameters and others), and analyzing loss and profit to manage internal and external issues without affecting the organization.

Similarly, this technique evolved a global rank order (validated through TOPSIS) for MAT variables, in which the top five ranked variables were process monitoring, virtual manufacturing, adaptive agents, automatic agents, and demand forecasting agents. This indicates that the MAT variables are active not only in their respective groups but also throughout the system. As a result, properly bridging the high-tech manufacturing process with high-tech product features and MAT variables can effectively maintain quality standards in high-tech manufacturing organizations.

Thereafter, the DEMATEL approach was used to measure the influential and cause-and-effect relationships between the factors and variables. High-tech manufacturing process measured the most influential factors and helps maintain high accuracy by choosing the right tool, machine, and process. Quality management systems and organizational factors belong to the effects group and are influenced by the remaining key causative variables. The effectiveness of the effect group factors or variables depends on the efficient working of the cause group factors or variables. This integration between high-tech manufacturing and the multi-agent system helps generate a high-tech production system in which manufacturing flaws should be eliminated automatically; maintenance goes simultaneously with production, decreases market risks, and increases product quality.

The implications of MAT in high-tech manufacturing organizations face various challenges, the most significant of which is the poor knowledge of MAT. Industrial automation can progress only if manufacturers are aware of this concept. However, the author's novel efforts may raise awareness among manufacturers, researchers, market experts, and professional. Nevertheless, the authors believe that the facts presented in their research can be helpful to those who want to understand the conceptual aspects of MAT and are working to expand the scope of the research being conducted in this area. The concepts and findings of this research may be useful to decision-makers who want to transform their current systems with novel innovations and those interested in establishing modern manufacturing businesses. This study may be used as a benchmark for the implication of MAT in high-tech manufacturing systems to increase quality standards in high-tech manufacturing systems.

Finally, this research concludes that effective execution of the 'high tech manufacturing process' followed by 'high tech product features' and others. It may offer process accuracy, innovative product features, a viable market for competition, and decreased maintenance costs, which may help to achieve the desired organizational standards. This study also developed a ranking order for MAT variables for HTM industries. Manufacturers must focus on the variables that achieve the highest ranking and belong to causal groups to meet quality standards. Such integration of MAT with a HTM system allows the system to become autonomous and provides the most appropriate alternative solutions to issues that may develop during production, maintenance, or marketing. This would help investors feel comfortable, stimulate economic activity, increase demand from neighboring nations, and boost productivity with minimal effort. Despite its many benefits, this strategy has some drawbacks, such as depending on expert judgments that are not always consistent due to human bias and the need for extra work to obtain consistent responses. In future studies, advanced MCDM techniques should be used to perform further analyses on the proposed framework. Moreover, techniques such as the Ordinal Priority Approach (OPA) and artificial intelligence network should be applied to refine the variables and validate the results to obtain more precise outcomes.

DISCLOSURE STATEMENT

No potential conflict of interest was reported by the author(s).

ETHICAL STATEMENT

Articles do not rely on clinical trials.

DATA AVAILABILITY STATEMENT

The authors confirm that the data supporting the findings of this study are available within the article.

FINDING

This research does not contain any financial funding from any source.

REFERENCES

Bhatia, M. S., Kumar Jakhar, S., & Dora, M. (2022). Analysis of Barriers to Closed-Loop Supply Chain: A Case of the Indian Automotive Industry. *IEEE Transactions on Engineering Management*, *69*(5), 1999–2013. DOI: 10.1109/TEM.2020.2998794

Bousdekis, A., Magoutas, B., Apostolou, D., & Mentzas, G. (2015). "A proactive decision making framework for condition-based maintenance", *Industrial Management and Data Systems*. *Industrial Management & Data Systems*, *115*(7), 1225–1250. DOI: 10.1108/IMDS-03-2015-0071

Burcher, P., Lee, G., & Sohal, A. (1999). Lessons for Implementing AMT Lessons for Implementing AMT Some Case Experiences with CNC in Australia, Britain and Canada [# MCB University Press.]. *International Journal of Operations & Production Management*, *19*(5/6), 515–527. DOI: 10.1108/01443579910260847

Chan, F. T. S., Swarnkar, R., & Tiwari, M. K. (2007). Infrastructure for co-ordination of multi-agents in a network-based manufacturing system. *International Journal of Advanced Manufacturing Technology*, *31*(9–10), 1028–1033. DOI: 10.1007/s00170-005-0115-9

Chang, B., Chang, C. W., & Wu, C. H. (2011). Fuzzy DEMATEL method for developing supplier selection criteria. *Expert Systems with Applications*, *38*(3), 1850–1858. DOI: 10.1016/j.eswa.2010.07.114

Chen, Y., Lee, J. Y., Sridhar, S., Mittal, V., McCallister, K., & Singal, A. G. (2020). Improving Cancer Outreach Effectiveness Through Targeting and Economic Assessments: Insights from a Randomized Field Experiment. *Journal of Marketing*, *84*(3), 1–27. DOI: 10.1177/0022242920913025

Chien, T. H., Lin, Y. I., & Tien, K. W. (2013). Agent-based negotiation mechanism for multi-project human resource allocation. *Journal of Industrial and Production Engineering*, *30*(8), 518–527. DOI: 10.1080/21681015.2013.861521

Coad, A., & Rao, R. (2008). Innovation and firm growth in high-tech sectors: A quantile regression approach. *Research Policy*, *37*(4), 633–648. Advance online publication. DOI: 10.1016/j.respol.2008.01.003

Cooke, A. D. J., & Zubcsek, P. P. (2017). The connected consumer: Connected devices and the evolution of customer intelligence. *Journal of the Association for Consumer Research*, *2*(2), 164–178. DOI: 10.1086/690941

Daabes, A. S. A., & Kharbat, F. F. (2017). Customer-based perceptual map as a marketing intelligence source. *International Journal of Economics and Business Research, 13*(4), 360. DOI: 10.1504/IJEBR.2017.084381

Huang, M. H., & Rust, R. T. (2018). Artificial Intelligence in Service. *Journal of Service Research, 21*(2), 155–172. DOI: 10.1177/1094670517752459

Huang, M. H., & Rust, R. T. (2021). Engaged to a Robot? The Role of AI in Service. *Journal of Service Research, 24*(1), 30–41. DOI: 10.1177/1094670520902266

Khandelwal, A. (2010). The long and short (of) quality ladders. *The Review of Economic Studies, 77*(4), 1450–1476. DOI: 10.1111/j.1467-937X.2010.00602.x

Kim, S. Y., Schmitt, B. H., & Thalmann, N. M. (2019). "Eliza in the uncanny valley: Anthropomorphizing consumer robots increases their perceived warmth but decreases liking", *Marketing Letters. Marketing Letters, LLC*(1), 1–12. Advance online publication. DOI: 10.1007/s11002-019-09485-9

Lavbič, D., & Rupnik, R. (2009). Multi-agent system for decision support in enterprises. *Journal of Information and Organizational Sciences, 33*(2), •••.

Lécuyer, C., & Brock, D. C. (2006). The materiality of microelectronics. *History and Technology, 22*(3), 301–325. DOI: 10.1080/07341510600803440

Leitao, P., Rodrigues, N., Turrin, C., & Pagani, A. (2015). "Multiagent System Integrating Process and Quality Control in a Factory Producing Laundry Washing Machines", *IEEE Transactions on Industrial Informatics. IEEE Transactions on Industrial Informatics, 11*(4), 879–886. DOI: 10.1109/TII.2015.2431232

Leung, E., Paolacci, G., & Puntoni, S. (2018). Man Versus Machine: Resisting Automation in Identity-Based Consumer Behavior. *JMR, Journal of Marketing Research, 55*(6), 818–831. DOI: 10.1177/0022243718818423

Lim, M.K. and Zhang, Z. (2003), "A multi-agent based manufacturing control strategy for responsive manufacturing", *Journal of Materials Processing Technology*, Vol. 139 No. 1-3 SPEC, pp. 379–384, .DOI: 10.1016/S0924-0136(03)00535-1

Liu, C., Li, Y., & Shen, W. (2014). "Integrated manufacturing process planning and control based on intelligent agents and multi-dimension features", *International Journal of Advanced Manufacturing Technology. International Journal of Advanced Manufacturing Technology, 75*(9–12), 1457–1471. DOI: 10.1007/s00170-014-6246-0

Liu, X., Cui, X., Song, G., & Xu, B. (2014). Development of a virtual maintenance system with virtual hand. *International Journal of Advanced Manufacturing Technology*, *70*(9–12), 2241–2247. Advance online publication. DOI: 10.1007/s00170-013-5473-0

Liu, Y., Eckert, C. M., & Earl, C. (2020). A review of fuzzy AHP methods for decision-making with subjective judgements. *Expert Systems with Applications*, *161*(April), 113738. Advance online publication. DOI: 10.1016/j.eswa.2020.113738

López, A., Casquero, O., Estévez, E., Armentia, A., Orive, D., & Marcos, M. (2023). "An industrial agent-based customizable platform for I4.0 manufacturing systems", *Computers in Industry. Computers in Industry*, *146*, 103859. Advance online publication. DOI: 10.1016/j.compind.2023.103859

Madrid-Guijarro, A., Auken, H. Van and García-Pérez-de-Lema, D. (2007), "An Analysis of Factors Impacting Performance of Spanish Manufacturing Firms", *Journal of Small Business and Entrepreneurship*, Taylor and Francis Ltd., Vol. 20 No. 4, pp. 369–386, .DOI: 10.1080/08276331.2007.10593406

Mahdjoub, M., Monticolo, D., Gomes, S., & Sagot, J. C. (2010). A collaborative Design for Usability approach supported by Virtual Reality and a Multi-Agent System embedded in a PLM environment. *Computer Aided Design*, *42*(5), 402–413. DOI: 10.1016/j.cad.2009.02.009

Mahesh, M., Ong, S. K., & Nee, A. Y. C. (2007). A web-based multi-agent system for distributed digital manufacturing. *International Journal of Computer Integrated Manufacturing*, *20*(1), 11–27. DOI: 10.1080/09511920600710927

Odeyinka, O. F., Okandeji, A. A., & Sogbesan, A. A. (2022). "A fuzzy TOPSIS model for selecting raw material suppliers in a manufacturing company", *Nigerian Journal of Technology* [AJOL]. *African Journals Online*, *41*(4), 797–804. DOI: 10.4314/njt.v41i4.17

Oroojlooy, A., & Hajinezhad, D. (2022). *"A review of cooperative multi-agent deep reinforcement learning", Applied Intelligence*. Springer., DOI: 10.1007/s10489-022-04105-y

Ottaway, T. A., & Burns, J. R. (2000). An adaptive production control system utilizing agent technology. *International Journal of Production Research*, *38*(4), 721–737. DOI: 10.1080/002075400189121

Ov, T. R. (2003). A Multi-Agent Architecture for Dynamic Scheduling of Steel Hot Rolling*. *Journal of Intelligent Manufacturing*, *14*, •••.

Pěchouček, M., Vokřínek, J., & Bečvář, P. (2005). ExPlanTech: Multiagent support for manufacturing decision making. *IEEE Intelligent Systems*, *20*(April), 67–74. Advance online publication. DOI: 10.1109/MIS.2005.6

Sahebjamnia, N., Tavakkoli-Moghaddam, R., & Ghorbani, N. (2016). Designing a fuzzy Q-learning multi-agent quality control system for a continuous chemical production line - A case study. *Computers & Industrial Engineering*, *93*, 215–226. DOI: 10.1016/j.cie.2016.01.004

Sharma, A. and Sharma, S.K. (2021), "Analyzing the role of multiagent technology in preventing airplane crash using AHP and DEMATEL approach", *International Journal of Crashworthiness*, .DOI: 10.1080/13588265.2021.2008739

Sharma, S. K., Vishwakarma, S., & Jha, N. (2017). "Prognosis agent technology: Influence on manufacturing organizations", *International Journal of Advanced Manufacturing Technology*. *International Journal of Advanced Manufacturing Technology*, *92*(1–4), 435–446. DOI: 10.1007/s00170-017-0025-7

Shavandi, A., & Khedmati, M. (2022). A multi-agent deep reinforcement learning framework for algorithmic trading in financial markets. *Expert Systems with Applications*, *208*, 118124. Advance online publication. DOI: 10.1016/j.eswa.2022.118124

Si, S. L., You, X. Y., Liu, H. C., & Zhang, P. (2018). *"DEMATEL Technique: A Systematic Review of the State-of-the-Art Literature on Methodologies and Applications", Mathematical Problems in Engineering*. Hindawi Limited., DOI: 10.1155/2018/3696457

Singh, A., Misra, S. C., Kumar, V., & Kumar, U. (2022). "Identification and ordering of safety performance indicators using fuzzy TOPSIS: A case study in Indian construction company", *International Journal of Quality and Reliability Management*. *International Journal of Quality & Reliability Management*, *39*(1), 77–114. DOI: 10.1108/IJQRM-02-2020-0051

Singh, V., Kumar, V., & Singh, V. B. (2023). A hybrid novel fuzzy AHP-TOPSIS technique for selecting parameter-influencing testing in software development. *Decision Analytics Journal, Elsevier Inc.*, *6*, 100159. Advance online publication. DOI: 10.1016/j.dajour.2022.100159

Singh, V., & Sharma, S. K. (2023). Critical factors of multi-agent technology influencing manufacturing organizations: An AHP and DEMATEL-oriented analysis. *International Journal of Computer Integrated Manufacturing*, •••, 1–23. DOI: 10.1080/0951192X.2023.2209857

Steenhuis, H. J., & De Bruijn, E. J. (2006), "High technology revisited: Definition and position", *ICMIT 2006 Proceedings - 2006 IEEE International Conference on Management of Innovation and Technology*, Vol. 2, pp. 1080–1084, DOI: 10.1109/ICMIT.2006.262389

Stornelli, A., Ozcan, S., & Simms, C. (2021). "Advanced manufacturing technology adoption and innovation: A systematic literature review on barriers, enablers, and innovation types", *Research Policy. Research Policy*, *50*(6), 104229. Advance online publication. DOI: 10.1016/j.respol.2021.104229

Sumrit, D. and Anuntavoranich, P. (2013), "Using DEMATEL Method to Analyze the Causal Relations on Technological Innovation Capability Evaluation Factors in Thai Technology-Based Firms", *International Transaction Journal of Engineering, Management, & Applied Sciences & Technologies*, Vol. 4 No. 2.

Sun, J., Zhang, Y. F., & Nee, A. Y. C. (2001). A distributed multi-agent environment for product design and manufacturing planning. *International Journal of Production Research*, *39*(4), 625–645. DOI: 10.1080/00207540010004340

Tang, H., Li, D., Wan, J., Imran, M., & Shoaib, M. (2020). A Reconfigurable Method for Intelligent Manufacturing Based on Industrial Cloud and Edge Intelligence. *IEEE Internet of Things Journal*, *7*(5), 4248–4259. Advance online publication. DOI: 10.1109/JIOT.2019.2950048

Tether, B. S., & Storey, D. J. (1998). Smaller Firms and Europe's High Technology Sectors: A Framework for Analysis and Some Statistical Evidence. *Research Policy*, *26*(9), 947–971. DOI: 10.1016/S0048-7333(97)00053-X

Timoshenko, A., & Hauser, J. R. (2019). "Identifying customer needs from user-generated content", *Marketing Science. Marketing Science*, *38*(1), 1–20. DOI: 10.1287/mksc.2018.1123

Unctad. (2023), *Technology and Innovation Report 2023*.

Vogel-Heuser, B., Lee, J., & Leitão, P. (2015). "Agents enabling cyber-physical production systems", *At-Automatisierungstechnik. Automatisierungstechnik*, *63*(April), 777–789. Advance online publication. DOI: 10.1515/auto-2014-1153

Vrba, P., & M rík, V. (2010). Capabilities of dynamic reconfiguration of multiagent-based industrial control systems. *IEEE Transactions on Systems, Man, and Cybernetics. Part A, Systems and Humans*, *40*(2), 213–223. DOI: 10.1109/TSMCA.2009.2034863

Yıldızbaşı, A., & Ünlü, V. (2020). "Performance evaluation of SMEs towards Industry 4.0 using fuzzy group decision making methods", *SN Applied Sciences. SN Applied Sciences*, *2*(3), 355. Advance online publication. DOI: 10.1007/s42452-020-2085-9

Yoon, K., & Hwang, C. L. (1985). Manufacturing plant location analysis by multiple attribute decision making: Part isingle plant strategy. *International Journal of Production Research, 23*(2), 345–359. DOI: 10.1080/00207548508904712

Zakhama, A., Charrabi, L., & Jelassi, K. (2019). Intelligent Selective Compliance Articulated Robot Arm robot with object recognition in a multi-agent manufacturing system. *International Journal of Advanced Robotic Systems, 16*(2), 1729881419841145. Advance online publication. DOI: 10.1177/1729881419841145

Zemlickienė, V. (n.d.). *ANALYSIS OF HIGH-TECHNOLOGY PRODUCT DEVELOPMENT MODELS.*

Zhang, J. (n.d.). Multi-Agent Based. *Production Planning and Control.*

Chapter 12
Artificial Intelligence in Detecting and Diagnosing Microbiome-Driven Cancer:
An Emerging Paradigm

Abhiram Kumar
https://orcid.org/0000-0002-3422-2117
Birla Institute of Technology and Sciences, Pilani, India

Anjali Rai
Jagannath University, Chaksu, India

Shivang Shukla
https://orcid.org/0009-0002-6395-6838
Jagannath University, Chaksu, India

Chhavi Dhiman
Birla Institute of Technology and Sciences, Pilani, India

Kumar Pranav Narayan
https://orcid.org/0000-0001-5198-2495
Birla Institute of Technology and Sciences, Pilani, India

ABSTRACT

This chapter aim to explore transformative role of artificial intelligence (AI) in detecting and diagnosis of cancer, addressing critical gaps in early detection, diagnosis, personalized treatment, and improved patient outcomes. AI based algorithms,

DOI: 10.4018/979-8-3373-3146-1.ch012

including machine learning and deep learning, are increasingly utilized to analysed complex microbiome data, identify cancer-associated signature, and predict therapeutics response. Moreover, we elucidate how AI facilitates the integration of multi-omics data (genomics, transcriptomics and metabolomics) with microbiome profiles, enhancing our understanding of cancer pathogenesis and progression. It also examines the application of AI in developing non-invasive diagnostic tools using microbiome biomarkers for various cancer, such as colorectal, lung, and breast cancer. Furthermore, the chapter addresses the challenges and opportunities in leveraging AI for microbiome-based caner immunotherapy. Conclusively, by highlighting future directions and the potential of AI revolutionize cancer care through microbiome-based precision medicine.

1. INTRODUCTION

Figure 1. Graphical Abstract

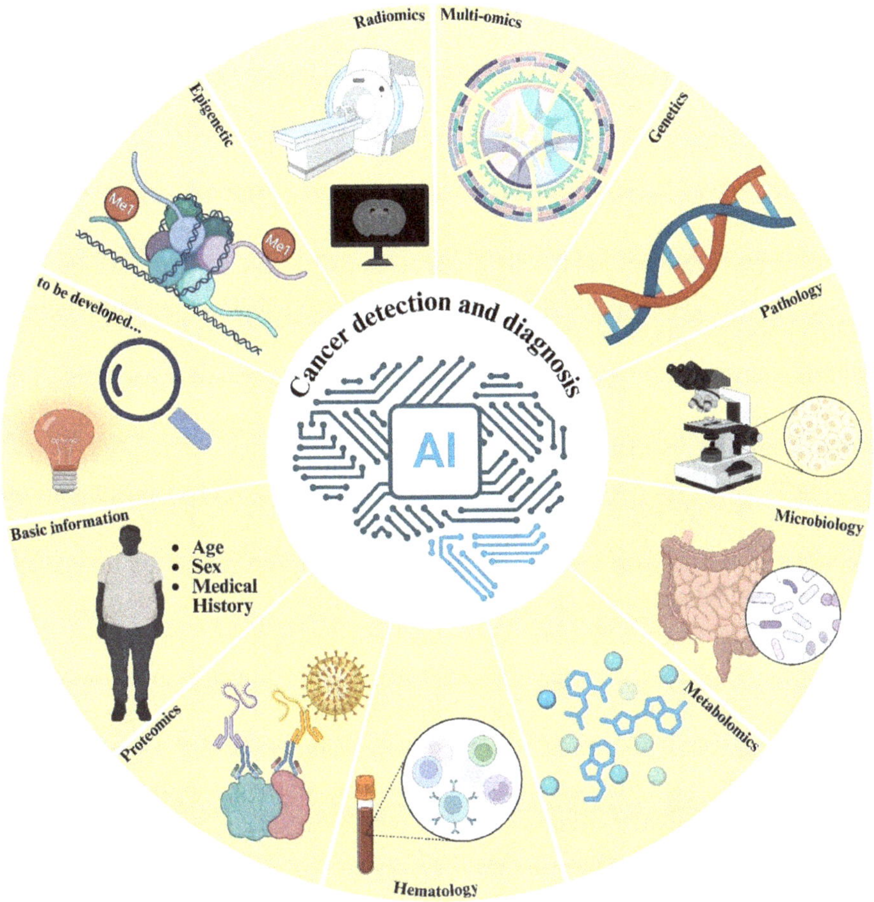

Microbiome-based cancer diagnosis is becoming an increasingly important supplement to the genomics approach in cancer diagnosis; however, current models for microbiome-based cancer diagnosis face general challenges: not only can diagnosis models not be adapted from one cancer to another, but models built on microbes from tissues cannot be adapted for diagnosis based on microbes from blood. As a result, a microbiome-based model appropriate for a wide range of cancer types is urgently required. We have introduced Deep Micro Cancer, an artificial intelligence-based diagnosis model for a wide range of cancer forms. Built on random forest models, it has demonstrated improved performance on tissue samples from over twenty different forms of cancer. By utilizing transfer learning approaches, better accuracies could be acquired, particularly for cancer types with only a few samples, which could meet clinical requirements. Furthermore, transfer learning approaches have permitted great diagnostic accuracy, which can be obtained with blood samples. These findings suggested that some types of microorganisms, when dug using modern artificial techniques, could show the detailed disparities between cancer patients and healthy people. Deep Micro Cancer has created a novel platform for reliable cancer diagnosis based on tissue and blood samples, which might be utilized in clinics (Xu et al., 2023). An estimated three trillion bacteria reside in the human body, where they coordinate a wide range of physiological functions and disease vulnerabilities. Despite having roughly, the same number of cells in the body as human cells, bacteria have 100 times more genetic diversity, which results in exceptional mechanistic and metabolic capabilities that affect not only their own microbial niche but also host tissue-specific and immune cell functions. In addition to bacteria, eukaryotic fungus, protozoa, and viruses make up the human microbiome (Dekaboruah et al., 2020). All things considered, in a healthy environment, the host and its microbiome coexist as a meta organism by exchanging a nutrient-rich milieu for assistance with digestion and metabolism (Cullin et al., 2021). The gut microbes and host form a complex ‘super-organism' where the host benefits from symbiotic interactions in many important areas of life. However, this symbiotic relationship may be disrupted, and disease may be promoted by changes in the microbiome brought on by environmental changes (infection, nutrition, or lifestyle) or by abnormalities in the host's regulatory circuits that govern bacterial sensing and homeostasis. There is growing evidence that bacterial microbiota plays a significant role in the development of cancer (Schwabe & Jobin, 2013). Numerous mucosal surfaces and low-biomass habitats within different cancer types have microbial populations. It is possible for microbial individuals to move back and forth between niches. For instance, it is hypothesized that oral-native Fusobacteria will travel from the mouth to the large intestine via the bloodstream or the digestive system, where they will settle and colonize. There is growing evidence that these microbial populations play a role in cancer. With an emphasis on immune responses, dysbiosis, genotoxicity, metabolism, and methods to target the microbiome for cancer prevention, we explore the connections between the bacterial microbiota and cancer in this opinion piece.

These days, AI is a new and quickly developing paradigm that considers several scientific domains, including those related to cancer patient care. The ability of a computer to learn and identify patterns and interactions from enough representative models and then use this information to improve the existing approach to the process of decision-making in a particular subject, can be viewed as a broad notion. AI is changing the landscape of precision oncology by combining the vast amounts of data from multi-omics analyses with the most recent developments in high-performance computing and innovative deep-learning techniques. Interestingly, AI is being used in a wider range of applications, such as new methods for cancer detection, screening, diagnosis, and classification; characterization of cancer genomics; analysis of the tumor microenvironment; evaluation of biomarkers for prognostic and predictive purposes; and follow-up and drug discovery strategies (Luchini et al., 2022).

By evaluating molecular data to find biomarkers, mutations, and therapeutic targets, AI in genomics enables precision oncology and individualized treatment plans. In order to minimize toxicity and improve therapeutic results, AI models also assist with radiotherapy treatment planning and forecast chemotherapy responses. Additionally, AI improves prognostication by integrating imaging data and electronic health records to predict survival, recurrence, and treatment-related problems. It cuts down on development time and expenses by speeding up drug discovery and enhancing clinical trial matching. By combining data from many sources, AI-based decision support systems help oncologists make difficult therapeutic decisions. Data uniformity, model interpretability, ethical issues, and regulatory approval are still obstacles in the way of its potential. AI is anticipated to better customize cancer treatment, enhance results, and lower medical expenses with more research (Farina et al., 2022).

1.1 Rationale for Integrating AI with Microbiome Research

Comprehensive profiling of microbial communities has been made possible by advancements in sequencing technologies; yet, the resulting datasets are large, varied, high-dimensional, and frequently challenging to interpret using conventional statistical methods. By finding patterns, forecasting disease outcomes, and deriving valuable insights from vast amounts of microbiome data, AI, in particular machine learning (ML) and deep learning (DL), provides strong capacities to solve these issues.

Scientists can differentiate between illness symptoms, find the microbial biomarkers, and group patients according to the microbial signature with the help of AI, which makes it possible to precisely classify, cluster and analyze the microbial profiles. These methods are effective in detecting the risk of diseases like hypertension, type 2 diabetes and colon cancer (Hernández Medina et al., 2022).

A network-wide comprehension of host-microbiome interactions is also made possible with the help of AI. AI helps in the discovery of finding disease mechanism, treatment reactions and drug microbiome interactions by evaluating multiomics data. The main advantage of AI is to improve the personalized treatment. Many environmental factors and host impact the highly personalized microbiome. According to the specific microbial makeup of the individuals, AI model can customize the intervention of food, probiotics and medication regimens. AI also facilitates the identification of new microorganisms and medicinal targets, improves functional annotation, and speeds up the assembly of microbial genomes. AI integration is fraught with difficulties, including insufficient standardization, interpretability problems, data heterogeneity, and small sample numbers, despite its potential. To integrate AI-driven microbiome research into clinical practice, efforts are being made to overcome these constraints through the use of federated learning, explainable AI, and standardized data protocols (Loganathan & Priya Doss C, 2022).

1.2 Artificial Intelligence in Cancer Detection and Diagnosis

Cancer treatments have evolved significantly during the last 15 years, recognizing scientists' understanding of the molecular mechanisms underlying malignancies. Nevertheless, out of all the main diseases, cancer medicines have the lowest chance of succeeding in clinical trials. Researchers are concentrating on certain aspects of the tumor's microenvironment and developing tailored medicines to treat cancer in the future. AI's powerful analytical capabilities have the potential to completely transform cancer research, diagnosis, and therapy. An unprecedented chance to integrate data with complex research findings is presented by big malignant research data, which in turn calls for powerful computing capabilities. AI is the application of mathematical algorithms that simulate how the human mind functions to address the incredibly complex problems that healthcare facilities face, like those brought on by biological system anomalies like cancer (Sekhar, 2024). In the future, digital health care for illness diagnosis and treatment is expected to be dominated by AI and ML, which are already having a big impact on daily life. Technology developments in AI and ML have made it possible for autonomic neuropathy disease diagnosis tools to use massive data sets to address the difficulties of human illness detection at a very early stage, especially in cancer. It manages learning, multiprocessing, mistake tolerance, and decision-making to enhance oncological treatment. AI may therefore be able to address the issue that expert systems aren't always universal or objective in addition to bringing together many aspects of clinical variability. Many research has demonstrated that AI that utilizes ML may produce accurate predictions and diagnoses (Iqbal et al., 2021).

1.3 Thematic Classification of Microbiome–Cancer Studies

To enhance clarity and originality, current research on microbiome–cancer interactions has been organized into three major themes: (i) multi-omics integration, (ii) immunotherapy prediction, and (iii) non-invasive diagnostics. These categories reflect the progression from mechanistic understanding to translational application. **Table 1** presents a comparative summary of over twenty representative studies, highlighting their methodology, cancer type, technological relevance, and limitations.

This thematic grouping offers a structured overview of how microbiome science is shaping precision oncology through integrated, predictive, and diagnostic approaches.

Table 1. Thematic grouping and comparative summary of representative studies on microbiome-driven cancer research.

Theme	Study (Author, Year)	Methodology (sample, assay, analysis)	Cancer type	Main finding	Key limitations
Immunotherapy prediction	(Routy et al., 2018)	Fecal metagenomics; clinical cohorts; survival/response analysis.	NSCLC, RCC, Urothelial cancers.	Gut microbiome composition *(Akkermansia spp.)* associates with PD-1 response; antibiotics reduce response probability supports microbiome as predictive biomarker and interventional target.	Cohort heterogeneity; observational (causality limited); limited sample size for some subgroups.
	(Gopalakrishnan et al., 2018)	16S + metagenomics of stool/oral samples; responder vs nonresponder comparison.	Metastatic melanoma.	Higher α-diversity and enrichment of *Ruminococcaceae/ Faecalibacterium* in responders; microbial signatures predict response.	Single cancer type (melanoma); confounders (diet/antibiotics) not fully controlled.
	(Matson et al., 2018)	Fecal shotgun metagenomics; human-to-germ-free mouse fecal transfer experiments.	Metastatic melanoma.	Identified species (e.g., *Bifidobacterium longum*) enriched in responders; fecal transfer improved ICI efficacy in micemechanistic support.	Small cohort; human mouse transfer may not fully recapitulate human host factors.
	(Sivan et al., 2015)	Gnotobiotic mouse models; probiotic administration.	Preclinical (melanoma models).	*Bifidobacterium spp.* promote antitumor immunity and improve anti-PD-L1 efficacy foundational mechanistic proof.	Preclinical only; species differences limit direct clinical translation.
	(Zheng et al., 2019)	Metagenomic sequencing longitudinal during ICI therapy.	Hepatocellular carcinoma (HCC).	Dynamic gut microbiome shifts correlate with anti-PD-1 response; candidate taxa for monitoring.	Sample size; tumor-specific immune milieu may differ from other cancers.
Multi-omics integration	(Nejman et al., 2020)	Large multi-tumor survey; 16S/amplicon + microscopy + culture from tumor tissues.	Multiple (breast, lung, ovary, pancreas, melanoma, brain, bone).	Demonstrated tumor-type specific intratumoral microbiomes; evidence of intracellular bacteria and tumor-type specificity justifies tumor microbiome as a distinct compartment.	Low biomass challenges (contamination risk); need rigorous controls for false positives.
	(Liang et al., 2024)	Cohort study: metagenomics + metabolomics + WES + RNA-seq (multi-omics).	Colon cancer (left/right)	Identified distinct microbes, metabolites and host genomic features shows power of integrated multi-omics to stratify tumor subtypes and mechanisms.	High cost and complexity; integration methods and reproducibility remain challenging.

continued on following page

Table 1. Continued

Theme	Study (Author, Year)	Methodology (sample, assay, analysis)	Cancer type	Main finding	Key limitations
	(Muller et al., 2024)	Multi-omic integration methods paper (computational).	Method development (disease-agnostic)	Presents robust pipelines and statistical frameworks for integrating microbiome multi-omic datasets improves cross-study signature discovery.	Computational assumptions; needs broad adoption and benchmarking on more cancer cohorts.
	(N. Zhang et al., 2023)	Review of multi-omics applications in tumor microbiome research.	Review	Summarizes methods (metagenomics, metatranscriptomics, metabolomics, host genomics) and application areas.	Review (synthesis) primary-data gaps remain.
	(Jayakrishnan et al., 2024)	Paired tissue microbiome + plasma metabolome + ML prediction.	Pan-cancer/ host-microbiome studies	Demonstrates ML on multi-omic features can reveal host–microbial interaction patterns and predictive signatures.	Overfitting risk; external validation needed.
Non-invasive diagnostics (fecal/other)	(Zeller et al., 2014)	Fecal metagenomics; LASSO classifier for CRC detection; cross-cohort validation.	Colorectal cancer (CRC).	Early demonstration that fecal microbiome profiles (metagenome) can classify CRC vs controls proof-of-concept for non-invasive screening.	Cohort/ population differences; need larger multi-ethnic validation for screening use.
	(Yuan et al., 2021)	Review/meta-analysis of fecal bacteria biomarkers for CRC.	CRC (review)	Summarizes multiple fecal bacterial biomarkers and diagnostic models; highlights promising taxa and classifiers.	Heterogeneous methods across studies limit pooled conclusions.
	(Yao et al., 2021)	Predictive model for CRC using stool microbiome.	CRC	New predictive model showing clinical potential for non-invasive diagnosis.	Needs prospective clinical validation.
	(Q. Ding et al., 2022)	Review of fecal biomarkers (microbes, VOCs, DNA/RNA/ proteins) for CRC.	CRC	Provides comprehensive assessment of non-invasive markers and combined multi-marker strategies.	Emphasizes technical variability and need for standardization.
	(Olovo et al., 2021)	Review of fecal microbial biomarkers and early diagnosis.	CRC	Discusses feasibility and limitations of microbiome-based early detection.	Similar: cohort heterogeneity, preanalytical variability.
Intratumoral & spatial microbiome / single-cell	(Ghaddar et al., 2022)	Single-cell level analysis (SAHMI) of tumor microbiome and host transcriptomes.	Pancreatic cancer (and others)	Links tumor-resident microbes with cellular programs and immune states adds spatial/ single-cell resolution to tumor–microbiome interactions.	Technical novelty; contamination and sensitivity concerns; needs replication.
	(Fu et al., 2022)	Mouse models + intratumor bacteria depletion experiments; metastasis assays.	Preclinical (lung metastasis models)	Intratumoral bacteria promote metastasis in models mechanistic link between tumor microbiome and progression.	Preclinical; human relevance needs confirmation.

continued on following page

Table 1. Continued

Theme	Study (Author, Year)	Methodology (sample, assay, analysis)	Cancer type	Main finding	Key limitations
Predictive models & computational approaches	(Yang et al., 2024)	ML models using gut microbiome to predict neoadjuvant therapy response.	Mixed (neoadjuvant immuno-chemo)	Demonstrates predictive power of microbiome-based machine learning for therapy response.	Requires external validation; model generalizability uncertain.
	(Sherwani et al., 2025)	Time-series multi-omics frameworks for microbiome research.	Methods	Time-resolved multi-omics improves causal inference and dynamic biomarker discovery.	Demands longitudinal sampling and cost.
Clinical/ Interventional evidence & reviews	(Petrelli et al., 2025)	Systematic narrative review of clinical trials and studies linking microbiome and ICI outcomes (2015–2025).	Multiple	Synthesis of clinical evidence for microbiome influence on ICI efficacy and interventions (FMT, diet, probiotics).	Many studies are small or heterogenous; standardized interventional trials still limited.
	(Almonte et al., 2025)	Review of microbiome interventions to boost cancer therapy.	Multiple	Discusses FMT, probiotics, dietary modulation as therapeutic adjuncts and issues for translation.	Safety, donor selection, regulatory and mechanistic gaps.
General/ Foundational epidemiology & CRC reviews	(Vogtmann & Goedert, 2016)	Epidemiologic and mechanistic reviews summarizing microbiome–CRC links.	CRC, multiple	Provide baseline epidemiologic evidence linking gut microbiome shifts to CRC risk and progression.	Many observational associations, need prospective and mechanistic work.

2.1 Machine Learning and Deep Learning Approaches

Advancements in the ML and DL have significantly enhanced the accuracy and efficiency of cancer detection and diagnosis. These tools help doctors with tumor identification, prognosis prediction, early diagnosis and customized treatment planning. While DL, a more advanced branch automatically extracts hierarchical features from the aw input data by stimulating the neural architecture of the human brain, ML uses statistical algorithms that learn from labeled data to produce predictions.

Various ML approaches, like logistic regression, k-Nearest Neighbors (k-NN), decision tree, Random Forests and Support Vector Machines (SVM), have been used extensively in cancer research. These algorithms have proven to be very useful in determining cancer subtypes, predicting patient survival, assessing treatment responses and distinguishing between benign and malignant lesions. For example, ML models trained on clinical and radiomic data from computed tomography (CT) and magnetic resonance imaging (MRI) scans have demonstrated encouraging outcomes in the classification of gliomas and the diagnosis of lung cancer. Locally Linear Embedding (LLE) and Principal Component Analysis (PCA) two feature reduction techniques, improved the effectiveness of classifiers like logistic regression and stochastic gradient descent in the diagnosis of ovarian cancer, reaching diagnostic accuracies of up to 99.8% (Sekhar, 2024).

Convolutional Neural Networks (CNNs) is a type of DL technique which has been highly efficient in image- based cancer diagnosis. CNNs are particularly good at evaluating thermoscopic pictures, histopathology slides, and mammograms because they automatically learn the spatial hierarchies of features from medical images. CNNs trained on more than 129,000 photos were able to detect skin cancer and even beat dermatologists in distinguishing between benign nevi and melanomas. DL models demonstrated high sensitivity and specificity in detecting microcalcifications and mass lesions in breast cancer. Furthermore, to help with accurate localization and characterization, DL algorithms like U-Net and ResNet have been used to segment tumor areas in colorectal and prostate malignancies.

By combining clinical information, tumor markers, and imaging characteristics, Artificial Neural Networks (ANNs) have also been effectively utilized for risk prediction in diseases such as lung and pancreatic cancer. Using extensive national datasets, an ANN model in one study was able to predict the risk of pancreatic cancer with over 87% sensitivity and 80% specificity. Furthermore, hybrid systems that integrate DL with conventional ML techniques or combine multiple DL models have shown improved diagnostic resilience and accuracy across a range of cancer types.

Although both ML and DL have the potential to be transformative, their levels of complexity and data requirements vary. ML techniques work best on organized, smaller datasets and are typically interpretable. On the other hand, DL needs a lot of computer power and massive annotated datasets, but it excels at processing unstructured data, such as photos. DL models, however, frequently lack transparency, which makes clinical adoption difficult. Enhancing model interpretability and clinician trust requires the development of explainable AI.

2.2 AI Algorithms for Microbiome Analysis

In microbiome research, AI is becoming a disruptive force by offering strong instruments for unraveling the intricate dynamics, structure, and operation of microbial communities. Human gut microbiota is made up of trillions of bacteria which is essential for the immunity, metabolism and the development of numbers of illness, including the autoimmune diseases, chronic inflammatory disorders (CIDs) and metabolic syndrome. The high-dimensional, dynamic character of microbiome data is frequently difficult to detect with traditional microbiological techniques. AI has changed this field by making it possible to understand data in a high-throughput, accurate and scalable manner.

Some prominent ML techniques like Random forests, support vector machines (SVMs) and k-means clustering used to classify microbial profiles and forecast disease state using metagenomic and metabolomic data. Supervised ML model can connect the microbiome fingerprints to clinical symptoms, while unsupervised techniques can assist identifying new microbial community structures. For example, the random classifier has been prominently used to differentiate between the healthy and pathological microbiome specimen with high predictive accuracy. These methods are useful for identifying biomarkers, classifying patients and identifying therapeutics targets (D'Urso & Broccolo, 2024).

DL techniques add an extra level of analytical complexity. Nonlinear patterns and temporal dynamics in microbiome data can be modeled using algorithms such as CNNs and recurrent neural networks (RNNs). Deep Micro, a CNN-based framework, has been validated to predict disease states directly from sequencing data, even for low-abundance or unculturable microbial species. Likewise, graph neural networks (GNNs) are used to simulate microbial interaction networks, allowing researchers to deduce co-occurrence correlations and functional connectivity among microbial species. AI algorithms are also having a big impact on multi omics integration in microbiome research. By merging the multi-omics data, AI models provide a complete understanding of microbial functions and their effects on host health. AI model enables monitoring of microbiome changes and can adapt the treatment tactics in real time, ushering in a new era of precision medicine in microbiome-targeted therapies (Sun et al., 2023).

Reinforcement Learning (RL) offers a strong platform for adaptive and personalized microbiome therapy, where an AI "agent" engages dynamically with the patient's biological system to learn to optimize treatment approaches. In this approach, the agent chooses interventions like particular probiotics, diet changes, or prebiotic mixtures based on the current microbiome composition and clinical condition, then gets a reward signal expressing therapeutic results (e.g., enhanced metabolite patterns or improved inflammation). Through repeated interactions, the RL model optimal policy learns to maximize long-term benefit to the patient, not short-term (K et al., 2025). Augmenting this, Online Learning models facilitate ongoing adaptation through incremental updating of parameters when new patient data are received without need for complete retraining. Through Bayesian or stochastic updating, the model updates its predictions in real time without losing responsiveness to the unique dynamics of individual microbiomes. In combination, RL and Online Learning form a closed-loop, self-optimizing system that is able to continually personalize interventions and enhance clinical outcomes in microbiome-focused precision medicine(S. si Zhang et al., 2021).

2.3 Identifying Cancer-Associated Microbiome Signatures

Researchers have highlighted the crucial function of microbiota in health and disease, especially in cancer. Whereas the relation of gut microbiota on colorectal cancer has been well established and the very less is known about the microbial environment within tumors of other cancer type. To close this gap, a large- scale study was performed using the data from The Cancer Genome Atlas (TCGA) conducted a brief study of intertumoral microbiota from 32 different types of cancer. Using the whole-transcriptome sequencing (RNA-seq) data and computational method like Kraken2 for the classification of microbe, this work evaluated more than 11,000 tumor samples to find cancer-specific microbial signatures and potential connection with molecular and clinical characteristics. Investigation found that each cancer type possesses a unique microbial composition, defining the unique features of tumor-microbiota interaction. For instance, breast invasive carcinoma frequently harbored *Prevotella, Streptomyces* and *Escherichia*, whereas colon adenocarcinoma was enriched with Fusobacterium and Bacteroides. Additionally microbial diversity was found to vary across immune subtypes (C1-C6), with taxa like Lactobacillus and *streptococcus* enriched in immune-active C3 and C4 subtypes, suggesting the dynamic role in shaping the tumor immune microenvironment. Significantly, microbial patters also vary according to the tumor stage as advanced stage breast cancer shows higher level of *Escherichia*, indicating a link between the microbial dysbiosis and cancer progression. These findings highlight the importance of tumor specific microbiota in influencing disease behavior and immune response (Yang et al., 2024).

2.4 Towards Autonomous and Adaptive Diagnostic Systems

The future of microbiome-based cancer diagnostics is being defined by the intersection of autonomous and adaptive artificial intelligence systems that can operate autonomously, learn in real-time, and continuously optimize. They represent a shift away from traditional, clinician-driven processes to smart, self-managing diagnostic systems that can examine biological data, discern disease states, and sharpen their predictions without human input. Autonomous systems refer to AI-based technologies capable of executing diagnostic and analytical tasks autonomously once activated (Jiang et al., 2024). In microbiome-oncology, autonomy is realized through technologically advanced applications like AI-enhanced lab-on-a-chip and point-of-care microfluidic biosensing systems that have the capability to autonomously acquire, process, and analyze biological samples (e.g., stool, saliva, or blood) for the detection of dysbiosis signatures or microbial biomarkers for tumorigenesis. Such systems integrate biosensors, microfluidic actuators, embedded machine vision, and cloud-connected AI modules to deliver end-to-end diagnostics from sample acquisition through result generation without human intervention. For instance, an autonomous diagnostic microdevice can employ embedded CNNs to analyze fluorescence or impedance-based sensor outputs and forecast cancer risk with negligible latency. The hallmark of autonomy here is the system's capacity to execute complex diagnostic workflows and make preliminary clinical inferences independently, guided by decision algorithms and pre-trained inference models (Shajari et al., 2023).

Adaptive systems, on the other hand, highlight the ability to learn and improve in real-time. These models adaptively change as they receive new patient data, modifying their internal parameters to capture developing biological and clinical trends. In cancer-microbiome studies, adaptive AI is able to iteratively optimize treatment regimens based on shifts in microbial diversity, metabolomic profiles, or immune responses. For example, reinforcement learning (RL) architectures have been suggested to iteratively maximize dietary therapy or probiotic treatment by assessing patient response as "reward signals." Likewise, online learning algorithms support real-time model updating as new sequencing or clinical information are learned, allowing for tailored treatment suggestions that are up-to-date and data-driven (Kabudi et al., 2021). This flexibility is a manifestation of a closed-loop intelligence cycle, wherein the physiological feedback from each patient drives the subsequent computational inference, resulting in a dynamic, changing decision system. The fusion of autonomy and flexibility terminates in closed-loop feedback structures, a characteristic property of intelligent control systems. In such systems, wearables, ingestible diagnostic devices, and IoT-enabled biosensors constitute the input layer, streaming multi-dimensional data (inflammatory markers, microbial signatures, metabolic profiles) to a diagnostic engine powered by AI. The AI model processes these inputs independently to make predictions or trigger therapeutic interventions like suggesting particular microbiome-modulating treatments. The results of these interventions are subsequently fed back into the AI model as feedback, allowing iteratively self-correcting and optimizing the model. This structure mirrors closely that of cyber-physical control systems found in engineering, wherein feedback controls adaptive performance in the long term (Jin et al., 2020; Shajari et al., 2023).

3. INTEGRATION OF MULTI-OMICS DATA USING AI

The integration of multi-omics data with AI represents a fundamental paradigm shift in biomedical research. While each omic delivers valuable data on its own, when combined, they can yield new and valuable insights. Integrating multi-omics data can reveal new cell subtypes, cell connections, and interactions across the many omic layers, resulting in gene regulation and phenotypic results. Because each omic layer is causally linked to the next, multi-omics integration assists to unravel this relationship and appropriately represent cell phenotypic. We are currently in the era of collecting information from millions of cells. The integration of this enormous, complex, multimodal data has the potential to disclose a lot about biological mechanisms and pathways, but it poses a significant difficulty for researchers. Ultimately, advanced computational tools and approaches are available to address this issue (Hossain et al., 2015).

Recent advancements in sequencing and high-throughput technologies have resulted in massive amounts of multi-omics data, including genomics, epigenomics, transcriptomics, proteomics, metabolomics, lipidomics, glycomics, cytomics/cellomics, microbiomics, metagenomics, radiomics, interactomics, and chemical genomics. Most omics data display the molecular landscape of distinct endophenotypes at multiple levels, with the exception of genotypes (genomics and epigenomics) and environmental information (microbiomics, metagenomics, chemical genomics). omics data is vital for linking genetic information to phenotypic outcomes and predicting environmental reactions. Few summarized biomarkers we have summarized in **Table 2,** which endophenotypes can be used as biomarkers to identify specific disease causes and design safe and effective therapies. Integrating all levels of omics data is crucial for understanding the complexity and interdependence of biological systems, as each type offers a distinct viewpoint on molecular events inside cells, tissues, or species. To comprehend how genetic information is turned into functional molecules and phenotypes, it's crucial to integrate many layers of omics data, from DNAs and RNAs to proteins and phenotypic consequences, as per molecular biology's core dogma. Integrating data across different omics levels identifies regulatory factors that influence biological pathways, exposing intricate relationships and feedback loops. Individual omics datasets provide limited information about a biological system. Integrating genetic and phenotypic data will improve the predictive capacity of computational models are illustrated in **Fig 1** (Wu & Xie, 2025).

Table 2. The table summarizes the various types of biological biomarkers and AI-algorithms that can be used for detection of cancer at an early stage.

Cancer Type	Biomarker	AI-Algorithm	Ref.
Colon cancer	Genetic mutations (e.g., KRAS G12D), protein markers (e.g., DEFA5, IL8)	Random Forest, Naive Bayes, Support Vector Machine (SVM), Gradient Boosted Trees, Decision Trees	(Graham-Knight et al., 2025)
Colon cancer	DEFA5, DEFA6, MMP7, IL8, SPP1, 16S RNA sequencing	Random forest (for biomarker identification)	(Yan et al., 2012)
Lung cancer	EGFR, KRAS, ALK, ROS1, BRAF, PD-L1, TMB, TME-related immune signatures (tumor microenvironment) and non-coding RNAs, lncRNAs, cfDNA, DELFI scores	Convolutional Neural Networks (CNNs) Random Forests (RF) Support Vector Machines (SVMs), Gradient Boosted Trees	(Çalışkan & Tazaki, 2023)
Breast cancer	ER (Estrogen Receptor)	CNN, Transfer Learning, AI-CAD, ProFound AI, ShuffleNet	((Vasanthapuram, 2025)
CNS Cancer	IDH mutation – found in astrocytomas and oligodendrogliomas TERT promoter mutation – associated with glioblastoma.	Radiomics + CNN, Random Forests, Deep Learning, Radiomic Signature Analysis	(Khalighi et al., 2024)
Gastric Cancer	HER2 overexpression, HER2/neu amplification, MSI-H, and PD-L1	CNN (Convolutional Neural Network)	(Cao et al., 2022)

DL integration allows for the analysis of many omics' layers at the same time to capture interactions between different biological components. Combining transcriptomics and proteomics data with DeepLIFT, a meta-learning method that incorporates Cox hazard loss, enhances survival analysis performance, and makes models more interpretable. Similarly, graph-based models such as multi-omics graph convolutional networks (MoGCN) and regularized hybrid multi-view coding (RHMC) enable comprehensive learning from many data modalities by matching latent representations. These models determine consistent and biologically meaningful variables for disease classification and subtype analysis (Wekesa & Kimwele, 2023).

Furthermore, integrated platforms such as Graph Omics enable users to investigate and generate hypothesis by integrating diverse omics datasets. Other methods, such as cust Omics, employ deep neural networks to incorporate many data types for biomarker identification and categorization. The use of attention processes, such as self-attention and coordinated attention, enhances model accuracy by focusing on biologically relevant aspects (Ballard et al., 2024).

Figure 2. Artificial Intelligence to integrate multiomics data (e.g., Transcriptomics, Metagenomics, Epigenomics, Genomics, Proteomics, and Metabolomics for cancer therapeutic target identification.) (Created in https://BioRender.com)

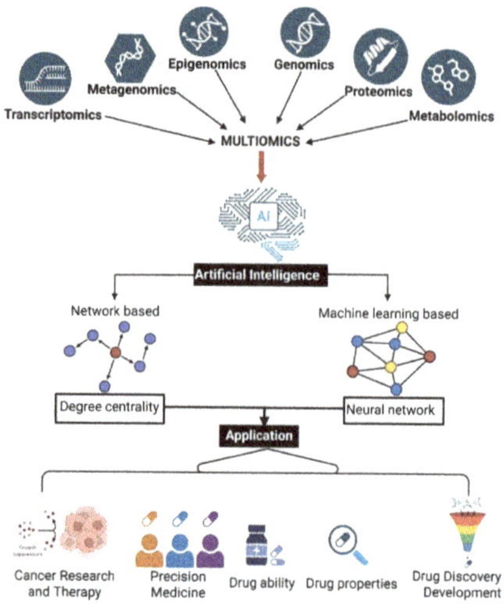

3.1 Genomics and Transcriptomics

In 1953, Francis Harry Compton Crick and James Dewey Watson discovered the DNA structure, sparking a revolution in biological sciences. These changes include the following milestones: a) Sanger and Coulson, and Maxam and Gilbert discovered the first sequencing technologies. b) Kary B. Mullis discovered the polymerase chain reaction in 1985. These accomplishments set the groundwork for precision medicine years later (Watson & Crick, 1953). Initially, precision medicine has mostly focused on genomes to identify illness disparities between groups.German botanist Hans Winkler coined the term "genome" to refer to the set of haploid chromosomes. The term "genome" now encompasses all genetic material present in any creature. The chromosomal and extrachromosomal DNA that makes up the genome includes both coding as well as non-coding genes that may or may not be involved in protein synthesis. Differentiating between non-coding genes and non-coding sections of the genome is crucial. While the latter lacks genes and frequently has repetitive sequences that are not transcripted, the former is a non-coding region with genes that are transcripted but not translated.

The multidisciplinary study of the entire genome is known as genomics. Genomics differs from genetics in that it focuses on identifying gene interactions and their synergistic effects, as well as the interaction between genes and the environment. It also studies non-coding portions of the genome. Since the Human Genome Project (HGP) was finished more than ten years ago, the field of genomic medicine has grown to include ideas like exomes, DAN, biomarkers, codons, and exons (Roth, 2019). Since its completion in 2003, HGR has greatly advanced our knowledge of how the genome contributes to disease mechanisms. Advancements in genomics have influenced the integration of genetics into clinical care. Genome investigations require specific tools and expertise for both sequencing and data interpretation (Collins et al., 1998). It is feasible to create a treatment that is specific to each patient's genetic and environmental features by examining the genetic variations linked to a problem. This lowers expenses and minimizes adverse effects. One important turning point in precision medicine was the human genome project. On the other hand, precision medicine applications have advanced significantly since the advent of next-generation sequencing (NGS) technologies. NGS is a major breakthrough in precision medicine (Yetgin, 2025).

RNA transcript found in a cell or tissue is referred to as the transcriptome. Alternative splice variants are part of the transcriptome. The transcriptome is important to investigate since the majority of human genes undergo alternative splicing. Many eukaryotic genes have exons and introns. After transcription, introns are removed, whereas exons are translated into proteins. The process of cutting off introns from mRNA and joining the exons is known as splicing. When a single mRNA goes through several rearrangements (removal of introns and reconnection of exons) in different tissues or stages of cell development, this is known as alternative splicing. A single mRNA can be used to create several proteins without compromising the connected gene. The fibronectin gene's nuclear pre-RNAs are transformed by alternative splicing into 20 mRNAs that encode 20 different proteins. Without altering its DNA, the fibronectin gene generates 20 mRNAs and 20 proteins. Most human gene functions are affected by alternative splicing (95% of multiexonic genes) (Rédei, 2008).

Errors at this level can lead to diseases such as Duchenne muscular dystrophy (mutations in the dystrophin gene and loss of function), β+-thalassaemia (reduced β-globin protein levels and anemia), and others. Consequently, any alteration to this path may result in disease. At the genomic level, these defects are not identified 13. The study of transcriptomics examines how the transcriptome affects the function of cells, tissues, and organisms in a variety of biological contexts. The relationship between the transcriptome and genome implies that the information of an organism is expressed through RNA (the transcriptome) and stored in DNA (the genome). Understanding how illness or environmental influences (such as drugs or hormones) impact the transcripts of cells, tissues, or organisms is the goal of transcriptomics. Non-coding RNA is another critical component of transcriptomics. These functional factors significantly impact illness occurrence and treatment response (Khodadadian et al., 2020).

Transcriptomics reveals post-transcriptional modifications that genomics cannot identify. Identifying potential pathogenic alterations can lead to more effective treatment options. In precision medicine, transcriptomics is a supplementary field that connects proteomics and genomics. Methodologies for transcriptome analysis differ according to whether a single gene or all of the transcripts of a cell, tissue, or organism are being evaluated (Khodadadian et al., 2020).

3.2 Metabolomics and Microbiome Synergy

The important discovery of "unknown unknowns" marked the beginning of modern microbiome study. When scientists started using targeted sequencing of ribosomal RNA (rRNA) genes to survey microbial populations in the mid-1980s, they discovered previously undiscovered branches of the tree of life. This was either because the organisms were not visible to the human eye or could not be grown in a lab. Thirty years ago, there were a lot of "unknown unknowns" regarding the bacteria that live in, on, and around us. The progress made in microbiome research is astounding (Stahl et al., 1985).

Metabolomics is a key technique for understanding how microbial community's function. Metabolomics is an emerging field that involves measuring hundreds or thousands of tiny molecules in a system. It includes metabolite profiling, lipidomics, and metabolic flux Metabolomics is a crucial method for comprehending the functioning of microbial communities. In the new dis. Metabolomicsstabolomics, thousands or even hundreds of small molecules within a system are measured. In addition to focused studies of small molecules including sex hormones, lipid mediators30, and amino acids, it encompasses metabolite profiling, lipidomics, and metabolic flux assessments. By measuring minute molecules, metabolomics offers a real-time view of dynamic changes, even while genomes, transcriptomics, and metadata offer useful information on phenotypes and molecular processes. The cell uses small molecules for several purposes, such as building blocks (amino acids), signaling (lipid mediators, neurotransmitters), and membrane construction (lipids). They also contribute to a number of exogenously affected processes, such as oxidative stress response (redox potential), energy metabolism, and inflammatory response, to name a few. Metabolomics is the general concept of measuring the functional properties of dozens to hundreds of molecules. This functional knowledge is necessary to comprehend the relationships between microbial communities and disease (Johnson et al., 2016).

Platforms based on nuclear magnetic resonance (NMR) or mass spectrometry can be used to perform metabolomics. About 125 tiny molecules, such as fatty acids, carbohydrates, and nucleic acids, can be quantitatively and reproducibly measured using NMR.NMR has a great degree of robustness, is non-destructive, and can be used in vivo, for example, to gather biochemical data on human body tissues without requiring a biopsy. However, clinical and research labs looking to perform metabolomics are increasingly using mass spectrometry-based platforms because of the comparatively poor sensitivity of NMR. Metabolomics' technical platform is independent of sample type and treatment group, making it a valuable tool for both hypothesis formulation (untargeted techniques) and testing (targeted procedures). Many "known unknowns" remain in the field of metabolomics. Because of their chemical diversity, it is hard to utilize a profiling approach to measure every species or even class of small molecule. As a result, the roles of individual molecules are quantitatively identified using focused small molecule analysis, which is based on reliable mass spectrometry experiments. For instance, assessing specific lipid mediators involved in the arachidonic acid pathway can help to clarify inflammatory (Anvarnia et al., 2019).

3.3 Dynamic Network Models as Digital Twins for Adaptive Intervention

AI-driven network models of the host-microbiome-tumor axis act as digital twins by creating a patient-specific virtual replica of biological interactions. These models integrate microbiome profiles, immune signaling pathways, tumor dynamics, and treatment responses to simulate how interventions such as targeted probiotics, dietary modifications, or chemotherapeutic regimens affect the system. By testing strategies in silico, the digital twin predicts outcomes and identifies optimal interventions before clinical application. This approach embodies adaptive control, enabling continuous refinement of personalized therapies based on simulated patient-specific responses, reducing trial-and-error and enhancing treatment precision (Ogunsakin et al., 2023).

3.4 From Reactive Diagnostics to Autonomous Microbiome-Guided Oncology

Historically, cancer diagnostics based on microbiomes have been working mainly in a reactive model detecting microbial dysbiosis only after the onset of clinical symptoms or molecular signs of malignancy. Although informative, it is still hampered by delayed treatment and the static nature of diagnostic models that do not account for the dynamic trajectory of the host–microbe ecosystem. Conversely, a closed-loop, autonomous microbiome-directed oncology platform sees a system that is in a state of continuous adaptation, sensing, learning, and responding to microbial and metabolic variations in real-time (Usman et al., 2024). Here, biosensor arrays or sequencing-based monitoring systems detect high-resolution microbial information from non-invasive sample streams (e.g., saliva, stool, blood). Such streams are input into an AI-based inference engine, which processes changes in microbial composition, metabolite profiles, and host immune interactions. The system automatically detects new pathogenic or oncogenic microbial patterns and produces predictive warnings well ahead of frank tumorigenesis. When combined with therapeutic modules e.g., targeted probiotics, bacteriophages, or microbiome-modulating nutraceuticals this system can trigger individualized interventions and track their efficacy, completing the loop (Green Buzhor et al., 2025).

A central aspect of this closed-loop design is its adaptive learning component, which regularly revises its predictive algorithms using longitudinal information. This enables the system to detect new microbial strains, antibiotic resistance patterns, and even diet-induced microbial changes that may impact cancer risk or therapeutic response. For example, dietary changes in fiber, probiotic, or xenobiotic exposure can quickly remold microbial networks. The adaptive layer utilizes reinforcement and transfer learning techniques to re-calibrate predictive thresholds so that diagnostic and therapeutic advice can stay current under shifting biological conditions (Sailer et al., 2024).

4 .AI-DRIVEN NON-INVASIVE DIAGNOSTIC TOOLS

4.1 Microbiome Biomarkers for Early Cancer Detection

Identifying accurate biomarkers for diseases like cancer is a tough task. Most current cancer tissue sampling approaches lack the Identifying accurate biomarkers for diseases like cancer is a tough task. Current cancer tissue sample techniques lack the ability to identify patients who will not respond to therapy and struggle to appropriately categorize cancer subtypes due to tumor variability among patients.

A biomarker should be easily measurable, non-invasive, and cost-effective. The gut microbiota provides a non-invasive way to identify disease biomarkers, potentially detecting infections early on. Combining microbiome-related indicators with clinical data and other biomarkers improves illness classification accuracy. Microbes have been linked to the progression of malignancies from adenoma to carcinoma, including CRC. These microorganisms can help assess the success of CRC therapies. Biomarkers obtained from the microbiome include mast cells (MCs), microRNAs (miRNAs), imaging approaches, and ML. Models are gaining popularity as non-invasive biomarkers for disease diagnosis and prognosis, providing insight into the future of Precision medicine (Zwezerijnen-Jiwa et al., 2023).

Occasionally, a number of genetic or chemical indicators interact with the human microbiome. The presence of particular bacteria is indicated by changes in small RNA patterns in the faeces of CRC patients. The effectiveness of current biomarkers can be increased by utilising a network of numerous interconnected biomarkers. Cancer and cardiovascular disorders are now the leading causes of death worldwide, replacing infectious diseases. There are 10 million cancer deaths and 19 million new cases worldwide, and these numbers are expected to rise by 50% over the next 20 years, especially in underdeveloped nations (Thomas et al., 2023).

In a number of cancers, including as colorectal, breast, lung, pancreatic, and prostate cancer, the human microbiome has become a crucial predictor of early cancer identification, prognosis, and response to treatment. Certain microbiological signatures, like those of *Fusobacterium nucleatum*, *Bacteroides fragilis*, and *Escherichia coli* (which include the pks pathogenicity island), have been closely linked to carcinogenesis in colorectal cancer (CRC). These microorganisms promote cancer growth through a variety of ways, including chronic inflammation, DNA damage, and immunological regulation. *F. nucleatum*, the most thoroughly investigated, activates the β-catenin signaling pathway and facilitates immune evasion. Its detectability in stool and tissue samples makes it an intriguing non-invasive biomarker. The combination of microbiome-based stool tests and traditional fecal immunochemical testing (FIT) has greatly improved early CRC detection accuracy. Furthermore, circulating microbial DNA (cmDNA) in plasma is being investigated as a liquid biopsy technique for CRC. Notably, microbiome makeup influences treatment response; *F. nucleatum* enrichment is related with resistance to fluoropyrimidine chemotherapy, whereas beneficial bacteria like *Akkermansia muciniphila* are associated with better outcomes in patients undergoing immune checkpoint inhibition (Kumar, Maurya, et al., 2025).

In breast cancer, changes in the gut microbiota, such as decreased *Lactobacillus/Bifidobacterium* and increased *E. coli and F. nucleatum*, impair estrogen metabolism and increase the likelihood of hormone receptor-positive tumors. Microbial DNA found in breast tissue and blood can distinguish cancer, and gut diversity appears to influence endocrine medication success (Fan et al., 2023).

Elevated levels of *Streptococcus, Veillonella*, and *Prevotella* in respiratory samples are diagnostic for lung cancer. Gut microbes such as *Akkermansia muciniphila* improve immunotherapy response, but dysbiosis indicates resistance. cmDNA and AI-assisted microbiome analysis have significantly improved diagnostic accuracy. Oral and intestinal dysbiosis in pancreatic cancer, including P. gingivalis, F. nucleatum, and altered *Bacteroides/Firmicutes* ratios, contribute to carcinogenesis and medication resistance. Some bacteria deactivate gemcitabine, emphasizing microbiome modification as a treatment option. Microbial changes in prostate cancer, particularly increased *Bacteroides, Clostridium, and E. coli*, promote tumor development and regulate androgen metabolism, affecting responses to androgen restriction therapy. Probiotics and nutritional modifications are being tested to improve treatment outcomes (Eslami et al., 2025).

4.2 Clinical Applications and Limitations

There is an urgent need to move away from small-scale pilot studies and towards prospective, suitably powered, multicenter trials that are quality verified through sampling and analytical methodologies if the microbiome is to be successfully mined for its biomarker potential in colorectal cancer. The best method for qualitatively or quantitatively assessing the gut microbiome for early cancer diagnosis in terms of function, species richness, variety, and absolute or relative abundance is still up for debate. It's still uncertain if the global microbiome is a good biomarker for early cancer because of its significant interindividual variation and evolutionary instability both within and on colorectal cancer. Future studies should therefore concentrate on biomarkers that have been demonstrated to exist in sensitive populations and are mechanistically connected to established biological pathways. However, given that it can provide future clinicians with noninvasive, actionable information on the anatomical location or development of the disease, this is possibly where the microbiome holds the greatest potential (Zwezerijnen-Jiwa et al., 2023).

4.3 IoT-Enabled Microbiome Sensing and Real-Time Data Acquisition

The intersection of the IoT and microbiome analytics is a decisive turning point in the development of autonomous and adaptive diagnostic ecosystems, making it possible to have a continuous dialogue between the biological environment of the human body and the computational acumen of AI systems. IoT-connected sensors like smart toilets, ingestible biosensors, and nano-patch wearables redefine passive biological monitoring as an active, self-sustaining process of real-time microbiome sensing and adaptive feedback. Smart toilets with multi-omics sensors and microfluidic analyzers quantitatively measure microbial diversity, metabolite flux, and inflammatory biomarkers from stool samples automatically, translating biological waste into a high-resolution diagnostic input stream. Ingestible sensors designed as biocompatible, transient electronic capsules detect and relay gut-specific parameters like pH, enzymatic activity, metabolite gradients, and volatile organic compounds, offering a dynamic picture of gut–microbiome interactions as they occur in vivo(Pandey et al., 2025). Supplementing these are wearable biosensor systems integrated into skin patches or fabrics that continuously monitor peripheral inflammation markers, cytokine levels, or microbial metabolites secreted via sweat or breath, correlating systemic host responses with microbial fluctuations. These diverse, real-time streams of data are bridged by cloud-based AI architectures that automatically analyze, correlate, and contextualize microbial and physiological dynamics through reinforcement learning and adaptive modeling algorithms. The result is a living data ecosystem a dynamic one that responds to changes in biology, optimizes its diagnostic parameters autonomously, and feedbacks predictive insights into individualized interventions. This synergy leads to the new paradigm of a "Digital Gut Twin" a virtual copy of an individual's microbiome-host network that is being updated continually and can simulate perturbations, predict disease courses, and self-direct clinical decision-making. Essentially, microbiome sensing using IoT converts the microbiome as a static lab specimen to a digitally interactive, computationally dynamic biosystem that bridges the physical world of human biology and digital diagnostics' autonomous intelligence, thus mapping the architecture of next-generation precision medicine (Sobhan et al., 2025).

4.4 Edge AI Deployment, Connectivity, and Implementation Roadmap

Edge AI deployment can make use of devices like Raspberry Pi and NVIDIA Jetson, with the former providing a power-efficient, low-cost solution for light models and the latter offering greater computational power and GPU acceleration for heavy real-time processing(A. Y. Ding et al., 2022). Connectivity is very important, with the 3G network providing low-end, low-bandwidth communication for non-real-time applications and the 5G network facilitating ultra-low-latency and high-bandwidth performance that is required for smooth real-time inference and cloud integration. The cost analysis should also factor in the purchase cost of hardware, power requirements, networking infrastructure, and upkeep, with Raspberry Pi being less expensive but possibly taking more units, while Jetson modules are more expensive but can carry heavier loads (Deng et al., 2020). A formal deployment plan entails initiating with a pilot phase to identify how devices perform and what reliability they offer, then implementing a scale phase for large-scale deployment and optimization, and ultimately a maintenance phase to provide long-term system stability, updates, and model retraining (Shankar, 2024).

Limitation

Age, gender, BMI, diet, and antibiotic use all have an impact on the macro and microscopic structural heterogeneity and gut microbiota, which differ greatly between people. Group matching should be used to incorporate these characteristics into investigations comparing colorectal patients to healthy controls. Regretfully, most of the included research did not compare the two groups; instead, they only addressed possible influences on microbiome makeup. Because the gut microbiota is dynamic and changes with the pathology, as well as because it is vulnerable to confounding environmental factors like diet, medication, smoking, and other lifestyle choices, pilot studies in colorectal cancer have also shown a great deal of variation among tumors. Most of these were often not recorded, and only a small percentage were included in the analysis. Geographically different populations and countries have different amounts of faecal microbiome richness at the phylum and strain levels. It is unclear how many of the mentioned papers may be translated because many of them were carried out in Asia, which may restrict translation to other regions (Artemev et al., 2022).

Lack of consistency in sample collection and processing has been another major issue in these studies. Temperature and time until long-term preservation at -80°C differed significantly between investigations, as did sample collection (for instance, before or after bowel-cleansing agents were employed for the colonoscopy). They have been demonstrated to change the composition of the microbiome (Breitwieser et al., 2018).

5.AI IN MICROBIOME-BASED CANCER IMMUNOTHERAPY

Recent research on the gut microbiome's impact on immunotherapy efficacy has identified unique microbial profiles linked with improved or decreased treatment responses. Studies on anti-CTLA-4 immunotherapy have found that the presence of gut bacteria such as *Bacteroides fragilis*, *Bacteroides thetaiotaomicron*, and *Burkholderia cepacia* can considerably improve treatment effects. In addition, FMT from high B. fragilis abundance patients reduced tumor sizes. Microbiota has also been linked to immunotherapy-related toxicities such as colitis. Anti-CTLA-4 treatment also led to the enrichment of the bacterial families *Bacteroidaceae, Rikenellaceae*, and *Barnesiellaceae* in colitis-free patients. Anti-CTLA-4 treatment led to a higher proportion of bacterial families including *Bacteroidaceae, Rikenellaceae* and *Barnesiellaceae* in colitis-target free patients. Specifically, ML models built from those microbiome sequencing reads exhibited both 100% specificity and 70% sensitivity to predict colitis outcomes, illustrating AI's potential to process complex microbiome data and predict therapy related toxicities (Hosny et al., 2018).

Similarly, anti-PD-1 and anti-PD-L1 immunotherapy responses have been linked to gut microbiota composition. Responders exhibited greater microbial diversity, particularly with increased abundance of *Clostridiales, Ruminococcaceae, Faecalibacterium, Akkermansia muciniphila, and Lachnospiraceae.* AI techniques have been employed to construct predictive models based on microbial abundance, which correlate with progression-free survival and immune cell recruitment. For instance, fecal microbiota transplantation (FMT) trials combined with anti-PD-1 therapy demonstrated post-treatment microbiota shifts toward *Enterococcaceae, Enterococcus*, and Streptococcus australis; responders also showed enhanced cytotoxic T-cell activity and altered cytokine profiles. Additionally, short-chain fatty acids (SCFAs), such as butyrate and propionate, produced by gut bacteria, were associated with improved therapeutic outcomes—though these effects varied between CTLA-4 and PD-1 blockade therapies. As illustrated in the accompanying **Fig.2**, the general workflow for AI-based prediction of PD-1/PD-L1 expression and tumor mutational burden (TMB) involves several key steps: feature collection, data preprocessing, feature extraction, model development, and performance evaluation. This systematic approach facilitates the development of accurate predictive models that integrate microbiome profiles with immunotherapy outcomes (Hu et al., 2021).

Furthermore, AI-enabled microbiome analysis has been applied to innovative immunotherapy techniques such as CpG oligonucleotide therapy and CAR-T therapy. Antibiotic-mediated microbiome depletion lowered the efficiency of CpG-oligonucleotide immunotherapy, but enriching Alistipes and *Ruminococcus* and depleting *Lactobacillus* boosted responses. An AI-driven analysis of microbiome sequencing data in CAR-T treatment recipients discovered that individuals given broad-spectrum antibiotics had lower survival rates. *Bacteroides, Ruminococcus, Eubacterium*, and *Akkermansia* genera all demonstrated a substantial association with CAR-T efficacy. Furthermore, bacterial metabolites such as butyrate and pentanoate, generated by *Megasphaera massiliensis* and *Megasphaera elsdenii*, improved CAR-T cytotoxic performance in murine and human models.

Collectively, these studies show that AI can help with the integration and interpretation of complex microbiome data to predict immunotherapy outcomes, stratify patients based on microbiome profiles, and potentially guide microbiota-targeted adjunct therapies to improve cancer immunotherapy efficacy. The successful application of AI-trained models to identify responders, predict side effects, and assess metabolite impact emphasizes AI's critical role in unlocking the clinical potential of microbiome-based cancer immunotherapy (Xie et al., 2023)

Figure 3. The general workflow for AI-based prediction of PD-1/PD-L1 expression and TMB typically involves several key steps: feature collection, data preprocessing, feature extraction, model development, and performance evaluation. This process ultimately leads to the creation of a predictive model. (Created in https://BioRender .com)

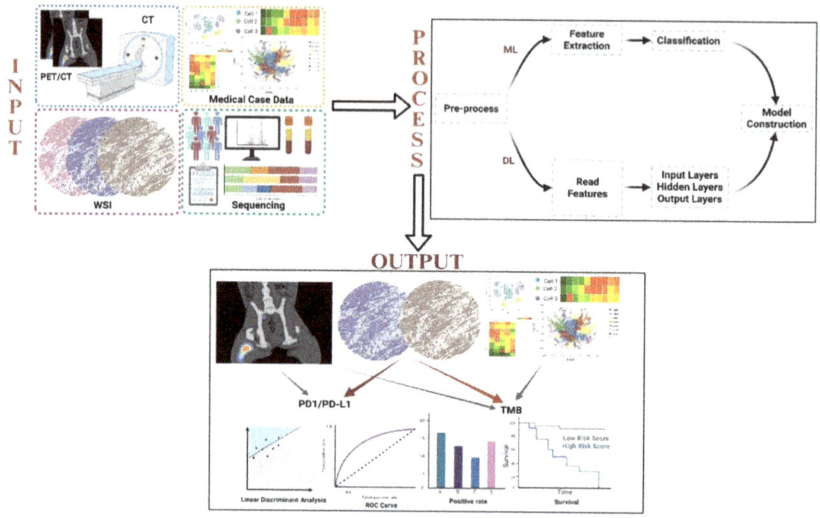

5.1 Predicting Immunotherapy Response

AI is now considered a successful tool for applying cancer immunotherapy. By filtering out only relevant phenotypic alterations and larger phenotypes, AI-based algorithms allow for the detection of phenotypic variants and the generation of data that can capture minute details that the human eye would miss. AI-based techniques have so far demonstrated excellent success in identifying and integrating a wealth of medical data, such as high-throughput data, medical imaging, and others, to directly predict treatment response in patients undergoing ICB (Yang et al., 2024).

Computed tomography (CT) and magnetic resonance imaging (MRI) are two examples of medical imaging, which is a technical procedure that represents the interior of a body and is frequently utilized in clinical practice and research. The application of medical images to monitor treatment outcomes is hampered by the time-consuming tasks done by radiologists and the detailed descriptions of tumors and lymph nodes. Combining high-dimensional imagery with the rapidly evolving AI methods in medical imaging can improve the radiological workflow (Raymond Geis et al., 2019).

This study found that NSCLC lesions with unique imaging patterns, such as compact borders and non-uniform density, and diverse morphological profiles were more likely to respond to immunotherapy. Furthermore, radiomics approaches have provided a viable method for predicting anti-PD-1 and anti-PD-L1 in lung cancer by comparing with genomic data such as tumour mutational burden (TMB) and RNA sequencing (Trebeschi et al., 2019). Given the foregoing, it is possible to predict a patient's response to ICB immunotherapy with accuracy and consistency by employing AI algorithms in medical imaging-based radiomics. This also applies to other solid tumours such as urinary tract cancer, esophageal carcinoma of the squamous cells, squamous cell cancer of the head and neck carcinoma, and melanoma. The development of radiomic signatures to forecast clinical outcomes in patients undergoing anti-PD-L1, anti-PD-1, and/or anti-CTLA-4 immunotherapy has also made extensive use of AI-based models. These investigations into various cancers showed consistent patterns across cancer types, pointing to links between immunotherapy response and radiomics biomarkers. As a result, AI-based medical imaging analysis can be helpful in predicting the therapeutic response and clinical results of patients receiving immunotherapy, which may help guide customized cancer immunotherapy in clinical settings (Wang et al., 2018).

Certain hematological indicators help identify patients with favorable treatment responses to identify those who are benefiting from immunotherapy by incorporating AI approaches. Nowadays, liquid biopsies provide a quick and flexible way to depict tumor heterogeneity. They can be used in conjunction with ML algorithms to forecast the therapeutic impact of ICB using hematological genetic indicators such microRNA and circulating tumor cell DNA (Mu et al., 2020). Additionally, AI-based algorithms found additional biomarkers in blood tests to forecast the results of cancer patients receiving ICB. Interestingly, several serum-derived proteins have been found by ML-based algorithms to stratify the outcomes of ICB patients. Furthermore, molecular signatures have been found and detected using AI in an automated manner to forecast immunotherapy response(Tosta et al., 2019). According to research, the ML-based framework uses multiplex spatial protein profiling to predict the ICB response by analysing the pathophysiological and molecular components in the tumour microenvironment and probing the protein expression in microscopic compartments. Previous studies have shown that ML-based models on histopathological imaging analysis enhance the efficient identification of appropriate patients for cancer immunotherapy, hence expanding the benefits of immunotherapy to a wider target group. Furthermore, integrating AI into cancer immunotherapy enables personalised and effective standard of care for each patient by developing accurate immune-associated biomarkers and controlling algorithms for predicting treatment response, medication resistance, and survival outcomes (Vandenberghe et al., 2017).

5.2 Modulating the Gut Microbiome to Enhance Efficacy

The gut microbiome has emerged as an important regulator of host immune responses and a moderator of immunotherapy results. Several studies have found that specific bacterial taxa increase responses to immune checkpoint inhibitors such as anti-CTLA-4 and anti-PD-1/PD-L1 treatments. The presence of Bacteroides fragilis and its recolonization after antibiotic treatment has been demonstrated to greatly improve anti-CTLA-4 medication efficacy. Furthermore, FMT (fecal microbiota transplantation) investigations demonstrated that a higher abundance of B. fragilis was associated with reduced tumor diameters, implying a function in boosting anti-tumor immunity. Similarly, patients who remained colitis-free following anti-CTLA-4 medication had higher levels of *Bacteroidaceae, Rikenellaceae,* and *Barnesiellaceae.* These data suggest that bacteria play a dual function in increasing therapeutic efficacy and perhaps decreasing immune-related side effects such as colitis (Simpson et al., 2023).

In addition to bacterial species, fungal microbiota has been involved. Non-responsiveness to checkpoint blocking in melanoma was linked to increased abundance of the fungus genus Cladosporium. Experimental investigations found that administering a combination of live *Clostridium butyricum* and checkpoint inhibitors dramatically enhanced clinical responses and progression-free survival in patients with metastatic renal cancer. Certain commensals, such as *Bifidobacterium pseudolongum, Lactobacillus johnsonii*, and Olsenella spp., have been demonstrated to improve immunotherapeutic efficacy in preclinical cancer models, thanks in part to microbial compounds such inosine, which promote anti-tumor T-cell responses (Maurya et al., 2021).

Overall, modulating the gut microbiota using probiotics, antibiotics, FMT, and nutritional interventions has great promise for improving immunotherapy efficacy. Inconsistencies between studies, likely due to patient heterogeneity, cancer type, geographic variation, and sequencing methods, highlight the need for standardized, large-scale, multicentric studies to identify reproducible microbial biomarkers and develop targeted microbiome-modulating strategies as adjuvant treatments in cancer immunotherapy (Kang et al., 2024).

5.3 Personalized Microbiome Interventions

Recent breakthroughs in microbiome research have highlighted the importance of commensal bacteria in affecting therapeutic results in cancer treatment, paving the path for tailored microbiome therapies. Personalized microbiome interventions, in contrast to standard precision oncology, which focuses primarily on tumor genomes to design medicines, acknowledge the gut microbiota as a dynamic and important component of the tumor microenvironment and systemic immunity. Each person has a distinct makeup of gut microorganisms, which influences not just baseline immune responses but also the host's ability to absorb chemotherapeutic drugs, regulate epigenetic marks, and develop certain immune cell profiles. For example, the gut microbiota is known to create short-chain fatty acids (SCFAs) like butyrate and propionate by fermenting food fibers. These SCFAs regulate histone post-translational modifications (HPTMs), specifically the suppression of histone deacetylase (HDAC), which regulates gene transcription in immune cells such as cytotoxic CD8+ T cells and regulatory T cells (Tregs). A healthy microbiome that produces SCFAs has been linked to increased anti-tumor immunity and immunotherapy response (**Fig. 3**).

Figure 4. AI-driven framework for personalized medicine in cancer therapy.

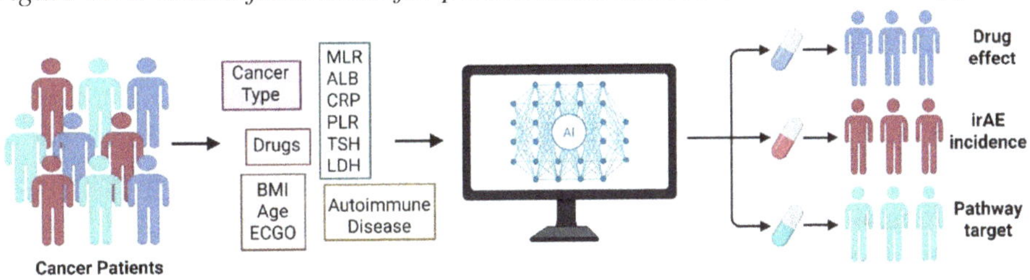

5.4 Human Microbiota and Personalized Cancer Treatments: Role of Commensal Microbes in Treatment Outcomes for Cancer Patients

Furthermore, the microbiome has a direct effect on the metabolism and absorption of several chemotherapy medicines. Gut microbial enzymes metabolize or modify more than forty medications, including frequently used medicines such as methotrexate and irinotecan (Kumar A, Sakhare K et al., 2022). Variations in microbial communities can thus result in differing drug detoxification, efficacy, and toxicity in patients. Personalized therapies may include altering the microbial environment to improve medication activation or lessen side effects. Specific microbial taxa have been identified in the context of immune checkpoint blockade therapies, such as anti-PD-1, anti-PD-L1, and anti-CTLA-4 antibodies, that correlate with better treatment outcomes. *Bifidobacterium spp.*, *Akkermansia muciniphila*, and *Bacteroides fragilis,* for example, have been shown to restore or enhance anti-tumor immunity in antibiotic-treated or germ-free mice when administered alone or in combination with checkpoint. This provides compelling evidence for microbiome analysis prior to immunotherapy to detect favorable or unfavorable microbial signatures and tailor therapies accordingly (Leffler & Lamont, 2015).

Fecal microbiota transplantation (FMT) is emerging as one of the most promising approaches to personalized microbiome-based cancer therapy. Studies have demonstrated that transferring gut microbiota from cancer patients who respond to immunotherapy into non-responders or preclinical animal models can significantly enhance treatment efficacy. These findings underscore the therapeutic potential of responder-derived microbiota or specialized microbial consortia in sensitizing tumors to immunotherapeutic agents. Additionally, the discovery of an intratumoral microbiome and its influence on local immune regulation introduces a further dimension of personalization by enabling targeted manipulation of site-specific microbial populations within tumors (Kumar, Kumar, et al., 2024).

This paradigm shift toward microbiome-informed cancer therapy is further strengthened by the integration of AI, which facilitates the analysis of complex microbial data throughout the entire cancer care continuum—from diagnosis to prognosis, therapy, remission, and recurrence (**Fig 4**). Microbial DNA derived from stool or tumor tissues is processed using AI-driven models to predict treatment responses, identify optimal therapeutic strategies (such as bispecific antibodies), and monitor therapeutic efficacy. AI also plays a crucial role in linking gut microbiota composition with tumor behavior and immune responses, supporting the development of personalized treatment protocols. Post-treatment, AI can be utilized to detect signs of remission or recurrence by analyzing shifts in microbial signatures, enabling real-time monitoring of disease progression. Personalized microbiome interventions such as dietary modulation, selective probiotic supplementation, short-chain fatty acid (SCFA) replacement therapy, and targeted histone deacetylase (HDAC) inhibition are rapidly emerging as integral components of cancer treatment. By combining genetic, immunological, and microbiological characterization, these innovative strategies aim to optimize host-microbiota synergy and ultimately improve patient outcomes (Singh et al., 2024).

Figure 5. Integrative Role of AI and Microbiome Across the Cancer Care Continuum. The role of Artificial Intelligence (AI) in integrating microbiome data throughout the cancer care span, including diagnosis, prognosis, therapy, remission and recurrence. In the first step during diagnosis, microbial DNA from patient sample such as blood, tissue, stool is analysed to detect the cancer. Then in the prognosis phase the Gut and tumour microenvironment (TME) microbiome profiles are in order to forecast clinical results and treatment response. AI models assist in guiding treatment strategies, including the immunotherapies like bispecific antibodies that activate effector T cells, leading to cancer cell death through perforin and granzyme release. In the remission stage, AI helps monitor the microbiome to detect micro-metastases or confirm disease-free status. Continuous surveillance through microbiome analysis also enables early detection of recurrence. Central to all stages, AI integrates and interprets complex microbial and clinical data to support personalized and predictive cancer management.

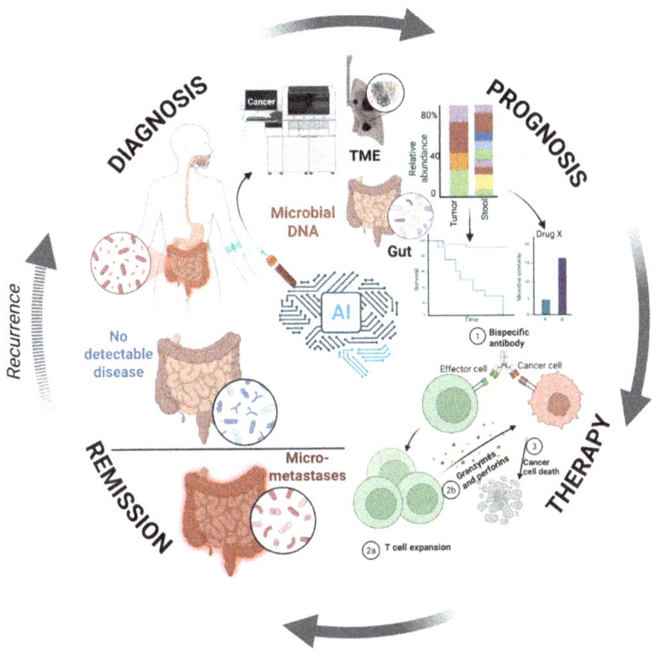

6. CHALLENGES AND ETHICAL CONSIDERATIONS

6.1 Data Privacy and Security

Since the time of Hippocrates, ethics has been regarded as the foundation of medical practice. The Belmont Report provides a modern framework that emphasizes four fundamental ethical principles autonomy, beneficence, nonmaleficence, and justice which continue to guide moral reasoning in medicine and biomedical research (Beauchamp & Childress, 2019). In the context of AI, these principles are equally vital. AI ethics has been defined as "a set of values, principles, and techniques that employ widely accepted standards of right and wrong to guide moral conduct in the development and use of AI technologies." Ethical oversight is imperative because the misuse, design flaws, or unintended consequences of AI systems may cause harm to individuals, healthcare institutions, and society at large. Therefore, the moral responsibility that once rested solely on physicians now extends to developers, data scientists, and institutions deploying AI-driven healthcare systems.

AI-driven healthcare systems especially those employing ML for predictive analytics, treatment optimization, and disease diagnosis rely heavily on access to large volumes of sensitive patient data, including genetic profiles, medical imaging, and electronic health records (EHRs). While such data enable remarkable progress in clinical accuracy and personalization, they also introduce profound privacy and security risks. Healthcare data are among the most targeted forms of digital information due to their intrinsic value. Cyberattacks can lead to identity theft, insurance fraud, blackmail, or deliberate manipulation of clinical data, undermining patient safety and public trust (Naik et al., 2022). Insider threats, where authorized personnel misuse data, further compound this risk. In AI-integrated clinical systems, such breaches could corrupt model performance, skew predictions, and erode the integrity of decision-making pipelines. The increasing use of cloud-based ML platforms for data storage and processing raises additional vulnerabilities. Without stringent safeguards such as end-to-end encryption, multi-factor authentication, and secure access protocols, patient data may be exposed to unauthorized third parties(Sheller et al., 2020). Even anonymized datasets are not immune advanced re-identification algorithms can potentially link de-identified medical data to individuals by correlating them with external datasets. This highlights the need for robust differential privacy techniques and continuous monitoring to ensure true anonymity.

Furthermore, regulatory compliance remains a cornerstone of ethical AI implementation. Failure to align with global data protection frameworks such as HIPAA (U.S.), GDPR (E.U.), and the Data Protection Act (U.K.) can lead not only to legal consequences but also to the erosion of patient trust, which is indispensable for successful AI integration in healthcare. The challenge is intensified by the limited explainability and auditability of many modern ML algorithms, often referred to as "black-box models." These limitations hinder regulatory verification and pose barriers to transparency and accountability.

A promising approach to addressing these issues is federated learning, which allows multiple institutions to collaboratively train AI models without transferring raw patient data to a central server. This decentralized model preserves patient confidentiality while leveraging the power of large, diverse datasets (Murdoch, 2021). However, federated learning is not immune to risk: data integrity breaches, model poisoning attacks, and gradient leakage can still compromise privacy and model reliability. Therefore, advanced secure aggregation protocols, blockchain-based auditing, and differentially private gradient updates are critical to ensure the resilience of such distributed AI systems. In the emerging field of autonomous microbiome–oncology systems, where IoT devices continuously collect microbiome and physiological data to guide adaptive interventions, data security becomes even more crucial. These systems operate at the intersection of human biology and digital intelligence, and thus any breach could have direct clinical consequences. Maintaining ethical compliance in such dynamic data environments demands continuous risk assessment, strong encryption of data streams, and real-time anomaly detection for cyber threats. Ultimately, safeguarding data privacy and security is not merely a technical challenge but a moral and legal imperative. Ethical AI in healthcare must be built upon secure digital infrastructures, transparent governance mechanisms, and the unwavering protection of patient autonomy and confidentiality. Only by embedding these principles into the design and deployment of AI-driven medical systems can we ensure that technological advancement serves humanity without compromising its core values.

6.2 Ethical Considerations in AI Deployment

AI implementation poses huge ethical issues that need to be resolved in order to attain international standards, such as algorithmic bias, digital divide, data sovereignty, and labor displacement. Algorithmic bias may occur when AI algorithms are trained on biased or non-representative data, resulting in unequal outcomes, especially in sensitive areas such as healthcare. The digital divide could extend inequalities since groups with limited exposure to AI technologies might not benefit from them. Sovereignty issues over data arise with cross-border usage, necessitating regulatory compliance within each jurisdiction and preserving personal privacy. Furthermore, job disruption through AI automation poses challenges, which necessitate reskilling and upskilling of workers. The solutions to these challenges are necessary for responsible and equitable AI development (Rivi -Tipei, 2023).

7. TOWARDS MICROBIOME-BASED PRECISION ONCOLOGY

Advances in high-throughput sequencing and ML have allowed for thorough analysis of the human gut microbiome, showing its significance in cancer development and treatment response. Microbiome-based algorithms can now help identify high-risk individuals by evaluating complex microbial datasets, however issues like data heterogeneity and sparsity persist. Improved bioinformatics tools are required to identify functional linkages between microbial populations and tumor biology

Parallel advances in next-generation sequencing (NGS) have improved the identification of predictive and prognostic biomarkers. Short-read systems aid in variant discovery for therapeutic applications, but long-read sequencing enables comprehensive transcriptome analysis. Integrating these techniques with AI enhances data interpretation and speeds up biomarker identification in precision oncology (Davenport & Kalakota, 2019).

Tools such as scPriorGraph, which uses single-cell RNA-seq data to map cell identities based on gene set priors, demonstrate how microbiome and cellular data might be linked to improve cancer diagnosis and therapy planning. Furthermore, AI-powered digital pathology and imaging improve diagnosis speed and accuracy, particularly in environments with few healthcare professionals. Automated solutions also improve electronic health record administration, facilitating clinical decision-making and decreasing administrative strain. As research progresses, microbiome signatures are likely to be integrated into cancer processes, assisting with patient categorization, tailored therapy selection, and treatment monitoring, alongside genomic and transcriptomic data. This convergence represents an encouraging step toward completely personalized cancer care (Davenport & Kalakota, 2019).

8. FUTURE PERSPECTIVES AND RESEARCH DIRECTIONS

Bioinformatics connects biology and computational analysis, unlocking patterns and insights from complex datasets. It provides tools and methods for analyzing, interpreting, and extracting meaningful knowledge from large amounts of data. Over the last decade, bioinformatics has evolved beyond the use of computational tools for biological and pharmaceutical data analysis. AI has revolutionized biological research and marked a new era of innovation (Kumar, Dhiman, Kumar, et al., 2025). The rapid increase of biological data (Big Data) through high-throughput sequencing and other cutting-edge technologies has presented a huge issue due to its time-consuming and complex nature. Machine methods like SVM, random forests, and neural networks are becoming essential tools in bioinformatics (Kumar et al., 1 C.E., 2022, 2024) These advanced tools enable researchers to handle complicated challenges in disciplines such as genetics, protein structure prediction, drug development, disease diagnosis, and biomarker identification.AI has the potential to alter bioinformatics, as demonstrated by the following instances. AI is rapidly altering bioinformatics, enabling better knowledge of biological processes, disease diagnostics, and individualized medicinal tactics. However, it's important to recognize that AI is not a panacea. Ethical considerations for data privacy, algorithmic bias, and misuse require careful attention. Embracing technology safely, ethically, and cooperatively can lead to tailored medicine, enhanced healthcare, and a greater understanding of life's complexities. AI has substantially impacted both developing and established bioinformatics applications. AI-driven tools and algorithms can accelerate research, improve data interpretation, and promote innovation in biomedical sciences (Gupta et al., 2024).

Drug development remains challenging due to the large number of potential drug-target combinations and the complexity of model prediction, resulting in time-consuming and costly processes.AI is transforming drug discovery across multiple applications. Graph neural networks (GNNs) anticipate pharmacological interactions with targets, guiding drug selection. Generative adversarial networks (GANs) use ML to create new medicinal compounds with desired attributes, speeding up drug discovery. AI models can anticipate drug outcomes, allowing for tailored medicine and focused therapy tactics.

It is anticipated that multiomics techniques would be crucial to the development of personalized medicine. By combining genomic, transcriptomic, proteomic, and metabolomic data, medical professionals can customize medications to each patient's unique traits, increasing effectiveness and lowering side effects. Whole exome sequencing (WES) and whole genome sequencing (WGS) have enormous promise for improving our knowledge of complicated illnesses like cancer, heart problems, and neurological conditions. By determining the molecular mechanisms behind these diseases, multiomics research can assist in the discovery of new biomarkers for early detection and possible therapeutic targets. Over the next ten years, there may be significant advancements in the integration of multiomics with other emerging technologies, which could result in the creation of customized virtual patient models. Before being used in actual clinical settings, these models would allow medicines and interventions to be tested in silico. Additionally, the ability of researchers and clinicians to understand complex datasets will improve with the development of multiomics data visualization tools, enabling the use of AI-driven genetic insights in clinical practice (Manzoni et al., 2018).

With numerous new trends, technological developments, and interdisciplinary approaches propelling innovation in precision medicine, the future of AI in genomics appears bright. Achieving the full potential of this research will require overcoming current challenges and creating new applications. AI-powered genomics will continue to be at the forefront of scientific and medical innovation with sustained funding and cooperation across other (Jamialahmadi et al., 2024). A details tree diagram is summarized in **Fig 5** by showing the advantages, future direction and current limitations.

Figure 6. A tree diagram represents the advantages, future direction, limitations and microbiome-based biomarkers for early stage of cancer detection and diagnosis.

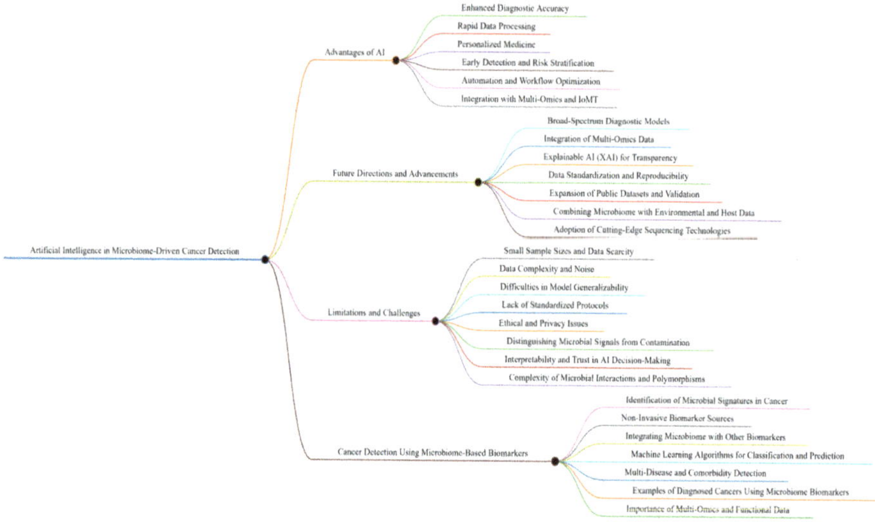

CONCLUSION

Convergence of microbiome science, AI, and the IoT is opening the doors to a game-changing paradigm in the management of cancer an adaptive, smart, and biologically integrated care system. The human microbiome, comprising an enormous catalog of genetic and metabolic data, is an active biosensor that mirrors the physiological as well as pathological status of the host. With incorporation into AI-based analytical models, this information becomes actionable clinical intelligence that can forecast therapeutic response, resistance mechanisms, and side effects well in advance of their clinically evident appearance.AI introduces autonomy and accuracy to this system. Deep learning architectures can incessantly examine high-dimensional biological data, discover hidden patterns of correspondence between microbial signatures and cancer phenotypes, and direct clinicians toward evidence-based, individualized treatment plans. In the background, IoT-enabled sensing and actuation systems offer the essential interface between the biological and digital spaces receiving real-time measurements on critical parameters, drug delivery processes, and treatment compliance while facilitating remote monitoring and responsive interventions.

Combined, these technologies form an auto-adaptive, closed-loop system one that improves along with the biology of the patient and dynamically changes therapeutic regimens. An integrated system such as this guarantees that cancer treatment advances beyond fixed treatment protocols towards a dynamically optimized model with balance of efficacy, safety, and quality of life. The synergy between microbiome-informed diagnostics, AI's cognitive capabilities, and IoT's seamless connectivity heralds the dawn of a new era in oncology one defined by intelligent automation, biological awareness, and human–machine collaboration. The future of medicine will no longer be based on human interpretation or digital computation alone but will arise from the synergy of human clinicians and autonomous, biology-conscious AI systems. This co-evolution of technology and medicine has the potential to revolutionize cancer care into a proactive, predictive, and personalized science where precision, flexibility, and empathy intersect to redefine patient outcomes.

REFERENCES

Almonte, A. A., Thomas, S., & Zitvogel, L. (2025). Microbiota-centered interventions to boost immune checkpoint blockade therapies. *The Journal of Experimental Medicine*, *222*(7), e20250378. Advance online publication. DOI: 10.1084/jem.20250378 PMID: 40261296

Anvarnia, A., Mohaddes-Gharamaleki, F., Asadi, M., Akbari, M., Yousefi, B., & Shanehbandi, D. (2019). Dysregulated microRNAs in colorectal carcinogenesis: New insight to cell survival and apoptosis regulation. *Journal of Cellular Physiology*, *234*(12), 21683–21693. DOI: 10.1002/jcp.28872 PMID: 31131450

Artemev, A., Naik, S., Pougno, A., Honnavar, P., & Shanbhag, N. M. (2022). The Association of Microbiome Dysbiosis With Colorectal Cancer. *Cureus*, *14*(2). Advance online publication. DOI: 10.7759/cureus.22156 PMID: 35174040

Ballard, J. L., Wang, Z., Li, W., Shen, L., & Long, Q. (2024). Deep learning-based approaches for multi-omics data integration and analysis. *BioData Mining*, *17*(1), 38. Advance online publication. DOI: 10.1186/s13040-024-00391-z PMID: 39358793

Beauchamp, T., & Childress, J. (2019). Principles of Biomedical Ethics: Marking Its Fortieth Anniversary. *The American Journal of Bioethics*, *19*(11), 9–12. DOI: 10.1080/15265161.2019.1665402 PMID: 31647760

Breitwieser, F. P., Lu, J., & Salzberg, S. L. (2018). A review of methods and databases for metagenomic classification and assembly. *Briefings in Bioinformatics*, *20*(4), 1125–1139. DOI: 10.1093/bib/bbx120 PMID: 29028872

Çalışkan, M., & Tazaki, K. (2023). AI/ML advances in non-small cell lung cancer biomarker discovery. *Frontiers in Oncology*, *13*, 1260374. DOI: 10.3389/fonc.2023.1260374 PMID: 38148837

Cao, R., Tang, L., Fang, M., Zhong, L., Wang, S., Gong, L., Li, J., Dong, D., & Tian, J. (2022). Artificial intelligence in gastric cancer: Applications and challenges. *Gastroenterology Report*, *10*, goac064. Advance online publication. DOI: 10.1093/gastro/goac064 PMID: 36457374

Collins, F. S., Patrinos, A., Jordan, E., Chakravarti, A., Gesteland, R., & Walters, L. R. (1998). New goals for the U.S. Human Genome Project: 1998-2003. *Science*, *282*(5389), 682–689. DOI: 10.1126/science.282.5389.682 PMID: 9784121

Cullin, N., Azevedo Antunes, C., Straussman, R., Stein-Thoeringer, C. K., & Elinav, E. (2021). Microbiome and cancer. *Cancer Cell*, *39*(10), 1317–1341. DOI: 10.1016/j.ccell.2021.08.006 PMID: 34506740

D'Urso, F., & Broccolo, F. (2024). Applications of Artificial Intelligence in Microbiome Analysis and Probiotic Interventions—An Overview and Perspective Based on the Current State of the Art. *Applied Sciences 2024, Vol. 14, Page 8627, 14*(19), 8627. DOI: 10.3390/app14198627

Davenport, T., & Kalakota, R. (2019). The potential for artificial intelligence in healthcare. *Future Healthcare Journal, 6*(2), 94–98. DOI: 10.7861/futurehosp.6-2-94 PMID: 31363513

Dekaboruah, E., Suryavanshi, M. V., Chettri, D., & Verma, A. K. (2020). Human microbiome: An academic update on human body site specific surveillance and its possible role. *Archives of Microbiology, 202*(8), 2147–2167. DOI: 10.1007/s00203-020-01931-x PMID: 32524177

Deng, S., Zhao, H., Fang, W., Yin, J., Dustdar, S., & Zomaya, A. Y. (2020). Edge Intelligence: The Confluence of Edge Computing and Artificial Intelligence. *IEEE Internet of Things Journal, 7*(8), 7457–7469. DOI: 10.1109/JIOT.2020.2984887

Ding, A. Y., Peltonen, E., Meuser, T., Aral, A., Becker, C., Dustdar, S., Hiessl, T., Kranzlmüller, D., Liyanage, M., Maghsudi, S., Mohan, N., Ott, J., Rellermeyer, J. S., Schulte, S., Schulzrinne, H., Solmaz, G., Tarkoma, S., Varghese, B., & Wolf, L. (2022). Roadmap for edge AI. *Computer Communication Review, 52*(1), 28–33. DOI: 10.1145/3523230.3523235

Ding, Q., Kong, X., Zhong, W., & Liu, W. (2022). Fecal biomarkers: Non-invasive diagnosis of colorectal cancer. *Frontiers in Oncology, 12*, 971930. DOI: 10.3389/fonc.2022.971930 PMID: 36119474

Eslami, M., Naderian, R., Bahar, A., Babaeizad, A., Rezanavaz Gheshlagh, S., Oksenych, V., & Tahmasebi, H. (2025). Microbiota as diagnostic biomarkers: Advancing early cancer detection and personalized therapeutic approaches through microbiome profiling. *Frontiers in Immunology, 16*, 1559480. Advance online publication. DOI: 10.3389/fimmu.2025.1559480 PMID: 40406094

Fan, J. Q., Zhao, W. F., Lu, Q. W., Zha, F. R., Lv, L., Ye, G.-L., & Gao, H.-L. (2023). Fecal microbial biomarkers combined with multi-target stool DNA test improve diagnostic accuracy for colorectal cancer. *World Journal of Gastrointestinal Oncology, 15*(8), 1424–1435. DOI: 10.4251/wjgo.v15.i8.1424 PMID: 37663945

Farina, E., Nabhen, J. J., Dacoregio, M. I., Batalini, F., & Moraes, F. Y. (2022). An overview of artificial intelligence in oncology. *Future Science OA, 8*(4), FSO787. Advance online publication. DOI: 10.2144/fsoa-2021-0074 PMID: 35369274

Fu, A., Yao, B., Dong, T., Chen, Y., Yao, J., Liu, Y., Li, H., Bai, H., Liu, X., Zhang, Y., Wang, C., Guo, Y., Li, N., & Cai, S. (2022). Tumor-resident intracellular microbiota promotes metastatic colonization in breast cancer. *Cell, 185*(8), 1356–1372.e26. DOI: 10.1016/j.cell.2022.02.027 PMID: 35395179

Ghaddar, B., Biswas, A., Harris, C., Omary, M. B., Carpizo, D. R., Blaser, M. J., & De, S. (2022). Tumor microbiome links cellular programs and immunity in pancreatic cancer. *Cancer Cell, 40*(10), 1240–1253.e5. DOI: 10.1016/j.ccell.2022.09.009 PMID: 36220074

Gopalakrishnan, V., Spencer, C. N., Nezi, L., Reuben, A., Andrews, M. C., Karpinets, T. V., Prieto, P. A., Vicente, D., Hoffman, K., Wei, S. C., Cogdill, A. P., Zhao, L., Hudgens, C. W., Hutchinson, D. S., Manzo, T., Petaccia De Macedo, M., Cotechini, T., Kumar, T., Chen, W. S., & Wargo, J. A. (2018). Gut microbiome modulates response to anti-PD-1 immunotherapy in melanoma patients. *Science, 359*(6371), 97–103. DOI: 10.1126/science.aan4236 PMID: 29097493

Graham-Knight, J. B., Liang, P., Lin, W., Wright, Q., Shen, H., Mar, C., Sam, J., & Rajapakshe, R. (2025). External Testing of a Commercial AI Algorithm for Breast Cancer Detection at Screening Mammography. Https://Doi.Org/10.1148/Ryai.240287, *7*(3). DOI: 10.1148/ryai.240287

Green Buzhor, M., Longobardi, G., Kandli, O., Krinsky, A., Avramoff, O., Katyal, A., Salomon, K., Miari, A., Venkert, D., Barnatan, T. T., García Alvarado, A., Greenberg, S., & Satchi-Fainaro, R. (2025). Harnessing Next-Generation 3D Cancer Models to Elucidate Tumor-Microbiome Crosstalk. *Advanced Healthcare Materials, 03198*, e03198. Advance online publication. DOI: 10.1002/adhm.202503198 PMID: 41045195

Gupta, A., Kumar, S., & Kumar, A. (2024). Big Data in Bioinformatics and Computational Biology: Basic Insights. *Methods in Molecular Biology (Clifton, N.J.), 2719*, 153–166. DOI: 10.1007/978-1-0716-3461-5_9 PMID: 37803117

Hernández Medina, R., Kutuzova, S., Nielsen, K. N., Johansen, J., Hansen, L. H., Nielsen, M., & Rasmussen, S. (2022). Machine learning and deep learning applications in microbiome research. *ISME Communications 2022 2:1, 2*(1), 1–7. DOI: 10.1038/s43705-022-00182-9

Hosny, A., Parmar, C., Quackenbush, J., Schwartz, L. H., & Aerts, H. J. W. L. (2018). Artificial intelligence in radiology. *Nature Reviews. Cancer, 18*(8), 500–510. DOI: 10.1038/s41568-018-0016-5 PMID: 29777175

Hossain, M. S., Joshi, T., & Stacey, G. (2015). System approaches to study root hairs as a single cell plant model: Current status and future perspectives. *Frontiers in Plant Science*, *6*(MAY), 1–7. DOI: 10.3389/fpls.2015.00363 PMID: 26042143

Hu, J., Cui, C., Yang, W., Huang, L., Yu, R., Liu, S., & Kong, Y. (2021). Using deep learning to predict anti-PD-1 response in melanoma and lung cancer patients from histopathology images. *Translational Oncology*, *14*(1), 100921. Advance online publication. DOI: 10.1016/j.tranon.2020.100921 PMID: 33129113

Iqbal, M. J., Javed, Z., Sadia, H., Qureshi, I. A., Irshad, A., Ahmed, R., Malik, K., Raza, S., Abbas, A., Pezzani, R., & Sharifi-Rad, J. (2021). Clinical applications of artificial intelligence and machine learning in cancer diagnosis: Looking into the future. *Cancer Cell International*, *21*(1), 1–11. DOI: 10.1186/s12935-021-01981-1 PMID: 34020642

Jamialahmadi, H., Khalili-Tanha, G., Nazari, E., & Rezaei-Tavirani, M. (2024). Artificial intelligence and bioinformatics: A journey from traditional techniques to smart approaches. *Gastroenterology and Hepatology from Bed To Bench*, *17*(3), 241–252. DOI: 10.22037/GHFBB.V17I3.2977 PMID: 39308539

Jayakrishnan, T. T., Sangwan, N., Barot, S. V., Farha, N., Mariam, A., Xiang, S., Aucejo, F., Conces, M., Nair, K. G., Krishnamurthi, S. S., Schmit, S. L., Liska, D., Rotroff, D. M., Khorana, A. A., & Kamath, S. D. (2024). Multi-omics machine learning to study host-microbiome interactions in early-onset colorectal cancer. *NPJ Precision Oncology*, *8*(1), 146. DOI: 10.1038/s41698-024-00647-1 PMID: 39020083

Jiang, H., Li, L., Bao, Y., Cao, X., & Ma, L. (2024). Microbiota in tumors: New factor influencing cancer development. *Cancer Gene Therapy*, *31*(12), 1773–1785. DOI: 10.1038/s41417-024-00833-0 PMID: 39342031

Jin, X., Liu, C., Xu, T., Su, L., & Zhang, X. (2020). Artificial intelligence biosensors: Challenges and prospects. *Biosensors & Bioelectronics*, *165*, 112412. DOI: 10.1016/j.bios.2020.112412 PMID: 32729531

Johnson, C. H., Spilker, M. E., Goetz, L., Peterson, S. N., & Siuzdak, G. (2016). Metabolite and microbiome interplay in cancer immunotherapy. *Cancer Research*, *76*(21), 6146–6152. DOI: 10.1158/0008-5472.CAN-16-0309 PMID: 27729325

K, B., Venkatesan, L., Benjamin, L. S., K, V., & Satchi, N. S. (2025). Reinforcement Learning in Personalized Medicine: A Comprehensive Review of Treatment Optimization Strategies. *Cureus*, *17*(4). DOI: 10.7759/cureus.82756

Kabudi, T., Pappas, I., & Olsen, D. H. (2021). AI-enabled adaptive learning systems: A systematic mapping of the literature. *Computers and Education: Artificial Intelligence*, *2*, 100017. DOI: 10.1016/j.caeai.2021.100017

Kang, X., Lau, H. C. H., & Yu, J. (2024). Modulating gut microbiome in cancer immunotherapy: Harnessing microbes to enhance treatment efficacy. *Cell Reports Medicine*, *5*(4), 101478. Advance online publication. DOI: 10.1016/j.xcrm.2024.101478 PMID: 38631285

Khalighi, S., Reddy, K., Midya, A., Pandav, K. B., Madabhushi, A., & Abedalthagafi, M. (2024). Artificial intelligence in neuro-oncology: Advances and challenges in brain tumor diagnosis, prognosis, and precision treatment. *NPJ Precision Oncology*, *8*(1), 1–12. DOI: 10.1038/s41698-024-00575-0 PMID: 38553633

Khodadadian, A., Darzi, S., Haghi-Daredeh, S., Eshaghi, F. S., Babakhanzadeh, E., Mirabutalebi, S. H., & Nazari, M. (2020). Genomics and transcriptomics: The powerful technologies in precision medicine. *International Journal of General Medicine*, *13*, 627–640. DOI: 10.2147/IJGM.S249970 PMID: 32982380

Kumar, A., Dhiman, C., Kumar, M., Kannappan, N., Kumar, D., Chourasia, M. K., & Narayan, K. P. (2025). Development of a quality-by-design-based hybrid RP-HPLC method for Glimepiride: Bioanalytical and industrial applications. *Journal of Applied Pharmaceutical Science*, *15*(4), 102–115. DOI: 10.7324/JAPS.2025.214654

Kumar, A., Dhiman, C., Narayan, K. P., Kumar, M., & Sharma, M. (2025). Nanotherapeutics: Restoring Gut-Dysbiosis and Inducing Cancer Apoptosis. Spatially Variable Genes in Cancer: Development, Progression, and Treatment Response, 177-204.

Kumar, A., Kumar, M., Singh, R., Upadhyay, P., & Mukherjee, A. (2024). Quality-by-design (QbD) aided formulation optimization of amlodipine besylate oral thin film. *Indian Journal of Pharmaceutical Education and Research*, *58*(2s, 2S), S444–S452. DOI: 10.5530/ijper.58.2s.48

Kumar, A., Maurya, P., & Narayan, K. P. (2025). Fabrication of polymeric matrix-type transdermal patch containing glimepiride for therapeutic management of type-2 diabetes. *Indian Journal of Pharmaceutical Education and Research*, *59*(3), 936–948. DOI: 10.5530/ijper.20250924

Kumar, A., Rana, R., Saklani, R., Kumar, M., Yadav, P. K., Tiwari, A., & Chourasia, M. K. (2024). Technology Transfer of a Validated RP-HPLC Method for the Simultaneous Estimation of Andrographolide and Paclitaxel in Application to Pharmaceutical Nanoformulation. *Journal of Chromatographic Science*, *62*(4), 356–363. DOI: 10.1093/chromsci/bmad070 PMID: 37674403

Kumar, A., Sakhare, K., Bhattacharya, D., Chattopadhyay, R., Parikh, P., Narayan, K. P., & Mukherjee, A. (2022). Communication in non-communicable diseases (NCDs) and role of immunomodulatory nutraceuticals in their management. *Frontiers in Nutrition*, *9*, 966152. DOI: 10.3389/fnut.2022.966152 PMID: 36211513

Leffler, D. A., & Lamont, J. T. (2015). Clostridium difficile infection. *The New England Journal of Medicine*, *372*(16), 1539–1548. DOI: 10.1056/NEJMra1403772 PMID: 25875259

Liang, L., Kong, C., Li, J., Liu, G., Wei, J., Wang, G., Wang, Q., Yang, Y., Shi, D., Li, X., & Ma, Y. (2024). Distinct microbes, metabolites, and the host genome define the multi-omics profiles in right-sided and left-sided colon cancer. *Microbiome*, *12*(1), 1–17. DOI: 10.1186/s40168-024-01987-7 PMID: 39731152

Loganathan, T., & Priya Doss, C. (2022). The influence of machine learning technologies in gut microbiome research and cancer studies - A review. *Life Sciences*, *311*, 121118. DOI: 10.1016/j.lfs.2022.121118 PMID: 36404489

Luchini, C., Pea, A., & Scarpa, A. (2022). Artificial intelligence in oncology: Current applications and future perspectives. *British Journal of Cancer*, *126*(1), 4–9. DOI: 10.1038/s41416-021-01633-1 PMID: 34837074

Manzoni, C., Kia, D. A., Vandrovcova, J., Hardy, J., Wood, N. W., Lewis, P. A., & Ferrari, R. (2018). Genome, transcriptome and proteome: The rise of omics data and their integration in biomedical sciences. *Briefings in Bioinformatics*, *19*(2), 286–302. DOI: 10.1093/bib/bbw114 PMID: 27881428

Matson, V., Fessler, J., Bao, R., Chongsuwat, T., Zha, Y., Alegre, M. L., Luke, J. J., & Gajewski, T. F. (2018). The commensal microbiome is associated with anti-PD-1 efficacy in metastatic melanoma patients. *Science*, *359*(6371), 104–108. DOI: 10.1126/science.aao3290 PMID: 29302014

Maurya, P., Kumar, A., Singh, S., Nisha, R., Pal, R. R., Chourasia, M. K., & Saraf, S. A. (2021). Bio-analytical method development for estimation of levofloxacin: Application in drug estimation in nano-formulations and pharmacokinetic studies. *Indian Journal of Pharmaceutical Education and Research*, *55*(3), S814–S824. DOI: 10.5530/ijper.55.3s.189

Mu, W., Tunali, I., Gray, J. E., Qi, J., Schabath, M. B., & Gillies, R. J. (2020). Radiomics of 18F-FDG PET/CT images predicts clinical benefit of advanced NSCLC patients to checkpoint blockade immunotherapy. *European Journal of Nuclear Medicine and Molecular Imaging*, *47*(5), 1168–1182. DOI: 10.1007/s00259-019-04625-9 PMID: 31807885

Muller, E., Shiryan, I., & Borenstein, E. (2024). Multi-omic integration of microbiome data for identifying disease-associated modules. *Nature Communications*, *15*(1), 1–13. DOI: 10.1038/s41467-024-46888-3 PMID: 38521774

Murdoch, B. (2021). Privacy and artificial intelligence: Challenges for protecting health information in a new era. *BMC Medical Ethics*, *22*(1), 1–5. DOI: 10.1186/s12910-021-00687-3 PMID: 34525993

Naik, N., Hameed, B. M. Z., Shetty, D. K., Swain, D., Shah, M., Paul, R., Aggarwal, K., Brahim, S., Patil, V., Smriti, K., Shetty, S., Rai, B. P., Chlosta, P., & Somani, B. K. (2022). Legal and Ethical Consideration in Artificial Intelligence in Healthcare: Who Takes Responsibility? *Frontiers in Surgery*, *9*, 862322. Advance online publication. DOI: 10.3389/fsurg.2022.862322 PMID: 35360424

Nejman, D., Livyatan, I., Fuks, G., Gavert, N., Zwang, Y., Geller, L. T., Rotter-Maskowitz, A., Weiser, R., Mallel, G., Gigi, E., Meltser, A., Douglas, G. M., Kamer, I., Gopalakrishnan, V., Dadosh, T., Levin-Zaidman, S., Avnet, S., Atlan, T., Cooper, Z. A., & Straussman, R. (2020). The human tumor microbiome is composed of tumor type–specific intracellular bacteria. *Science*, *368*(6494), 973–980. DOI: 10.1126/science.aay9189 PMID: 32467386

Ogunsakin, R., Mehandjiev, N., & Marin, C. A. (2023). Towards adaptive digital twins architecture. *Computers in Industry*, *149*, 103920. DOI: 10.1016/j.compind.2023.103920

Olovo, C. V., Huang, X., Zheng, X., & Xu, M. (2021). Faecal microbial biomarkers in early diagnosis of colorectal cancer. *Journal of Cellular and Molecular Medicine*, *25*(23), 10783–10797. DOI: 10.1111/jcmm.17010 PMID: 34750964

Pandey, S., Chaudhary, M., & Tóth, Z. (2025). An investigation on real-time insights: Enhancing process control with IoT-enabled sensor networks. *Discover Internet of Things*, *5*(1), 1–17. DOI: 10.1007/s43926-025-00124-6

Petrelli, F., Ghidini, A., Dottorini, L., Ghidini, M., Zaniboni, A., & Tomasello, G. (2025). Clinical Evidence for Microbiome-Based Strategies in Cancer Immunotherapy: A State-of-the-Art Review. *Medicina (Lithuania)*, *61*(9), 1595. https://doi.org/DOI: 10.3390/MEDICINA61091595/S1

Raymond Geis, J., Brady, A. P., Wu, C. C., Spencer, J., Ranschaert, E., Jaremko, J. L., Langer, S. G., Kitts, A. B., Birch, J., Shields, W. F., van den Hoven van Genderen, R., Kotter, E., Gichoya, J. W., Cook, T. S., Morgan, M. B., Tang, A., Safdar, N. M., & Kohli, M. (2019). Ethics of artificial intelligence in radiology: Summary of the joint European and North American multisociety statement. *Radiology*, *293*(2), 436–440. DOI: 10.1148/radiol.2019191586 PMID: 31573399

Rédei, G. P. (2008). Transcriptome. *Encyclopedia of Genetics, Genomics, Proteomics and Informatics*, 1999–1999. DOI: 10.1007/978-1-4020-6754-9_17231

Rivi -Tipei, I. (2023). Ethical Considerations in Artificial Intelligence Development and Deployment. *SCIENTIA MORALITAS - International. Journal of Multidisciplinary Research*, *8*(2), 13–26.

Roth, S. C. (2019). What is genomic medicine? *Journal of the Medical Library Association: JMLA*, *107*(3), 442–448. DOI: 10.5195/jmla.2019.604 PMID: 31258451

Routy, B., Le Chatelier, E., Derosa, L., Duong, C. P. M., Alou, M. T., Daillère, R., Fluckiger, A., Messaoudene, M., Rauber, C., Roberti, M. P., Fidelle, M., Flament, C., Poirier-Colame, V., Opolon, P., Klein, C., Iribarren, K., Mondragón, L., Jacquelot, N., Qu, B., & Zitvogel, L. (2018). Gut microbiome influences efficacy of PD-1–based immunotherapy against epithelial tumors. *Science*, *359*(6371), 91–97. DOI: 10.1126/science.aan3706 PMID: 29097494

Sailer, M., Ninaus, M., Huber, S. E., Bauer, E., & Greiff, S. (2024). The End is the Beginning is the End: The closed-loop learning analytics framework. *Computers in Human Behavior*, *158*, 108305. DOI: 10.1016/j.chb.2024.108305

Schwabe, R. F., & Jobin, C. (2013). The microbiome and cancer. *Nature Reviews Cancer 2013 13:11*, *13*(11), 800–812. DOI: 10.1038/nrc3610

Sekhar, G. (2024). *Advanced AI and Machine Learning Techniques and Algorithms in Cancer Research*. DOI: 10.31219/osf.io/7skz3

Shajari, S., Kuruvinashetti, K., Komeili, A., & Sundararaj, U. (2023). The Emergence of AI-Based Wearable Sensors for Digital Health Technology: A Review. *Sensors (Basel)*, *23*(23), 9498. DOI: 10.3390/s23239498 PMID: 38067871

Shankar, V. (2024). Edge AI: A Comprehensive Survey of Technologies, Applications, and Challenges. *2024 1st International Conference on Advanced Computing and Emerging Technologies, ACET 2024*. DOI: 10.1109/ACET61898.2024.10730112

Sheller, M. J., Edwards, B., Reina, G. A., Martin, J., Pati, S., Kotrotsou, A., Milchenko, M., Xu, W., Marcus, D., Colen, R. R., & Bakas, S. (2020). Federated learning in medicine: facilitating multi-institutional collaborations without sharing patient data. *Scientific Reports 2020 10:1*, *10*(1), 1–12. DOI: 10.1038/s41598-020-69250-1

Sherwani, M. K., Ruuskanen, M. O., Feldner-Busztin, D., Nisantzis Firbas, P., Boza, G., Móréh, Á., Borman, T., Putu Erawijantari, P., Scheuring, I., Gopalakrishnan, S., & Lahti, L. (2025). Multi-omics time-series analysis in microbiome research: A systematic review. *Briefings in Bioinformatics*, *26*(5), bbaf502. Advance online publication. DOI: 10.1093/bib/bbaf502 PMID: 41052276

Simpson, R. C., Shanahan, E. R., Scolyer, R. A., & Long, G. V. (2023). Towards modulating the gut microbiota to enhance the efficacy of immune-checkpoint inhibitors. *Nature Reviews Clinical Oncology 2023 20:10, 20*(10), 697–715. DOI: 10.1038/s41571-023-00803-9

Singh, D., Dhiman, V. K., Pandey, M., Dhiman, V. K., Sharma, A., Pandey, H., Verma, S. K., & Pandey, R. (2024). Personalized medicine: An alternative for cancer treatment. *Cancer Treatment and Research Communications, 42*, 100860. DOI: 10.1016/j.ctarc.2024.100860 PMID: 39827574

Sivan, A., Corrales, L., Hubert, N., Williams, J. B., Aquino-Michaels, K., Earley, Z. M., Benyamin, F. W., Lei, Y. M., Jabri, B., Alegre, M. L., Chang, E. B., & Gajewski, T. F. (2015). Commensal Bifidobacterium promotes antitumor immunity and facilitates anti-PD-L1 efficacy. *Science, 350*(6264), 1084–1089. DOI: 10.1126/science.aac4255 PMID: 26541606

Sobhan, A., Hossain, A., Wei, L., Muthukumarappan, K., & Ahmed, M. (2025). IoT-Enabled Biosensors in Food Packaging: A Breakthrough in Food Safety for Monitoring Risks in Real Time. *Foods 2025, Vol. 14, Page 1403, 14*(8), 1403. DOI: 10.3390/foods14081403

Stahl, D. A., Lane, D. J., Olsen, G. J., & Pace, N. R. (1985). Characterization of a Yellowstone hot spring microbial community by 5S rRNA sequences. *Applied and Environmental Microbiology, 49*(6), 1379–1384. DOI: 10.1128/aem.49.6.1379-1384.1985 PMID: 2409920

Sun, T., Niu, X., He, Q., Chen, F., & Qi, R. Q. (2023). Artificial Intelligence in microbiomes analysis: A review of applications in dermatology. *Frontiers in Microbiology, 14*, 1112010. DOI: 10.3389/fmicb.2023.1112010 PMID: 36819026

Thomas, A. M., Fidelle, M., Routy, B., Kroemer, G., Wargo, J. A., Segata, N., & Zitvogel, L. (2023). Gut OncoMicrobiome Signatures (GOMS) as next-generation biomarkers for cancer immunotherapy. *Nature Reviews. Clinical Oncology, 20*(9), 583–603. DOI: 10.1038/s41571-023-00785-8 PMID: 37365438

Tosta, T. A. A., de Faria, P. R., Neves, L. A., & do Nascimento, M. Z. (2019). Color normalization of faded H&E-stained histological images using spectral matching. *Computers in Biology and Medicine, 111*, 103344. Advance online publication. DOI: 10.1016/j.compbiomed.2019.103344 PMID: 31279982

Trebeschi, S., Drago, S. G., Birkbak, N. J., Kurilova, I., C lin, A. M., Delli Pizzi, A., Lalezari, F., Lambregts, D. M. J., Rohaan, M. W., Parmar, C., Rozeman, E. A., Hartemink, K. J., Swanton, C., Haanen, J. B. A. G., Blank, C. U., Smit, E. F., Beets-Tan, R. G. H., & Aerts, H. J. W. L. (2019). Predicting response to cancer immunotherapy using noninvasive radiomic biomarkers. *Annals of Oncology : Official Journal of the European Society for Medical Oncology*, *30*(6), 998–1004. DOI: 10.1093/annonc/mdz108 PMID: 30895304

Usman, I., Anwar, A., Shukla, S., & Pathak, P. (2024). Mechanistic Review on the Role of Gut Microbiota in the Pathology of Cardiovascular Diseases. *Cardiovascular & Hematological Disorders Drug Targets*, *24*(1), 13–39. DOI: 10.2174/01187152 9X310857240607103028 PMID: 38879769

Vandenberghe, M. E., Scott, M. L. J., Scorer, P. W., Söderberg, M., Balcerzak, D., & Barker, C. (2017). Relevance of deep learning to facilitate the diagnosis of HER2 status in breast cancer. *Scientific Reports*, *7*(1), 45938. Advance online publication. DOI: 10.1038/srep45938 PMID: 28378829

Vasanthapuram, G. (2025). Multi-Modal Generative AI and LLM-Driven Clinical Interpretation Pipeline for Early Detection of Oncology Markers in Radiological Imaging. *International Journal of Computer Science and Engineering Research and Development*, *15*(3), 34–44. DOI: 10.63519/IJCSERD_15_03_005

Vogtmann, E., & Goedert, J. J. (2016). Epidemiologic studies of the human microbiome and cancer. *British Journal of Cancer*, *114*(3), 237–242. DOI: 10.1038/bjc.2015.465 PMID: 26730578

Wang, D. Y., Salem, J. E., Cohen, J. V., Chandra, S., Menzer, C., Ye, F., Zhao, S., Das, S., Beckermann, K. E., Ha, L., Rathmell, W. K., Ancell, K. K., Balko, J. M., Bowman, C., Davis, E. J., Chism, D. D., Horn, L., Long, G. V., Carlino, M. S., & Johnson, D. B. (2018). Fatal Toxic Effects Associated With Immune Checkpoint Inhibitors: A Systematic Review and Meta-analysis. *JAMA Oncology*, *4*(12), 1721–1728. DOI: 10.1001/jamaoncol.2018.3923 PMID: 30242316

Watson, J. D., & Crick, F. H. C. (1953). Molecular structure of nucleic acids: A structure for deoxyribose nucleic acid. *Nature*, *171*(4356), 737–738. DOI: 10.1038/171737a0 PMID: 13054692

Wekesa, J. S., & Kimwele, M. (2023). A review of multi-omics data integration through deep learning approaches for disease diagnosis, prognosis, and treatment. *Frontiers in Genetics*, *14*, 1199087. Advance online publication. DOI: 10.3389/fgene.2023.1199087 PMID: 37547471

Wu, Y., & Xie, L. (2025). AI-driven multi-omics integration for multi-scale predictive modeling of genotype-environment-phenotype relationships. *Computational and Structural Biotechnology Journal, 27*, 265–277. DOI: 10.1016/j.csbj.2024.12.030 PMID: 39886532

Xie, J., Luo, X., Deng, X., Tang, Y., Tian, W., Cheng, H., Zhang, J., Zou, Y., Guo, Z., & Xie, X. (2023). Advances in artificial intelligence to predict cancer immunotherapy efficacy. *Frontiers in Immunology, 13*, 1076883. DOI: 10.3389/fimmu.2022.1076883 PMID: 36685496

Xu, W., Wang, T., Wang, N., Zhang, H., Zha, Y., Ji, L., Chu, Y., & Ning, K. (2023). Artificial intelligence-enabled microbiome-based diagnosis models for a broad spectrum of cancer types. *Briefings in Bioinformatics, 24*(3), bbad178. Advance online publication. DOI: 10.1093/bib/bbad178 PMID: 37141141

Yan, Z., Li, J., Xiong, Y., Xu, W., & Zheng, G. (2012). Identification of candidate colon cancer biomarkers by applying a random forest approach on microarray data. *Oncology Reports, 28*(3), 1036–1042. DOI: 10.3892/or.2012.1891 PMID: 22752057

Yang, Z., Ma, J., Han, J., Li, A., Liu, G., Sun, Y., Zheng, J., Zhang, J., Chen, G., Xu, R., Sun, L., Meng, C., Gao, J., Bai, Z., Deng, W., Zhang, C., Su, J., Yao, H., & Zhang, Z. (2024). Gut microbiome model predicts response to neoadjuvant immunotherapy plus chemoradiotherapy in rectal cancer. *Med (New York, N.Y.), 5*(10), 1293–1306.e4. DOI: 10.1016/j.medj.2024.07.002 PMID: 39047732

Yao, Y., Ni, H., Wang, X., Xu, Q., Zhang, J., Jiang, L., Wang, B., Song, S., & Zhu, X. (2021). A New Biomarker of Fecal Bacteria for Non-Invasive Diagnosis of Colorectal Cancer. *Frontiers in Cellular and Infection Microbiology, 11*, 744049. DOI: 10.3389/fcimb.2021.744049 PMID: 34976850

Yetgin, A. (2025). Revolutionizing multi-omics analysis with artificial intelligence and data processing. *Quantitative Biology, 13*(3), e70002. Advance online publication. DOI: 10.1002/qub2.70002

Yuan, B., Ma, B., Yu, J., Meng, Q., Du, T., Li, H., Zhu, Y., Sun, Z., Ma, S., & Song, C. (2021). Fecal Bacteria as Non-Invasive Biomarkers for Colorectal Adenocarcinoma. *Frontiers in Oncology, 11*, 664321. DOI: 10.3389/fonc.2021.664321 PMID: 34447694

Zeller, G., Tap, J., Voigt, A. Y., Sunagawa, S., Kultima, J. R., Costea, P. I., Amiot, A., Böhm, J., Brunetti, F., Habermann, N., Hercog, R., Koch, M., Luciani, A., Mende, D. R., Schneider, M. A., Schrotz-King, P., Tournigand, C., Tran Van Nhieu, J., Yamada, T., & Bork, P. (2014). Potential of fecal microbiota for early-stage detection of colorectal cancer. *Molecular Systems Biology*, *10*(11), 766. Advance online publication. DOI: 10.15252/msb.20145645 PMID: 25432777

Zhang, N., Kandalai, S., Zhou, X., Hossain, F., & Zheng, Q. (2023). Applying multi-omics toward tumor microbiome research. *iMeta*, *2*(1), e73. DOI: 10.1002/imt2.73 PMID: 38868335

Zhang, S., Liu, J., & Zuo, X. (2021). Adaptive online incremental learning for evolving data streams. *Applied Soft Computing*, *105*, 107255. DOI: 10.1016/j.asoc.2021.107255

Zheng, Y., Wang, T., Tu, X., Huang, Y., Zhang, H., Tan, D., Jiang, W., Cai, S., Zhao, P., Song, R., Li, P., Qin, N., & Fang, W. (2019). Gut microbiome affects the response to anti-PD-1 immunotherapy in patients with hepatocellular carcinoma. *Journal for Immunotherapy of Cancer*, *7*(1), 193. Advance online publication. DOI: 10.1186/s40425-019-0650-9 PMID: 31337439

Zwezerijnen-Jiwa, F. H., Sivov, H., Paizs, P., Zafeiropoulou, K., & Kinross, J. (2023). A systematic review of microbiome-derived biomarkers for early colorectal cancer detection. *Neoplasia : an International Journal for Oncology Research*, *36*, 100868. Advance online publication. DOI: 10.1016/j.neo.2022.100868 PMID: 36566591

Chapter 13
Quantum-SVM (SVM) for Parkinson's Disease Prediction Using Wearable IoT Sensors

N. Jagadish Kumar
https://orcid.org/0000-0001-7677-9872
SRM Institute of Science and Technology, India

Chin-Shiuh Shieh
https://orcid.org/0000-0003-3187-458X
National Kaohsiung University of Science and Technology, Taiwan

Mong-Fong Horng
National Kaohsiung University of Science and Technology, Taiwan

G. Elangovan
SRM Institute of Science and Technology, India

K. Rajkumar
SRM Institute of Science and Technology, India

ABSTRACT

The amalgamation of wearable IoT sensors with QSVM technology is transforming the early identification and ongoing surveillance of Parkinson's Disease (PD). This chapter examines the enhancement of PD diagnostics through quantum machine learning by evaluating real-time kinematic data from various sensors. Traditional models such as Classical SVM, RF, and DNN encounter difficulties due to the intricacy and high dimensionality of time-series sensor data. QSVM, employing

DOI: 10.4018/979-8-3373-3146-1.ch013

Variational Quantum Circuits, optimizes feature extraction and classification, thereby diminishing computational requirements while improving accuracy. In benchmark evaluations, QSVM attained an accuracy of 97.2%, exceeding that of Classical SVM, RF, and DNN, and concurrently decreasing computational time by 35%. The efficacy of QSVM, in conjunction with edge computing and federated learning, has the potential to revolutionize personalized, privacy-centric healthcare. Notwithstanding the constraints of quantum hardware, QSVM presents significant promise in enhancing PD diagnosis and advancing intelligent healthcare.

1. INTRODUCTION

The evolution of wearable Internet of Things (IoT)-based sensors has significantly impacted contemporary healthcare by facilitating the acquisition of real-time sensor data for the ongoing monitoring of patients. Instruments such as accelerometers, gyroscopes, and electromyography (EMG) sensors are extensively utilized in the surveillance of neurodegenerative disorders, particularly in the context of predicting conditions such as Parkinson's disease. These sensors meticulously monitor movement patterns, tremors, and gait irregularities, thereby enabling healthcare practitioners to identify symptoms at a nascent stage. IoT-enabled intelligent healthcare solutions offer customized monitoring and predictive analytics, thereby minimizing hospital admissions and improving the quality of patient care. Nevertheless, the efficient management of high-dimensional, sensor-generated time-series data necessitates the application of sophisticated ml methodologies for precise classification and informed decision-making.

Ml with in the domain of healthcare assumes a pivotal function in the analysis of extensive volumes of real-time sensor data acquired from wearable Internet of Things (IoT) devices. The Quantum Support vector machine (QSVM), in conjunction with traditional ml methodologies such as Classical SVM, RF, and Deep Neural Networks (DNN), is extensively implemented for the prognostication of Parkinson's disease through the examination of movement anomalies. The processes of feature extraction, classification accuracy, and computational efficiency are paramount for facilitating accurate and trustworthy diagnostic outcomes. Adaptive healthcare systems harness ml models to diminish computational duration while concurrently enhancing accuracy, precision, recall, and F1-score. Nevertheless, conventional ml models encounter significant challenges when addressing intricate sensor data, thereby necessitating the exploration of quantum-enhanced artificial intelligence techniques.

Quantum computing possesses the capacity to transform wearable IoT-enabled medical diagnostics by harnessing quantum parallelism to enhance classification precision and diminish latency. The Quantum Support vector machine (QSVM), which employs Variational Quantum Circuits (VQC), significantly advances the real-time monitoring of Parkinson's disease by facilitating expedited and more precise decision-making processes. The enhancement of feature selection through quantum methods, specifically utilizing Grover's algorithm, markedly improves the speed and efficacy of classification in comparison to traditional Classical SVMs (SVM). Moreover, quantum cryptography guarantees the secure transmission of patient information within IoT-based intelligent healthcare frameworks. Although the constraints of quantum hardware present a significant obstacle, the amalgamation of QSVM, federated learning, and edge computing in forthcoming smart healthcare innovations provides a hopeful outlook for the evolution of intelligent and adaptive medical diagnostic systems.

1.1 Problem Statement:

Parkinson's Disease (PD) constitutes a progressive neurodegenerative condition that profoundly influences motor capabilities, necessitating prompt diagnosis and ongoing surveillance to facilitate effective therapeutic interventions. Conventional ml methodologies, encompassing Classical SVM, Random Forest (RF), and Deep Neural Networks (DNN), frequently encounter difficulties when addressing the high-dimensional, intricate time-series data produced by wearable Internet of Things (IoT) sensors, resulting in computational inefficiencies and less than optimal accuracy. Moreover, the challenge of real-time processing persists due to the substantial computational expenses related to feature extraction and classification within extensive sensor datasets.

Quantum Support vector machine (QSVM) presents a potentially transformative solution by utilizing Variational Quantum Circuits (VQC) to enhance the processes of feature extraction and classification, thereby diminishing computational burdens while augmenting diagnostic accuracy. Nonetheless, the practical implementation of QSVM for real-time monitoring of Parkinson's Disease is impeded by challenges associated with quantum hardware limitations, noise disturbances, and the integration into pre-existing healthcare infrastructures.

This research endeavors to rectify these deficiencies by examining the effectiveness of QSVM in analyzing wearable IoT sensor data for PD prediction, contrasting its performance with traditional ml frameworks, and investigating its viability for integration with edge computing and federated learning, thereby fostering scalable and privacy-preserving healthcare solutions.

1.2 Objectives of the Proposed Chapter:

Evaluate QSVM Performance – Assess QSVM's efficacy in wearable IoT sensor data processing for Parkinson's prediction regarding accuracy, efficiency, and noise resilience.
Address Challenges and Solutions – Recognize QSVM's challenges, including hardware constraints and noise, and propose solutions for effective implementation.
Compare with Traditional Models – Conduct a comparative analysis of QSVM with SVM, RF, and DNN on accuracy, feature extraction, and computational expense.
Future Directions – Investigate QSVM's applicability in healthcare, focusing on quantum hardware innovations, edge computing integration, and its relevance to neurodegenerative diseases.

2. BACKGROUND

2.1 Need for IoT Sensors in Parkinson's Disease Monitoring

Parkinson's Disease (PD) is a progressive neurodegenerative disorder characterized by motor impairments such as tremors, slowness of movement, and gait difficulties. Conventional diagnostic methods often depend on irregular clinical evaluations and subjective judgments, leading to delayed detection and limited accuracy. The integration of Internet of Things (IoT)-enabled wearable sensors introduces a transformative approach to PD management by enabling real-time, continuous, and objective monitoring of patients. Devices such as accelerometers, gyroscopes, and electromyography (EMG) sensors can detect subtle movement abnormalities, rigidity, or tremors that may not be evident during clinical examinations. Smart rings and pressure-sensing footwear further enhance the capacity to analyze hand tremors and walking patterns, providing detailed insight into disease progression.

Remote patient monitoring (RPM) through IoT systems allows clinicians to supervise patients from afar, making timely therapeutic decisions without frequent hospital visits. This not only lessens the burden on healthcare systems but also improves patient comfort and safety through features like automated fall detection. The immense data generated from IoT sensors—including time-series measurements of tremors, gait irregularities, and muscle activity—require advanced analytical techniques for meaningful interpretation, enabling data-driven and adaptive treatment plans tailored to the unique physiological needs of each patient with Parkinson's Disease.

2.2 Parkinson's Disease Prediction Using Traditional Ml Algorithms

Traditional ml techniques have been widely utilized for predicting Parkinson's Disease (PD) through sensor-based data. These methodologies evaluate time-series information from wearable IoT sensors to identify movement irregularities linked to PD.

Commonly Used Ml Algorithms

SVM (SVM):

SVM (SVM) represents a formidable supervised learning algorithm that is extensively employed for classification endeavors, particularly esteemed for its capacity to efficiently process high-dimensional data. In the realm of predicting Parkinson's Disease (PD), SVM delineates an optimal hyperplane that differentiates individuals diagnosed with PD from those devoid of the affliction, predicated on derived features such as gait attributes, hand tremors, or vocal variances. The algorithm demonstrates superior performance in contexts where the data possesses distinct margins of separation and is capable of functioning effectively even with datasets of limited to moderate size. SVM employs kernel functions—such as linear, polynomial, or radial basis function (RBF)—to transmute non-linearly separable data into a higher-dimensional realm wherein separation becomes feasible. Nonetheless, when engaged with large-scale, time-series sensor data produced by wearable devices such as accelerometers and gyroscopes, SVM may encounter significant performance impediments. The duration of training and memory consumption escalates markedly with the increase in dataset size, thereby rendering it less appropriate for real-time or resource-limited environments. Furthermore, the optimization of SVM parameters, including kernel type, cost function (C), and gamma, necessitates meticulous calibration to avert underfitting or overfitting. In spite of these constraints, SVM continues to serve as a robust baseline model owing to its resilience, especially when integrated with dimensionality reduction methodologies such as Principal Component Analysis (PCA). Prospective developments may encompass hybrid models or quantum-enhanced variations to surmount these computational obstacles while maintaining classification precision.

Random Forest (RF):

Random Forest (RF) constitutes a formidable ensemble learning methodology that augments classification efficacy through the construction and aggregation of outputs derived from multiple decision trees. Each individual tree within the ensemble is trained on a randomly selected subset of both data instances and features, thereby fostering diversity and mitigating the propensity for overfitting that is frequently associated with solitary decision trees. This methodology demonstrates particular efficacy in addressing intricate datasets characterized by heterogeneous features, exemplified by those procured from wearable sensors utilized in the monitoring of Parkinson's Disease (PD). RF exhibits intrinsic resilience to noise and missing data, rendering it particularly advantageous for practical medical applications wherein data inconsistencies are prevalent. Furthermore, it generates feature importance scores, thereby facilitating the identification of the most pertinent biomarkers for the detection of PD. Nonetheless, the performance of the algorithm can be significantly impacted by various hyperparameters, including the number of trees, the depth of each tree, and the quantity of features evaluated at each bifurcation. The optimization of these parameters is imperative for attaining elevated levels of accuracy and efficiency, particularly within contexts that are time-sensitive or constrained by available resources. Although RF typically demonstrates superior scalability compared to SVMs (SVM) when applied to extensive datasets, it may still encounter obstacles concerning model interpretability and computational expenses during the inference phase.

Deep Neural Networks (DNN):

Deep Neural Networks (DNNs) exhibit remarkable efficacy in modeling intricate, non-linear correlations within sensor data, attributable to their multi-tiered architecture, which encompasses several hidden layers that facilitate the acquisition of complex patterns and characteristics. In the realm of Parkinson's Disease (PD) detection, DNNs possess the capability to autonomously extract pertinent features from unprocessed sensor data, including accelerometer and gyroscope measurements, obviating the necessity for extensive manual feature engineering. They demonstrate superior performance when engaged with large datasets, capitalizing on their potential to assimilate vast quantities of information to enhance model precision. Nevertheless, this proficiency necessitates substantial computational resources, as the training of deep networks demands considerable processing power and memory, particularly when addressing high-dimensional time-series data. Furthermore, DNNs may experience protracted training durations due to the intricacy involved in optimizing the multitude of weights and biases dispersed across the layers. The meticulous process of fine-tuning hyperparameters, including the number of layers, learning rate, and batch size, is imperative to attain optimal performance. Although DNNs frequently surpass traditional models such as SVMs (SVM) and Random Forests (RF) with regard to predictive accuracy, they may encounter challenges such as overfitting in smaller datasets and necessitate extensive labeled data for effective training.

K-Nearest Neighbors (KNN):

K-Nearest Neighbors (KNN) constitutes a straightforward yet efficacious non-parametric classification methodology that allocates labels to novel data instances by evaluating the predominant class among the 'k' most analogous examples within the training dataset. The degree of similarity is typically quantified through various distance metrics, including but not limited to Euclidean, Manhattan, or Minkowski distances. Within the realm of Parkinson's Disease (PD) prediction utilizing wearable sensor data, KNN is applicable for identifying anomalies by juxtaposing a patient's sensor measurements against those from previously classified cases. Its inherently intuitive framework facilitates ease of implementation and interpretation. Nonetheless, the efficacy of KNN is substantially impacted by the selection of distance metric, the specified value of 'k', and the existence of irrelevant or redundant features. It exhibits particular susceptibility to the "curse of dimensionality," wherein an increase in feature dimensions diminishes the significance of distance assessments, subsequently resulting in diminished classification accuracy. Furthermore, KNN can incur considerable computational expenses during inference, as it necessitates the retention of the entire training dataset and the computation of distances to all instances for each novel prediction.

Logistic Regression (LR):

Logistic Regression (LR) constitutes one of the most essential and interpretable statistical frameworks employed for binary classification endeavors, encompassing the detection of Parkinson's Disease (PD). It quantifies the likelihood that a specific input is associated with a designated class by utilizing a sigmoid function applied to a linear amalgamation of input variables. Owing to its inherent simplicity, LR affords lucid insights into the impact of individual features on the predictive outcome, rendering it profoundly advantageous in clinical environments where interpretability is of paramount importance. In the context of PD prediction, LR can elucidate the significance of particular sensor-derived attributes, such as tremor frequency or gait velocity. Nonetheless, its efficacy is constrained when confronted with intricate, non-linear patterns frequently encountered in physiological and time-series sensor datasets. LR posits a linear decision boundary, which may inadequately encapsulate the complexities inherent in authentic medical data collections. Furthermore, LR exhibits heightened sensitivity to multicollinearity and outliers, which can compromise its predictive consistency. While regularization methodologies such as L1 (Lasso) and L2 (Ridge) can assist in alleviating overfitting and enhancing generalization, LR continues to underperform in comparison to more sophisticated models like Random Forests or Deep Neural Networks in high-dimensional, non-linear contexts.

2.3 Challenges of Traditional Ml for Parkinson's Disease Prediction

The utilization of conventional ml methodologies for predicting Parkinson's Disease via wearable IoT sensor data encounters several obstacles. These obstacles arise from the characteristics of PD symptoms, the intricacies of sensor-derived time-series data, and limitations in computational resources. The following section provides a comprehensive theoretical analysis of these obstacles.

High Dimensionality and Curse of Dimensionality

The monitoring of Parkinson's Disease (PD) frequently hinges upon a diverse array of sensor-derived characteristics, including acceleration, angular velocity, gait dynamics, tremor intensity, and muscle activity, which are acquired through wearable technological devices. Collectively, these attributes facilitate a thorough analysis of motor symptoms, thereby contributing to precise diagnosis and the monitoring of disease progression. Nevertheless, the incorporation of a multitude of features culminates in high-dimensional datasets, which may invoke the curse of dimensionality—a phenomenon characterized by the sparsity of data within the feature space, ultimately impairing the efficacy of various ml algorithms. This sparsity exacerbates the challenges faced by distance-based models, such as K-nearest neighbors (KNN), while simultaneously augmenting computational complexity, resulting in overfitting and diminished generalization capabilities. Techniques for dimensionality reduction, including Principal Component Analysis (PCA) and t-distributed Stochastic Neighbor Embedding (t-SNE), are frequently employed to alleviate this challenge by converting the original feature space into a lower-dimensional framework. Although these methodologies enhance model efficiency and facilitate visualization, they also carry the potential risk of information loss, particularly if critical discriminative features are not preserved throughout the transformation process. Furthermore, approaches such as t-SNE, while exemplary for exploratory analysis, are not invariably appropriate for subsequent predictive modeling. Consequently, meticulous preprocessing, judicious feature selection, and rigorous validation are paramount to achieving an equilibrium between data simplification and diagnostic precision within ml workflows pertinent to PD.

Noise and Variability in Sensor Data

Wearable Internet of Things (IoT) sensors are integral to the continuous monitoring of Parkinson's Disease (PD) as they facilitate the real-time acquisition of physiological and motion-related data, encompassing tremors, gait irregularities, and limb movements. Nonetheless, these sensors are susceptible to noise contamination stemming from environmental disturbances, inaccuracies in device calibration, motion artifacts, and electromagnetic interference. Such noise has the potential to obscure the underlying signal patterns, thereby impairing model performance. Conventional ml models, such as RF and SVM, exhibit particular vulnerability to noise, as they may erroneously assimilate spurious variations present in the data, culminating in diminished classification accuracy and generalization capabilities. In response to this challenge, a variety of preprocessing techniques are implemented to mitigate noise within the input signals prior to the training of models. Approaches such as wavelet transforms facilitate multiresolution analysis of time-series data, adeptly isolating noise components from pertinent patterns. Kalman filtering provides a dynamic and recursive methodology for estimating true signal values in evolving systems, while moving average filters effectively attenuate short-term fluctuations.

Class Imbalance in PD Datasets

Parkinson's Disease (PD) datasets are often characterized by class imbalance, wherein the quantity of non-PD instances considerably surpasses that of PD cases. This disparity results in the formulation of biased ml models that preferentially represent the majority class, frequently culminating in diminished recall and erroneous classification of actual PD instances—an outcome that assumes particular significance in clinical environments where timely and precise detection is imperative. Imbalanced datasets present challenges to algorithms such as Logistic Regression, SVMs (SVM), and Random Forest, which typically prioritize overall accuracy over performance metrics specific to individual classes. In response to this issue, methodologies such as the Synthetic Minority Oversampling Technique (SMOTE) and cost-sensitive learning are routinely employed. SMOTE produces synthetic examples of the minority class through interpolation among existing instances, thereby aiding in the equilibrium of class distributions. Conversely, cost-sensitive learning modifies the training regimen to impose greater penalties for the misclassification of minority class instances, consequently enhancing sensitivity. While these approaches demonstrate efficacy, they are not devoid of limitations. SMOTE has the potential to introduce synthetic noise, particularly when utilized in contexts with overlapping class boundaries or within datasets characterized by noise, which may adversely affect the generalization capabilities of the model.

Temporal Dependencies and Feature Engineering Complexity

Parkinson's Disease (PD) is distinguished by the gradual progression of both motor and non-motor manifestations over time, which encompass tremors, rigidity, gait abnormalities, and cognitive deterioration. This inherent dynamism mandates the employment of models that are adept at comprehending temporal patterns for the purposes of accurate diagnosis, monitoring of disease progression, and individualized treatment strategies. Traditional ml algorithms such as SVMs and RF possess an intrinsic static quality that fails to capture the temporal dependencies present within the data. These methodologies necessitate manual feature engineering to derive time-related characteristics from sensor data, frequently resulting in the attenuation of temporal context and restricted adaptability to real-time fluctuations in symptoms. In contrast, Long Short-Term Memory (LSTM) networks, a subset of recurrent neural networks (RNN), have been specifically engineered to manage sequential and time-series data by preserving long-term dependencies via memory gates. This capability enables LSTMs to effectively model variations in sensor readings over time, thereby rendering them particularly proficient for the ongoing monitoring of PD.

Computational Complexity and Scalability Issues

Deep Neural Networks (DNNs) excel in complex pattern recognition for Parkinson's Disease (PD) sensor data but demand large datasets and intensive computational resources, limiting deployment in real-time edge healthcare environments. Similarly, Support Vector Machines (SVMs) manage high-dimensional data effectively yet face quadratic computational complexity, reducing scalability for large volumes of streaming data. Cloud–edge computing architectures help reduce processing demands by distributing workloads between remote servers and local devices, balancing performance with cost and responsiveness. Nonetheless, limited parallel processing capabilities, communication overheads, privacy concerns, network instability, and synchronization delays remain significant barriers to adoption.

Generalization and Overfitting Risks

ML models for PD prediction often struggle to generalize across varied patient populations due to differences in sensor modalities, placement, calibration, and demographic attributes. This data heterogeneity increases overfitting risks, where models perform well in training but poorly on unseen data. Strategies to improve generalization include cross-validation for thorough evaluation, dropout in deep learning to reduce reliance on specific features, and data augmentation to simulate realistic variations such as noise, signal shifts, or sensor inconsistencies.

Limited Interpretability of ML Models in Healthcare

Complex models like DNNs lack transparency, posing challenges in medical contexts where explainability is vital. Regulatory bodies such as the FDA and HIPAA require interpretable AI systems. Tools like SHAP and LIME enhance understanding but add complexity to models. Beyond interpretability, traditional ML faces issues of noise, class imbalance, and temporal dependencies, addressed through feature selection and optimization. Quantum-based approaches, such as Quantum Support Vector Machines (QSVM), leverage quantum parallelism to improve PD classification accuracy and scalability.

3. QUANTUM ML FOR PARKINSON'S DISEASE PREDICTION

Quantum Ml is transforming the domain of neurological disorder identification, particularly in the prognostication of Parkinson's Disease (PD). Conventional ml techniques frequently encounter challenges in processing the extensive, high-dimensional datasets produced by wearable Internet of Things (IoT) sensors that observe symptoms such as tremors, gait irregularities, and muscular rigidity. Quantum computing provides a substantial benefit through its capacity to analyze immense volumes of data concurrently employing principles such as superposition and entanglement. By utilizing Quantum Support vector machines (QSVM) and quantum kernel methodologies, QML improves pattern recognition, facilitating more precise and timely identification of PD.

Wearable Internet of Things (IoT) devices incessantly gather physiological information, which necessitates preprocessing prior to its application for forecasting. Quantum feature mapping methodologies convert this unrefined sensor information into higher-dimensional quantum states, enhancing classification precision. In contrast to classical models, quantum algorithms can adeptly discern subtle patterns within the data that may signify early-stage Parkinson's Disease (PD), thereby offering a vital benefit in early intervention. Furthermore, the capacity of quantum computing to execute computations concurrently diminishes processing duration, rendering real-time PD detection more practicable for clinical applications.

Quantum Ml possesses the capacity to revolutionize the prognostication of Parkinson's Disease (PD) by augmenting the efficacy and precision of medical diagnostics. In contrast to conventional ml frameworks, which depend on classical computational principles, QML employs quantum states to process extensive volumes of high-dimensional data concurrently. This proficiency is particularly advantageous in scrutinizing intricate patterns within physiological signals, such as tremors, gait irregularities, and heart rate fluctuations, obtained from wearable

Internet of Things (IoT) sensors. Through the implementation of quantum feature mapping and the utilization of quantum kernels, QML is capable of discerning subtle disease indicators that may be disregarded by classical models. Moreover, quantum algorithms can improve the accuracy of classification endeavors, diminishing false positives and enhancing early-stage PD identification.

3.1 Comparative Overview: Traditional ML vs. Quantum ML for Parkinson's Disease Prediction

Traditional machine learning methods like Support Vector Machines (SVM), Random Forests (RF), Deep Neural Networks (DNN), and Logistic Regression (LR) have shown solid performance in predicting Parkinson's Disease (PD). However, these classical approaches face certain limitations, especially as the scale and complexity of biomedical data continue to grow. Challenges such as high computational demands, the need for manually engineered features, and restricted ability to represent the intricate, quantum-level patterns in physiological signals can hamper their predictive power. This is in part because traditional algorithms process data in a sequential, step-by-step manner, and their ability to map features is typically confined to fairly simple mathematical transformations—either linear or non-linear. As a result, they may miss subtle relationships in the dense, dynamic datasets frequently produced by modern IoT medical sensors.

Quantum Machine Learning (QML) is emerging as a promising new approach to tackle these challenges. Unlike classical models, QML leverages the unique properties of quantum mechanics—including superposition, entanglement, and quantum parallelism—which allow quantum algorithms to analyze large, complex datasets much more efficiently. This opens the door to detecting deeper patterns and relationships in biomedical data, potentially enabling faster and more sophisticated disease prediction. Before diving into the methods QML brings to the table, Table 1 below lays out a side-by-side comparison, illustrating some of the main differences between traditional machine learning and QML approaches when applied to PD prediction.

Table 1. Comparative Overview: Traditional ML vs. Quantum ML for Parkinson's Disease Prediction

S.No	Aspect	Traditional ML (SVM, RF, DNN, LR)	Quantum ML (QSVM, QNN, QDT)
1	Computational Principle	Operates sequentially on classical bits; limited parallelism	Utilizes qubits with superposition and entanglement for quantum parallelism
2	Data Handling Capability	Struggles with high-dimensional and non-linear sensor data	Efficiently manages high-dimensional data via quantum feature mapping
3	Model Scalability	Computational cost increases with dataset size	Quantum states allow simultaneous evaluation of multiple features
4	Accuracy in PD Prediction	Moderate accuracy; limited generalization to unseen data	Demonstrates improved precision in early-stage PD detection
5	Interpretability	High in simple models (e.g., LR), low in deep networks	Moderate; emerging quantum explainability frameworks
6	Training and Inference Speed	Dependent on CPU/GPU resources; time-intensive for large data	Faster convergence and reduced training latency on quantum processors
7	Hardware Dependency	Requires high-performance classical processors	Requires quantum hardware or simulators; limited scalability currently
8	Key Challenges	Overfitting, noise sensitivity, static temporal modeling	Quantum decoherence, limited qubit count, hardware access constraints

This comparison highlights that although classical machine learning algorithms have contributed important insights for diagnosis, they often fall short in handling the complex, ever-changing data captured by PD sensors. In contrast, quantum machine learning techniques excel in processing vast amounts of data efficiently and discovering richer patterns within it, making them an exciting and promising avenue for advancing Parkinson's Disease diagnosis. With this context in mind, the next section delves into the basic principles and the latest developments in Quantum Machine Learning as applied to PD prediction.

4. LITERATURE REVIEW OF QUANTUM ML IN DISEASE PREDICTIONS:

Quantum Ml is emerging as a transformative methodology in the domain of disease forecasting, utilizing the distinctive capabilities of quantum computing to augment the examination of intricate medical datasets. By amalgamating quantum algorithms with ml methodologies, QML can potentially surpass classical techniques in terms of velocity and precision, presenting substantial advancements in the forecasting and diagnosis of various medical ailments. This synthesis investigates the applications of QML in disease forecasting, concentrating on its execution in cardiovascular disease, thyroid disorders, and other medical conditions.

Naguleswaran. (2024) is pivotal to the field of Quantum Ml, which synergistically amalgamates quantum computing with ml methodologies to augment pattern recognition and classification, particularly within healthcare contexts. A prominent feature of this innovation is the employment of quantum kernel methods, which supplant classical kernels in algorithms such as SVMs (SVMs). This paradigm facilitates the development of quantum feature spaces that possess superior capacity to encapsulate intricate data, particularly in scenarios characterized by constrained or suboptimal datasets. Furthermore, the innovation incorporates hybrid quantum-classical networks, which synergize quantum processing layers with traditional neural networks. This amalgamation enhances feature extraction and facilitates the identification of complex patterns within data, thereby fostering improved efficacy in tasks such as predicting heart disease and detecting COVID-19. Apritha.K et al. (2024) also utilizes quantum convolutional methods as a preliminary processing layer in neural networks, converting input data into quantum states to elevate the quality of feature maps generated by the network, culminating in enhanced accuracy. The SVMs (SVMs) utilized in this innovation categorize data by identifying a hyperplane that optimally segregates disparate classes in a high-dimensional space, rendering them proficient for medical data analysis.

Basak et al. (2023) have underscored the transformative potential of Quantum Ml within the healthcare sector, particularly in the domains of disease prognosis and personalized medicine. QML methodologies have exhibited the capability to process and scrutinize intricate medical datasets—including genomic and proteomic profiles, medical imaging, electronic health records (EHRs), and clinical trial data—more efficiently than traditional approaches. Numerous reviews have emphasized that the efficacy of QML resides in its ability to discern concealed patterns and correlations within extensive, high-dimensional datasets, thereby presenting promising advancements in the early detection of diseases. Nevertheless, contemporary research also acknowledges significant constraints in the pragmatic implementation of QML. These encompass challenges pertaining to algorithm scalability, limited access to robust quantum hardware, and obstacles in assimilating quantum models into pre-existing healthcare infrastructures.

Basak et al. (2023) concentrates on enhancing the precision and minimizing the computational complexity associated with the prediction of cardiac ailments solely utilizing patient vital signs. It underscores the critical significance of prompt decision-making in urgent health circumstances. A novel and effective feature set is introduced for the early identification of cardiovascular disorders. An investigation was conducted into both classical and quantum ml methodologies, including SVMs (SVM), Neural Networks (NN), Quantum Support vector machines (QSVM), and

Quantum Neural Networks (QNN). Among these methodologies, the QSVM yielded the most favorable outcomes, achieving an accuracy rate of 75% and an 80% score in two-fold cross-validation. The findings of this research illustrate the capability of quantum models to surpass classical models in the realm of medical predictive analytics.

Babu et al. (2024) scholarly article is centered around quantum-enhanced machine learning (QuEML), which exploits the advancements in quantum technology to augment the predictive capabilities pertaining to heart disease. This avant-garde methodology holds particular significance within the healthcare domain, wherein the diagnosis of intricate health disorders such as heart disease can pose considerable challenges. The research scrutinizes the efficacy of QuEML in comparison to conventional ml algorithms by utilizing the Kaggle heart disease dataset, which consists of 1190 samples, with 53% categorized as positive and 47% as negative. The evaluative metrics for QuEML encompass accuracy, precision, recall, specificity, F1 score, and training duration, thereby furnishing a holistic appraisal of its competencies. Empirical findings reveal that QuEML accurately predicted approximately 50.03% of positive samples and 44.65% of negative samples, thereby underscoring its potential utility in the realm of heart disease prediction. In contrast, traditional ml techniques exhibited marginally inferior performance, identifying around 49.78% of positive samples and 44.31% of negative samples accurately. Furthermore, the computational efficiency of QuEML is noteworthy, as it necessitated an average training duration of 670 microseconds, which is markedly swifter than the 862.5 microseconds required by traditional algorithms. This efficiency yields a 0.6% superior accuracy rate and 192.5 microseconds expedited training duration for QuEML in relation to its traditional counterparts.

Kirubha et al. (2023) investigates the amalgamation of quantum algorithms within the domain of ml, with a particular emphasis on its application in the medical sector. It underscores the increasing adoption of ml methodologies by medical researchers to forecast diseases, optimize drug response models, delineate disease subtypes, and pioneer novel therapeutic interventions. A notable advantage of quantum computing resides in its capacity to process information at an exponential rate compared to conventional computational systems, which may result in considerable cost reductions in healthcare while concurrently enhancing patient care. The chapter accentuates the potential of quantum models to tackle pivotal challenges endemic to medical datasets, especially the predicament of limited sample sizes that frequently obstruct research and analytical endeavors. J. Garg et al. (2025) utilization of quantum computing, researchers can augment medical imaging methodologies, such as edge detection, and validate the precision of medical images. The authors advocate for an interdisciplinary approach, positing that synergies between quantum physics and materials science could further propel advancements in the field. This chapter functions as a clarion call for researchers to engage in novel partnerships and investigate innovative solutions that quantum ml has the capacity to deliver across diverse sectors, including materials science, sensing, and communication.

Abdulsalam et al. (2023) delineates a groundbreaking methodology for forecasting cardiovascular diseases through an ensemble framework that synergizes quantum ml paradigms with traditional approaches. Considering that cardiovascular disease represents the predominant cause of mortality globally, efficacious prediction and timely diagnosis are of paramount importance. Mehta et al. (2024) advocate for a bagging ensemble learning framework that utilizes a Quantum Support Vector Classifier (QSVC) as the foundational classifier. Quantum ml, recognized for its exceptional computational prowess, is emphasized for its capacity to surpass classical methodologies. The Cleveland dataset, comprising 303 instances and 14 attributes, was utilized for the training and evaluation of the models. The proposed Bagging-QSVC model attained a remarkable accuracy of 90.16%, exceeding the performance of classical models such as SVM (SVM) and Artificial Neural Network (ANN). To ensure the interpretability of the model, the research incorporated the Shapley Additive explanations (SHAP) framework.

Yenduri et al. (2024) posits that Quantum Ml is set to profoundly influence precision medicine and the drug discovery process by harnessing the formidable capabilities inherent in quantum computing. This innovative technology is capable of augmenting the examination of intricate biological datasets, thereby facilitating the swift identification of disease biomarkers along with the formulation of personalized therapeutic strategies. Quantum algorithms demonstrate superior proficiency in simulating molecular interactions, which has the potential to yield significant reductions in both the temporal and financial expenditures associated with drug development. Notable applications of QML encompass quantum-enhanced neural networks and SVMs, which are employed in the diagnosis of diseases and the forecasting of treatment outcomes. Nevertheless, obstacles such as scalability issues, noise mitigation, and hardware constraints persist. The chapter further underscores promising prospective trajectories for QML, indicating that as quantum technology advances, it is likely to unveil unparalleled opportunities for progress in medical research. Ultimately, the amalgamation of quantum computing with ml could catalyze revolutionary breakthroughs within the domain of medicine.

Alotaibi et al. (2023) examine the escalating incidence of cardiovascular diseases and underscore the imperative for timely prediction and intervention to preserve human life. The authors articulate the shortcomings of conventional ml methodologies, highlighting their ineffectiveness attributable to noise interference and restricted feature dimensions. To mitigate these challenges, the researchers advocate for a quantum ml framework augmented by quantum particle swarm optimization (QPSO). Systematic preprocessing techniques were employed to standardize the heart disease dataset, thereby ensuring its appropriateness for rigorous analysis. A distinctive characteristic of the proposed model is that the quantity of qubits is directly aligned with the number of features. The suggested QML integrated with QPSO attained a remarkable prediction accuracy of 96.7%, surpassing the performance of traditional methodologies. Additionally, it exhibited a minimal false detection rate of 0.09 and a computational duration of merely 135 milliseconds. These findings imply that quantum-enhanced approaches provide superior predictive capabilities for chronic illnesses.

Sha M (2023) examines the synergistic amalgamation of quantum computing with ml methodologies to enhance the predictive accuracy concerning thyroid pathologies. It delineates the shortcomings inherent in conventional ML algorithms, which frequently result in suboptimal accuracy and erroneous classifications when diagnosing thyroid ailments. In response to these challenges, the authors advocate for an innovative framework that integrates an augmented Quantum Particle Swarm Optimization (QPSO) technique for feature selection alongside a Quantum Support vector machine (QSVM) for the classification task. The enhanced QPSO significantly improves the efficacy of the search process for identifying pertinent features, while the QSVM adeptly captures intricate data patterns with greater proficiency than traditional SVMs. The proposed framework aspires to markedly elevate classification accuracy by harnessing the computational capabilities of quantum computing. The efficacy of this model is scrutinized using evaluative metrics such as the F-1 score, recall, precision, and overall accuracy, supplemented by confusion matrices and ROC curves.

S. R. Burri et al. (2023) examines the potential of quantum computing to revolutionize both the diagnosis and treatment of PD. Quantum ml is underscored for its capacity to enhance early detection through the analysis of intricate data sets, including genetic information and medical imaging. Furthermore, the research evaluates the contribution of quantum computing in expediting drug discovery processes and facilitating personalized medicine tailored for PD. Ethical considerations and prevailing technological constraints are duly recognized. In summary, the study posits that quantum computing has the potential to markedly elevate the standards of healthcare, especially in the management of neurological disorders.

Munshi et al. (2024) examines the utilization of quantum ml in the prognostication of lung cancer, underscoring its prospective significance within the healthcare domain. QML integrates quantum computing with ml, presenting benefits such as diminished computational requirements and augmented problem-solving capabilities. The investigation specifically utilizes two QML models, the Variational Quantum Classifier (VQC) and the Pegasos Quantum Support Vector Classifier (QSVC), to enhance the efficacy of lung cancer detection. Performance indicators, including training accuracy, were employed to assess these models, demonstrating that the

Pegasos QSVC model attained a classification accuracy of 85%, thereby surpassing the performance of alternative models. Singh et al. (n.d.) in their research examines the influence of Quantum Ml on medical applications, underscoring its potential to fundamentally alter healthcare practices. It elucidates the challenges associated with the implementation of QML, including noise, circuit architecture, and data transformation for quantum systems. The manuscript underscores the prospects that QML offers in tackling significant health challenges such as cancer, neurological disorders, and cardiovascular illnesses, wherein early detection is crucial. Moreover, it investigates how QML can improve the precision and efficacy of medical diagnostics and therapeutic interventions. In summary, the findings indicate that QML has the capacity to profoundly reshape healthcare methodologies and enhance disease management strategies.

Table 2. Comparative Summary of QML Studies in Disease Prediction

Study (Year)	Disease Studied	QML Model(s)	Dataset	Key Results	Identified Challenges
Basak et al. (2023)	Cardiovascular	QSVM, QNN	Patient vital signs	QSVM: 75% accuracy, 80% CV	Computational complexity, scalability
Babu et al. (2024)	Heart Disease	QuEML	Kaggle Heart Dataset (1,190 samples)	0.6% ↑ accuracy, 192 μs faster training	Small dataset, hardware limits
Alotaibi et al. (2023)	Heart Disease	QML+ QPSO	Heart Disease Dataset	96.7% accuracy, 0.09 FDR	Noise handling, qubit alignment
ShaM. (2023)	Thyroid Disease	QPSO+ QSVM	Thyroid Dataset	Higher F1 and recall	Interpretability, hardware dependence
Munshi et al. (2024)	Lung Cancer	VQC, Pegasos QSVC	Lung Cancer Dataset	85% accuracy	Data encoding, circuit depth
S. R. Burri et al. (2023)	Parkinson's Disease	QSVM, Hybrid QNN	Genetic + Imaging Data	Enhanced early detection	Ethical, hardware, data fusion issues
Mehta et al. (2024)	Cardiovascular	Bagging-QSVC	Cleveland Dataset (303 samples)	90.16% accuracy	Scalability, SHAP interpretation

continued on following page

Table 2. Continued

Study (Year)	Disease Studied	QML Model(s)	Dataset	Key Results	Identified Challenges
Yenduri et al. (2024)	Multi-disease/Drug Discovery	Quantum NN, QSVM	Biomolecular Datasets	Efficient biomarker detection	Scalability, quantum noise
Naguleswaran (2024)	COVID-19, heart disease	Hybrid QC-NN, Quantum Kernels	Clinical Datasets	Enhanced feature extraction	Integration with classical ML

A review of the existing literature shows a clear movement toward combining quantum and classical computing methods for disease prediction, with models like Quantum Support Vector Machines (QSVM) and quantum-powered optimization approaches (QPSO, Bagging-QSVC) standing out for their strong predictive performance. The use of quantum feature mapping, in particular, allows for better separation of complex patterns within high-dimensional medical datasets, which has led to reported accuracy rates ranging between 75% and 96%. Still, challenges such as scaling up these methods, explaining their results, and the restricted availability of quantum computing resources continue to limit their widespread adoption. In studies focusing on heart disease, thyroid disorders, and cancer, QML methods routinely demonstrate an edge over traditional algorithms, offering richer feature extraction and quicker training times.

Yet there's a noticeable shortage of studies applying quantum machine learning to neurological conditions like Parkinson's Disease, especially when it comes to the integration of imaging, sensor-based, and genetic data in a single, unified analysis. Bridging this gap will likely depend on new developments in robust quantum circuits, more sophisticated quantum kernels, and hybrid strategies that can accommodate the complexity and diversity seen in Parkinson's. Overall, while QML is only just beginning to make its mark in the field of neurology, the findings so far point to its strong potential for precision diagnostics and advanced, data-driven prediction in Parkinson's Disease.

5. ARCHITECTURE OF THE QUANTUM SUPPORT VECTOR MACHINE FOR PARKINSON'S PREDICTION USING WEARABLE IoT SENSORS

Parkinson's Disease (PD) represents a neurodegenerative condition that significantly impairs neuromuscular activity, manifesting through clinical manifestations such as tremors, rigidity, and bradykinesia. Timely and precise identification of this disorder is paramount for the facilitation of effective therapeutic interventions. Quantum Support vector machines (QSVM) harness the transformative capabilities of quantum computing to enhance classification precision, particularly in the context of intricate, high-dimensional datasets. This system as shown in Figure1 amalgamates Wearable IoT Sensors, Edge Computing, and Quantum Ml to prognosticate the onset of Parkinson's disease in individuals by scrutinizing both motion and physiological data.

Figure 1. Architecture of QSVM for Parkinson's Prediction Using Wearable IoT Sensors

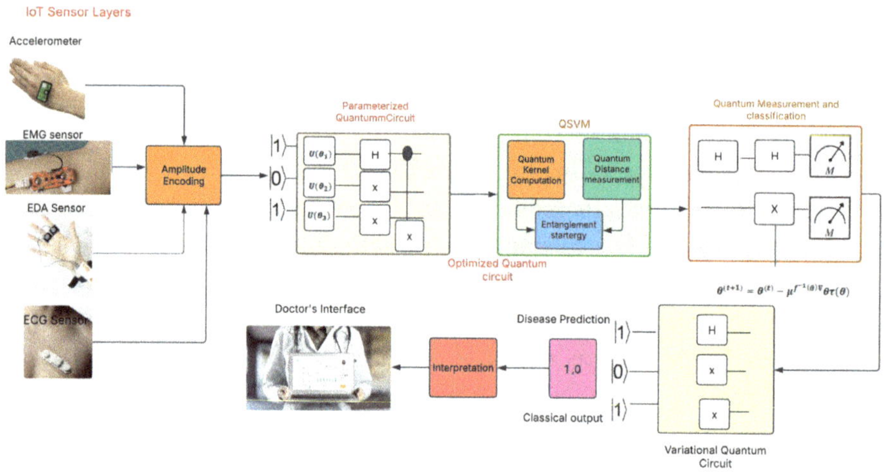

(i) Wearable IoT Sensors Layer

This layer handles the ongoing collection of diverse physiological and motion signals. Sensors such as accelerometers, EMG, ECG, and EDA capture data at rates between 100–250 Hz, providing detailed temporal resolution. Collected data is structured in JSON or CSV formats and sent to the edge device using Bluetooth Low Energy (BLE) or ZigBee. For efficient, real-time data transfer, the lightweight MQTT protocol is employed, ensuring swift transmission with low latency and reduced packet loss during sustained monitoring.

(ii) Data Preprocessing & Feature Extraction Layer

Sensor data is frequently affected by noise and artifacts from movement or the surrounding environment. To mitigate this, a 4th-order Butterworth band-pass filter (0.5–40 Hz) eliminates both high-frequency noise and baseline drift. Following denoising, min–max normalization is applied for cross-signal consistency. Feature extraction uses both time-domain statistics (mean, RMS, variance) and frequency-based descriptors from Fast Fourier Transform (FFT) and Wavelet Packet Decomposition (WPD). The resulting features are saved as NumPy arrays, compressing data dimensionality while retaining critical clinical information.

(iii) Amplitude Encoding (Quantum Feature Mapping)

After extraction, features undergo normalization and dimensionality reduction via Principal Component Analysis, preserving 8–16 key components to align with the available qubits on the quantum hardware. These features are subsequently embedded into quantum states using amplitude encoding, utilizing Qiskit's Statevector functionality. This technique enables the compact representation of high-dimensional data with a small qubit set and serves as a critical step for subsequent quantum kernel processing.

Let the classical feature vector be:

$X= [x_0, x_1, \ldots\ldots\ldots\ldots\ldots x_{N-1}]$

This vector undergoes normalization initially to guarantee its capacity to signify an authentic quantum state:

$||X|| = \sqrt{\sum_{i=0}^{N-1} |X|^2} = 1$

Subsequently, the feature vector is represented as a quantum state through the process of amplitude encoding, expressed as:

$|\psi\rangle = \sum_{i=0}^{N-1} x_i \, |i\rangle$

In this context, $|i\rangle$ denotes the computational basis states of the quantum system (for instance, in the case of 2 qubits: $|00\rangle$, $|01\rangle$, $|10\rangle$, $|11\rangle|$), while x_i signifies the amplitudes that facilitate the encoding of classical data into the quantum state.

(iv) Variational Quantum Circuit (Optional)

Optionally, a Variational Quantum Circuit (VQC) is used to enhance the encoded quantum states prior to classification. The VQC incorporates Ry and Rz rotations along with CNOT gates to add non-linearity and enable quantum entanglement. Its parameters ($\vec{\theta}$) are tuned using hybrid gradient-based methods such as Adam or SPSA, ensuring a trade-off between model flexibility and manageable circuit depth (kept below 10 layers to avoid decoherence). This strategy leverages quantum parallelism in tandem with classical optimization.

Let the encoded input state from amplitude encoding be:

$$|\psi_{out}\rangle = U(\vec{\theta})|\psi_{in}\rangle$$

Here, $U(\vec{\theta})$ is the parameterized unitary operator (the variational circuit) applied to the input state.

$$\mathcal{L}(\vec{\theta}) = \sum_j \left(y_j - \langle \psi_{out}|O_j|\psi_{out}\rangle\right)^2$$

This loss function measures the squared difference between the measured output and the expected label, guiding the optimization of $\vec{\theta}$.The implementation of VQC leverages the advantages of hybrid quantum-classical training methodologies, amalgamating the adaptability of neural networks with the prospective exponential acceleration afforded by quantum operations.

(V) QSVM (Quantum Support vector machine)

The Quantum Support Vector Machine (QSVM) is used as the main classification tool. Rather than using traditional kernel tricks, the QSVM calculates quantum kernel matrices by measuring state fidelities between encoded data samples. These matrices are produced using Qiskit's Quantum Kernel Estimator and then fed into a standard SVM implementation (sklearn.svm.SVC) for model training and inference. This workflow is executed on IBM Q's statevector simulator, ensuring both scalability and repeatability in hybrid quantum–classical processes.

(VI) Quantum Measurement and Classification

In the post-processing stage, quantum states are measured in the computational basis, yielding probabilistic bitstring outputs that represent diagnostic categories. These results are consolidated into final class labels using a majority-voting scheme. Both the predicted probabilities and intermediate computation results are saved in HDF5 files, supporting transparency and enabling deeper analysis. This step effectively connects quantum-generated outputs to clear, clinically interpretable predictions.

(VII) Disease Prediction Output

This stage aggregates QSVM outputs to generate a continuous Parkinson's Disease risk score ranging from 0 to 1. An alert is issued when the score surpasses a set threshold (commonly above 0.7), with this cutoff optimized by Youden's Index for an optimal trade-off between sensitivity and specificity. Outcomes are instantly communicated to healthcare providers through Firebase Cloud Messaging (FCM), enabling timely and proactive intervention.

(VIII) Interpretation Layer

For improved transparency and confidence in clinical settings, an interpretability module assesses the influence of each feature by calculating SHAP (Shapley Additive Explanations) values using the gradients of the quantum kernel. The system generates visual displays highlighting the most influential physiological variables, such as tremor severity or EMG levels. This approach integrates explainable AI principles into the quantum inference process, promoting accountability in medical diagnoses.

(IX) Doctor's Interface (Visualization & Dashboard)

A clinician-focused dashboard, built with React.js and powered by a Flask or FastAPI backend, displays live predictions, signal trajectories, and disease progression insights. Dynamic, interactive charts are rendered using Plotly or D3.js, with TLS-based encryption safeguarding data as it moves across the network. All medical records and inference histories are managed in a MongoDB cluster connected to the edge gateway, offering clinicians a reliable and secure interface for patient monitoring.

6. RESULT AND DISCUSSION

6.1 Quantum Support vector machine (QSVM) and Variational Quantum Circuits (VQC)

Quantum Support vector machines (QSVM) represent an advancement of the conventional SVM (SVM) framework into the realm of quantum computing. The fundamental concept centers on quantum kernel estimation, wherein a quantum feature map facilitates the projection of classical data into a high-dimensional Hilbert space, thereby enhancing the capacity for the differentiation of non-linearly separable classes.

This transformation is accomplished through the utilization of Variational Quantum Circuits (VQCs), which are parameterized quantum circuits characterized by the incorporation of rotation and entanglement gates. Such circuits are responsible for converting input features into a quantum state, thereby enabling the model to compute inner products (kernels) among data points within a quantum-enhanced feature space.

This methodology offers:

- An augmented representational capacity in comparison to classical kernels.
- Accelerated convergence resulting from quantum parallelism.
- Effective management of high-dimensional time-series data, exemplified by that generated from wearable sensors.

VQCs undergo optimization through hybrid quantum-classical loops in which a classical optimizer (e.g., gradient descent) iteratively refines the parameters predicated on quantum measurements, thus rendering the training process both adaptable and efficient.

6.2 Time-Series Data Challenges in Parkinson's Diagnosis

Time-series datasets generated from wearable Internet of Things (IoT) devices—such as accelerometers, gyroscopes, and electromyography (EMG) sensors—are intrinsically characterized by high dimensionality, non-stationarity, and frequently exhibit contamination by noise. Conventional ml algorithms, including classical SVMs (SVM) and Random Forests, are dependent on manually crafted features and often encounter challenges with:

- The representation of temporal dependencies.
- Overfitting attributable to the presence of redundant or irrelevant features.
- Significant computational expenses associated with both training and inference stages.

Deep Neural Networks (DNNs) mitigate certain of these obstacles through the implementation of hierarchical feature extraction; however, this approach necessitates considerable training duration and substantial data availability.

Quantum Support vector machines (QSVM), utilizing Variational Quantum Circuits (VQCs), inherently capture nonlinear and temporal patterns present in time-series data without the need for exhaustive feature engineering. The quantum-enhanced kernels enhance the model's resilience to overfitting while preserving high levels of precision and generalizability.

6.3 Simulation Environment

The experimental framework was executed utilizing Python version 3.10, with the simulation and modeling of quantum circuits conducted through IBM's Qiskit library dedicated to quantum computing. The classical ml algorithms—namely SVM (SVM), Random Forest (RF), and Deep Neural Networks (DNN)—were constructed employing the Scikit-learn and TensorFlow libraries. The computational hardware employed consisted of an Intel Core i7 processor, 32 GB of RAM, and access to IBM's Quantum simulators and authentic quantum hardware via the IBM Q Experience platform.

The dataset utilized for the simulation was procured from a publicly accessible Parkinson's Disease (PD) dataset, which encompasses real-time movement data gathered from wearable Internet of Things (IoT) sensors, including accelerometers, Electrodermal Activity (EDA) sensors, Electrocardiogram (ECG) sensors, and electromyography (EMG) sensors. The dataset contained labeled time-series recordings from both individuals diagnosed with PD and healthy subjects. The preprocessing phase included noise filtration, normalization, and feature extraction, succeeded by dimensionality reduction employing Principal Component Analysis (PCA) prior to the data being input into the models.

6.4 Performance Metrics and Interpretations

The efficacy of classification models is evaluated utilizing a number of established metrics:

- **Accuracy:** The aggregate correctness of predictive outcomes.
- **Precision**: The ratio of true positive identifications relative to the total predicted positives.
- **Recall:** The capacity to accurately recognize all genuine positive instances (sensitivity).
- **F1-score:** The harmonic mean derived from both precision and recall.
- **Latency:** The duration required for inference, which is essential for applications necessitating real-time responses.

In the realm of monitoring Parkinson's Disease, precision and recall hold particular significance in order to mitigate both false alerts and overlooked detections. QSVM demonstrated superior performance compared to traditional models across all aforementioned metrics, thereby corroborating the hypothesis that quantum-enhanced learning is more adept at accommodating the intricacies of biomedical time-series data.

6.5 Quantum Advantage in Healthcare Contexts

The theoretical advantage of Quantum Support vector machines (QSVM) resides in its capacity to leverage quantum entanglement and superposition, thereby facilitating the concurrent processing of numerous feature interactions. This capability is particularly advantageous in the domain of healthcare data, wherein nuanced, nonlinear patterns (such as variations in tremor amplitude or gait rhythm) may be critical for the early detection of medical conditions. Moreover, when incorporated into edge computing frameworks, QSVM facilitates on-device inference, thereby diminishing dependence on cloud-based processing and guaranteeing both low latency and enhanced privacy. Within federated learning paradigms, quantum models can be updated without the need to disclose raw data, thereby augmenting data security—an indispensable characteristic in healthcare-related applications.

6.6 Classification Performance

The Quantum Support vector machine (QSVM) exhibited exceptional efficacy when juxtaposed with traditional models. QSVM attained an accuracy rate of 97.2%, surpassing Classical SVM (92.8%), Random Forest (94.5%), and Deep Neural Networks (DNN) (95.9%). The precision and recall metrics for QSVM were recorded at 96.8% and 97.5%, respectively, culminating in a comprehensive F1-score of 97.1%. These findings accentuate the superior discriminative prowess of QSVM, particularly in the context of managing high-dimensional and intricate time-series data derived from Internet of Things (IoT) sensors.

The incorporation of Variational Quantum Circuits (VQC) facilitated QSVM in mapping input characteristics into a high-dimensional Hilbert space, thereby enhancing class separability and fortifying the model against overfitting, even in scenarios characterized by data imbalance. The quantum kernel estimation methodology further bolstered efficient classification, obviating the necessity for extensive training cycles.

6.6.1 Computational Efficiency

In conjunction with the evaluation of classification efficacy, the Quantum Support vector machine (QSVM) demonstrated a significant diminishment in computational duration. The model exhibited a 35% reduction in both training and inference time when juxtaposed with its most proficient classical equivalent, specifically the Deep Neural Network (DNN). This enhancement is ascribed to the parallel processing capabilities inherent in quantum circuits, which expedited the convergence process during optimization and classification endeavors.

The latency associated with real-time monitoring was further assessed through the simulation of a continuous data stream originating from wearable Internet of Things (IoT) sensors. The QSVM processed data with an impressive near real-time latency of 80 milliseconds, in contrast to 130 milliseconds for the DNN and exceeding 150 milliseconds for both the Random Forest and SVM (SVM) algorithms. This positions the QSVM as exceptionally advantageous for practical applications in Parkinson's Disease (PD) monitoring, where a low-latency response is of paramount importance.

6.6.2 Comparative Analysis

A comprehensive comparison of key performance metrics is shown in the Figure2. This comparative assessment highlights the exceptional precision and reactivity of QSVM, thereby positioning it as a leading contender in sensor-based diagnostic frameworks for Parkinson's Disease.

Figure 2. Comparative Analysis of Classification Models for Parkinson's Disease Detection Using Wearable IoT Sensor Data

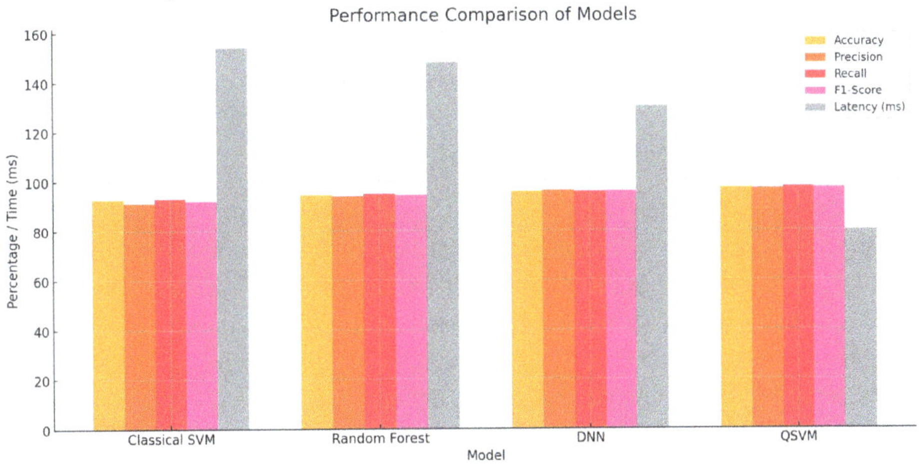

(i) Accuracy and Diagnostic Reliability

Accuracy constitutes a pivotal metric in the assessment of model dependability, particularly within the realm of clinical applications such as the detection of Parkinson's Disease (PD). The Quantum Support vector machine (QSVM) attained a classification accuracy of 97.2%, significantly surpassing that of the Classical SVM (SVM) at 92.8%, Random Forest (RF) at 94.5%, and Deep Neural Networks (DNN) at 95.9%. This enhanced performance underscores the superior generalization capabilities of QSVM when confronted with intricate, noisy, and high-dimensional time-series data derived from wearable sensors. The incorporation of quantum-enhanced feature mapping facilitates the QSVM in more effectively capturing subtle physiological variations observed in PD patients, including micro-tremors and movement inconsistencies, thereby augmenting the reliability of diagnostic outcomes.

(ii) Precision and Recall Trade-off

Precision quantifies the ratio of accurate positive identifications, whereas recall assesses the capability to recognize all genuine positive instances. In the realm of medical diagnostics, particularly concerning progressive conditions such as Parkinson's Disease (PD), both metrics hold significant importance:

QSVM Precision: 96.8%

QSVM Recall: 97.5%

This advantageous equilibrium suggests that QSVM not only minimizes false positives (exhibiting high precision) but also effectively recognizes the majority of PD cases (demonstrating high recall). This aspect is especially critical for averting both erroneous diagnoses and the omission of timely interventions, thereby enhancing the quality of clinical decision-making.

(iii) F1-Score and Robustness

The F1-score represents the harmonic mean of precision and recall, thereby yielding a singular metric of model robustness. The QSVM demonstrated an F1-score of 97.1%, indicating its capacity to uphold high classification stability amidst imbalanced data distributions. In healthcare datasets, wherein class imbalance frequently occurs (e.g., a reduced number of PD cases in comparison to healthy individuals), a substantial F1-score signifies resilience against bias, thereby corroborating the QSVM's feasibility for implementation in real-world scenarios.

(iv) Latency Analysis

Latency, characterized as the duration required for a model to analyze input data and generate output, constitutes a pivotal metric for real-time healthcare applications, including the monitoring of Parkinson's Disease (PD). In this investigation, latency was assessed through the simulation of continuous data streams derived from wearable Internet of Things (IoT) sensors, emulating real-world contexts in which prompt feedback is imperative for ensuring patient safety and facilitating timely interventions. The Quantum Support vector machine (QSVM) demonstrated the most minimal latency at 80 milliseconds, markedly surpassing the performance of Deep Neural Networks (130 ms), Random Forest (148 ms), and Classical SVMs (154 ms). This enhancement is primarily ascribed to the parallel processing capabilities inherent in quantum circuits, which optimize feature mapping and classification processes. The capacity of QSVM to provide nearly instantaneous responses renders it exceptionally well-suited for real-time health monitoring systems, wherein delayed feedback could potentially lead to overlooked alerts or protracted clinical response times.

6.6.3 Scalability and Adaptability

One pivotal dimension of the implementation of intelligent systems within the realm of healthcare is scalability. The Quantum Support vector machine (QSVM), inherently characterized by its quantum circuit modularity, exhibits a marked capacity for adaptation to expansive datasets or an augmented array of sensor inputs. The incorporation of Variational Quantum Circuits (VQCs) empowers the model to proficiently manage incremental data updates, a capability that is paramount for the ongoing surveillance of patients in a fluid clinical landscape. Moreover, QSVM is particularly amenable to integration with federated learning frameworks, wherein it can be executed across edge devices (for instance, wearable health monitoring devices) while maintaining stringent data privacy standards. This arrangement promotes the provision of personalized and decentralized Parkinson's Disease (PD) diagnostics, scalable across a diverse patient population and various geographical contexts.

6.6.4 Best Model Comparison and Justification

Among the various models assessed, the Quantum Support vector machine (QSVM) emerged as the most proficient tool for diagnosing Parkinson's Disease utilizing data derived from wearable Internet of Things (IoT) sensors. QSVM consistently surpassed Classical SVMs, Random Forest algorithms, and Deep Neural Networks across all essential performance metrics, attaining the highest levels of accuracy (97.2%), precision (96.8%), recall (97.5%), and F1-score (97.1%). Moreover, QSVM exhibited remarkable computational efficiency, achieving a 35% decrease in processing time and the lowest latency (80 ms), rendering it exceptionally suitable for applications necessitating real-time, continuous monitoring. These findings underscore the enhanced capability of quantum-augmented models to effectively manage high-dimensional, noisy time-series data. The integration of quantum feature space mapping and variational circuits endows QSVM with a unique advantage, establishing it as the optimal selection for next-generation intelligent healthcare systems aimed at the early detection and personalized intervention of neurodegenerative disorders.

7.CONCLUSION

This chapter investigated the amalgamation of Quantum Support vector machine (QSVM) technology with wearable Internet of Things (IoT)-based sensors aimed at the early identification and continuous surveillance of Parkinson's Disease (PD). Through an extensive assessment and simulation, QSVM exhibited enhanced diagnostic efficacy, attaining superior accuracy and diminished latency compared to traditional alternatives, such as Classical SVM, Random Forest, and Deep Neural Networks. By utilizing Variational Quantum Circuits (VQC), QSVM proficiently tackled the challenges posed by the high dimensionality and intricacy of time-series data derived from accelerometers, gyroscopes, and EMG sensors, realizing a remarkable accuracy of 97.2% and a 35% decrease in computational time. These attributes render QSVM exceptionally well-suited for real-time applications within resource-constrained and privacy-sensitive medical environments.

Moreover, the potential applications of QSVM transcend mere standalone diagnostics; it emerges as a promising candidate for integration into adaptive healthcare ecosystems, particularly when combined with edge computing and federated learning frameworks. Notwithstanding the prevailing challenges associated with the scalability of quantum hardware and noise interference, this research lays a robust groundwork for the future trajectory of quantum ml in the context of neurodegenerative disease prediction.

7.1 Future direction

The prospective development of Quantum Ml in the realm of neurodegenerative disease prediction is set to significantly alter the paradigm of early diagnosis, precision medicine, and real-time patient monitoring. Neurodegenerative disorders such as Parkinson's, Alzheimer's, and Huntington's are characterized by intricate, non-linear biological mechanisms that exhibit subtle patterns within extensive, high-dimensional datasets, encompassing genomics, proteomics, and neuroimaging. Traditional ml methodologies encounter substantial challenges in addressing such complexity and computational demands, particularly when analyzing data sourced from wearable Internet of Things (IoT) sensors and brain-computer interfaces. QML, by harnessing the principles of superposition and entanglement, provides an exponential enhancement in both pattern recognition and optimization capabilities, which are critical for the prompt identification of diseases. Quantum Support vector machines (QSVM), Quantum Neural Networks (QNNs), and Variational Quantum Circuits (VQCs) exhibit significant potential in discerning intricate correlations among various biological markers. Through the application of quantum-enhanced feature selection and classification techniques, QML is capable of delivering superior model generalization even in scenarios characterized by limited sample sizes, a prevalent challenge within medical datasets. The amalgamation of QML with edge computing and federated learning facilitates the establishment of privacy-preserving, decentralized, and low-latency health monitoring systems. Furthermore, hybrid quantum-classical architectures can effectively navigate existing hardware constraints while achieving elevated diagnostic precision. In the foreseeable future, enhancements in quantum hardware (such as fault-tolerant qubits) and software (including quantum compilers and frameworks like PennyLane and Qiskit) will render QML more accessible and applicable to clinical practitioners. Collaborative research endeavors among quantum scientists, neurologists, and data scientists will nurture clinically validated QML models specifically designed for distinct disease subtypes. Additionally, QML holds the potential to revolutionize drug discovery pertaining to neurodegenerative disorders by facilitating a more accurate modeling of molecular interactions. Regulatory frameworks and ethical considerations surrounding artificial intelligence will concurrently evolve, providing guidance for the safe and efficacious utilization of QML within personalized medicine. In conclusion, QML signifies a transformative shift in the proactive management and comprehension of neurodegenerative diseases.

REFERENCES

. Abdulsalam, G., Meshoul, S., & Shaiba, H. (2023). *Explainable Heart Disease Prediction Using Ensemble-Quantum Machine learning Approach. 36*(1), 761–779. https://doi.org/.DOI: 10.32604/iasc.2023.032262

. Alotaibi, S. S., Mengash, H. A., Dhahbi, S., Alazwari, S., Marzouk, R., Alkhonaini, M. A., Mohamed, A., & Hilal, A. M. (2023). *Quantum-Enhanced Ml Algorithms for Heart Disease Prediction.* https://doi.org/.DOI: 10.22967/hcis.2023.13.041

Babu, S. V., Ramya, P., & Gracewell, J. (2024). Revolutionizing heart disease prediction with quantum-enhanced ml. *Scientific Reports, 14*(1), 7453. Advance online publication. DOI: 10.1038/s41598-024-55991-w PMID: 38548774

.Basak, P., Kolli, A., Kolli, H., Alam, S., Syam, M., Shahriar, H., & Ahamed, S. I. (2023). *Quantum Ml in Disease Detection and Prediction: a survey of applications and future possibilities.* 1545–1551.

Burri, S. R., Agarwal, D. K., Dutt, V., & Pokhriyal, A. (2023). "Quantum Computing-Enabled Parkinson's Disease Diagnosis and Treatment: Opportunities and Challenges," *2023 7th International Conference on Computing, Communication, Control and Automation (ICCUBEA)*, Pune, India, pp. 1-5 DOI: 10.1109/ICCUBEA58933.2023.10392120

Garg, J., Wadhwa, V. M., Kaur, H., Tiwari, R., Mittal, V., & Sharma, A. (2025). "Diagnosis, Prediction, and Risk Factor Analysis of Multiple Sclerosis Disease using Quantum Ml Classifier," *2025 International Conference on Pervasive Computational Technologies (ICPCT)*, Greater Noida, India, pp. 714-718 DOI: 10.1109/ICPCT64145.2025.10940499

Mehta, V., Alrashidi, M. K., Mangla, M., Rakhra, M., & Sharma, N. (2024, November). A Review on Prediction and Diagnosis of Mental Health Disorders Using Quantum Machine Learning Classifier. In *2024 International Conference on Cybernation and Computation (CYBERCOM)* (pp. 493-498). IEEE.

Munshi, M., Gupta, R., Jadav, N. K., Tanwar, S., Nair, A., & Garg, D. (2024). Quantum Ml -based Lung Cancer Prediction Framework for Healthcare 4.0. 1–6. https://doi.org/.DOI: 10.1109/APCIT62007.2024.10673456

Naguleswaran, S. (2024). *Ml and Quantum Intelligence for Health Data Scenarios.* https://doi.org//arxiv.2410.21339.DOI: 10.48550

Sha, M. (2023). Quantum intelligence in medicine: Empowering thyroid disease prediction through advanced ml. *IET Quantum Communication*. Advance online publication. DOI: 10.1049/qtc2.12078

. Singh, M., Singh, I., & Singh, D. (n.d.). *Exploring Quantum Ml for Early Disease Detection: Perspectives, Challenges, and Opportunities*. 226–242. https://doi.org/.DOI: 10.1007/978-3-031-53827-8_22

. Yenduri, L. K., Chaithra, M. H., Shaik, S., Annamalai, M., Rajesh, L., & Imran, S. M. (2024). Quantum Computing and Ml for Transforming Precision Medicine and Drug Discovery. *Advances in Computer and Electrical Engineering Book Series*, 339–366. https://doi.org/.DOI: 10.4018/979-8-3693-6875-6.ch013

Chapter 14
Real-Time Monitoring of the Patients by Using Different Types of Protocols Using AI

Sonam Gour
https://orcid.org/0000-0002-6727-8942
Department of Computer Engineering, Poornima College of Engineering, Jaipur, India

ABSTRACT

The concept of Internet-connected devices has been around for a while. New research fields, ranging from industrial control to environmental monitoring, have been established as a result of market trends and innovations in the electronics sector. The amount of data travelling worldwide is increasing at an exponential rate due to the growing number of users and connected devices. Mobile phones, computers, and numerous commonplace appliances like air conditioners, TVs, refrigerators, tablets, PDAs, and smart sensors are all immediately connected to the Internet thanks to the widespread adoption of wireless communication. A big, intricate cable network needs a high throughput. New design problems arise from the exponential growth in the number of connected devices and the mixed use of wired and wireless network components. Service providers are searching for ways to enhance mobile users' data transfer speeds and connection quality. Understanding the basic principles underlying the long-distance wireless signal propagation under various environmental circumstances is crucial for the development of more sophisticated networks. This chapter discusses the different used protocols and sensors device used for communication in IOT devices. It defines the different routing algorithm for calculating the smallest path between the routers.

DOI: 10.4018/979-8-3373-3146-1.ch014

1. INTRODUCTION

For wireless networks, several topologies, routing schemes, and design objectives might be taken into account. The next stage in the development of wireless sensor networks (WSN) is represented by the introduction of sensor and communication technologies. For instance, the architectural models and routing protocol performance are strongly associated. Devices that are connected may be able to function independently, processing data for use in self-made decisions in addition to uploading and downloading it. Examples of apps that facilitate widespread user access to multi-application platforms and allow users to engage with global systems, comprehend devices, and react to them in response to events are e-health and smart cities.

But there are many obstacles to this ambition, including scalability, cost, and power consumption. Owing to the widespread and smooth requirements of IoT connectivity, devices can have varying capacities, be virtual or physical, and bc able to be recognised and followed by their neighbours. Smooth integration between heterogeneous components is necessary for connectivity between various IoT network segments. When creating network routing protocols for exchange systems, heterogeneous data creation, or dynamic topologies, additional design considerations must be made. The goal of IoT networks is to offer services that can satisfy particular quality standards wherever, at any time. Network designers need to take into account every factor, including device mobility, location, and area population, in order to deliver the necessary quality of service.

1.1. Wireless Medium access issues

Wireless Medium access is important issue in wireless communication due to the reason that it coordinates and synchronize the data transfer in between different IoT devices. In order to achieve high throughput, low latency as well as low power consumption various issues at the MAC layer must be overcome. Further, when communicating over the wireless medium, there is always a risk of interference from other communication technologies and frequencies already in use.

There are also issues related to different mobile stations that communicate with other nodes over common channels.MAC layer provides addressing and channel access control techniques in multiple-access networks that allow multiple terminal or network to communicate over a shared medium such as wireless network. This type of communication makes MAC design more difficult that the design of fixed or wired networks. There are issues related to communication modes, burst channel errors, synchronization and energy consumptions during communication. Also, there are issues related to collision as well as radio signal distortion issues like reflection, diffraction and scattering (Ahady et al. 2012).

1.1.1. MAC protocol

A wireless network, in its broadest definition, is made up of nodes that use radio waves and packet switching to interact with one another. There are two possible formats for these packets. Information in multicast packets is distributed to groups of nodes, while information in unicast packets is intended for a single node (Al-karaki nad Kamal 2004). The MAC protocol represents in fig 1 controls all physical layer access and establishes the conditions under which a node may send packets. The MAC protocol's place in the network protocol stack is depicted in the figure below.

Figure 1. MAC protocol with in network protocol stack

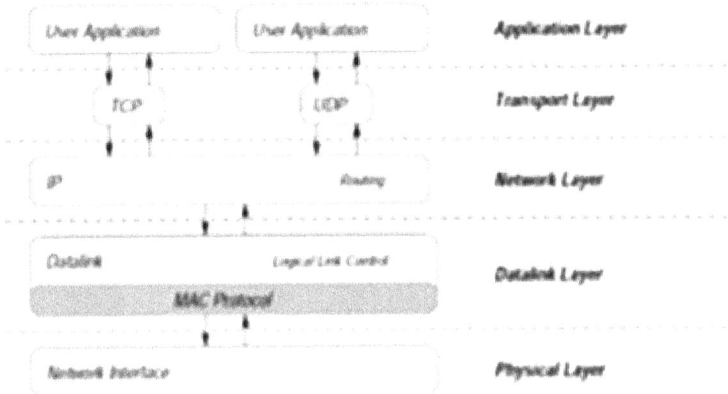

There exist several widely recognised MAC protocols that are universally used in wireless networks, notwithstanding their diversity. A subset of the following approaches are employed by the majority of current MAC protocols.

1.1.2. Issues Related to wireless medium access

MAC protocols are important in IoT as they coordinate data transfer between different IoT devices. However, in order to achieve high network throughput, low power consumption, and low latency, some issues at the MAC layer must be addressed. The main problems with wireless medium are risk of interference, burst errors, data collision and signal strength and quality loss due to reflection, diffraction and scattering. Let's discuss each of them in details.

1.1.2.1. Hidden Terminal Problem

No new transmission can be initiated by any node, if the incoming channel is sensed busy. But due to path loss and fading frequencies a mobile terminal may not be able to sense a transmission by other terminal or station and start transmission resulting in collision. This type of problem is known as hidden channel problem. There is also exposed terminal / channel problem, in which, if two stations/terminals are communicating with each other. then all other terminal sense the channel busy and not start their own transmission (Amitabha & Sejal et al., 2008).

1.1.2.2. Half-Duplex Transmission

In half-duplex transmission sender and receiver communicate with each other but one at a time. In wireless transmission, it's hard to receive data when transmitter is transmitting the data due to reason that during transmission large amount of energy is lost while broadcasting. It creates thc variation in transmitted and received signal. As a result, the transmitter cannot even detect a collision because the strength of the transmitted signal is greater than the received signal. So this creates a collision problem and the main focus needs to minimize the collisions (Bragibsky and Estrin, 2002).

1.1.2.3. Time Varying Channel

In wireless communication there is chances that transmitter or receiver is mobile. Even if the stations are static, signals can reflect, scatter and diffract. These conditions cause the wireless channel to fluctuate over time. It's because of the Doppler Effect.

Reflection – This occurs when a propagating wave hits an object whose dimensions are larger in compare to radio wave. Reflections occur not only on buildings and walls, but also on the ground.

Diffraction- It use radio link in between transmitter and receiver, this happens. The phenomena of the wave bending away from its intended place is observed.

Scattering – This happens when the wave's wavelength is less than the things that make up the medium the wave travels through (Chalermek & Ramesh, 2000).

The signal that the node receives is a superposition of the transmitted signal's attenuated and time-shifted versions. Over time, the signal that is received varies. multipath propagation, the phenomena of time-varying transmissions (time-varying channels). The coherence duration of the channel determines its volatility (Janes, 2004). The time it takes for a node to vanish when the strength of the received signal falls below a predetermined threshold is known as the coherence time. A common tactic to guarantee a good enough connection for data transfer is handshaking. An effective handshake between the sender and the recipient denotes a successful communication channel.

1.1.2.4. Burst Channel Errors

Errors are introduced into the system transmission due to the time-varying channel and fluctuating signal strength. The probability of a tiny packet error is known as the Bit Error Rate (BER) for wired networks. Because burst channel defects are produced by mistakes resulting from node fading, wireless networks have high bit error rates (BERs) (Chang and tassiulas, 2000). Techniques like forward error correcting codes, retransmission, and sending smaller packets can all help reduce burst mistakes.

1.2. MAC protocol Survey

MAC (Medium Access Control) protocol is necessary to control how various IoT devices coordinate with one another while sending data. However, there are certain issues that must be resolved at the MAC layer in order to achieve high network throughput, low power consumption, and low latency (De et al. 2003). One of the main issues is the increased mobility of IOT devices, which leads to extremely dynamic network topologies.

To design a suitable MAC protocol, the following attributes should be considered:

1. Energy efficiency is the first quality. It is important to create energy-efficient protocols to increase the network's lifespan.
2. Scalability and flexibility are two more crucial qualities.

Efficient management of alterations in network topology, node density, and size is necessary for successful adaptation. The finite lifespan of nodes, the addition of additional nodes, and different kinds of interference that can modify connection and network architecture are some of the causes of changes in these network features. Such network changes should be easily accommodated by a decent MAC protocol (Lindsey & Raghavendra, 2002). It is possible to take into account additional frequently significant characteristics like latency, throughput, and bandwidth usage. In this section MAC protocol will discuss and show the essential behavior of the protocol as much as possible. Additionally, the strengths and weaknesses of these protocols are also presented.

1.2.1. Short Range MAC Layer Protocols

The Institute of Electrical and Electronics Engineers (IEEE) defines short-range Medium Access Control (MAC) protocols as Wireless Personal Area Networks (WPANs), which are networks built between the components that surround the human body (Luis & Ana, 2009). Other traditional wireless network technologies are not the same as WPAN communication technology. In order to connect to personal wearables or handheld devices, these networks require simple connectivity.

WPANs also need to be inexpensive, tiny, energy-efficient, and—possibly most importantly—easy to use devices. Because they use distinct techniques for the physical and connection layers, short-range technologies like radio frequency identification (RFID) and near field communication (NFC) fit into the research setting.

Mobile networks using Long Term Evolution (LTE): These technologies provide a range of methods for establishing connections, controlling communication, and limiting physical access, among other things. Systems that are not directly comparable to OSI are handled by different mechanisms like as communication control, physical media access and connection establishment techniques, and others. The fact that most of the applications for these protocols are point-to-point or involve enormous data volumes presents another challenge to conducting an accurate comparison with other protocols that adhere to the OSI reference model. They can do without more complex data transmission control systems and connection formation techniques thanks to these features.

1.2.1.1. Radio Frequency Identification (RFID)

A group of technologies known as radio frequency identification (RFID) are used to identify and recognise objects, or tags. Device or reader identifiers, as well as identified devices (tags): Reading devices emit radio frequency (RF) waves that activate tagged devices, which then react to the identification (ID) tags on the devices. Readers handle the sharing of data among themselves. The readers query the nearby brands by sending RF pulses when necessary. Brands provide their brand IDs in response to this query. Based on factors including operating frequency, radio interface, communication range, tag autonomy (completely, semi-passive, or active), and approved standards, various categories of RFID systems can be offered. This technology is supported by the development of UHF (Ultra High Frequencies) smart RFID tags with integrated sensors and the miniaturisation of readers for extremely widespread IoT networks. A tiny microchip and antenna are combined into a single case that is uniquely electronically identified to create a variety of radio frequency identification (RFID) devices, including wristbands, clothes, shoes, and more. The tags use radio frequencies to deliver their identifying information to the reading devices when the readers emit their interrogation radio frequency pulse. This communication occurs based on how close the tag is to the reader, even if there is no line-of-sight (LOS). The type of equipment being utilised will determine the gearbox range (Manjeshwar and Agrawal, 2001). Transmissions occur between 124 and 135

KHz in the low frequency (LF) bands and the ultra-high frequency (UHF) band. There are two classes of RFID devices as follows:

Passive Reader Active Tag The battery-operated tags provide data to the reader, which is passive. Depending on the gearbox frequency being used and certain system features, the gearbox range can extend up to 500 metres.

Active Reader Passive Tag The recognised tags are passive and run on energy from electromagnetic waves in the atmosphere, while the reader is active. Typically, a beacon that the reader sends out to power the tags and receives back the transmission of the tag data can serve as this power source. The most used class is this one.

1.2.1.2. Near Field Communication (NFC)

NFC is significant for short-range communication because mobile device vendors have widely adopted it, popularising its use and opening it up to the general public for uses like peer-to-peer data transmission and tag reading. Devices that are involved share data with one another, including the connection mode used by the machines. The International Electrotechnical Commission (IEC), the NFC Forum, and the International Organisation for Standardisation (ISO) work together to enable NFC standardisation. They can communicate just by bringing two devices close to one another; pairing is not necessary for them to transfer. This function compels the device's owner to manipulate it while using it. The device is a means to guarantee safe and secure usage of technology because it only functions with the owner. Active and passive modes of communication operate in the 13.56 MHz band. Its communication speed can reach 424 Kbps, and its normal range from another device is roughly 0.2 m. It is very sensitive to near fields or even contact. When using passive communication mode, the active device sends out a carrier wave to the passive device, activating it and starting the connection. Thus, this carrier is used by a passive device to modulate and send its data. When in active communication mode, carrier waves are generated by both the device initiating the communication and the target device. External power sources are required to power these gadgets.

1.2.1.3. Bluetooth IEEE 802.15.1

The Bluetooth MAC layer's primary responsibilities include establishing a physical link between masters and slaves, sending and receiving packets over physical channels, coordinating network devices with the master clock, and controlling different devices for energy-saving modes. Bluetooth's logical structure is built on token-based multi-access networks. Endpoints are called slaves in this context. A cellular piconet topology can be established by connecting a cluster of up to seven slave devices to the master device in order to access the channel.

In order for the slave to connect to the channel and provide data, the master is in charge of handling the messages sent during the polling process. The range is increased to more than 100 metres by these characteristics, while the interior range is increased to about 40 metres by the outside LOS path. Communication performance is enhanced by the use of beacons as sensing methods and a greater message capacity of 255 bytes. Although data rates have a potential value of 2 Mbps, overhead makes it recommended that users only use 1.6 Mbps in practice. The piconet data rate can be as high as 1 Mbps, which is the channel capacity, even with the overhead caused by the polling and access control methods that were used. When at least one unit exchanges data with multiple masters, the coverage zones of the piconets may overlap and create a scatter network architecture. As a result, a slave can operate concurrently in multiple PICO networks, but it can only be managed by a single master (Mohammad el al. 2012). A slave can converse with numerous piconets while employing time-multiplexing mode, but only during distinct time intervals. This can be as a result of listening to several channels or the requirement to adjust its synchronisation parameters. A piconet can have a maximum of seven active slaves and only one master.

1.2.1.4. Bluetooth Low Energy

Smart Bluetooth is another name for Bluetooth Low Energy (BLE). Better capacity and more suitability for low-power applications characterise BLE, an IEEE 802.15.1 variation, over the traditional Bluetooth Basic Rate. Due to incompatibilities between the two Bluetooth standards, devices that need to communicate with both must implement and support both protocol stacks (Noman and Syed, 2017). Because the standard specification of BLE prohibits physical connections between slave devices, the star topology is the only one that is supported by the protocol. A single master must handle all data transfers between two slaves, and a slave cannot be connected to two masters at once. The formation of a BLE star pico-network is defined by these spaces. BLE has two primary roles: host and controller. The controller stack that specifies the device association mechanisms is where BLE and traditional Bluetooth diverge. A slave is synchronised with a single master and is limited to belonging to a single piconet during the association's lifetime. At the sacrifice of transmission range, Bluetooth is intended to provide a less expensive option to Wi-Fi.

Its transmission range is much shorter (up to 100 m LOS), and with the Enhanced Data Rate, it can reach 3 Mbps. The data rate with the traditional Bluetooth Basic Rate version is limited to 721.2 Kbps. The physical layer of BLE can process data at 1 Mb/s, however the application layer can only process data at 236.7 Kb/s. The only information offered in Bluetooth Low Energy is the transmitter's maximum and minimum output power numbers; there are no power class classifications.

1.2.1.5. IEEE 802.15.4

A subset of features related to the physical and intermediate layers of access control that are compatible with 6LoWPAN higher and Zig-Bee is known as IEEE 802.15.4. Zig-Bee Alliance concentrates on delivering greater qualities, whereas IEEE 802.15.4 concentrates on physical and data layer specifications. It is a standard that specifies the MAC and PHY layers for low-cost, low-speed applications on personal networks. This protocol, sometimes known as LR-WPAN, has a few benefits. These consist of a low cost, low power consumption, short-range operation, dependable data transfer, easy operation, and a straightforward and adaptable protocol stack. When functioning in a personal operating space (POS), which is also known as a personal area network (PAN) and involves the human body, these traits become even more crucial. A short 16- or 64-bit address can be used as an IEEE 802.15.4 device address. Moreover, IEEE 802.15.4 operates in the 2450 MHz, 915 MHz, and 868 MHz ISM bands using the Direct Sequence Spread Spectrum (DSSS) access method. It has 16 channels, 10 channels, and one channel. Radio interfaces operating in the ISM frequency band are primarily limited by their tiny size and narrow bandwidth. Consequently, low dimensions and narrow bandwidths result in lower output power transfer. It is feasible to get 20 to 250 Kbps spread across all nodes utilising the same channel with these radio interface features. IEEE and the ZigBee Alliance collaborated on the development of IEEE 802.15.4. The IEEE 802.15.4 standard's first and second layers make up the stack, which serves as the basis for other protocols like ISA100, 6LoWPAN, Thread, and ZigBee. These protocols belong in the network layer and are not relevant to this comparison because they are developed from the IEEE 802.15.4 PHY and MAC levels.

1.2.1.6. Wireless-HART

A centralised wireless network is what Wireless-HART (Highway Addressable Remote Transducer Protocol), a variation of IEEE 802.15.4, effectively does. IEEE 802.15.4 is made to meet the stringent timing requirements, important security requirements, and high levels of interference from barriers of industrial wireless applications. While the IEEE 802.15.4 PHY and the Wireless-HART protocol have the same requirements, the latter creates its own MAC layer using the TDMA method. There is never a guarantee of value delay when Bluetooth is used for wireless communication between endpoints (Paolo 2005). Bluetooth's quasi-static star topology and lack of a channel hopping method hinder its scalability. They are not appropriate for usage in industrial settings because of these features. IEEE 802.11b/g channels are numbered 11 through 26 and have a 5 MHz spacing between them, enabling transmission speeds of up to 250 Kbps. At the media access control (MAC) layer, Wireless-HART employs proprietary Time Division Multiple Access (TDMA), which includes 10ms synchronised timeslot capabilities. These characteristics enable the routing of messages through interference and barriers related to network topology. The application of self-organizing and self-healing network techniques made feasible by the network layer allows for this (Qiangfeng 2004). Despite being primarily a centralised wireless network, Wireless-HART has a network manager within its stack to facilitate communication and routing. This can fulfil wireless industrial applications and ensure network performance. Wireless-HART is primarily concerned with single-hop communication, and the network layer is in charge of allocating neighbouring network devices. In contrast to IEEE 802.15.4, Wireless-HART combines frequency hopping at the MAC layer with time-synchronized TDMA to enable many devices to send data simultaneously across various channels. They all facilitate the establishment of minimal dependability, data routing and signalling capabilities, maintenance issues, and network development features.

1.2.1.7. IEEE 802.11 a/b/g/n/ah

IEEE 802.11 is without a doubt one of the most talked about and utilised protocols in terms of its features and applications. The necessity for high data transfer rates drove its design. Its technology has advanced to satisfy ever-more-specific criteria since it was standardised by the IEEE as a WLAN protocol. The IEEE 802.11 set of standards was started by this development, and they later merged and adopted the moniker Wireless Fidelity (Wi-Fi). The Wi-Fi Alliance is this organisation that certifies Wi-Fi products. This gadget, called WLAN System Toolbox, ensures that Wi-Fi products are compliant with standards by ensuring that market items are compatible with PHY layer settings. Furthermore, it facilitates the utilisation of diverse regional implementations, thereby aiding in the protocol's evolution.

According to the standard, communication devices are called stations (STAs) and are capable of acting on their own. Direct communication between two devices creates an ad hoc topology. An Access location (AP) is a STA that is defined as a traffic concentration location of other STAs. A STA-AP can connect to many STAs to create a Basic Service Set (BSS) by using its defined coverage area, known as a Basic Service Area (BSA). Typically, a wired connection is used to link the STA-AP to a WAN or the Internet. is superior to IEEE 802.11a because of less interference in the spectrum.

A distributed system (DS) that links various STA-APs within the same LAN can also be established by constructing an Extended Service Set (ESS) transport backbone architecture. With 52 orthogonal frequency division multiplexing (OFDM) rates of less than 54 Mbps, 802.11a operates in the 5 GHz spectrum. There are fragments in its transmission speed: 48, 36, 24, 18, 12, 9, or 6 Mbps. Due to reduced spectrum interference, the 5 GHz band has an advantage over the IEEE 802.11a standard because the 2.4 GHz band is more widely used and has greater spectrum pollution. When it comes to handling obstacles, IEEE 802.11a devices do less well in terms of transmission than IEEE 802.11b devices. A network that can accommodate a big number of nodes running low power policies is also necessary. The IEEE 802.11ah Task Group worked to satisfy these demands and modified the IEEE 802.11 protocol's PHY and MAC levels to provide for this situation. IEEE 802.11ah can be utilised as a backhaul infrastructure to link sensors to data collectors in addition to being a WSN. Its broad coverage and quick data transfer make this possible.

1.2.2. Long Range Protocol

1.2.2.1. Long Term Evolution

The Next Generation Wireless, or 4G, stands for Long Term Evolution, and it is the latest national broadband standard for public safety. This standard will give public safety personnel in the field improved capabilities and access to digital technologies. To completely integrate public safety into the digital era, LTE is required. Every day, new tech products and applications are brought onto the market, competing with those that were limited to office servers and desktop computers just a few years ago. This network will facilitate the ongoing creation of public safety-specific applications and contribute to the improvement of first responder operations' effectiveness and efficiency.

Although the LTE standard offers fast speeds and is very promising for first responders, there are restrictions on the application of associated technologies in public safety. It won't be as easy as flicking a switch to switch to LTE. Over the next few years, a comprehensive and intricate construction and implementation procedure will take place. In order to do this, present public safety broadband users will need to coordinate and adapt to a great extent. Currently, these users operate in fragmented spectrum bands across numerous commercially and publically backed networks. In contrast to the existing wireless environment, which lacks real country-wide interoperability, a statewide network based on the LTE standard will allow for interoperability both between jurisdictions and between public safety devices. This network will facilitate the ongoing creation of public safety-specific applications and contribute to the improvement of first responder operations' effectiveness and efficiency. LTE technology is highly versatile, allowing it to be implemented across a broad range of frequency bands, making it as simple to run and deploy as feasible. Less than 5 MHz to 20 MHz scalable bandwidths are provided by LTE, which also supports TDD unpaired and FDD paired spectrum (Rabiner et al., 1999).

1.2.2.2. Narrow-Band IoT (NB-IoT)

Another LTE standard derivative that is specified in 3GPP is called Narrow-Band IoT (NB-IoT). It is intended for even more restricted IoT applications. It operates on a 180 kHz bandwidth and is based on narrowband communication. This leads to a significant decrease in the data transfer rate. NB-IoT has a longer range and uses less power than enhanced machine-type communication.

Three modes are available for NB-IoT use:

in-band
guard-band
Standalone and LTE.

The stand-alone mode uses the designated spectrum (such as GSM bands), the guard band uses the portion of the LTE band that is not in use, and the in-band mode uses the LTE band NB-IoT necessitates a hardware upgrade of the current LTE infrastructure, making its deployment more difficult than that of the enhanced Machine Type Communication.

1.2.2.3. Long Range

The proprietary technology of Semtech is called Long Range (LoRa). Depending on the user's location, it can utilise many ISM sub-GHz spectrum bands using Chirp Spread Spectrum (CSS) modulation. LoRa communications are immune to Doppler shift and can withstand detection and jamming to a reasonable extent. LoRa has various configurable settings, such the spreading factor, to fine-tune the trade-off between data throughput and range (ranging from 0.3 to 50 kbps). Although LoRa is a physical layer technology, the LoRa Alliance supports LoRaWAN, an open protocol for the network and MAC levels. Three device classes are described by LoRaWAN.

Class-A have severe power constraints
Class-B have moderate power constraints
Class-C are always-on gadgets.

While feasible, FotA updates over LoRa may be challenging to implement. They might only need to use the more power-hungry Class C multicast devices. According to a number of studies, LoRa has a good communication range when line-of-sight is accessible, but its range is significantly reduced when it is not, like in an urban setting.

1.3. Survey of Routing Protocols

Routing in WSNs is an extremely difficult task due to certain complexity involved in deployment, node setup, and node addressing schemes. For a huge quantity of sensor nodes building a global addressing scheme is not feasible, hence traditional IP based addressing scheme cannot be applied in a wireless sensor network environment. The dynamic nature as well as, unattended deployment area makes the node configuration difficult. Sensor nodes are tightly resource-constrained and need careful resource management techniques. It's very difficult to find and maintain the route in sensor network due to energy restrictions and sudden changes in the network topology. In the sensor networks mainly data-centric routing is preferred over other techniques. If the node deployment area is small then single-hop communication is preferred, in contrast to multi-hop communication where deployment area is large. Multi-hop communication not only delivers the sensed data but also give them information about the routes from source to destination. The procedure of finding the appropriate routes from source to destination is known as routing and traditionally network layer is involved in this process but in the sensor networks, the data link layers have a major role for the routing process.

1.3.1. Routing Constraints in WSNs

The development of routing techniques in a wireless sensor network is influenced by many factors while designing efficient communication and routing strategies. The routing challenges and design constraints must be solved before the deployment of a sensor network in the fields. In the following section, there is a summarization of some of the challenges involved in routing techniques are.

The universal identification for each sensor node is impossible to implement where the quantity of nodes in the deployment area is large. So, classical IP based addressing scheme is not feasible in such an environment.

The data flow-based routing techniques are generally applied in the sensor network which is dissimilar from conventional network routing techniques.

There is large data redundancy in the sensor network due to the reason that several sensor nodes can produce the same type of data. The elimination of data redundancy is an important task in the sensor network routing strategies.

There are severe constraints on bandwidth, node data storage capacity as well as on transmission energy.

1.3.2. Design challenges in Routing FOR WSNs

There are many design constraints in wireless sensor networks e.g. energy, bandwidth, storage and processing power limitations. While designing the routing procedures, following issues must be resolved.

1.3.2.1. Energy Efficiency

The energy storage is a major issue in the wireless sensor network. The nodes in the sensor networks are mostly battery-powered and consume more power during their operating time. The performance of network operation, as well as efficiency of data transmission, degrades if the battery power of each node falls below their threshold level, so there is a need for routing protocols that must be energy efficient.

1.3.2.2. Complexity

The routing protocols are more complex in wireless sensor networks which affect its performance. The main reason behind this is that there is massive hardware complexity in the sensor nodes due to its small size and limited battery life.

1.3.2.3. Scalability

The scalability is another issue that must be taken into concern while designing the routing protocols. By deploying the additional nodes or expanding the deployment area the routing protocol must work efficiently with any expansion modifications in the network.

1.3.2.4. Delay

For real-time applications, there must be no delay in the data transmission in sensor networks. So there is a requirement of routing protocol that must offer minimum delay in data transmission.

1.3.2.5. Robustness

The wireless sensor nodes are generally deployed in hostile environments, thus routing protocol must be designed that is proficient to accept all sorts of the environment including hostile and loss environment.

1.3.2.6. Fault Tolerance

During network operation, few sensor nodes may stop working or may encounter physical damage. If some nodes fail, then there should be options to change the routing path or to recover from nodes failures, so that the normal operation of the sensor network will not be affected (Rao et al. 2003)

1.3.2.7. Data Transmission

The data transmission in the wireless sensor networks varies depending on the type of application. In general, the data transmission may be flow-based, event-driven, and query-driven or it may be a continuous flow. To design an effective routing protocol the constraints related to the transmission medium and nature of data flow are taken into consideration. The problem linked with a wireless medium like fading, attenuation, error rate, data loss must be resolved effectively.

1.3.3. Routing Protocol in WSNs

The routing protocols in wireless sensor networks can be divided in different categories based on nodes functionality, network topology, nodes behavior, location-aware as well as the services the network provides. The routing protocols can also be classified into reactive as well as pro-active categories. In reactive routing, the routing path is decided based on the events or when any query is initiated, whereas, in the pro-active routing, the path is decided in advance and remains the same for all time. Another classification in routing protocols is done based on the routing initiated by either source or destination node known as source or destination-initiated routing. In general, the routing algorithms induced in wireless sensor networks fall into three categories known as centralized, distributed and location-based algorithms. The centralized algorithms are executed at the base station which possesses the knowledge about the whole sensor network. The centralized algorithms are difficult to implement due to its complexity.

In the distributed routing algorithms, all nodes with the sensor network communicate with each through message passing to decide the best path from source to destination. The localized routing algorithms are executed in a few nodes, locations or in the devices that belong to a particular location. The type of routing algorithm is a major factor during the deployment of sensor nodes in the field. The localized algorithms are designed to optimize the communication between adjacent nodes whereas the distributed algorithms optimizc the communication between all the sensor nodes (Sadagopan et al. 2003). The centralized algorithms play a major role when the command and control of the sensor network are with a centralized authority known as a base station. The algorithms characteristics are an essential parameter to be taken into account while deciding the routing strategies. In case of localized algorithms, the routing protocol must reinforce and optimize the transmission between neighbors. For centralized algorithms, combining the messages that concurrently go to central node (even when their source of origin may be different) could be an advantage. The distributed algorithms should efficiently support the transmission between any two pairs of nodes. In conclusion, localization-based algorithms depend on a solution that delivers geographic coordinates like GPS making the solution more costly.

1.3.4. Classification of Routing Protocols in Wireless Sensor Networks

The routing techniques in wireless sensor network can be classified in to following categories

1.3.4.1. Attribute-Based or Data-Centric Routing Protocols

The following protocol falls in these categories

1.3.4.1.1. Sensor Protocols for Information via Negotiation (SPIN)

The sensor node that uses the SPIN follows the data naming convention by using high-level data descriptors, known as Metadata. The sensor node uses meta-data negotiations to eliminate the broadcasting of redundant data in the entire network. These nodes make transmission choices based on application-specific information or based on the resources available to them at a particular time. This capability allows sensor nodes to efficiently disseminate information among themselves with minimum energy requirements. In comparison to other protocols, SPIN can deliver 60% higher data with a given amount of energy in point-to-point networks. Another advantage' of this protocol is that it is a single-hop protocol i.e. node communicates with its neighbours only.

1.3.4.1.2. Directed Diffusion

In the directed diffusion the data in sensor networks is collected based on interest message. The process of directed diffusion involves several parameters like gradient, interests, data message and reinforcement. The sink node broadcasts interest message which is a query to access the specific data from the network. Each interest message consists of the narrative of a sensing task to be performed by sensor nodes for acquiring data. The gradient is data that matches the event and is specifically designed to draw the event towards the sink node.

1.3.4.1.3. Rumour Routing

Rumour routing creates paths to the nodes where events occur and later to route the queries along these paths. All the nodes inject packets with higher TTL in the network known as mobile agents. These agents propagate a long distance in the network so that the distant node has awareness about which node initiated certain events. The agents keep track of each visited node to avoid loop. If the query matches the demanded event, then data is transmitted back through the learned path from the event node to the query node. Rumour routing is more energy-efficient than flooding based system especially when geographic information is not available represents in fig 2 (Takahiro et al. 2010).

Figure 2. Rumour Routing

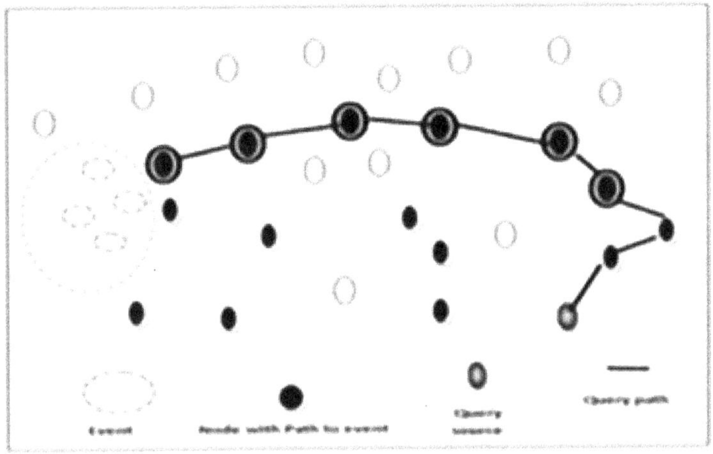

1.3.4.1.4. COUGAR

It is another data-centric routing protocol. It uses the cluster head node to perform data aggregation and these cluster head nodes transmit the data back to the base station. The base stations generate the queries and plan for data flow, network computation and then transmit them to respective nodes. These query plans also describe how to select the cluster head. This protocol provides in-network computation capabilities for energy efficiency in a situation when data generation is huge. This protocol has some limitations like need more storage space, consumes more computation power as well as synchronization problems (Wan et al. 2008).

1.3.4.1.5. ACQUIRE (Active Query Forwarding in Sensor Networks)

In this technique, a node fires an active query packet in the sensor network. Here network is considered as a distributed database where composite queries further sub-divided into sub queries. As the query propagates through the network each node when encountering the query, it tries to resolve it and if the resolution is not found then it further subdivides the query into the smaller and smaller sub-query and forwards to different nodes until it is completely resolved. After resolution, the result is returned back to the base station or to the querying node as a complete result.

1.3.4.2. Geographical Routing Protocols

This algorithm uses the location information in order to enhance the efficiency of routing. In this technique, each node can easily identify its neighbours. The location information can be derived using GPS or can be modelled using a virtual coordinate system. The popular routing techniques in this category are described below (Yao et al. 2002).

1.3.4.2.1. Geographic Adaptive Fidelity

This scheme optimizes the performance of routing in a wireless sensor network by selecting equivalent nodes. The nodes are said to be equivalent if there is the same set of neighbours and belongs to the same routing path. In order to identify the equivalent nodes, the position of the nodes is identified using a virtual grid and periodically updates their position information by the exchange of information between them. In order to save energy, the nodes set their sleep and wake up time and select the nodes which will take part in routing at a particular interval of time.

1.3.4.2.2. Geographic and Energy Aware Routing

This technique used energy awareness as well as location, informed nodes selection heuristic techniques to route the packets towards destination region. The main idea of this technique is to consider the only certain region of interest rather than forwarding the packets toward all sensor nodes, by using this approach energy can be conserved. This protocol keeps the estimated cost as well as learning cost for reaching a packet from source to destination. The calculated cost is transmitted back each time a packet reaches the destination so that the routing path for the next packet can be adjusted.

1.3.4.3. ROUTING Protocols based on Protocol Operation

Multipath Routing Protocols

There are certain techniques that uses multiple routes between source and destination for enhancing the network performance. Using this approach resilience of the network is enhanced as if one path fails then data can be routed on an alternate path. But the energy consumption of the network increases due to the maintenance overhead of multiple paths. There are procedures for the selection of primary and secondary paths. If the energy of the primary path falls below a certain threshold then secondary paths can be chosen for routing. Multipath routing is used to improve the reliability of wireless sensor networks.

SAR (Sequential Assignment Routing)

This protocol is based on priority level of each packet. It is the first protocol that considers the quality of service in routing criteria. Each route is related to a metric that classifies the quality of service, this metric dependent on delay and energy cost for each path. This technique periodically recalculates packet priority, energy level, and QoS metric.

Maximum Lifetime Routing

The maximum lifetime routing uses the technique in which energy consumption is optimized with the help of multiple routes. The algorithm uses the energy consumption rate during data transmission and the residual energy level at the source and receiver. The active path is examined to control the remaining energy at the same time alternate routes in between source and destination are discovered. This technique achieves an optimum network lifetime in routing.

Hierarchical Routing Protocols

The hierarchical routing simplifies the process of managing the network as well as it minimizes the energy consumption by classifying nodes at different levels into a group of nodes known as a cluster. The nodes with higher residual energy are selected as a cluster head. The classification of techniques in hierarchical routing depends on the method of selecting cluster head and behaviour of nodes in inter and intra-cluster domains.

LEACH (Low Energy Adaptive Clustering Hierarchy)

LEACH is a self-adaptive as well as a self-organizing clustering method where energy load is distributed equally in all nodes within a cluster. During setup phase cluster head is declared which depends on residual energy. The role of cluster head nodes is periodically changed and is transferred to other nodes in to allocate the energy. The cluster head is accountable for the data aggregation and acts as a router to the sink node. During the setup phase, all nodes send the signal to the base station about their position and energy level, based on the information received the base station decides the structure of cluster and corresponding cluster heads.

PEGASIS (Power-Efficient Gathering in Sensor Information Systems)

It is a hierarchical routing algorithm that uses a greedy algorithm in a chain-based approach, all sensor nodes structured in such a way that they form a chain and if any nodes die then the chain is reorganized to evade the dead node. It performs data fusion at every node except the last node in the chain. Every node infuses data from neighbours to its own data to produce a packet of the same length and then forward it to the next node. It improves energy saving in comparison to LEACH, by removing the overhead of dynamic clustering; minimize the distance between nodes, restraining the number of transmissions by using only one transmission per round to the base station.

TEEN (Threshold Sensitive Energy Efficient Sensor Network Protocol)

TEEN is a hierarchical protocol for the reactive networks, there is an immediate response based on the change in threshold parameters. It uses two types of threshold known as Hard and soft threshold. In case of hard threshold, the sensed attribute immediately transmitted if it reaches its upper limit by switching on its transmitter, whereas in soft threshold sensed value is transmitted in the interval. The hard threshold transmits only when sensed data value is in the range under consideration by this procedure it cuts the number of transmissions. The soft threshold further minimizes the number of transmission by not transmitting the data if there is no change in sensed data. The main limitation of this technique is that the sensor node never transmits data if there is no modification in sensed value and the base station doesn't know about the health of sensor nodes.

1.4. Efficiency enhances by AI

Through the development of more autonomous, adaptive, and predictive communication systems, artificial intelligence (AI) improves protocol efficiency. AI improves network speed by dynamically modifying configurations based on real-time data analysis, doing away with strict, static rules.

1.4.1. Adaptive and dynamic routing

The rigid, predefined logic that underpins traditional routing protocols like OSPF and AODV makes it difficult for them to adjust to the changing needs of contemporary networks. AI offers intelligent routing, which increases efficiency by using machine learning to make decisions based on shifting network conditions.

1.4.2. Predictive resource allocation

AI's predictive capabilities enable networks to move from a reactive to a proactive management strategy. By analyzing historical and real-time data, AI can forecast future network demands and allocate resources accordingly.

1.4.3. Automated network management

By automating complicated and repetitive network management operations, artificial intelligence (AI) lowers the possibility of human error and empowers systems to self-optimize and self-heal.

1.4.4. Enhanced security and privacy protocols

AI makes it possible for more efficient threat identification and mitigation, which greatly improves security protocols.

1.4.5. Optimized wireless communications

In wireless systems, AI helps overcome traditional limitations related to energy consumption and resource constraints by creating more adaptive protocols. Role of AI in the health care represented in figure 3.

Figure 3. Role of AI in healthcare

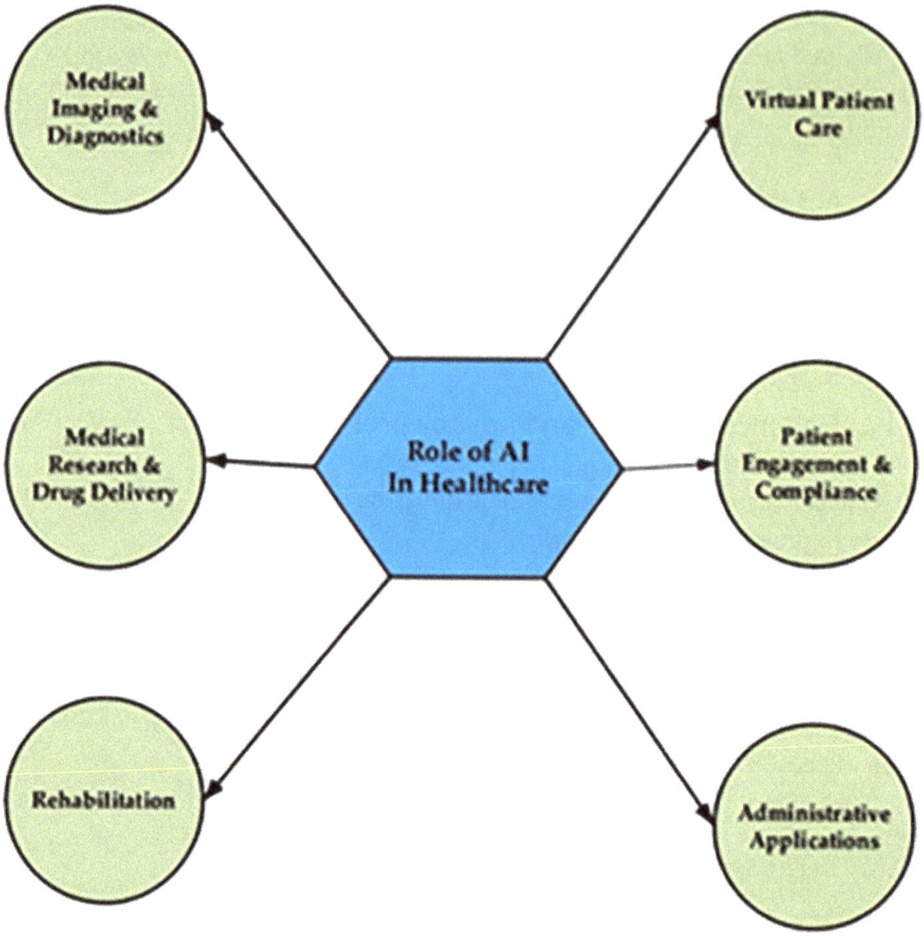

1.5. Data aggregation & Dissemination

1.5.1. Data Aggregation

The technique of using one or more sensor nodes to find information from other sensor nodes is known as data aggregation. In a wireless sensor network, data aggregation aims to reduce data redundancy and lengthen energy lifespan decrease in the quantity of data packets delivered over the network since aggregation uses less energy than sending several packets containing the same information.

1.5.1.1. Data Aggregation Approaches

Data aggregation comes in four varieties.

Cluster-based method

A hierarchical strategy known as the "cluster-based approach" divides the network into several clusters. Every cluster has a cluster head and selects a cluster from among its members. Data received from cluster members is aggregated locally by the primary cluster head role, which subsequently sends the result to the base station. Information can be immediately shared between a cluster head and a sink via remote transmission or multi-hop communication utilising additional cluster heads.

Tree based approach

An aggregate tree is made using an aggregation idea that is defined by the Tree Based Approach. This tree can be defined as a minimal skeleton where the source node functions as the leaves and the sink node as the root. From the output nodes up to the root nodes, data starts to flow. The primary drawback of this strategy is the potential for data packet losses at any tree node, which might bring down the network as a whole.

Multipath method

The drawbacks of the tree-based method are addressed by this strategy. By using many neighbours in the aggregation tree, each node might deliver data packets across multiple paths according to this method. Thus, with the aid of an intermediate Node, a data packet is transmitted along several channels from the source to the destination. The drawback of this strategy is overhead, and a ring topology is an example of it (Zunguru et al. 2012).

Hybrid Strategy

A hybrid approach combines elements of the multipath, tree, and cluster approaches. This method is primarily employed to adaptively get the best possible data aggregation performance.

Information Sharing

Sensor nodes gather data and communicate with the base station or any other interested node during the data dissemination process. The data is generated by the source node, and the event is the information that has to be reported. A sink is a node that seeks the information and is interested in the event. Thus, the sensor network is used to route data throughout the entire process. There are two steps to this process: interested nodes broadcast to their neighbouring nodes in the network in the first phase, and nodes submit the request data in the second step after receiving the request.

Data Dissemination Approach

There are various ways to disseminate data, including the following:

Inundation

If the data packet is not received by the destination node or if the required number of hops is not completed. The gathered data is then disseminated by each node until the packet reaches its intended node. Flooding's primary benefit is that it doesn't require expensive topology maintenance or route discovery; nonetheless, it has a number of drawbacks, including resource blindness, overlap, and implosion.

Rumours

One use of the flooding strategy is gossip. Instead of broadcasting each packet to every neighbour, this method sends a packet to a single neighbour that is randomly chosen from the neighbour tables. The completion of this process may take a lengthy time. Gossip steers clear of the issues face at the time of implementation. Body area network is represented in figure 4.

Figure 4. Body area network

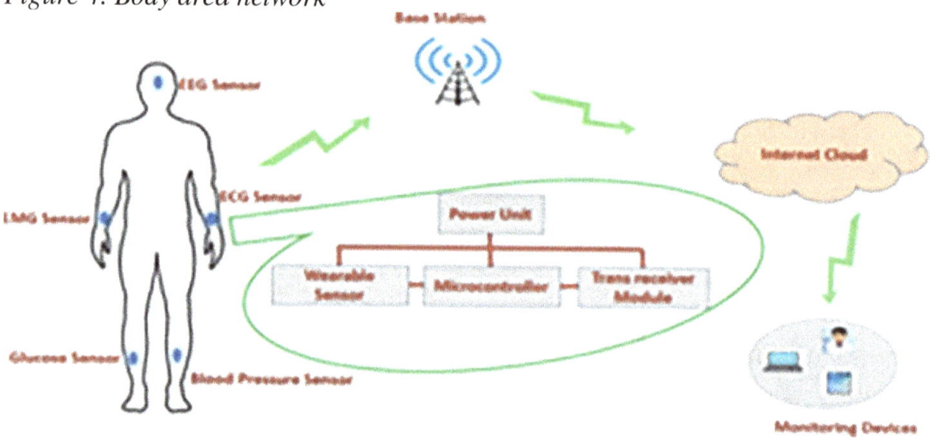

1.6. Case Study: Patient monitoring following surgery using a BLE Body Area Network

1.6.1. Objectives & scope

Early detection of post-operative problems (such as sepsis, DVT, respiratory distress, and surgical site infection) allows clinicians to take action and save costs, readmissions, and morbidity.

1.6.2. Hardware and device layer

Sensors:

Continuous or episodic single-lead ECG patch: HR, HRV, and arrhythmia detection.
PPG wrist/patch: perfusion index, pulse rate, and SpO_2.
Continuous trend in the skin/axillary temperature patch.
chest acceleration (respiratory rate) or a respiratory effort sensor.
Accelerometer/IMU: falls, gait, mobility (DVT risk), and activity.
Cuffless blood pressure measurement (pulse transit time) and wound cameras (patient-initiated photos) are optional.

Device Choices:

Sensor-to-gateway link using BLE Low Energy (saves power).
In order to recover missed uplinks, sensors retain a brief circular buffer.
Battery life goal: depending on design, 3–7 days for patches or 24–72 hours for continuous high-fidelity ECG.

1.6.3. Communication and data format

For metrics, use standardised GATT characteristics; if bespoke, use a version that includes semantic metadata and UUIDs. Sensor_id, timestamp (ISO 8601 + timezone), sequence number, firmware version, battery, and sample type are among the data packets.

1.7. Conclusion

Protocols are very useful for real time monitoring for exchanging the health information in sensors, monitoring devices and healthcare providers. It provides the interoperability between different IoT devices. The different protocols are used for like HL7, MQTT, DICOM and Zigbee. The inventions and early diagnosis are providing the better result for the unauthorized access. Protocols are essential for improving the patient outcomes, data integrity and maintaining the security of the patients.

REFERENCES

Ahady, A. El-KadeR, A., Sherine, & Hussein, E. (2012). *A Comparative Analysis of Hierarchical Routing Protocols in Wireless Sensor Networks*. Cairo: IGI Global.

Al-Karaki, J. N., & Kamal, A. E. (2004). Routing techniques in wireless sensor networks: A survey. *IEEE Wireless Communications*, *11*(6), 6–28. DOI: 10.1109/MWC.2004.1368893

. Amitabha, G., & Sajal, K. D. (2008). Coverage and connectivity issues in wireless sensor networks: A survey. *Elsevier Journal of pervasive and mobile computing*, 303-334.

Chalermek, I., & Ramesh, G. (2000). Directed diffusion: a scalable and robust communication paradigm for sensor networks. *Sixth Annual International Conference on Mobile Computing and Networking* (pp. 56–67). Boston: ACM.

Chang, J., & Tassiulas, L. (2000). Energy Conserving Routing in Wireless Ad Hoc Networks. *In Proceedings of the 19th Conference of the IEEE Communications Society*, (pp. 22–31). Tel-Aviv Israel.

De, S., Qiao, C., & Wu, H. (2003). Meshed Multipath Routing with Selective Forwarding: An Efficient Strategy in Wireless Sensor Networks. *Computer Networks*, *43*(4), 481–497. DOI: 10.1016/S1389-1286(03)00355-4

Lindsey, S., & Raghavendra, C. (2002). PEGASIS: Power-Efficient Gathering in Sensor Information Systems. *In Proceedings of the Aerospace Conference* (pp. 1125–1130). Big Sky, MT: IEEE. DOI: 10.1109/AERO.2002.1035242

Luis, J. G., & Ana, L. (2009). Routing Protocols in Wireless Sensor Networks. *Sensors (Basel)*, •••, 8400–8420. PMID: 22291515

Manjeshwar, E., & Agrawal, D. (2001). A Routing Protocol for Enhanced Efficiency in Wireless Sensor Networks. *In Proceedings of the 15th International Parallel and Distributed Processing Symposium (IPDPS)* (pp. 2009–2015). San Francisco, CA, USA: ACM. DOI: 10.1109/IPDPS.2001.925197

Mohammad, H. A., Abdul, H. A., Shukor, A. R., & Md, A. (2012). An Overview of Data Routing Approaches for Wireless. *Sensors (Basel)*, •••, 3965–3995. PMID: 23443040

Noman, S., & Syed, R. H. (2017). *Wireless Sensor Networks-Insights and Innovations*. IntechOpen.

Paolo, S. (2005). Topology control in wireless ad hoc and sensor networks. *ACM Computing Surveys*, •••, 164–194.

Qiangfeng, J., & D., M. (2004). Routing protocols for sensor networks. *First IEEE Consumer Communications and Networking Conference* (pp. 93-98). Las Vegas: IEEE. DOI: 10.1109/CCNC.2004.1286839

Rabiner, W., Kulik, J., & Balakrishnan, H. (1999). Adaptive Protocols for Information Dissemination in Wireless Sensor Networks. *In Proceedings of the Fifth Annual International Conference on Mobile Computing and Networking (MOBICOM)* (pp. 174–185). Seattle, WA, USA: ACM.

Rao, A., Ratnasamy, S., Papadimitriou, C., Shenker, S., & Stoica, I. (2003). Geographic Routing without Location Information. *In Proceedings of the Ninth Annual International Conference on Mobile Computing and Networking (MOBICOM)* (pp. 96–108). San Diego, CA, USA: ACM. DOI: 10.1145/938985.938996

Sadagopan, N., Krishnamachari, B., & Helmy, A. (2003). The ACQUIRE Mechanism for Efficient Querying in Sensor Networks. *In Proceedings of the First IEEE International Workshop on Sensor Network Protocols and Applications (SNPA)* (pp. 149–155). Anchorage: IEEE. DOI: 10.1109/SNPA.2003.1203365

Takahiro, H., Vladimir, I., & Erik, B. (2010). *Wireless Sensor Network Technologies for the Information Explosion Era*. Springer Berlin Heidelberg.

Wan, J., Yuan, D., & Xu, X. (2008). A Review of Routing Protocols in Wireless Sensor Networks. *4th International Conference on Wireless Communications, Networking and Mobile Computing* (pp. 1-4). Dalian: IEEE. DOI: 10.1109/WiCom.2008.946

. Yao, Y., & Gehrke, J. (2002). The Cougar Approach to In-Network Query Processing in Sensor Networks. *SIGMOD*, 9–18.

Zungeru, A. M., & Kah, P. S. (2012). Review: Classical and swarm intelligence based routing protocols for wireless sensor networks: A survey and comparison. *Journal of Network and Computer Applications*, *35*(5), 1508–1536. DOI: 10.1016/j.jnca.2012.03.004

Chapter 15

A Deep Learning-Based Lip Reading Module Applied for Forensic Analysis and Silent Speech Decoding

Vijayalakshmi G. V. Mahesh
https://orcid.org/0000-0002-1917-7506
BMS Institute of Technology and Management, India

H. P. Akshay Gowda
BMS Institute of Technology and Management, India

S. Badari Narayana
BMS Institute of Technology and Management, India

S. Dhedheepya Gowda
BMS Institute of Technology and Management, India

Suhas Gowda M. S.
BMS Institute of Technology and Management, India

ABSTRACT

Machine learning and Deep learning algorithms in digital forensics play a significant role in crime detection aiding forensic analysis through identifying patterns, trends based on data. Audio and Video data serve as a great electronic evidence in digital forensics for investigation purpose. In forensic investigations, crucial video

DOI: 10.4018/979-8-3373-3146-1.ch015

footage often lacks audio or suffers from poor sound quality, making it difficult for analysts to fully grasp the conversations that took place and traditional speech analysis methods depend on audio data, leaving a significant void when only silent footage is available. Several existing systems struggle with contextual understanding, that can lead to misinterpretations, and also raise privacy and ethical issues when applied in surveillance and forensic contexts. Thus chapter proposes to provide a solution employing advanced computer vision and deep learning methods called Lip reading module for forensic analysis and silent speech decoding that seeks to fill this void by offering a way to extract speech information from lip movements, potentially revealing insights that would otherwise remain hidden. The modular design of the project makes it a potential forensic analysis assistant tool for silent speech decoding.

1. INTRODUCTION

In the age of digitization, the society relies more on digital systems where most of the devices such as IoT devices, computer systems, mobile phones, tablets etc., are all cloud connected (Vermesan & Friess (Eds.), 2022) and three is rapid increase in the enormous amount of information flow (digital data) that is available to everyone. The digital era is witnessing a growth with the rapid increase in IoT devices such as smart cameras and edge computing systems and is also paving way to the expansion of digital forensics domain through the creation of crime scenes. These pose threats towards accessibility and ethical considerations and risks of data breaches, privacy and cyberattacks which needs to be taken care of.

Digital forensics involves acquisition, examination, analysis and interpretation of digital data using advanced tools and technologies and also preserving it in its original form as required to be submitted to the court of law.

Digital forensics could be any of these types (Jones & Winster, 2022 ; Irons & Lallie, 2014) as displayed in Figure 1.

Figure 1. Digital Forensics types

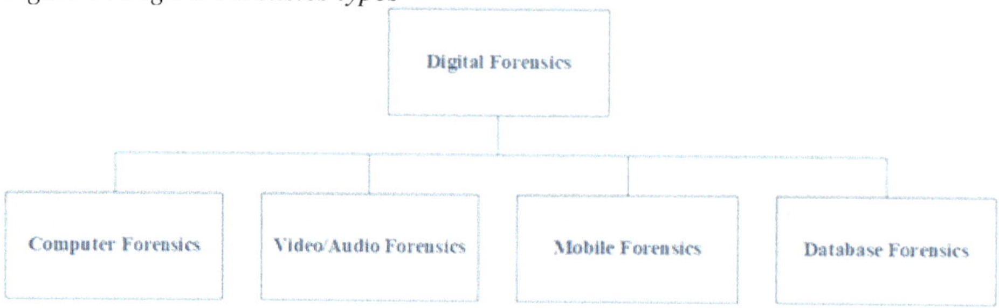

(i)Computer Forensics: Computer Forensics, widely known as the cyber forensics involves acquisition of the data from the computer systems and digital devices. Computer Forensics focuses on utilizing techniques such as reverse steganography, disk imaging, file carving and cross drive analysis to discover the evidence, to prevent the originality of the data from unnecessary tampering or modification in the data and identifying patterns as applicable to the crime investigation. It finds use prominently in identifying fraudulent cases, theft of intellectual property or employee misconduct.

(ii)Video/Audio Forensics: This refers to the analysis of digital audio and video data that serves as digital evidence to support the investigation for legal purposes. Video/Audio Forensics involves exploration and enhancement of audio and video data acquired from various available sources such as surveillance cameras, voice mail recordings, telephonic conversations etc., to derive required information. The key aspect here is to establish the integrity and authenticity of the evidence. Audio and Video forensics relies on digital signal processing utilizing the techniques of audio and video authentication, enhancement to improve the intelligibility, voice and video analysis and recovery/restoration of the data from the corrupted sources. They are used in speaker identification, tampering and fake news detection, accident reconstruction and deep fake

(iii)Mobile Forensics: It involves the process of deriving and analyzing the information stored in the mobile devices such as tablets, smart phones or any hand held devices. These devices store large amount of text data, phone call records, chat conversations, media files, app data, online search history and current location data and history. Experts extract specific information from the device that includes its storage, SIM card information or cloud storage. It finds applications in criminal investigations (identify theft, fraud, violence), corporate security and legal proceedings.

(iv)Database Forensics: The database forensics acquires and examines the electronic data stored in the database to identify clues to detect crimes. The prime factor here is the review and validation of timestamp associated with the database. It finds evidence in the following sources such as stored tables & records, transaction logs, audit logs, database server logs, backup files, metadata, network traffic etc.,. Database forensics faces challenges when databases are too large and time consuming, volatility in data and complexity in the systems. It is utilized to identify data breaches, unauthorized access, data modification or deletion, suspicious activities, denial of service attacks, unauthorized transfer of information and many more.

Though the process of digital forensics varies with distinct situations or circumstances, but the core remains same and it includes the following phases (Flaglien, 2017).) as illustrated in Figure 2

Figure 2. Digital Forensics process

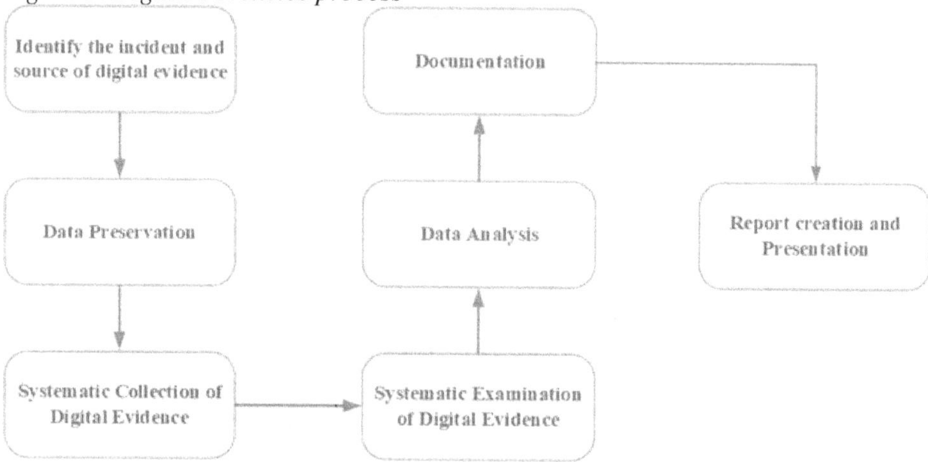

The process at the beginning involves identification of the incident and source of digital evidence. This phase also includes determination of the relevant location information and the scope of investigation.

Once the source of evidence is identified, it is important to protect the data by shielding it from any damage, alteration or deletion for its reliability. Further the data needs to systematically analyzed to find the appropriate information. This may include recovering the files deleted, metadata and keywords assessment, data deciphering etc.,, Data analysis helps in reconstructing the events, recognize the hidden patterns and draw conclusions appropriate to the investigation process.

The entire process of digital evidence collection, its analysis and the conclusions drawn are to be recorded to become acceptable in the court of law. A report needs to be submitted to end the digital forensic process. It should include the abstract of the case, digital evidence collected, methods employed for analysis and conclusions in simple. of the investigation process carried.

The digital forensics is a domain that combines the components of forensic science, computer science and law to identify, preserve, analyze the digital evidence and to extract the required information to support digital investigation. Developing and implementing effective and efficient frameworks is required to apply for the entire Digital Forensics process Advancement in the signal processing algorithms, artificial intelligence(AI) with machine and deep learning algorithms are transforming digital forensics by making it easier in processing large volumes of digital data and effectively trace the evidence.

The existing surveillance systems with smart cameras generate large amount of video data where most of the data is left passively recorded that lacks real-time intelligence necessary for modern forensic investigation. Further, certain scenarios that involve underground operations or noisy environments 'verbal communication' or audio is lost which might be the key evidence. For the investigations based on digital video evidence, sometimes it is important to consider the speech or audio in the footage. but the poor quality of audio or lack of audio (silent speech) or inaccessibility to the audio makes it difficult to analyze or understand the events that took place.

Also, manual review or the use of traditional methods for forensic analysis will not be able to handle large volume and the speed at which the data arrives posing problems towards efficient addressing of the need which necessitates the shift towards automatic systems.

Thus this chapter presents a framework that overcomes the above mentioned challenges by integrating visual speech recognition(VSR)/ silent speech decoding(-SiSpD) with advanced deep learning methods under IoT enabled edge AI. With the implementation of VSR with IoT the system will be able to achieve real time silent speech decoding. This framework applies video forensics that analyses video frames using image processing algorithms and deep learning to detect lip movements and identify phonemes or words. The presented work bridges the gap between silent video footage and meaningful speech interpretation. The objectives of the work on application of video forensics include (i) Creation of a system that can detect and monitor lip movements in real-time using computer vision techniques, (ii) Develop a deep learning model trained on visual data (lip movements) to identify phonemes or words without relying on audio input, (iii) Link the visual features to phonetic units or directly to words and sentences, enhancing the system's vocabulary and accuracy over time and finally alert the law enforcement to take necessary action.

Further, the chapter is organized into four sections, where section 2 provides a survey on the available methods, techniques, datasets and challenges. Section 3 describes the methodology adopted in the work in detail. The implementation of the work using 3D convolutional neural networks and bidirectional recurrent neural networks and the corresponding results is presented in section 4. Finally, section 5 concludes the chapter providing future scope and limitations of lip reading.

2. LITERATURE SURVEY

The domain of lip-reading, silent speech decoding or visual speech recognition has seen significant advancements, thanks to improvements in IoT, computer vision, signal processing, machine learning, deep learning and natural language processing. The work presented here builds on this foundation, aiming to create a reliable, real-time lip-reading system for forensic analysis, silent communication, and assistive applications. Here, this section explores key literature covering the major developments and challenges in the field of digital forensics, visual speech recognition/ silent speech decoding.

A qualitative review on the digital forensics, its challenges in terms of evidence collection & preprocessing, report preparation & presentation and ethical aspects are provided by (Casino et al. 2022). In the digitization era, the role of IoT and its contribution towards digital forensics is significant as it provides required infrastructure for extraction, analysis and detection of forensic evidence through IoT framework. Consequently, the risks and threats of IoT cyber-attack is also increasing which needs attention through the utilization of advanced digital forensic investigation techniques to track the cybercrimes. (Yaacoub et al. 2022) has provided a review of forensic and anti-forensic methods, types and techniques that can be applied to IoT domain that would aid the investigators in their process. The authors have also reported about the challenges and limitations of the forensic techniques. (Chango, X., Flor-Unda, O., Gil-Jiménez, P., & Gómez-Moreno, H. (2024)), provided a systematic review on utilizing the advancements in technology by the forensic investigators, its impact and the challenges. The review identified that the inclusion of technology benefits the investigation in terms of remote analysis of the data, digitizing and constructing the crime scene, biometric trait examination etc., that improves the accuracy of the investigation.

(Ghasemi & Mahmoudi, 2024), investigated the usage of IoT in forensic analysis and law enforcement for evidence collection from the crime scene. The results emphasized on the importance of developing forensic strategies, frameworks for working on digital forensics in IoT devices. Further the work also stressed on the importance of collecting the evidence from the IoT devices as they provide précised real time data in its real form with authentication as required by the court of law. (Bakar, 2025), studied about the perseverance and buried forensic evidence in smart IoT frameworks. Further it presents the challenges in recovering unclear, deleted data from the cloud storage and the measures to be taken in encrypting the data and utilizing advanced tools for improved digital forensics in IoT systems.

(Ghanem et al., 2025), conducted work to simulate cyberattacks on PC and Raspberry pi platforms investigating system logs, network traffic and memory storage for their forensic analyzing abilities in weaponized IoT scenario. The results indicated that the PCs are able are more advantageous with respect to memory, support from tools and system log whereas Raspberry pi is facing challenges in terms of their architecture which emphasizes the need for specialized tools for forensic analysis in the IoT framework. (Lutta, P. 2024), presented a framework to overcome the technical challenges of IoT in forensic analysis for smart home applications. The work highlighted on the usage of HI-SDR: hash indexed sparse distributed representation and machine learning algorithms in improving the feature representation and anomaly detection capabilities in noisy environments that contributes towards the progression in digital forensics. (Sawarkar and Hazari, 2024), developed cyber physical system with blockchain technology that used a Hadoop distributed file system with ML algorithm for improved security. The results indicated the cost reduction with parallel processing IoT forensics. (Salem and Hamarsheh, 2024), provided guidelines through their work to investigate network forensics on the system with smart cameras through MAoIDFFT IoT framework.

The strengths of Artificial Intelligence(AI) is increasingly used in digital forensics to automate various process, analyze large volume and complex datasets and aid in criminal investigations. AI with its subsets machine learning and deep learning algorithms is exceptionally effective in identifying the patterns, traces and predictive analytics leading to effective and efficient investigations. A systematic analysis on the significance of AI in digital forensics is provided by (Dunsin et al. 2024). The work explores the use of AI as cutting edge initiative for data acquisition, analysis, pattern recognition and develop strategies to cyberattacks. The research also investigates the contributions, challenges and limitations of AI being effectively utilized in digital forensics and incident responses to detect and manage the attacks.

(Renukadevi et al, 2024), conducted a study to explore the possibility of using AI and its key to develop methods for forensic science analysis. The study focused on analyzing audio and video data to study the crime scene for improved investigation and also presents the challenges of using AI in forensic analysis. It also stresses on AIs contribution (Cărcăle, 2025)(Bhawna and Mahajan, 2024)., towards improving the forensic analysis process. (Rajasekar and Vezhaventhan, 2024), explored the influence of AI in forensic investigation for law enforcement for various applications, challenges, future directions and ethical concerns. (Nayak, 2024), discusses about utilizing automated methods for evidence collection and forensic investigation. The work presents details about the development in forensic analysis ranging from traditional methods to AI driven procedures as per the varying characteristics in digital crime. Further it explores the application of AI framework in forensic investigation for data collection, analysis and evaluation considering various performance metrics. The paper also highlights on the ethical consideration in using AI reliably and effectively to support the investigators in achieving improved performance.

(Akeiber, 2025), presents a work that discusses about the significance of AI in identifying hidden patterns in large volumes of data and anomalies to support digital forensics for crime investigation while it also highlights on the challenges such as ethical concerns related to data privacy, bias in the ML models, data transparency, deficiency in regulatory standards. The study also emphasizes on integrating blockchain and quantum computing for the advancements in digital forensics.

Visual speech recognition or speech decoding, where the speech is decoded and converted into text plays a crucial role in digital forensics especially in video forensics for evidence extraction and speaker identification. This utilizes computer vision, pattern recognition and AI for automating the process and improve the efficacy of detection. Several works have been identified in the literature corresponding to this domain. A methodical survey on the various existing techniques for processing and analyzing speech data for digital forensic applications is made available by (Kruthika & Nagavi, 2023). The work also covers information on various processes such as speech enhancement, speaker recognition and speech synthesis and a discussion about the challenges and future directions. (Battista et al., 2025), presents a detailed review on examining the visual cues to read the lips for speech recognition. The study describes the importance and the development of lip reading methods for its application across various domains. The study discovers its significance in the scenarios with linguistic deficiencies and noise.

(Chowdhury et al., 2025), did a detailed study on the existing ML based silent speech recognition (SSR)techniques, data collection methods, datasets available, integration of multimodal cues to improve SSR. The study also investigated the applying transfer learning for management of cross subject variability and further provided suggestions and future directions of SSR.

(Oghbaie et al, 2025) presented a survey on the application of deep learning methods in visual speech recognition. The study highlighted the key challenges, its solutions, important components of visual speech recognition such as features to be considered, performance metrics etc., The review also explored in identifying the advanced methods that include lightweight and fast deep learning models, few shot learning techniques, combination of pretrained models with fine tuning to be deployed for visual speech recognition.

(Pu and Wang, 2023) provided information on lip reading datasets available to conduct research on analyzing lip motion patterns for speech decoding. Further the review article also discusses about the feature extractions and provides their comparison to identify the suitable method that could be applied as required. (Hao et al., 2020) explored various deep learning architectures for lip reading and lip reading databases available. The article provides the information on risks, challenges and future directions of lip reading.

(Liu et al., 2024), presented event based lip reading module combining spiking neural network (SNN) framework with spatial temporal attention block and triplet loss to identify and discriminate thc phonetically similar words. The work proved its effectiveness in decoding the visually comparable words. (Li et al., 2024), utilized AI for visual speech recognition in metaverse for developing realistic avatars which can identify the words by analyzing lip movements and facial expressions. The work integrated several deep learning architectures that include 3D –CNN, ShuffleNet and densely connected Temporal CNN. The methodology was validated on the LRW and GRID Corpus datasets that provided an accuracy of 99.5% and 98.8% respectively laying the foundation for developing metaverse for future.

(Zhou et al., 2014) provides a survey on automatic speech recognition system based on visual data. This survey focuses on the points related to feature extraction from the visual cues such as speaker dependency, variation in pose and temporal information. The article also provides information about the facial landmark localization and audio visual databases that can be used for further research. (Fenghour et al., 2021) provided a brief survey on automated lip reading approaches based on deep learning approaches. The survey provided insights on a comparison between convolutional neural network with other networks, strengths of attention transformers, temporal convolutional neural networks over recurrent neural networks for classification purpose and state of the art methods available for lip reading for speech recognition. (Oghbaie et al., 2025), presented a broad survey on utilizing deep learning architectures for visual speech recognition throwing light on the entire process that includes attribute extraction, architectures, evaluation criteria with measures, available datasets and the key challenges of the domain. The study also directs on using strategies in selecting the appropriate models for improving the performance.

(Exarchos et al., 2024), integrated 3D convolutional neural network(CNN) and long short term memory(LSTM) deep learning architectures to recognize the words from lip movement considering the MobLip dataset. The work successfully combined the spatial and temporal information to identify the words at an accuracy of 87.5% indicating the strength of the method even with speaker diversity and varying lighting conditions. A similar work was conducted by (Wang et al., 2025), who proposed a light weight deep learning model termed Mini 3DCvT that combines 3D convolution and visual transform to derive spatio temporal information at local and global level from the images of LRW lip reading dataset. The performance of the model was acceptable that provided an accuracy of 88.3%.

(Negrão & Domingues, 2021) presented SpeechToText module that combines Voice activity detection and Automatic speech recognition to detect voice and further convert it to text to aid forensic investigation. The module when tested on LibriSpeech, found to be 100% accurate in detecting voice from unencrypted video files and a word error rate of 27.2% when transliterating English voice messages by nonnative speakers and a word error rate of 7.80% from native speakers. (Patilkulkarni, S. (2021) worked on detecting words from the analysis of lip movements from video data using deep learning. The work employed VGG16 pre trained Convolutional Neural Network for identifying and detecting the words that proved to be effective providing an accuracy of 76%.

(Shashidhar et al., 2022) proposed audio visual speech recognition for English language. The work was carried under three stages where audio was processed separately using Mel frequency cepstral coefficients(MFCC) for speech recognition, then Long short term memory (LSTM)model was used to detect speech from visual data. Finally, audio and video methods were integrated using neural network. The experimental results indicated highest performance accuracy of 91% from the integrated method whereas MFCC provided an accuracy of 90% and LSTM delivered 71%.

Understanding the movement of the lips during pronunciation of a word or character is challenging as certain movements overlap for different words. (Peymanfard et al., 2022) have come up with Viseme decoding that maps the lip movements to characters pronounced. Here the video data is initially converted to Viseme and later Viseme is converted to characters. This method has proven to be successful through an improved absolute error rate by 4%. (Ma et al., 2022) investigated the importance of hyper parameter tuning and data augmentation for visual speech recognition. The results demonstrated that the models that used additional training data generated using data augmentation and hyper parameter tuning works better on the on the non-publicly available data and also the overall performance of the model improves.

(Fernandez-Lopez& Sukno, 2022), demonstrated that the automatic lip reading systems can be trained competitively to be applied for various languages with small scale data provided certain relevant constraints are applied during the training process. Additionally, the work also stresses on utilizing data augmentation for improving temporal context. The results indicate the contending nature of the proposed methods as compared to other state of the art methods. (Jishnu & Antony, 2024), developed an end to end lip reading system to that uses 3D convolutions and recurrent neural network with connectionist temporal classification(CTC) loss to decode the text from the video frames that are sequential in nature. The method proved to be 95.3% accurate on the GRID corpus dataset.

(Yeo et al., 2025), suggested a speaker-adaptive lip reading technique that targets speakers at both the linguistic and visual levels by modifying a previously trained model. In particular, we employ the LoRA method and prompt tuning to a pre-trained lip reading model in order to successfully adjust the model to target speakers. It is shown that the proposed speaker-adaptive approach enhances sentence-level performance in the wild as well. In addition, we demonstrate that the suggested approach outperforms the earlier studies.

(Cao, 2024), proposed a lip reading architecture integrating 3D convolutional neural networks and transformers. The model was excellent in understanding the lip movements and it exhibited better performance with multiple background and speakers and reduced word error rate. To build a robust lip reading system that does not vary under different lighting conditions (Lee, 2024) built lip reading system based on deep neural network utilizing the infrared and depth images. The results indicated that recognition rate of the system under IR and depth images was 95.76% and 92.34% respectively under different lighting conditions.

(Sompalyam and Raveendran, 2024) developed LipCraft, a visual speech recognition system coupling with the advantages of convolutional neural networks and long short term memory with the connectionist temporal classification decoder. The project also considers an application developed with streamlit framework making the developed work accessible and the overall performance of the project was successful with accurate transcriptions. A system to understand the lip movements and transform the movements to speech is proposed by (Sahipjohn, 2024) considering self-supervised learning and sequence to sequence networks. Here the lip movements mapped to speech representations and then provided to vocoder to generate speech. This system demonstrated effectiveness in lip to speech synthesis proving its versatile applications.

(Nainan and Kulkarni, 2024) designed a dynamic audio visual speaker recognition system combining the handcrafted features and convolutional neural networks. The system derived Histogram of Gradient features from the lip movements from video frames and fused with audio features and provided to SVM CNN for classification that provided an accuracy of 91.4%. The system proved to be successful in liveliness detection and automatic speaker recognition. (Awan et al., 2025), tried bridging the gap between the audio speech recognition and visual speech recognition using deep learning with attention mechanism. The work was successful in translating the visual cues into intelligent speech segments with a promising accuracy of 97%.

(Chandrabanshi and Domnic, 2024), further proposed a multilayered hybrid network that combined the 3D-CNN and two bidirectional recurrent neural networks for visual speech recognition applied for liveliness detection. The method was found to be efficient as it provided a very less error rate of 1.3%. (Li et al., 2023) developed a methodology to improve the performance of speech recognition system in the noisy environment. The work aimed at enhancing the cross-modal fusion effect and lip reading ability to increase the accuracy of speech recognition in noisy conditions. The methodology showed a reduction in the word error rate by 4% in a -15dB SNR environment. (Wang et al., 2024) explored the application of depth sensing for locating the lips and analyzing its movements for effective speech recognition under varying light conditions. This method eliminated the use of intensity images and proved to be accurate and reliable.

(Chandrabanshi and Domnic, 2025), utilized visual speech recognition to overcome face spoofing by liveliness detection in facial image based authentication systems. The method decodes the words analyzing the lip movements from the video data along with facial image match providing dual authentication standard utilizing the capabilities of 3D-CNN, bi directional gated recurrent unit and LSTM. The results indicated the reduction in word error rate demonstrating the best performance of the method. The survey found certain works on lip reading applied to specific languages. (Ashour et al., 2024) considered ArabCeleb dataset to identify the speaker analyzing lip movements using deep learning architecture. The method found to be 98.6% accurate and proved to be reliable for Arabic speaker identification.

The survey indicated the depth of the work on visual speech recognition applied to various domains, the utilization of IoT, multiple signal processing methods, machine learning and deep learning algorithms/architectures. In the same line, this chapter presents a deep learning based lip reading module applied for forensic analysis and silent speech decoding analyzing the spatial temporal data from the video data. Further, the lip reading module generates the text which is further converted to speech and alert the law enforcement authorities to take necessary action subsequently.

3. METHODOLOGY

This section provides the methodological details of the lip reading system designed for forensic applications, which aims to utilize visual speech recognition techniques to enhance the accuracy of speaker identification, lip-to-speech transcription, and evidence analysis as indicated in Figure 3. The methodology is structured into several key stages: Video data acquisition, video processing, build and train deep learning model, character prediction, text generation, text to speech conversion and evaluation. Each of these stages contributes to ensuring that the system meets the high standards of forensic investigations where accuracy, reliability, and robustness are paramount. This methodology is tailored to meet the rigorous demands of forensic investigations, providing a powerful tool for evidence analysis and contributing to enhanced outcomes in criminal investigations and security surveillance.

The proposed approach lays a strong foundation for creating a highly effective lip reading system that can be seamlessly integrated into forensic workflows.

Figure 3. Methodology of visual speech recognition system

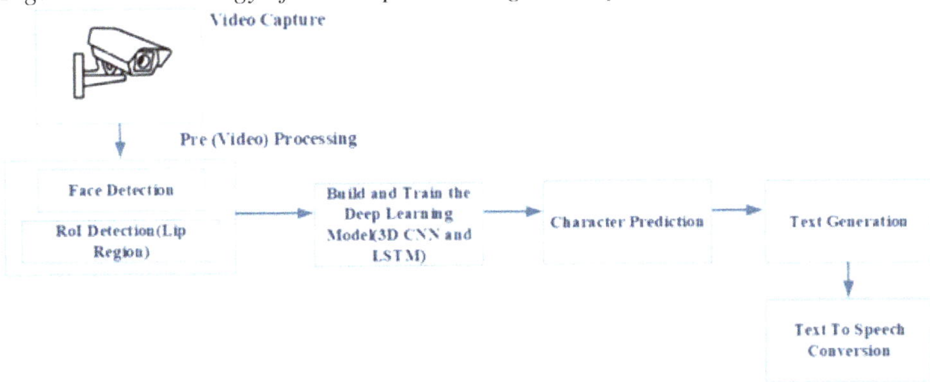

The first stage of the system involves capturing and preparing video data for further analysis. This starts with the camera module that records the video of the people speaking. The captured video is presented to preprocessing stage (illustrated in Figure 4)that involves: frame capture phase wherein the individual frames are extracted from the sequence of frames in video data, rgb frame conversion to gray scale, later from the individual frames, region of interest needs to be detected(RoI) where only the most relevant information is retained from the rest of the frame and here the RoI are the lip regions that are aimed and isolated to ensure that the system process and analysis only the lip region for its movements and shapes. Lastly, frame normalization is applied on the pixels in all the frames to have same scale.

Once the lip region is extracted, the next phase involves deep learning-based feature extraction using 3D Convolutional Neural Networks (CNNs) (Yuvaraj et al., 2023). Here the features are extracted using CNN where the important and key attributes such as lip shape, lip movement and changes in the texture are identified from the lip regions to map to the character or word pronounced.

Figure 4. Pre processing stage

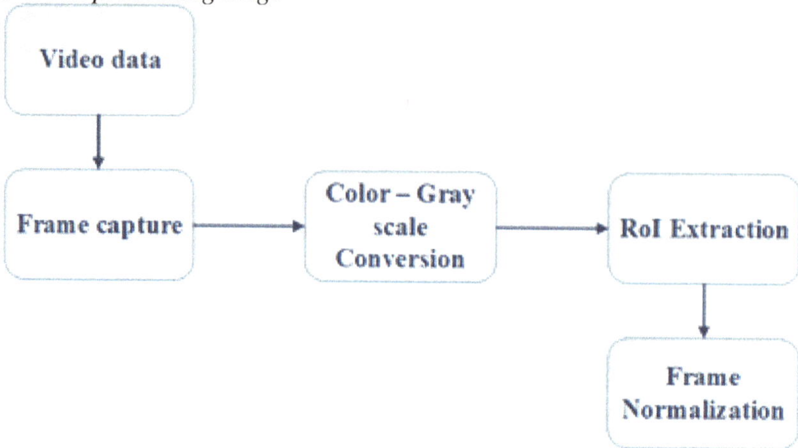

Since speech is a time series data, the time dependent information or temporal information that exists between the frames is analyzed by the temporal CNN. This is crucial as the model aids the system in understanding the evolution of lip shapes and movements over the time capturing the important phoneme transitions. Bothe the spatial and temporal CNNs focuses on deriving the spatio-temporal information from the video data utilizing the operations such as 2D convolution, max pooling, batch normalization and activation functions. The spatio-temporal information derived will contribute towards successful visual speech recognition.

Later, a fusion module is utilized after CNN processing to combine the spatial and temporal features in to a single representation that utilized the strengths of both frame based and time based data to make accurate character predictions. After the extraction of spatio-temporal features, Long short term memory (LSTM) to model the sequential dependencies present in the data (Siami-Namini et al., 2019). These networks are designed in a way to retain the long term dependencies in the speech related data. This deep hierarchical learning is predominantly effective in recognizing complex speech patterns.

The work presented integrates three-dimensional convolutional neural networks (3D CNNs) for processing spatial and temporal features, and bidirectional recurrent networks (BRNN) for sequence analysis based on (Assael et al., 2016).

The deep learning architecture with 3D CNN and BRNN utilized in the present work is illustrated in Figure 5. This architecture has initial feature extraction stage, where a 3D convolutional layer (Conv3D-1) analyzes the input video frames and transforms the data into a shape of (75, 46, 140, 128) - Here 128 represents the feature maps created. This first layer learns to detect basic patterns of motion and spatial features across consecutive frames. After activation, a max pooling layer

(MaxPooling 3D-1) reduces the spatial dimensions by half, resulting in an output shape of (75, 23, 70, 128), while preserving the temporal dimension of 75 frames.

The next stage of the architecture strengthens the feature extraction with a larger convolutional layer(Conv3D-2) that produces an output shape of (75, 23, 70, 256), doubling the feature maps to capture more complex patterns. After max pooling (MaxPooling 3D-2), the spatial dimensions are further reduced to (75, 11, 35, 256). The third convolutional block(Conv3D-3) then compresses these features into a shape of (75, 11, 35, 75) feature channels, followed by max pooling (MaxPooling 3D-3) that results in (75, 5, 17, 75), creating a more focused representation of the most relevant patterns.

Figure 5. Architecture of the deep learning model

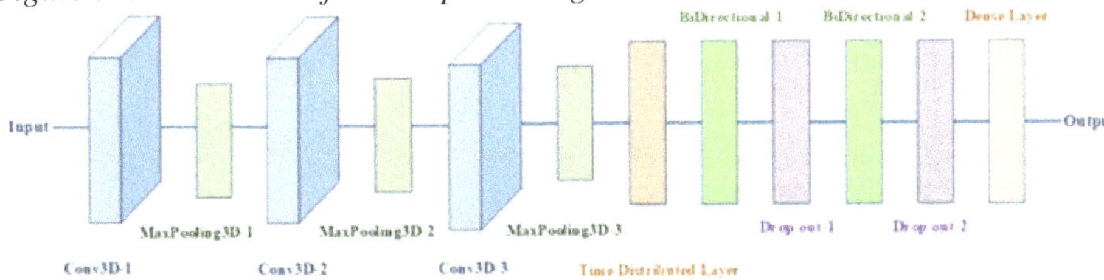

Next, the architecture transits from convolutional processing to sequence analysis through a time-distributed layer that reshapes the features into (75, 6375), flattening the spatial and channel dimensions. This is followed by two bidirectional layers (BiDirectional-1, BiDirectional-2) with dropout mechanisms(Dropout-1, Dropout-2). The first bidirectional layer transforms the data to 75, 256), and after dropout and the second bidirectional layer, maintains this shape while refining the features. The BiDirectional layers play an important role in this architecture as the sequence analysis is done forward and also backward making the network understand the context in both the directions.

The last stage of the architecture has a dense layer that transforms the output to a dimension (75,41), here 41 indicating the different character probabilities needed to generate the text. The output is further processed by the connectionist temporal classification(CTC) decoder that converts the prediction probabilities into text.

Finally, the last phase considers the processed data and predicts the most probable uttered characters or words, which is then converted to text and later speech respectively. The speech is further used to alert the law enforcement to take necessary action

4. IMPLEMENTATION AND RESULTS

The implementation of Deep Learning based Lip reading module applied for forensic analysis and silent speech decoding aims to create a cutting-edge AI-powered lip-reading module designed for forensic analysis and silent speech decoding by translating lip movements into text or synthesized speech. In forensic applications, it seeks to assist law enforcement and legal professionals by decoding speech from silent video footage, such as surveillance recordings, where audio is absent or unusable. This capability can provide critical insights during investigations, enabling the extraction of valuable speech content that may otherwise be inaccessible. The development of the lip-reading module required a powerful programming environment to handle deep learning and computer vision tasks. Python was chosen for carrying the silent speech decoding task or its rich ecosystem of machine learning and image processing libraries. Development was carried out in Jupyter Notebook, which enabled an interactive and step-by-step workflow.

Data acquisition and preprocessing

To develop an accurate lip reading model, the GRID CORPUS (https://www.kaggle.com/datasets/mohamedbentalb/lipreading-dataset) dataset was chosen as the primary source of training data. The GRID CORPUS dataset is frequently used in speech processing and visual speech recognition due to its clarity in recordings, a structured format and aligned text transcriptions. The data has short video clips of the speakers speaking(pronouncing) pre-defined sentences that make it suitable for working with deep learning models to map the lip movements with the words pronounced.

The GRID CORPUS is an audio visual dataset made available for studies in speech recognition (visual speech recognition). The dataset has recordings including audio and video of 1000 sentences pronounced by 18 male and 16 female speakers. The audio recordings(files) are down sampled at 25KHz and they have maximum amplitude of one. The video recordings are also sampled at 25 frames per second with two quality formats: normal quality with a resolution of 360 x 288 pixels and higher quality with a resolution of 720 x 576 pixels. Every recorded video clip from the dataset provides essential information about the lip movement and the underlying speech and is labelled with an identifier correlating to the spoken sentence. This ensures the mapping between the visual and the text data. The sample frames from the dataset is indicated in Figure 6.

Figure 6. Sample frames from the video of one speaker - Grid Corpus dataset

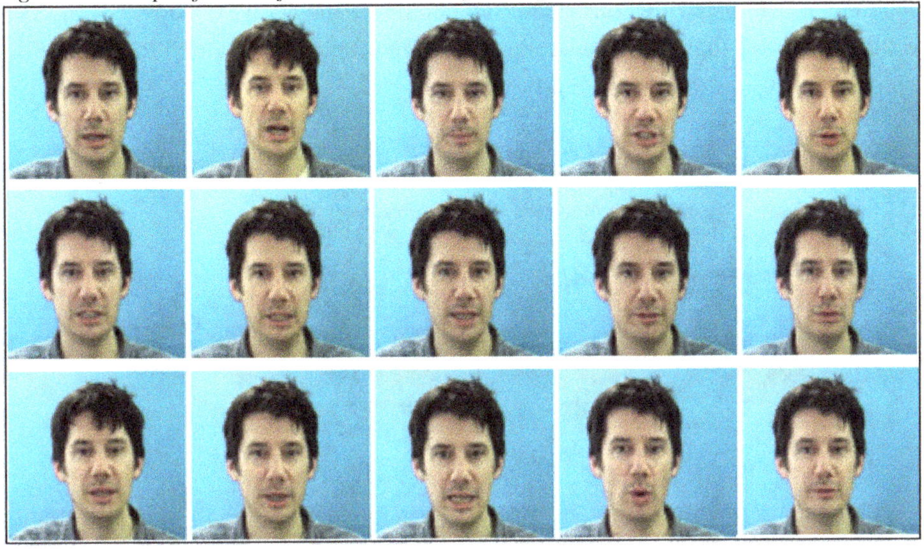

Together with the videos, the dataset also includes alignment files, which provide precise time-stamped transcriptions of the spoken content. These alignment files serve as ground truth labels, that help the deep learning model to learn the underlying correlation between visual features and phonetic elements. The alignments specify the exact timing of each word within the video, allowing the model to synchronize visual cues with speech patterns. This structured format ensures that the dataset can be effectively used for supervised learning, where each video frame is linked to its corresponding phoneme or word.

The acquired videos from the dataset is further preprocessed, thus it is provided to the video processing stage as illustrated in Figure 6, where the videos are broken down into smaller frames to be analyzed by the model. Each video was processed to extract 75 frames and each frame is further converted to grayscale to as the work is around lip reading. The frames are later cropped to focus on the lip region by creating a rectangle of dimension 46 x 140 pixels (as indicated in Figure 7), specifically indicating the horizontal and vertical pixel coordinates. This ensures that the model only processes the most relevant visual information.

Later the pixel values across all the frames are normalized using Z-score normalization such that the entire data to have zero mean and a standard deviation of 1 as required by the deep neural network.

$$\text{Frame}_{\text{norm}} = \frac{\left(\text{Current}_{\text{frame}} - \text{Mean}_{\text{across all frames}}\right)}{\text{Standard deviation}_{\text{across all frames}}} \quad (1)$$

One of the key features of the video processing function is the sliding window mechanism, which divides the video into smaller manageable windows. If the video clip is shorter than the number of frames required, it is padded to achieve the desired length.

Figure 7. Lip region(RoI) extraction

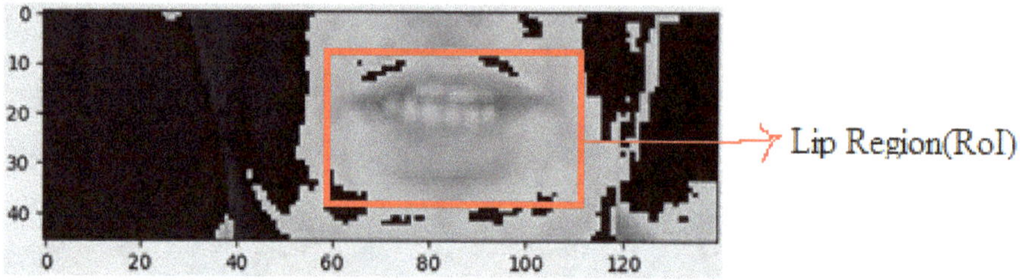

Once the videos are processed, the data is prepared for its use in the following lip reading module. As the work aims at speech decoding, the pronounced characters are to be predicted and the corresponding text has to be generated thus it is required to perform character mapping to support double way conversion. In this regard a vocabulary is built taking into account the English alphabet characters from {a→z}, numbers from {1→9} and few special characters. Consequently, every element from the vocabulary list is assigned token/unique id/sequence of numbers that aid in mapping these elements to numbers and vice versa to enhance model's training performance.

Training and Evaluation of the Deep Learning Model

For visual speech recognition, the deep learning model with 3D CNN and BRNN is trained and tested. To start, L2 regularization with a regularization parameter of 0.001 is applied to all the kernels of 3DCNN and BRNN to take care of overfitting. 3DCNN extracts the features and it is converted to 1 dimensional 75 feature vectors as required by BRNN. To prevent the problem of vanishing and exploding gradients, orthogonal kernel initializer is used in BRNN and at the output dense layer he_normal kernel initializer which is derived from the normal distribution is used that keeps the variance of gradients and activation functions in the network stable.

For training and testing, the dataset is shuffled and batched from which the first 450 batches form the training data and the rest are considered for testing the model. Adam optimizer is used with the learning rate set to 0.0001 such that it is stable for the first thirty epochs and above 30 it is decreased exponentially with a rate 9.5%/ epoch to prevent the loss function from oscillating near the minimum point of the loss function. For training CTC loss function is used with the number of epochs set to 100. During training, the model is trained for with a total of 84,7, 1924 parameters to create the final model. The model built for the task of lip reading and text generation has considered both the spatial and temporal aspects of the video data ensuring a complete understanding of the patterns in spatial and temporal domains of the video data.

After the completion of training process and once the model is built, the testing video data undergoes the process of preprocessing and is fed to the model for predicting the characters and further the generation of text. Since the predictions are encoded as numbers rather than readable text, the predictions are decoded using a mapping function using a function that converts the numeric predictions into their corresponding characters by iterating over each sequence and transforming it into a readable text format and then concatenating the characters for each word, ensuring the predicted output matches the ground truth text.

Finally, the performance of the model is evaluated using Character Accuracy(Acc-Char), Word Accuracy(Acc-Word) and Cumulative accuracy(Acc-Cum) computed using the equations (1), (2) and (3).

$$\text{Acc} - \text{Char} = \frac{\text{Number of correct characters predicted by the model}}{\text{Total number of characters}} \tag{2}$$

$$\text{Acc} - \text{Word} = \frac{\text{Number of correct words predicted by the model}}{\text{Total number of words}} \tag{3}$$

$$\text{Acc} - \text{Cum} = \frac{\sum_{j=1}^{M} I(\hat{y}_j - y_j)}{M}$$ re $\hat{y}_j$ i e predicted label of the jth instance

y_j s t e ground truth label of jth instance

$I(\hat{y}_j - y_j)$ s an in cator function that is '1' if the prediction is right and '0' if the predictions are wrong M is the total number of test samples (4)

The visual speech recognition model demonstrated an exceptional performance on short-duration video sequences, achieving a peak accuracy of 96% in controlled testing environments. Finally, a brief comparison is made between the proposed method and other approaches presented on lip reading (visual speech recognition) to find the reliability of the proposed method. The comparison is provided in Table.1

Table 1. Performance comparison with other deep learning architectures

Deep Learning Model	Dataset Used	Accuracy(%)
LipNet (Assael et al., 2016)	GRID corpus	95.2
Convolutional Neural Network with Temporal convolutional networks. (Kastaniotis et al., 2020)	Self developed	50
3D convolutional vision transformer (3DCvT) (Wang et al., 2022)	LRW lip reading dataset	88.5
Convolutional neural networks (CNNs) and deep neural networks (DNNs) (Rane et al., 2023)	GRID dataset	93
EMOLIPS (Ryumin et al., 2023)	RAVDESS CREMA-D	91.9 90.9
TCS-LipNet (Chen et al., 2023)	LRW	92.2
Combination of handcrafted features and convolutional neural networks. (Nainan and Kulkarni, 2024)	VidTIMIT Dataset	91.4
3D convolutional neural network(CNN) and long short term memory(LSTM) (Exarchos et al., 2024),	MobLip dataset	87.5
Mini 3DCvT – combination of 3D convolution and visual transformer (Wang et al., 2025)	LRW lip reading dataset	88.3
STDNet (Wu et al., 2025)	LRW LRW-1000	90.2 53.56
3DCNN and Bidirectional LSTMs with Multi-Head Attention mechanisms (Chaudhari et al., 2025)	GRID Audio-Visual Sentence Corpus	96
Proposed method 3DCNN and BRNN	GRID corpus	96

5. CASE STUDIES

This section presents some of the real world documented case studies that has implemented lip reading (visual speech recognition) for forensic investigation. Few cases presented here indicate the use of manual method and also the application of lip reading algorithms

[a]Case study i: Arlene Fraser murder case (UK, 1998)

The case: Arlene Fraser's disappearance from her home in 1998 was treated as missing person's case. Her husband Nat Fraser was the prime suspect but the legal case on him was weak as the Arlene's body was not found, there was no forensic nor eyewitness. During her disappearance Nat Fraser was on bail for the case -attempt to murder his wife which had happened before her actual disappearance.

In 2003, Nat was found guilty in the first trial and was sentenced to life imprisonment of 25 years. But his appeal to the court of law in the same year granted him leave. But during his second trial in 2012 he was again found to be guilty and was sentenced to 17 years' imprisonment.

Role of Lip reading: During Nat's stay in jail, his conversations were recorded on CCTV camera which was forwarded to the police for analysis. The camera recordings had silent speech which were analyzed using lip readings by the experts.

Conclusion: Some of the words extracted from the lip reading such as "They can't find her-it's impossible", "cut her up" charged Nat with murder of his wife.

[b]Case study ii: R v Luttrell (UK, 2003)

The case: R v Luttrell along with eight defendants were found guilty of committing armed robbery and disposal of the goods worth £11m. But R v Luttrell pleaded not guilty and he was under police surveillance.

Role of Lip reading: The CCTV camera recordings of the conversation during the selling of goods by R v Luttrell were analyzed with lip reading where the lip reading expert identified 452 words correctly out of 890 words.

Conclusion: The words decoded by the expert identified the name of the co-conspirator of Luttrell and the court accepted the lip reading evidence.

[c]Case Study iii: John Terry v Anton Ferdinand (UK, 2012)

The case: The English footballer John Terry was accused of making a racist remark against opposite player Anton Ferdinand during a football match between the teams Chelsea FC and Queen's Park Rangers FC.

The defence of John terry argued that the racist remark by Terry was in response to a previous accusation by Anton Ferdinand.

The role of Lip reading: The Television recording was provided for forensic analysis. The lip reading experts analyzed the words spoken by the accused.

Conclusion: Initially the magistrate found Terry not guilty of the charge due to lack of evidence. Terry also defended that he was just repeating what Ferdinand had earlier said. But the Football association later found him guilty.

[d]Case Study iv: Bruce Lehrmann defamation trial (Australia, 2023)

The case: Bruce Lehrmann an employee of a political organization sued a media company with a defamation case. The media had broadcast an interview with his former colleague who alleged Lehrmann for sexual assault.

Role of Lip reading: The CCTV footage with silent speech that recorded the conversation between Bruce Lehrmann and his colleague on the day of alleged assault was analysed using Lip reading algorithms to decode the speech. The transcription was submitted to the court of law as evidence.

Conclusion: Though there were objections initially, the lip reading evidence was considered and accepted as a reliable evidence. Through the evidence, Bruce Lehrmann was found guilty and lost his defamation case.

6. CONCLUSION

The Deep Learning based Lip reading module applied for forensic analysis and silent speech decoding presented in this chapter has successfully demonstrated the potential of advanced lip-reading in forensic analysis and silent speech decoding. By leveraging cutting-edge computer vision and machine learning techniques, the work provides a powerful tool for interpreting silent video recordings and decoding speech from lip movements with remarkable accuracy of 96%. In forensic analysis, the project can assist law enforcement and legal experts in extracting critical information from surveillance footage or silent video evidence, enhancing investigative capabilities and also providing required alerts in a closed loop system.

Challenges in Lip reading

Though Lip reading is found successful which is event from the results and also the case studies presented, it also faces certain challenges that needs immediate attention. The effectiveness of the lip reading system depends on various factors that include

i. Visibility of speech: Some characteristics of the speech like those involve tongue features (velvar consonants, alveolar consonants, palatal consonants) are not visible on lips.
ii. Homophenes: Many sounds produced inside the mouth during speech look similar on lips which are visually confusing for speech decoding.

iii. Context in speech: Knowing the context in speech is much important to generate a transcription based on lip reading which otherwise might be unclear. The context sometimes can also help in eliminating homophenes.
iv. Quality of video: Lip reading requires a clear video footage, the quality determines the reliability of the decoded speech. Resolution of the video, occlusions, variations in lighting, contrast and shadow are some of the factors that affect lip reading.

Future directions

The future directions of lip reading would be adapting explainable AI(XAI), light weight computational models and its incorporation for multiple languages. The current deep learning models work as black boxes which could be made transparent with XAI and it can help improving the models performance and its reliability by providing the required information regarding the occurrence of errors and revealing the constraints and biases of the model. Implementing the task of visual speech recognition using Light weight deep learning models can reduce the computational cost by balancing with improved accuracy.

The work will be more acceptable if implemented considering multiple languages specially the regional for several applications. However, it also challenges of developing the dataset, building vocabulary taking into account variations in tone, viseme ambiguities, diverse accents of the language.

REFERENCES

Akeiber, H. J. (2025). A comprehensive study of Cybercrime and Digital Forensics through Machine Learning and AI. Al-Rafidain Journal of Engineering Sciences, 369-395.

Ashour, M. I., Abbas, A. S., Ahmed, S. R., Haydar, N., Ghaffoori, A. J., Hussain, A. S. T., & Tawfeq, J. F. (2024, May). Enhancing Arabic speaker identification through lip movement analysis and deep representation learning. In *Proceedings of the Cognitive Models and Artificial Intelligence Conference* (pp. 335-340). DOI: 10.1145/3660853.3660938

Assael, Y. M., Shillingford, B., Whiteson, S., & De Freitas, N. (2016). Lipnet: End-to-end sentence-level lipreading. arXiv preprint arXiv:1611.01599.

Awan, A. B. Y., Zhang, T., Younas, N., Zhao, X., Fahad, M., & Geng, Y. (2025). Lips movement detection and conversion into text using deep learning-based CNN model. *Signal, Image and Video Processing*, *19*(9), 691. DOI: 10.1007/s11760-025-04178-5

Bakar, J. (2025). Capturing the Digital Scene: Applying Routine Activity Theory to IoT Smart Frames.

Battista, M., Collesei, F., Orzan, E., Fantoni, M., & Bottari, D. (2025). Lip-Reading: Advances and Unresolved Questions in a Key Communication Skill. *Audiology Research*, *15*(4), 89. DOI: 10.3390/audiolres15040089

Bhawna, & Mahajan, S. (2024). Artificial intelligence perspective on digital forensics. Applying Artificial Intelligence in Cybersecurity Analytics and Cyber Threat Detection, 33-44.

Cao, Y. (2024). Lips Reading Using Deep Learning Architecture (Doctoral dissertation, Auckland University of Technology).

Cărcăle, V. A. (2025). *The Future of Artificial Intelligence (AI).* Applications in Forensics.

Casino, F., Dasaklis, T. K., Spathoulas, G. P., Anagnostopoulos, M., Ghosal, A., Borocz, I., Solanas, A., Conti, M., & Patsakis, C. (2022). Research trends, challenges, and emerging topics in digital forensics: A review of reviews. *IEEE Access : Practical Innovations, Open Solutions*, *10*, 25464–25493. DOI: 10.1109/ACCESS.2022.3154059

Chandrabanshi, V., & Domnic, S. (2024). HNet: A deep learning based hybrid network for speaker dependent visual speech recognition. *International Journal of Hybrid Intelligent Systems*, *20*(4), 385–401. DOI: 10.3233/HIS-240014

Chandrabanshi, V., & Domnic, S. (2025). A deep learning approach for strengthening person identification in face-based authentication systems using visual speech recognition. *The European Physical Journal. Special Topics*, *234*(15), 1–15. DOI: 10.1140/epjs/s11734-025-01586-z

Chango, X., Flor-Unda, O., Gil-Jiménez, P., & Gómez-Moreno, H. (2024). Technology in forensic sciences: Innovation and precision. *Technologies*, *12*(8), 120. DOI: 10.3390/technologies12080120

Chaudhari, A., Abdullah, M., Deshpande, V., Zanke, T., Wath, S., Mulgir, S., & Jagtap, S. (2025). Deep learning based 3D residual convolutional and Multi-Head Attention (3D-RMA) for lip-reading. Results in Control and Optimization, 100608.

Chen, H., Li, W., Cheng, Z., Liang, X., & Zhang, Q. (2023, September). TCS-LipNet: Temporal & Channel & Spatial Attention-Based Lip Reading Network. In International Conference on Artificial Neural Networks (pp. 413-424). Cham: Springer Nature Switzerland. DOI: 10.1007/978-3-031-44201-8_34

Chowdhury, A. T., Newaz, M., Saha, P., AbuHaweeleh, M. N., Mohsen, S., Bushnaq, D., Chabbouh, M., Aljindi, R., Pedersen, S., & Chowdhury, M. E. (2025). Decoding silent speech: A machine learning perspective on data, methods, and frameworks. *Neural Computing & Applications*, *37*(10), 6995–7013. DOI: 10.1007/s00521-024-10456-z

Dunsin, D., Ghanem, M. C., Ouazzane, K., & Vassilev, V. (2024). A comprehensive analysis of the role of artificial intelligence and machine learning in modern digital forensics and incident response. *Forensic Science International Digital Investigation*, *48*, 301675. DOI: 10.1016/j.fsidi.2023.301675

Exarchos, T., Dimitrakopoulos, G. N., Vrahatis, A. G., Chrysovitsiotis, G., Zachou, Z., & Kyrodimos, E. (2024). Lip-reading advancements: A 3D convolutional neural network/long short-term memory fusion for precise word recognition. *BioMedInformatics*, *4*(1), 410–422. DOI: 10.3390/biomedinformatics4010023

Fenghour, S., Chen, D., Guo, K., Li, B., & Xiao, P. (2021). Deep learning-based automated lip-reading: A survey. *IEEE Access : Practical Innovations, Open Solutions*, *9*, 121184–121205. DOI: 10.1109/ACCESS.2021.3107946

Fernandez-Lopez, A., & Sukno, F. M. (2022). End-to-end lip-reading without large-scale data. *IEEE/ACM Transactions on Audio, Speech, and Language Processing*, *30*, 2076–2090. DOI: 10.1109/TASLP.2022.3182274

Flaglien, A. O. (2017). The digital forensics process. Digital forensics, 13-49.

Ghanem, M. C., Almeida Palmieri, E., Sowinski-Mydlarz, W., Al-Sudani, S., & Dunsin, D. (2025). Weaponized iot: A comprehensive comparative forensic analysis of hacker raspberry pi and pc kali linux machine. *IoT*, *6*(1), 18. DOI: 10.3390/iot6010018

Ghasemi, J., & Mahmoudi, N. (2024, October). The application of IoT devices in evidence collection and preservation In legal investigations In the common law legal system of Western countries. In 2024 11th International Symposium on Telecommunications (IST) (pp. 120-124). IEEE. DOI: 10.1109/IST64061.2024.10843627

Hao, M., Mamut, M., Yadikar, N., Aysa, A., & Ubul, K. (2020). A survey of research on lipreading technology. *IEEE Access : Practical Innovations, Open Solutions*, *8*, 204518–204544. DOI: 10.1109/ACCESS.2020.3036865

Irons, A., & Lallie, H. S. (2014). Digital forensics to intelligent forensics. *Future Internet*, *6*(3), 584–596. DOI: 10.3390/fi6030584

Jishnu, T. S., & Antony, A. (2024). LipNet: End-to-End Lipreading. [IJDM]. *Indian Journal of Data Mining*, *4*(1), 1–4. DOI: 10.54105/ijdm.A1632.04010524

Jones, G. M., & Winster, S. G. (2022). An insight into digital forensics: History, frameworks, types and tools. Cyber Security and Digital Forensics, 105-125.

Kastaniotis, D., Tsourounis, D., & Fotopoulos, S. (2020, October). Lip Reading modeling with Temporal Convolutional Networks for medical support applications. In 2020 13th International Congress on Image and Signal Processing, BioMedical Engineering and Informatics (CISP-BMEI) (pp. 366-371). IEEE. DOI: 10.1109/CISP-BMEI51763.2020.9263634

Kruthika, S. G., & Nagavi, T. C. (2023). Speech Processing and Analysis for Forensics and Cybercrime: A Systematic Review. Cybercrime in Social Media, 191-224.

Lee, K. S. (2024). Improving the Performance of Automatic Lip-Reading Using Image Conversion Techniques. *Electronics (Basel)*, *13*(6), 1032. DOI: 10.3390/electronics13061032

Li, D., Gao, Y., Zhu, C., Wang, Q., & Wang, R. (2023). Improving speech recognition performance in noisy environments by enhancing lip reading accuracy. *Sensors (Basel)*, *23*(4), 2053. DOI: 10.3390/s23042053

Li, Y., Hashim, A. S., Lin, Y., Nohuddin, P. N., Venkatachalam, K., & Ahmadian, A. (2024). AI-based visual speech recognition towards realistic avatars and lip-reading applications in the metaverse. *Applied Soft Computing*, *164*, 111906. DOI: 10.1016/j.asoc.2024.111906

Liu, Q., Ge, M., & Li, H. (2024). Intelligent event-based lip reading word classification with spiking neural networks using spatio-temporal attention features and triplet loss. *Information Sciences*, *675*, 120660. DOI: 10.1016/j.ins.2024.120660

Lutta, P. (2024). An Internet of Things forensics framework validated by machine learning (Doctoral dissertation, Staffordshire University).

Ma, P., Petridis, S., & Pantic, M. (2022). Visual speech recognition for multiple languages in the wild. *Nature Machine Intelligence*, *4*(11), 930–939. DOI: 10.1038/s42256-022-00550-z

Nainan, S., & Kulkarni, V. H. (2024). Multimodal Speaker Recognition using voice and lip movement with decision and feature level fusion.

Nayak, M.Anushka Deepak Kadage. (2024). Ai-enhanced digital forensics: Automated techniques for efficient investigation and evidence collection. J. *Electrical Systems*, *20*(1s), 211–229. DOI: 10.52783/jes.766

Negrão, M., & Domingues, P. (2021). SpeechToText: An open-source software for automatic detection and transcription of voice recordings in digital forensics. *Forensic Science International Digital Investigation*, *38*, 301223. DOI: 10.1016/j.fsidi.2021.301223

Oghbaie, M., Sabaghi, A., Hashemifard, K., & Akbari, M. (2025). When deep learning deciphers silent video: A survey on automatic deep lip reading. *Multimedia Tools and Applications*, *84*(32), 1–43. DOI: 10.1007/s11042-024-20156-4

Patilkulkarni, S. (2021). Visual speech recognition for small scale dataset using VGG16 convolution neural network. *Multimedia Tools and Applications*, *80*(19), 28941–28952. DOI: 10.1007/s11042-021-11119-0

Peymanfard, J., Mohammadi, M. R., Zeinali, H., & Mozayani, N. (2022, February). Lip reading using external viseme decoding. In *2022 International Conference on Machine Vision and Image Processing (MVIP)* (pp. 1-5). IEEE.

Pu, G., & Wang, H. (2023). Review on research progress of machine lip reading. *The Visual Computer*, *39*(7), 3041–3057. DOI: 10.1007/s00371-022-02511-4

Rajasekar, K. P., & Vezhaventhan, D. (2024, June). Artificial intelligence revolutionizing legal and forensic practices: A comprehensive analysis. In 2024 15th International Conference on Computing Communication and Networking Technologies (ICCCNT) (pp. 1-12). IEEE. DOI: 10.1109/ICCCNT61001.2024.10724685

Rane, M., Chaudhari, V., Vasant, M., Nambiar, B., Umak, S., & Kulkarni, A. (2023, October). Seeing the Unheard: Lip Reading with Deep Learning. In *2023 International Conference on Evolutionary Algorithms and Soft Computing Techniques (EASCT)* (pp. 1-6). IEEE. DOI: 10.1109/EASCT59475.2023.10393197

Renukadevi, P., John, S., & Shivani, N. (2024, September). Forensic Science: AI-Powered Image and Audio Analysis. In 2024 5th International Conference on Smart Electronics and Communication (ICOSEC) (pp. 1519-1525). IEEE.

Ryumin, D., Ryumina, E., & Ivanko, D. (2023). EMOLIPS: Towards Reliable Emotional Speech Lip-Reading. *Mathematics*, *11*(23), 4787. DOI: 10.3390/math11234787

Sahipjohn, N. (2024). Beyond Text: Expanding Speech Synthesis with Lip-to-Speech and Multi-Modal Fusion (Doctoral dissertation, International Institute of Information Technology Hyderabad).

Salem, Y., & Hamarsheh, M. M. (2024). Forensically analyzing IoT smart camera using MAoIDFF-IoT framework. *Forensic Science International Digital Investigation*, *51*, 301829. DOI: 10.1016/j.fsidi.2024.301829

Sawarkar, A. D., & Hazari, A. D. (2024). IoT Forensic Cyber Activities Detection and Prevention with Automated Machine Learning Model. Journal of Sensors. *IoT & Health Sciences*, *2*(02), 1–15. DOI: 10.69996/jsihs.2024006

Shashidhar, R., Patilkulkarni, S., & Puneeth, S. B. (2022). Combining audio and visual speech recognition using LSTM and deep convolutional neural network. *International Journal of Information Technology : an Official Journal of Bharati Vidyapeeth's Institute of Computer Applications and Management*, *14*(7), 3425–3436. DOI: 10.1007/s41870-022-00907-y

Siami-Namini, S., Tavakoli, N., & Namin, A. S. (2019, December). The performance of LSTM and BiLSTM in forecasting time series. In 2019 IEEE International conference on big data (Big Data) (pp. 3285-3292). IEEE. DOI: 10.1109/BigData47090.2019.9005997

Sompalyam, S., & Raveendran, L. S. (2024). LipCraft: Building a Visual Speech Recognition System with Machine Learning and Streamlit.

Vermesan, O., & Friess, P. (Eds.). (2022). *Digitising the industry Internet of Things connecting the physical, digital and VirtualWorlds*. CRC Press.

Wang, H., Cui, B., Yuan, Q., Pu, G., Liu, X., & Zhu, J. (2025). Mini-3DCvT: A lightweight lip-reading method based on 3D convolution visual transformer. *The Visual Computer*, *41*(3), 1957–1969. DOI: 10.1007/s00371-024-03515-y

Wang, H., Pu, G., & Chen, T. (2022). A lip reading method based on 3D convolutional vision transformer. *IEEE Access : Practical Innovations, Open Solutions*, *10*, 77205–77212. DOI: 10.1109/ACCESS.2022.3193231

Wang, X., Su, Z., Rekimoto, J., & Zhang, Y. (2024, May). Watch your mouth: Silent speech recognition with depth sensing. In *Proceedings of the 2024 CHI Conference on Human Factors in Computing Systems* (pp. 1-15). DOI: 10.1145/3613904.3642092

Wu, X., Tan, Z., Cheng, Z., & Ru, Y. (2025). STDNet: Improved lip reading via short-term temporal dependency modeling. *Virtual Reality & Intelligent Hardware*, *7*(2), 173–187. DOI: 10.1016/j.vrih.2024.07.003

Yaacoub, J. P. A., Noura, H. N., Salman, O., & Chehab, A. (2022). Advanced digital forensics and anti-digital forensics for IoT systems: Techniques, limitations and recommendations. *Internet of Things : Engineering Cyber Physical Human Systems*, *19*, 100544. DOI: 10.1016/j.iot.2022.100544

Yeo, J. H., Kim, C. W., Kim, H., Rha, H., Han, S., Cheng, W. H., & Ro, Y. M. (2025, April). Personalized lip reading: Adapting to your unique lip movements with vision and language. *Proceedings of the AAAI Conference on Artificial Intelligence*, *39*(9), 9472–9480. DOI: 10.1609/aaai.v39i9.33026

Yuvaraj, R., Baranwal, A., Prince, A. A., Murugappan, M., & Mohammed, J. S. (2023). Emotion recognition from spatio-temporal representation of EEG signals via 3D-CNN with ensemble learning techniques. *Brain Sciences*, *13*(4), 685. DOI: 10.3390/brainsci13040685

Zhou, Z., Zhao, G., Hong, X., & Pietikäinen, M. (2014). A review of recent advances in visual speech decoding. *Image and Vision Computing*, *32*(9), 590–605. DOI: 10.1016/j.imavis.2014.06.004

ENDNOTES

[a1] https://www.bbc.co.uk/pressoffice/pressreleases/stories/2005/10_october/19/fraser.shtml

[a2] https://www.telegraph.co.uk/news/uknews/1420816/Lip-reader-saw-Frasers-incriminating-conversations.html retreived 13 October 2025.

[a3] http://news.bbc.co.uk/2/hi/uk_news/scotland/2707855.stm

[b1] https://www.bailii.org/cgi-bin/markup.cgi?doc=/ew/cases/EWCA/Crim/2004/1344.html&query=luttrell&method=boolean

[b2] http://news.bbc.co.uk/2/hi/uk_news/england/2806155.stm

[c1] https://www.bbc.com/sport/football/19842795

[c2] https://www.theguardian.com/football/2012/oct/05/john-terry-judgment-commission

[d1] https://www.theguardian.com/media/live/2023/dec/19/bruce-lehrmann-defamation-trial-updates-day-19-fiona-brown-evidence-brittany-higgins-lisa-wilkinson-the-project-ntwnfb#:~:text=Lip%2Dreader%20spent%20three%20days,night%20of%20the%20alleged%20rape.

[d2] https://www.theguardian.com/media/2024/mar/20/bruce-lehrmann-timeline-defamation-trial-brittany-higgins-rape-allegation-network-ten-lisa-wilkinson-ntwnfb

Chapter 16
Policy and Regulatory Considerations for Blockchain and Federated Learning in IoV

Yashika Vashisht
Sharda University, India

Keshav Bisht
Sharda University, India

Renu Mishra
https://orcid.org/0000-0002-5211-8751
SSCSE, Sharda University, India

Mamta Narwaria
https://orcid.org/0000-0001-8941-5824
SSCSE, Sharda University, India

ABSTRACT

The union of Blockchain and Federated Learning (FL) technologies in the Internet of Vehicles (IoV) domain has ushered in new possibilities for privacy-preserving, decentralized, and cyber-resilient financial use cases. As autonomous and connected vehicles (CAVs) are integrated with edge computing and 5G/6G networks, these smart nodes increasingly engage in real-time financial transactions spanning from usage-based insurance, automated tolls, electric vehicle charging fee payments, to intelligent vehicle leasing and decentralized vehicular identity management. Such a dynamic scenario is best addressed by Blockchain technology, which provides a

DOI: 10.4018/979-8-3373-3146-1.ch016

decentralized ledger system that promises immutability, transparency, and trust, which are perfect for authenticating transactions like insurance claims, micropayments, and contractual fulfilment in mobility finance. This enables IoV stakeholders to build predictive systems with data sovereignty and low latency. Even so, the joint employment of blockchain and FL in financial environments poses important regulatory and ethical challenges.

1.INTRODUCTION

The Internet of Vehicles (IoV) is quickly transforming from a networked automotive communications platform into a hi-tech cypher-physical system (Lin et al., 2017). Cars can now compute and network, allowing them to perceive, process, and respond to huge volumes of data. However, in the effort of making the automobile safe, more navigational, and fun, IoV systems are now changing the financial interaction paradigm of vehicles giving much softer IoV systems automated and real-time functions from use-based insurance to tolling, vehicle micropayments, leasing, and even autonomous energy trading (Xu et al., 2021). Cars are increasingly becoming financial agents that complete transactions, execute, contracts, and interact with digital infrastructure. This shift from being passive carrier objects to active autonomous financial nodes creates new challenges, most importantly, data privacy issues, regulatory compliance, transparency, security, and managing boundries lability (Mishra et al., 2022). The financial system architecture centralized, highly regulated and audit-based contrasts with the mobile, decentralized, and privacy-driven nature of IoV operations.

To address these challenges, two enabling technologies emerged: Blockchain and Federated Learning (FL). The decentralized system of Blockchain can offer financial trust by enforcing obligation without being mutable or transparent.

Federated learning provides decentralized machine learning between edge devices such as vehicles without sharing raw data to a central point, maintaining user privacy and respecting present data privacy paradigms.

Based on this, this chapter formulates three closely related goals:

- To explore how blockchain and FL can be integrated together to enable autonomous financial decision-making in vehicles.
- To look at the policy and regulatory implications of this integration, with a focus on privacy, accountability, and cross-border compliance.
- To design a Federated Governance Framework that implements hybrid on-chain/off-chain monitoring for financial IoV ecosystems.

Finally, this chapter positions blockchain and federated learning as the pillars of autonomy of action and adaptivity of intelligence in next-gen mobility finance. Blockchain brings autonomy of action by means of self-executing smart contracts, while federated learning brings adaptivity of intelligence by virtue of ongoing model evolution. Combined, under a suitable regime of governance, they allow vehicles and financial infrastructures to behave as responsible, autonomous economic agents—representing an important step toward the attainment of autonomous and adaptive systems in global transport finance.

1.1 Blockchain: Trust, Transparency, and Auditability in Vehicular Finance

Blockchain, or Distributed Ledger Technology (DLT), is uniquely suited to provide secure, decentralized recordkeeping in environments where multiple parties engage in financial and contractual interactions (Zyskind et al., 2015; NIST, 2021). In IoV systems, these parties may include vehicle manufacturers, drivers, insurance companies, regulatory authorities, cloud service providers, leasing firms, and roadside infrastructure providers (Xu et al., 2021).

1.1.1 Applications in Financial Transactions

One of the most valuable applications of blockchain in IoV lies in its ability to secure financial transactions and contractual data across organizational boundaries (Kaur et al., 2023). In a decentralized vehicle network, blockchain can serve as the trusted layer for:

- Vehicle Identity and Ownership: Vehicles can be assigned blockchain-based identities to manage digital credentials such as ownership certificates, registration, and roadworthiness data. This supports tamper-proof leasing and resale transactions, even across jurisdictions.
- Insurance Claims Automation: In the event of a crash, blockchain can log the incident with sensor-based evidence and trigger an automatic smart contract to process claims. The immutability of this data ensures that the record is accepted as proof by insurance companies.
- Dynamic Tolling and Micro-Payments: Blockchain wallets linked to vehicles enable real-time toll payments, EV charging billing, or vehicle-to-vehicle service payments using cryptocurrencies or stablecoins (MOBI Consortium, 2021; Xu et al., 2021).
- Maintenance and Warranty Logs: Keeping maintenance history, repairs, recalls, and warranty details on a blockchain makes it easy to track everything.

This helps reduce fraud and builds trust when buying, selling, or insuring a vehicle, since everyone can see the real condition and service record of the car.

- Fleet Finance and Usage-Based Contracts: Commercial Fleet operators can use blockchain to make leasing smarter, enable pay-as-you-go financing, and track carbon offsets more accurately. With verified emission records stored on the blockchain, everything stays transparent and easy to audit.

There are various use cases to highlight blockchain's ability to reduce fraud, simplify audits, and eliminate intermediaries in complex vehicular finance environments(Kaur et al., 2023).

1.1.2 Legal and Regulatory Implications

Despite its advantages, blockchain introduces substantial regulatory and policy challenges when used in financial services, especially in the IoV domain:

- Conflict with Data Erasure Laws: Blockchain's immutability directly conflicts with privacy rights such as the "right to be forgotten" under the General Data Protection Regulation (GDPR) and similar provisions under the California Consumer Privacy Act (CCPA) and India's Digital Personal Data Protection (DPDP) Act.
- Jurisdictional Uncertainty: Blockchain works across borders, but laws don't always keep up. Rules about smart contracts, taxes on digital assets, and where data can be stored differ from country to country. For example, a leasing agreement made on the blockchain in Japan might not be legally recognized in the U.S. or Europe unless the laws between those regions are aligned.
- Lack of Legal Framework for Smart Contracts: Smart contracts enable automated execution of lease terms, insurance policies, or payment flows, but legal ambiguity exists regarding their enforceability and how liability is attributed in cases of failure.
- Governance of Permissioned Blockchains: In permissioned blockchain systems used by automotive consortiums or financial regulators, questions arise regarding who controls node access, who maintains consensus protocols, and how malicious participants are managed.

To make sure blockchain is used responsibly, regulators need to work closely with standardization groups like ISO, ITU, and IEEE, along with car industry experts. Together, they can create clear guidelines that help blockchain-based financial systems and heathcare systems (Mahesh et al., 2024) stay compliant with the law

From Automation to Autonomy: Blockchain's Role in Adaptive Systems

In the general context of Autonomous and Adaptive Systems, blockchain supports autonomy of execution and adaptivity of compliance. Autonomy occurs from deterministic smart contracts that can make decisions independently; adaptivity occurs when contract logic interacts with changing data streams, AI predictions, or policy updates.

For instance, a vehicle-leasing system based on blockchain can automatically change payment cycles using FL-derived wear-and-tear forecasts while still being inspectable by regulators. The blockchain ledger is the substrate of governance to guarantee that adaptation of whatever form is open, reversible, and compliant.

Therefore, blockchain technology in IoV finance is not only a decentralization tool; it is the institutional memory and governance system that makes machine decisions accountable, auditable, and legally interpretable. Coupled with federated learning it turns the financial Internet of Vehicles into a self-regulating, policy-responsive economic system.

1.2 Federated Learning: Privacy-Preserving Intelligence for Finance-Driven Mobility

Federated learning (FL) is an artificial intelligence technique that trains machine learning models collaboratively across distributed nodes without moving raw data to a central server. Instead, each participant such as a vehicle or roadside unit trains a model locally and transmits only the updated parameters or gradients. These updates are then aggregated into a global model.

1.2.1 Use Cases in Financial Services

In finance-driven IoV systems, FL enables vehicles and service platforms to develop intelligent financial models without exposing raw user data. Common use cases include:

- Usage-Based Insurance (UBI): Vehicles contribute model updates that analyze braking, speed, distance, and time-of-day driving patterns. These help insurers set dynamic premiums without having access to the driver's full driving history or biometric data.
- Credit Risk Assessment: Federated credit scoring models can be trained across mobile apps, vehicle dashboards, and embedded finance platforms without violating customer confidentiality or consent laws.
- Fraud Detection Models: Federated architectures allow insurers and leasing platforms to collaboratively train fraud detection systems that monitor ab-

normal behavior (e.g., odometer tampering or false damage claims) across thousands of vehicles.

- EV Charging and Microfinance Prediction: FL models can optimize energy pricing, demand forecasting, and real-time finance eligibility across smart charging stations without sharing user identity.

1.2.2 Challenges and Governance Needs

The application of FL in financial contexts introduces new regulatory and governance challenges, especially when these models influence real-time pricing, risk scoring, or eligibility decisions:

- Data Poisoning and Model Inversion: Malicious nodes may inject poisoned data to skew financial predictions, or extract sensitive information from shared model updates using gradient inversion or membership inference attacks.
- Non-IID Data and Model Bias: IoV data is often non-IID (non-independent and identically distributed). For instance, rural and urban driving patterns differ significantly. Without proper handling, FL models may show systemic bias against certain geographic, socio-economic, or demographic groups.
- Accountability and Legal Liability: When a federated AI model adjusts an insurance premium or denies a financial service, who is legally liable the car manufacturer, the model aggregator, or the fleet operator? Regulatory clarity on fault attribution in decentralized decision-making is urgently needed.
- Auditability and Model Explainability: For compliance, especially in finance, models must be transparent and explainable. Regulators may require audit logs for training data lineage, update history, and performance benchmarks. This is particularly important under GDPR's Article 22, which grants users the right to an explanation for automated decisions affecting them.

To support these requirements, regulatory sandboxes, privacy-preserving audits, and federated governance protocols must be introduced.

1.3 Integration Potential: The Synergy Between Blockchain and FL in Finance

The fusion of blockchain and federated learning presents a unique opportunity to construct finance-ready IoV ecosystems that are secure, privacy-preserving, and legally compliant

(Rajamanickam et al., 2025)

- Blockchain provides verifiable logs of model updates, participation history, and smart contract execution.
- FL ensures that financial insights are generated without raw data exposure, maintaining user privacy.
- Smart contracts can automate compliance enforcement, such as rejecting updates from unauthorized or non-compliant vehicles.
- Regulatory authorities can join the blockchain network as observer nodes, enabling transparent oversight of FL model development and deployment.

Together, these technologies open the door to exciting innovations in the Internet of Vehicles (IoV), like decentralized insurance, global leasing platforms, and AI audits powered by blockchain all designed to make connected vehicles smarter and more secure as depicted in *Figure 1.*

Figure 1. Overview of Policy and Regulatory Considerations for Blockchain and Federated Learning in IoV

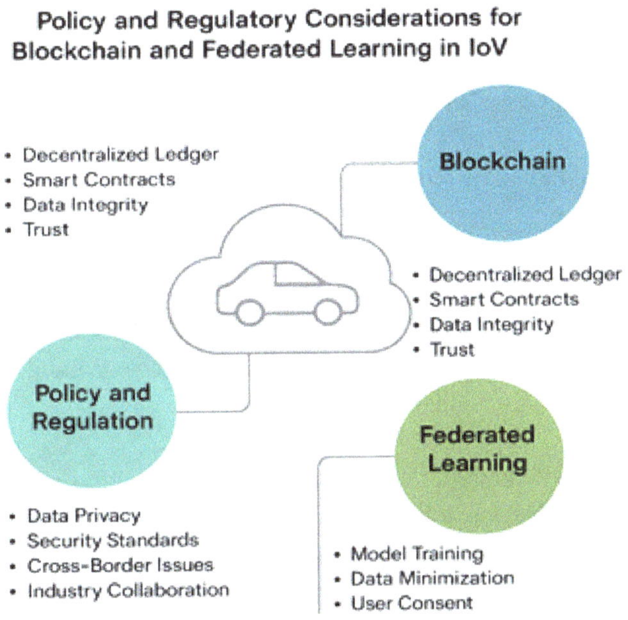

2. BLOCKCHAIN IN FINANCIAL IoV APPLICATIONS

2.1 Overview of Blockchain Capabilities in IoV Finance

Blockchain is a foundational technology for building trust, traceability, and transactional integrity in finance-focused IoV systems. Unlike centralized ledgers managed by banks or authorities, blockchain operates as a decentralized and immutable record, allowing multiple stakeholders such as insurers, fleet operators, leasing companies, governments, and vehicle owners to maintain synchronized access to shared financial and operational data.

Within the IoV domain, blockchain enables:

- Distributed identity verification, allowing vehicles to authenticate themselves securely across jurisdictions.
- Secure audit trails for insurance claims, toll records, vehicle leasing, and service history.
- Tamper-proof storage of compliance data for emissions, road taxes, or financing contracts.
- Token-based payments, enabling microtransactions between vehicles, infrastructure, and services.

2.2 Smart Contracts for Automated Financial Services

Smart contracts are self executing programs deployed on a blockchain that automatically enforce predefined terms and conditions without human intervention. In IoV finance, smart contracts reduce operational delays, eliminate the need for intermediaries, and minimize the possibility of disputes or fraud.

Common applications include:

- Insurance settlements: A vehicle involved in an accident could trigger a smart contract that collects telemetry data (e.g., speed, impact location), verifies it through roadside units or third-party oracles, and initiates automated claim settlement.
- Leasing and rental agreements: Vehicles can enforce time-bound leases by automatically logging usage metrics and issuing payments or late fees based on contract terms.
- Toll payments and congestion pricing: Smart contracts enable real-time, distance-based toll collection that adjusts based on traffic conditions and location.

- EV charging transactions: Charging sessions can be logged and billed through blockchain-integrated stations, with dynamic pricing models implemented in code.

Smart contracts add an additional layer of financial automation and enforcement, particularly where vehicles operate autonomously or cross regional boundaries.

2.3 Privacy, Compliance, and the Right to Be Forgotten

While blockchain immutability ensures transparency and auditability, it also introduces major regulatory tensions. Financial services often involve personally identifiable information and sensitive behavioral data that are subject to strict legal protections under privacy laws.

Key legal conflict include:

- Right to erasure (GDPR Article 17): Blockchain irreversible records contradict data deletion rights.
- Consent and purpose limitation: Once data is placed on-chain, it becomes difficult to restrict access or reuse.
- Cross-border data flow restrictions: Data recorded in one jurisdiction may be subject to legal scrutiny in another.

  To navigate these tensions, blockchain systems in IoV must adopt privacy-enhancing designs, including:
- Off-chain data storage: Only non-identifiable data hashes or pointers are recorded on-chain.
- Zero-knowledge proofs (ZKPs): These allow validation of data or claims (e.g., driving behavior scores) without revealing the underlying data.
- Selective disclosure mechanisms: Access control protocols limit what different users (e.g., insurers, governments, OEMs) can see.

Laws need to evolve to fit how blockchain works. This might include officially recognizing things like cryptographic deletion, acceptable levels of anonymization, and systems that are built to follow legal rules from the start.

2.4 Governance Models for Permissioned Blockchains

Financial systems in the Internet of Vehicles (IoV) usually run on permissioned blockchains, where only trusted parties like automakers, insurance companies, regulators, and fleet operators can manage the network or approve transactions. In these setups, figuring out who controls what and how decisions are made becomes a key challenge.

Challenges include:

- Node control and access rights: Who gets to read, write, or validate transactions?
- Consensus mechanisms: Are Proof-of-Authority (PoA), Practical Byzantine Fault Tolerance (PBFT), or other consensus models used?
- Dispute resolution: What happens when data is incorrect, fraudulent, or contested?
- Consortium blockchains: Multiple automotive and financial entities form a shared network under a charter and neutral governance body.
- Regulatory observer nodes: Public authorities are granted read-only or audit rights to ensure legal oversight.

With smart contracts, rules like spending limits, access tracking, and data retention can be built directly into the blockchain(Mahesh et al., 2024). But as vehicles start handling financial activities across borders, strong governance becomes crucial to ensure everything stays legal, transparent, and accountable.

2.5 Interoperability and Standardization Challenges

Since IoV networks operate globally, blockchain systems need to work smoothly with these interoperability gaps:

- Data formats and APIs: Lack of standardized schemas for logging tolls, energy usage, insurance events, etc.
- Cross-chain communication: Smart contracts or digital identities issued on one chain must be verifiable on others.
- Policy enforcement divergence: Different regions often have their own rules about how smart contracts should work or how long data must be kept. To tackle this, global groups like ISO, IEEE, and ITU are working to create common standards for using blockchain securely and legally in the Internet of Vehicles. Still, getting everyone to adopt these standards and adapt them locally is a big challenge.

2.6 Blockchain Use Cases in Financial IoV

Multiple real world initiative are exploring blockchain in mobility finance:

- Volkswagen's Digital Ledger for Vehicle Histories: Uses blockchain to track maintenance, repairs, and leasing milestones to improve resale trust.
- Dubai's Smart Mobility Initiative: Leverages blockchain to integrate car registration, tolling, and insurance in a unified mobility ledger.
- MOBI Consortium: A group of automakers including Ford, BMW, and GM exploring decentralized mobility identities and smart contracts for tolling and payments.
- Tata Elxsi's Autonomous Mobility Sandbox (India): Blockchain is used to secure vehicle data exchanges for leasing and finance evaluation.

These examples tell of blockchain's growing role in enabling fair, auditable, and decentralized finance within smart mobility ecosystems.

3. FEDERATED LEARNING IN FINANCIAL IoV

Federated Learning lets multiple devices like vehicles work together to train AI models without sharing their raw data with a central server. This is especially useful in the Internet of Vehicles, where sensitive information like driving habits, location history, and biometric data needs to be kept private while still being analyzed.

In finance-oriented IoV environments, FL enables the training of predictive and decision-support systems for insurance pricing, fraud detection, leasing risk assessment, credit evaluation, and usage-based mobility payments. These systems can learn from real-world patterns while remaining compliant with regional data protection regulations such as GDPR, CCPA, and India's DPDP Act.

This section tells of the operational structure of FL in IoV finance, its use cases, compliance challenges, and the emerging governance frameworks required to safely integrate it into vehicle-centred financial systems.

3.1 Architecture of FL in IoV Financial Applications

In the Internet of Vehicles, Federated Learning usually works like this: each vehicle or roadside unit trains a local AI model using its own data. Instead of sharing the raw data, they send updates like model improvements to a central system. This system combines all the updates to improve a global model and then sends the updated model back to the vehicles. The key point is that no raw data ever leaves the vehicle, keeping privacy intact.

In finance-focused deployments, this architecture supports:

- Real-time on-board model training for individual driver risk profiles.

- Edge-based anomaly detection to identify fraudulent financial behaviors.
- Collaborative model evolution across a fleet of vehicles from the same leasing firm, without centralized exposure of driver histories.

This process enhances data sovereignty and minimizes the risk of surveillance or misuse of private financial and behavioral data.

3.2 Regulatory and Legal Considerations

While federated learning aligns well with the principles of data minimization and privacy, it remains entangled in a range of unresolved legal and regulatory challenges especially when applied to financial services. One major area of concern lies in jurisdictional complexity. IoV systems often involve vehicles crossing national borders, training models on data collected in different territories. This raises difficult questions about which region's data protection laws apply when model updates are transmitted across borders. For example, if a vehicle in Germany contributes updates to a model that is later deployed in India, does the European Union's GDPR continue to apply to how that model is used? Similarly, if updates collected in California influence financial predictions in another U.S. state or country, must those predictions remain subject to California's stricter privacy standards.

The uncertainty around how federated learning works across borders makes it harder for banks, insurers, and fleet operators to follow the rules. It's tricky to enforce important legal requirements like getting user consent, limiting how long data is kept, and managing who can access information. More importantly, federated learning challenges the usual definition of 'data processing' under privacy laws. Even though raw data isn't shared, the model updates sent around might still be seen as personal data if they can be used to uncover sensitive details. As laws catch up, regulators will need to decide if these model updates fall under current data protection rules or if new rules are needed to cover these AI-specific cases.

Equally important are the security risks associated with FL in financial applications. Contrary to the assumption that federated learning is inherently secure, multiple attack vectors have emerged in recent years that threaten both data integrity and confidentiality. Gradient inversion attacks can be used to reconstruct portions of the original training data from shared model updates. Similarly, membership inference techniques may reveal whether a user's data was included in model training. This could be especially problematic in sensitive financial contexts, such as health-based insurance pricing or creditworthiness modeling, where the mere presence or absence of certain behavior can carry reputational or monetary consequences. Additional threats like poisoning attacks where a malicious client intentionally sends corrupted updates can skew the global model, potentially to favor or harm certain individuals or financial profiles.

3.3 Governance and Liability in Federated Financial AI

One of the most complex challenges associated with federated learning in financial IoV systems is the issue of accountability. As FL introduces multiple participants across jurisdictions, platforms, and service roles, assigning liability when things go wrong becomes less straightforward. For instance, when an insurance premium is unfairly calculated due to a flawed federated model, several actors may have contributed to that outcome: the aggregator who combined the updates, the edge clients who provided biased or inaccurate data, the developers of the algorithm, or even the financial entity relying on the model's outputs. Determining which party bears legal or financial responsibility in such scenarios is still an open question.

This ambiguity is particularly problematic when financial outcomes have legal consequences. Incorrect decisions about creditworthiness, insurance eligibility, or fraud detection can lead to lawsuits, regulatory penalties, or reputational damage. To mitigate these risks, governance structures need to be established within federated learning networks. These structures must define roles, responsibilities, and fallback mechanisms. Clear policies should identify which participants are authorized to train models, under what conditions training can take place, and what standards must be followed during aggregation. Institutions may also need to establish independent oversight boards or compliance committees to review FL deployments and certify models prior to launch.

Beyond legal liability, governance also encompasses incentive structures. Federated learning systems in finance must ensure that participating nodes such as vehicles, roadside units, or service centers are contributing accurate and representative updates. If nodes can benefit from the global model without meaningfully contributing, the system becomes vulnerable to free-riding and decline in performance. Similarly, malicious participants could attempt to corrupt model outcomes for financial gain. To prevent these behaviors, governance frameworks may include mechanisms such as contribution validation, update scoring, or token-based rewards. These can be enforced either through smart contracts or through regulatory rules, depending on the system architecture.

Institutions must also prepare for the long-term maintenance and updating of federated models. Financial data is highly dynamic; what is accurate today may be misleading tomorrow. Continuous learning systems must be equipped with versioning controls, rollback options, and ongoing evaluation metrics. At each stage of model evolution, there must be a mechanism for ensuring fairness, robustness, and compliance. These requirements highlight the need for a federated learning governance model that is not only technical in nature but also legal, ethical, and operational.

3.4 Toward International Harmonization

The final challenge and perhaps the most significant for financial IoV systems is achieving cross-border regulatory harmonization for federated learning. Vehicles do not operate in isolation. International fleets, cross-border logistics providers, and global auto manufacturers routinely move across jurisdictions. A federated model that begins training in one country may end up serving financial decision-making purposes in another. Yet today's regulatory frameworks are not designed for this kind of distributed, transnational data interaction.

This regulatory fragmentation increases the compliance burden for institutions, which must simultaneously adhere to overlapping or even contradictory privacy, AI, and financial laws. For example, consent requirements for data processing in Europe may not match those in the United States or India. Rules about algorithmic transparency in the EU may exceed those required elsewhere. Without coordinated standards, this creates uncertainty, discourages innovation, and limits the scalability of FL-based financial services.

One potential solution lies in the development of global regulatory frameworks modeled after existing initiatives such as the OECD Principles on Artificial Intelligence, the ISO/IEC standards for AI systems, and the European Commission's AI Act. By establishing minimum thresholds for fairness, accountability, transparency, and safety, these frameworks can provide a basis for interoperable compliance. Governments may also collaborate through regulatory sandboxes, in which experimental technologies like FL are tested under regulatory supervision. These controlled environments offer a valuable way to study risks and benefits while giving institutions room to innovate responsibly.

In the long term, achieving safe and ethical deployment of federated learning in financial IoV systems will require not just technical improvements but also cross-sectoral and international collaboration. Only through shared standards, mutual trust, and proactive governance can we unlock the full potential of this transformative technology while ensuring that users remain protected, services remain fair, and systems remain resilient.

4. INTEGRATION OF BLOCKCHAIN AND FEDERATED LEARNING IN FINANCIAL IoV SYSTEM

The evolution of intelligent, connected vehicles has paved the way for data-driven financial services that rely on real-time behavior analysis, secure digital identity, and decentralized access control. However, achieving the necessary balance between trust, automation, privacy, and regulatory compliance requires more than a single technological innovation. It demands a fusion of multiple technologies that together can provide a resilient and ethically aligned foundation. The integration of blockchain and federated learning (FL) stands out as a promising solution, particularly in the context of the Internet of Vehicles (IoV) and its growing intersection with financial technologies.

Both federated learning and blockchain emerged from different problem spaces FL to solve the challenge of training AI models without sharing raw data, and blockchain to secure and decentralize data storage and value exchange. Their integration addresses the weaknesses of each technology while amplifying their strengths, particularly in decentralized financial systems. In FL, the major limitation is trust there is no built-in way to verify whether participants are honest, whether the model update process is transparent, or whether the system is being manipulated by adversarial actors. Blockchain, by contrast, ensures immutability, transparency, and auditability, allowing the creation of secure records of all FL transactions, model updates, and contributor behaviors.

In the financial dimension of IoV, this integration is transformative. Vehicles constantly generate and act on data that can influence financial decisions: determining insurance premiums, approving leases, processing in-vehicle purchases, or verifying cross-border transactions. Federated learning can be used to develop context-aware financial models predicting a driver's risk profile, for example, or optimizing vehicle usage billing based on region, road type, and behavior. Yet these models, without secure and verifiable provenance, could be manipulated or suffer from poisoned data, ultimately impacting the financial system's integrity. Blockchain addresses this issue by recording every training update, submission, and aggregation step on a verifiable and immutable ledger.

By anchoring FL training events to the blockchain, developers and regulators gain access to a tamper-proof history of the model's development. Each vehicle's contribution can be cryptographically signed, timestamped, and verified, providing a transparent trail of who trained what, when, and how often. In the context of usage-based insurance, for instance, a vehicle's local model may estimate driver risk and send updates to a central model. The blockchain would record the metadata about that update such as contribution score, confidence level, and anonymized device ID ensuring that only legitimate updates are included. This prevents both data falsification and model manipulation. If a dispute arises say, a user contests their insurance rate the system can be queried to verify what data contributed to that pricing model and whether it complied with policy constraints.

Moreover, this integration supports programmable trust through the use of smart contracts. These are self-executing pieces of code deployed on the blockchain that automatically enforce rules. Smart contracts can be used to validate whether a participant is eligible to contribute to a federated training round, to automatically reward honest participation with tokens, or to reject poisoned or underperforming updates. In EV charging payments, for instance, a smart contract could combine vehicle data from an FL model (predicting preferred charging stations and demand) with blockchain-based tokens or digital assets to process payments in real time, without requiring third-party intermediaries.

The integration also benefits cross-organization financial collaboration, especially among insurers, banks, mobility providers, and manufacturers. In traditional centralized settings, these entities often hesitate to share data due to concerns over privacy, intellectual property, or competitive advantage. However, with federated learning, data remains local, and only model updates are shared. When combined with blockchain, even those updates can be verified, traced, and governed through shared consensus mechanisms. This makes it possible for a consortium of stakeholders to co-train a fraud detection model or lease pricing model, while using blockchain to ensure the process remains fair, accountable, and auditable for all parties involved.

A major value proposition of blockchain-FL integration is its capacity to resolve trust gaps in decentralized systems. In a federated learning network spanning hundreds or thousands of vehicles, it is difficult to determine whether all participants are contributing in good faith. Some may attempt to free-ride, downloading the improved global model without contributing meaningful updates themselves. Others may engage in model poisoning, deliberately injecting biased or malicious gradients into the system to disrupt its accuracy. Blockchain provides the enforcement mechanism to detect and prevent such behavior. By logging all actions and enforcing reputation systems through smart contracts, bad actors can be flagged, excluded, or penalized automatically. This creates a self-regulating ecosystem where participation is tied to measurable trustworthiness.

In the context of vehicular leasing, this trust model becomes essential. Leasing agreements often depend on predictions about how a vehicle will be used, how quickly it will depreciate, and how well it will be maintained. These factors can now be inferred through FL models trained on localized usage patterns. Blockchain can be used to maintain a history of updates to these predictive models, store the leasing terms, and even trigger smart contracts for payments, damage claims, or contract renewal based entirely on validated data. For example, if an FL model predicts that a vehicle has been under high operational stress, triggering early wear and tear, a smart contract could automatically adjust the lease payment structure, deducting penalties or initiating a service check.

Yet integrating blockchain and FL is not without its complexities. Technically, one of the main hurdles is throughput and latency. Federated learning can require frequent updates especially in time-sensitive environments such as vehicular networks while blockchain networks, particularly public ones, are relatively slow. Writing every training interaction to the chain is neither scalable nor efficient. Hybrid solutions are often proposed, where the majority of operations happen off-chain, and only key events or summaries are recorded on-chain. This allows for auditability and trust without overwhelming the network or inflating transaction costs.

To enhance scalability, permissioned blockchain systems are often preferred in IoV settings. These are blockchains where only vetted participants such as automakers, insurers, or governments can validate blocks or submit transactions. While this reduces decentralization compared to public chains, it increases performance and control, which are crucial in regulated environments. These systems can be designed to integrate seamlessly with federated learning frameworks, using predefined governance rules to manage model access, update verification, and participation scoring.

Another critical issue is data leakage through metadata. Even if raw data is not transmitted, the metadata associated with FL and blockchain transactions such as update times, location tags, or node identifiers could potentially be used to infer private information. To mitigate this, advanced cryptographic tools must be incorporated. Techniques such as zero-knowledge proofs, secure multi-party computation, and homomorphic encryption allow systems to validate model contributions or enforce rules without revealing underlying data. For example, a vehicle might prove that it has contributed a high-quality model update that meets all regulatory criteria, without disclosing any behavioral or biometric data. These privacy-preserving cryptographic techniques are vital for compliance with international data protection laws, particularly in finance.

From a legal standpoint, the fusion of blockchain and federated learning in financial IoV contexts introduces novel governance challenges. Smart contracts, though operationally reliable, still exist in a legal gray area. Many jurisdictions do not yet recognize them as enforceable contracts, particularly when they involve dynamic pricing or automated financial decisions. For example, an insurer that relies on a smart contract to adjust premiums based on FL-derived scores may face legal scrutiny if users contest the fairness of the model or the decision process. Ensuring legal validity may require hybrid systems where smart contracts operate under human supervision or are supplemented by legal fallback clauses coded into their logic.

Moreover, questions of jurisdictional conflict arise in multinational deployments. A federated model may involve vehicles in Europe, infrastructure in Asia, and service providers in North America, each governed by different data, AI, and financial regulations. Blockchain's decentralized nature only compounds this complexity, as ledgers are distributed and borderless by design. To address this, regulatory frameworks must evolve to support federated compliance, where system-level behaviors are evaluated for compliance across multiple regions. Standards organizations and regulatory bodies may need to collaborate to define compliance schemas, shared vocabularies, and testbeds for blockchain-FL systems in finance.

As the Internet of Vehicles continues to expand and financialization becomes embedded in mobility itself, this convergence will likely define the backbone of future transportation economies. However, its success will depend not only on technological innovation but also on cross-sector collaboration, ethical foresight, and regulatory adaptability. Governments, developers, insurers, and automakers must work together to ensure that blockchain-FL ecosystems promote fairness, privacy, interoperability, and legal accountability at scale.

5. CASE STUDIES AND GLOBAL REGULATORY LANDSCAPE

The integration of blockchain and federated learning in the financial Internet of Vehicles is not merely a conceptual or experimental idea it is being actively explored and implemented across various regions of the world. Governments, regulators, automotive firms, and technology companies are beginning to adopt, test, and regulate these technologies in different ways based on local legal frameworks, infrastructural readiness, and economic priorities. These global case studies offer critical insights into both the technical feasibility and the policy alignment necessary to enable compliant, secure, and effective adoption of decentralized, intelligent financial systems in IoV environments.

In this context, analysing the global landscape becomes essential not only to understand the opportunities and risks in different jurisdictions but also to inform international standardization efforts that could facilitate interoperability and trust.

5.1 Overview of Global Regulatory Trends

Globally, three dominant trends are emerging in how blockchain and FL integration in IoV finance is being governed. First, there is a strong push toward data sovereignty and decentralization, often driven by growing mistrust of large data monopolies and international platforms. Governments are increasingly requiring that personal and financial data be processed and stored within national borders. Federated learning naturally aligns with this principle by allowing data to remain on edge devices. Second, there is a rising emphasis on AI accountability and fairness. Countries are introducing regulations to ensure that automated decision-making systems used in finance are transparent, explainable, and free from bias particularly when decisions affect credit, insurance, or mobility rights.

5.2 Regional Case Studies

In the European Union, regulatory alignment around data privacy and AI fairness is highly developed. The General Data Protection Regulation (GDPR) sets strict limits on how personal data, including inferred data used in FL models, may be processed. It also guarantees individuals the right to understand and challenge automated decisions, such as those used in dynamic insurance pricing. The proposed AI Act further tightens requirements for high-risk AI systems including financial AI tools demanding risk assessments, human oversight, and robust documentation. Nevertheless, several EU-funded projects have been piloting blockchain-FL integrations. For example, the GAIA-X initiative envisions a federated cloud infrastructure for Europe where mobility data and financial services can be built in a compliant, decentralized architecture. Automotive companies in Germany and France have collaborated on early FL models for vehicle diagnostics and behavior-based insurance, often anchored by permissioned blockchains for audit and data lineage tracking.

In Japan, a different model has emerged. The country has been relatively progressive in regulating both blockchain and AI, though it has maintained a hands-off approach to centralized control. The Financial Services Agency of Japan (FSA) has issued guidelines that support innovation while emphasizing risk mitigation, particularly in fintech. Automotive firms like Toyota have invested heavily in blockchain and FL experiments for insurance, safety, and fraud detection. Toyota's Blockchain Lab and its Mobility Platform have incorporated blockchain identity systems and federated machine learning models to improve both privacy and fraud resistance in vehicle data. Japan's regulatory authorities have shown interest in using blockchain-FL systems to enable secure, auditable micro-insurance markets for autonomous vehicles and regional mobility providers.

In the United Arab Emirates (UAE), the government's ambition to become a global leader in blockchain has led to rapid implementation. The Dubai Blockchain Strategy aims to have all government-related documents secured via blockchain, with mobility and transport as key focus areas. The UAE Insurance Authority and the Roads and Transport Authority have been experimenting with smart contract-based vehicle leasing, blockchain-backed driver scoring systems, and localized FL-based accident prediction models. While the regulatory regime is more centralized, the government has created well-structured sandboxes and issued licenses for fintech experimentation, allowing companies to develop FL systems without violating local data laws provided that proper encryption and auditing mechanisms are in place.

In India, the emergence of the Digital Personal Data Protection (DPDP) Act has brought both opportunity and caution. On one hand, federated learning offers a method to comply with data localization and consent principles enshrined in the Act. On the other hand, the lack of clear legal definitions for AI liability and smart contract enforceability limits the rapid deployment of integrated blockchain-FL systems. Nevertheless, Indian automotive and insurance companies have begun investing in pilot projects. Tata Elxsi, for instance, has proposed FL-based fleet monitoring tools that support dynamic risk pricing, while startups are exploring token-based driving reward systems secured by blockchain. Regulatory support is emerging through initiatives by NITI Aayog and the Reserve Bank of India's fintech innovation hub, but full-scale implementation is still in early phases.

5.3 Comparative Policy Analysis

Comparing these case studies reveals distinct regulatory philosophies and deployment strategies. The EU leads in privacy-first regulation and high standards for AI accountability, which makes FL an attractive model but introduces challenges for blockchain's immutability. Japan and the UAE, while technologically ambitious, take more flexible, innovation-friendly approaches that allow faster piloting of integrated systems but may lack robust mechanisms for handling unintended outcomes or long-term legal implications. India represents a balancing model combining rising data protection awareness with a strong push for domestic technological innovation. This makes it a potential testbed for context-sensitive governance models that align blockchain and FL with both national sovereignty and financial inclusion goals.

Notably, one of the most challenging aspects across regions is the cross-border operability of federated learning models.

5.4 Lessons and Patterns from Global Implementation

Several important lessons emerge from global regulatory practice. First, trust infrastructure is essential. Without immutable and verifiable records, regulators hesitate to accept automated models as valid in financial settings. Blockchain fills this role, especially when linked to clear governance and audit procedures. Second, privacy is a non-negotiable feature, not an optional add-on. FL's ability to keep data local satisfies privacy regulations across all examined jurisdictions, but only if encryption, participant verification, and metadata minimization are properly enforced.

Third, regulatory bodies strongly prefer explainable systems. Where AI is used in insurance, credit scoring, or leasing, users must be able to contest outcomes and receive explanations. Blockchain-FL systems must be designed to enable this through built-in logging, model versioning, and human-readable summaries of model decisions. Finally, the most successful pilot projects are those operating within regulated sandboxes that allow both innovation and oversight. These controlled environments not only de-risk experimentation but also enable policymakers to observe how emerging technologies behave in real-world conditions before enacting permanent rules.

6. PROPOSED FRAMEWORK AND ROADMAP FOR POLICY HARMONIZATION

The convergence of blockchain and federated learning in the financial Internet of Vehicles offers a transformative pathway to secure, decentralized, and privacy-preserving data and financial ecosystems. However, as established in the previous sections, this convergence also introduces significant legal, technical, and governance complexities. Fragmented regulatory frameworks, diverse technological readiness levels, and inconsistent interpretations of data privacy, AI accountability, and digital contract enforceability all threaten to slow or even halt the adoption of such integrated systems. To address these barriers and ensure scalable, responsible innovation, there is a pressing need for a unified policy framework a set of structured guidelines that harmonizes legal, technical, and ethical principles for the combined deployment of blockchain and FL in financial IoV applications.

6.1 Need for a Unified Policy Approach

Current policy structures were designed in an era of centralized data, manual auditing, and siloed services. They do not fully account for the distributed, real-time, and autonomous nature of blockchain and FL systems. Financial applications especially those embedded in vehicles operate at the edge, beyond traditional enterprise boundaries, and often across jurisdictions. This demands a governance model that can provide clarity on questions of consent, liability, transparency, and interoperability. Without such a framework, institutions may face legal uncertainty, reduced user trust, and resistance from regulatory authorities.

Moreover, the overlap between financial regulations, AI ethics, and mobility law creates a multi-layered compliance environment. A federated learning model that dynamically adjusts an insurance premium may need to comply simultaneously with AI explainability standards, data protection requirements, and financial fairness laws. Blockchain's role in storing audit logs, automating payouts, or verifying identity may place it under payment processing laws or even crypto-asset regulation. A unified framework must address these interdependencies holistically rather than through disjointed, domain-specific rules.

6.2 Layered System Architecture for Blockchain–FL Integration

To transform these challenges into an implementable system, the integration architecture can be organized into four interdependent layers, each serving distinct functions within the hybrid governance ecosystem. This structure is illustrated conceptually in *Figure 2* described below.

Figure 2. Multi-Layer Federated Governance Framework Integrating Blockchain and FL in Financial IoV Systems

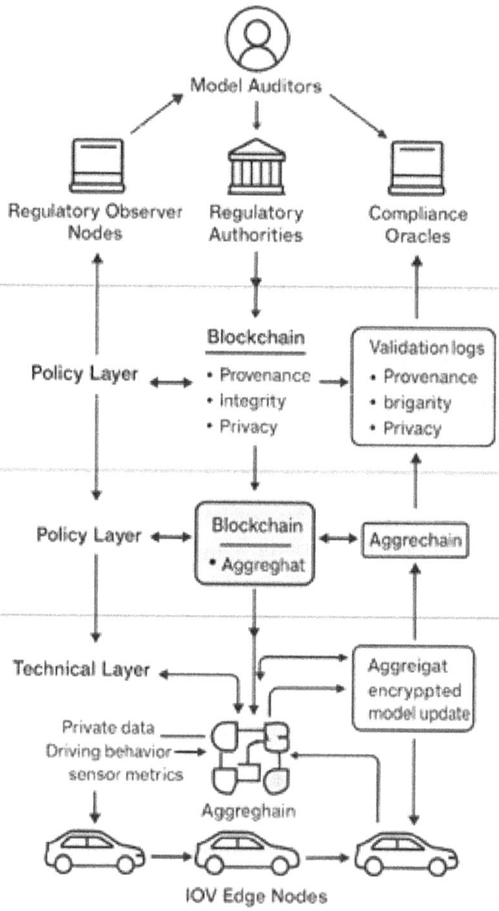

Layer 1: Data and Model Layer (Edge Intelligence)

At the lowest level, vehicles and roadside units perform local data collection and model training.

- **Function:** Secure on-device learning with privacy-preserving mechanisms.
- **Outputs:** Encrypted gradient updates, metadata on data provenance, and model confidence scores.
- **Relevance:** This layer embodies *adaptivity*—vehicles evolve their intelligence based on real-world behavior.

Layer 2: Blockchain Ledger Layer (Trust Fabric)

This layer records cryptographic proofs of model updates, contributor identities, and compliance metadata.

- **Function:** Provides tamper-proof logs, tokenized incentive systems, and automated smart- contract enforcement.
- **Relevance:** This layer embodies *autonomy*—actions are self-executing and auditable without central control.

Layer 3: Governance and Compliance Layer

At this level, institutional actors (insurers, regulators, auditors) interact through permissioned validator nodes.

- **Function:** Monitor on-chain transactions, verify model fairness certificates, and resolve compliance discrepancies.
- **Mechanisms:** Regulatory observer nodes and compliance oracles that synchronize legal changes (e.g., new data retention limits or AI Act provisions) with contract logic.
- **Relevance:** Ensures *institutional accountability* and bridges technical autonomy with legal oversight.

Layer 4: User Interaction and Policy Interface Layer

The final layer exposes application-level functionalities—smart dashboards, consent management tools, and audit interfaces for end-users and authorities.

- **Function:** Visualize transparency, facilitate consent revocation ("right to be forgotten"), and provide human interpretability of automated actions.
- **Relevance:** Aligns *human rights and ethical values* with machine-driven decision ecosystems.

6.3 Ethical and Legal Safeguards

A federated governance framework cannot succeed without embedding ethical safeguards directly into the system. This includes provisions for fairness, accountability, transparency, and contestability. In financial IoV systems, where decisions about pricing, leasing, and service eligibility can significantly impact users, it is essential that outcomes are not only technically correct but also socially acceptable.

First, fairness audits must be integrated into both federated model training and blockchain logging. These audits can detect discriminatory behavior, such as premium rates that disproportionately penalize certain demographics or geographic regions. Automated fairness metrics can be calculated as part of the model update process, and if thresholds are violated, contributions can be flagged or discarded. These audit results should be written to the blockchain for regulatory visibility.

Second, explainability mechanisms must be built in from the ground up. Where models are used to make financial decisions, users should have the right to receive an understandable explanation of why a particular decision was made and what data influenced it. Federated learning complicates this by abstracting the model training across thousands of devices. Therefore, explainability must include model provenance trails, summary dashboards, and visualizations that communicate how the system reached its decision.

Third, the framework must include redress mechanisms for contested outcomes. If a vehicle owner disagrees with an insurance pricing decision or lease adjustment, they must be able to challenge it and receive timely resolution. Blockchain provides the necessary transparency to support such investigations, while platform governance policies define how appeals are handled. These mechanisms must be embedded into smart contracts or platform protocols not merely offered as post-hoc manual support.

Finally, legal harmonization policies must be incorporated. This includes clauses about jurisdiction, legal recognition of smart contracts, and mutual compliance certifications for cross-border deployments. The governance framework should reference international standards wherever possible, such as ISO/IEC for AI governance, FATF recommendations for digital identity, and OECD AI Principles for ethical design.

6.4 Strategic Collaboration and International Alignment

No single government or organization can define or enforce global rules for blockchain and FL in financial IoV systems. Therefore, the framework must encourage strategic collaboration across borders and sectors. Automotive consortia, insurance coalitions, financial regulators, and standardization bodies must work together to develop shared reference architectures, legal templates, and technical APIs that support interoperability.

International alignment begins with shared compliance protocols for instance, a standard for federated model risk grading that can be used across insurers and nations, or a template for smart contracts that encode local tax and service laws. These protocols should be published openly and updated regularly through multi-stakeholder dialogue. Governments can also contribute by enabling bilateral regulatory sandboxes cross-border experimentation environments where companies from different jurisdictions can test integrated blockchain-FL solutions under joint supervision.

6.5 Roadmap for Implementation

To bring this framework into reality, a structured roadmap is necessary. Phase one involves pilot deployment and benchmarking. Institutions must identify test cases such as federated insurance pricing or blockchain-based vehicle leasing and deploy them in controlled environments. Metrics for performance, compliance, user satisfaction, and fairness should be monitored rigorously.

Phase two focuses on governance infrastructure. This includes forming federated oversight committees, developing model audit tools, creating open governance smart contracts, and setting up communication channels between nodes, users, and regulators. Educational campaigns are critical at this stage to inform participants about their rights, roles, and responsibilities.

Phase three involves regulatory integration and standardization. Successful pilots must be documented and translated into formal policy proposals, legal draftings, and compliance toolkits. Governments and regulators must be engaged directly to validate these frameworks and identify points of alignment or conflict with national laws.

Finally, phase four is scale and optimization. Systems that pass regulatory review and demonstrate ethical robustness can be deployed at scale first within jurisdictions, then across borders. Ongoing improvement mechanisms, such as federated community voting or adaptive regulation clauses, ensure that the framework remains relevant in the face of technological and legal change.

7. RISKS, LIMITATIONS, AND FUTURE RESEARCH DIRECTIONS

While the integration of blockchain and federated learning (FL) in financial Internet of Vehicles (IoV) environments holds vast transformative potential, it is not without inherent risks, open limitations, and areas where further research is urgently required. The pursuit of decentralization, automation, and privacy-preserving intelligence must be balanced with caution, oversight, and continuous innovation. This section critically reflects on the technological, legal, ethical, and operational gaps that must be addressed to realize secure, equitable, and scalable financial IoV ecosystems.

One of the most pressing technical limitations is the computational and communication overhead involved in deploying federated learning at scale. IoV environments consist of diverse vehicle types, with varying hardware capacities and communication bandwidths. While newer models may be equipped with on-board processing units capable of executing FL tasks, many legacy systems are not. Furthermore, federated learning requires frequent, high-volume exchange of model parameters especially in time-sensitive financial applications like dynamic pricing or fraud detection. This results in significant communication costs, which, when combined with blockchain logging operations, may exceed the capabilities of vehicular networks, particularly in low-bandwidth or high-latency environments. Future work must focus on lightweight FL algorithms, adaptive update mechanisms, and bandwidth-aware optimization strategies to make such systems practically deployable in heterogeneous and resource-constrained vehicle fleets.

In addition to technical vulnerabilities, legal ambiguity presents a critical obstacle to widespread adoption. Despite growing regulatory interest, there is still limited clarity on how FL-based AI decisions should be evaluated under existing financial and AI regulations. There is also no uniform international agreement on whether encrypted model updates are considered personal data, and whether cross-border transfer of such updates violates data localization mandates. Researchers and policymakers must therefore collaborate to develop compliance-aware architectures, model certification standards, and jurisdiction-sensitive privacy protocols that reflect the realities of global mobility.

Ethical considerations must also remain central to the conversation. FL and blockchain systems can unintentionally reinforce algorithmic bias, particularly if training data reflects existing inequalities. For instance, insurance models trained disproportionately on urban driving data may penalize rural drivers; lease pricing systems may replicate biases based on historical usage or geography. Without active correction, such models risk undermining equity and fairness particularly in financially sensitive domains. Future research should prioritize bias mitigation in decentralized systems, fairness-aware training objectives, and participatory design approaches that include end users, especially from marginalized communities.

Moreover, there is a need to explore the human factors of system use and acceptance. While blockchain-FL systems are technologically compelling, they may remain unintelligible or inaccessible to most vehicle users. If individuals are unable to understand how decisions are made or how to appeal them the system's legitimacy may be compromised. Research into human-computer interaction (HCI) for federated finance, explainable user interfaces, and legal-tech integration will be essential for improving usability, trust, and accessibility.

A promising avenue for future development is the concept of self-sovereign vehicular finance identities (SSVFI). These would allow vehicles and users to own and control their financial data, selectively share FL model contributions, and negotiate contracts using smart contracts based on local data contexts. Combining verifiable credentials, digital wallets, and reputation systems, such identities could enable a new era of user-controlled mobility finance. However, technical standards and legal definitions around digital identity, consent, and contract enforcement must evolve to support this vision.

Finally, the path ahead will benefit from interdisciplinary research and cross-sector collaboration. No single domain whether AI, law, automotive engineering, or finance can independently resolve the complex web of challenges posed by decentralized financial systems in IoV. Cross-cutting research initiatives, public-private partnerships, and regulatory sandboxes must be expanded to ensure that system design reflects not only technological innovation but also ethical reasoning, legal soundness, and user-centered values.

8. STANDARDIZATION AND INTEROPERABILITY FOR BLOCKCHAIN-FEDERATED LEARNING IN FINANCIAL IoV

The successful deployment of blockchain and federated learning (FL) in financial Internet of Vehicles (IoV) applications depends not only on technological maturity and regulatory alignment but also on the presence of widely accepted standards and interoperability frameworks. As vehicles cross jurisdictions, interact with multiple financial institutions, and switch between manufacturers, leasing providers, and insurers, the systems supporting them must be able to operate seamlessly across technological and legal boundaries. Without standardization, even the most secure and innovative systems risk becoming siloed, incompatible, and difficult to scale.

Several international organizations have already begun laying the groundwork for such standards. The International Organization for Standardization (ISO), through its ISO/TC 307 committee, has published guidelines on blockchain terminology, reference architectures, smart contract governance, and interoperability. These documents provide foundational language for describing and comparing systems across vendors and sectors. Meanwhile, the IEEE P7000 series addresses AI ethics, privacy, and explainability including frameworks for ensuring fairness and transparency in algorithmic systems used in critical decision-making. These are particularly relevant to federated financial models that influence creditworthiness, leasing eligibility, and insurance pricing.

In the European Union, the AI Act and Data Governance Act are creating legal incentives for standardization. Under these laws, high-risk AI systems including those used in transport, finance, and insurance must conform to specific technical and operational requirements. To assist with compliance, the EU is funding standardization bodies like CEN-CENELEC, which are working alongside ISO and ETSI to define trustworthy AI protocols. For FL systems used in mobility finance, this means that model updates, training processes, and outcome transparency must follow agreed-upon data formats, audit structures, and documentation templates. Blockchain systems supporting these AI applications must also support interoperable identity schemes and cross-chain data validation, especially for vehicles operating across EU member states.

In Asia, similar efforts are underway. Japan's Robot and AI Technology Strategy Council and India's Bureau of Indian Standards (BIS) have launched working groups to define technical standards for privacy-preserving machine learning, smart contracts, and decentralized identities. China has invested in blockchain standardization through its Blockchain-based Service Network (BSN), which aims to unify blockchain access for commercial and government services. While each of these initiatives is locally governed, they recognize the importance of global harmonization, particularly for transnational financial services.

The automotive sector itself has also recognized the need for consistent interfaces and shared vocabularies. Projects like MOBI (Mobility Open Blockchain Initiative) have proposed standards for vehicle digital identity (VID), decentralized insurance, and trip data monetization. These standards aim to allow cars from different manufacturers to interact with a shared mobility ledger, regardless of brand or infrastructure provider. Likewise, efforts by the AUTOSAR consortium and the Car Connectivity Consortium support interoperable vehicle data formats and communication protocols, which are essential for feeding FL training and blockchain event logs with reliable, standardized data.

Despite this progress, several gaps remain. Most FL frameworks are developed by private companies or research groups and often lack cross-platform compatibility. Differences in encryption, model serialization, and communication layers can prevent FL nodes from collaborating across brands or jurisdictions. Therefore, research into middleware layers, cross-chain bridges, and federated model translators is increasingly important. These tools would allow heterogeneous nodes to participate in a unified federated training round, anchored to a blockchain that is either universal or bridgeable across systems.

Another pressing issue is the lack of semantic interoperability. In financial contexts, terms like "premium adjustment," "risk score," or "vehicle eligibility" may have different meanings depending on the institution or region. Without shared ontologies and financial taxonomies, smart contracts and FL models may misinterpret data, leading to incorrect decisions. Work by the Financial Stability Board, ISO 20022 (financial messaging), and XBRL International may offer useful starting points for aligning language and expectations across blockchain-FL ecosystems in finance.To bridge these gaps, several recommendations can be made. First, governments should incentivize the adoption of open standards by requiring them in public procurement or fintech licensing. Second, industry alliances must continue to fund joint standardization efforts, particularly in the automotive, insurance, and AI sectors. Third, interoperability must be tested through collaborative pilots and shared testbeds, where vendors can validate whether their systems meet baseline compatibility metrics. Finally, open-source reference implementations and compliance certification schemes should be developed, allowing developers and regulators alike to benchmark systems and promote trust in emerging infrastructures.

9. CONCLUSION

The convergence of blockchain and federated learning (FL) within the financial Internet of Vehicles (IoV) landscape marks a profound evolution in how mobility data, artificial intelligence, and decentralized finance intersect. This chapter has explored the technological foundations, policy implications, and regulatory opportunities of this integration, presenting a comprehensive vision for privacy-preserving, scalable, and trustworthy mobility finance systems. In doing so, it has recognized the vehicle not merely as a transport entity, but as an intelligent economic agent capable of autonomous data generation, learning, and transaction.

Through the deployment of FL, vehicles can collaboratively train models that support real-time, context-aware decisions for insurance pricing, leasing management, fraud detection, and personalized financial services all without compromising user privacy. Blockchain, in turn, provides the infrastructure for immutable logging, verifiable trust, smart contract enforcement, and decentralized governance. Together, these technologies offer a secure foundation for the automated, ethical, and compliant financial services that tomorrow's smart mobility ecosystems demand.

Yet, as this chapter has emphasized, the path to adoption is not linear. Technical challenges related to scalability, energy efficiency, and adversarial robustness persist. Legal uncertainties regarding data jurisdiction, AI accountability, and smart contract enforceability remain unresolved. Ethical imperatives around fairness, explainability, and redress mechanisms must be addressed with care. These complexities require thoughtful policy harmonization, international standardization, and proactive governance.

By proposing a federated governance framework and examining real-world case studies, the chapter has laid the groundwork for responsible integration. It also highlights the critical role of interoperability, ethical design, and collaborative policy-making. The future of financial IoV systems hinges not solely on technical innovation, but on the collective ability of stakeholders governments, technologists, researchers, and end-users to build infrastructures that are intelligent, inclusive, and aligned with the public good.

REFERENCES

International Organization for Standardization. (2021). *ISO/TC 307 – Blockchain and distributed ledger technologies – Overview of governance*. [Online]. Available: https://www.iso.org/committee/6266604.html

Kaur, I., Mishra, R., Narwaria, M., & Saxena, S. (2023). Opportunities and Applications of Blockchain for Empowering Tele-Healthcare. In Khan, R., & Kumar, I. (Eds.), *Advancements in Bio-Medical Image Processing and Authentication in Telemedicine* (pp. 111–126). IGI Global Scientific Publishing., DOI: 10.4018/978-1-6684-6957-6.ch007

Lin, J., Yu, W., Zhang, N., Yang, X., Zhang, H., & Zhao, W. (2017). A survey on Internet of Vehicles: Architectures, enabling technologies, security and privacy, and applications. *IEEE Internet of Things Journal*, *5*(5), 3776–3797. DOI: 10.1109/JIOT.2018.2849609

Mahesh, G., & Mishra, R. (2024). Implementing a Secured Blockchain-Based EHR Management System With Remix IDE. In Mishra, R., Dwivedi, V., & Saxena, S. (Eds.), *Computational Convergence and Interoperability in Electronic Health Records (EHR)* (pp. 265–276). IGI Global Scientific Publishing., DOI: 10.4018/979-8-3693-3989-3.ch014

Mishra, R., Kaur, I., Sharma, V., & Bharti, A. (2022). Computational Intelligence and Blockchain-Based Security for Wireless Sensor Networks. In Tyagi, A. (Ed.), *Handbook of Research on Technical, Privacy, and Security Challenges in a Modern World* (pp. 324–336). IGI Global Scientific Publishing., DOI: 10.4018/978-1-6684-5250-9.ch017

National Institute of Standards and Technology (NIST). *Blockchain for IoT Security and Privacy*, 2021. [Online]. Available: https://www.nist.gov/publications

Rajamanickam, R., Mishra, R., & Chaturvedi, S. (2025). A Hybrid Cryptographic Framework for Privacy-Preserving and Securing Blockchain Transactions. *International Journal of Performability Engineering*, *21*(6).

World Economic Forum. *Blockchain Deployment Toolkit – Regulatory Guidance*, 2021. [Online]. Available: https://www.weforum.org/projects/blockchain-toolkit

Xu, X., Liu, C., Zhang, L., Qiu, M., & Chen, M. (2021). Blockchain-based federated learning for IoV: A secure and decentralized framework. *IEEE Transactions on Intelligent Transportation Systems*, . Advance online publication. DOI: 10.1109/TITS.2021.3083061

Zyskind, G., Nathan, O., & Pentland, A. (2015). *"Decentralizing privacy: Using blockchain to protect personal data," in 2015 IEEE Security and Privacy Workshops.*, DOI: 10.1109/SPW.2015.27

About the Contributors

Ben Othman Soufiene is an Assistant Professor of computer science at Applied College, King Faisal University, Saudi Arabia. He received his Ph.D. degree in computer science from Manouba University in 2016 for his dissertation on "Secure data aggregation in wireless sensor networks. He also holds M.S. degrees from the Monastir University in 2012. My research interests focus on the Internet of Medical Things, Wireless Body Sensor Networks, Wireless Networks, Artificial Intelligence, Machine Learning and Big Data. Dr. Ben Othman has published more than 120 papers at reputed international journals, conferences, and book chapters. He is an Editorial Board Member in the different Journals and Conferences. He serves as an associate editor/academic editor for international journals including IEEE Access, IEEE Sensors, IEEE Internet of Things, Elsevier, Springer, Taylor & Francis, IGI, IET, Telecommunication Computing Electronics and Control, and Wiley. Dr. Ben Othman is a Technical Program Committee Member for more than a dozen of international conferences.

***

Yamini B is a faculty member in the Department of Networking and Communications at the SRM Institute of Science and Technology, Kattankulathur, Tamil Nadu, India. Her academic and research interests include computer networks, wireless communication, and emerging technologies in network security. She is committed to fostering innovation in networking research and guiding students toward excellence in the field of communication technologies.

C. Thiyagarajan is an Assistant Professor in the Department of Artificial Intelligence and Machine Learning at Panimalar Engineering College, Chennai. He is dedicated to teaching and research in emerging AI and ML technologies, fostering innovation among students.

Senthil Kumar is an Assistant Professor in CSE (Cyber Security) at R.M.K. College of Engineering and Technology, committed to fostering technical excellence among students. His areas of interest include cybersecurity, emerging technologies, and industry-oriented research initiatives. He actively contributes to academic growth through teaching, mentoring, and participation in professional development activities.

Jaishri Gothania is an academician, author, researcher, and data scientist with over 9 years of experience. She has published 12 research articles, presented 6 conference papers, and holds multiple patents. Her work bridges data science and innovation, contributing significantly to academia, research, and technological advancement.

Pooja Jaiswal received the B.Sc. degree in Science from J.K. Institute of Applied Physics & Technology, University of Allahabad (U.P.) in 2007. She received the B.Tech & M.Tech degree in Electronics and communication from J.K. Institute of Applied Physics & Technology, University of Allahabad (U.P.) in 2010 & 2012. Her research interest includes Quantum computing and Image Processing with Deep learning and their applications. Presently she is pursuing PhD from J.K. Institute of Applied Physics & Technology, University of Allahabad, India.

K. Raju is an Associate Professor in the Department of Artificial Intelligence and Machine Learning at Rajalakshmi Engineering College, Chennai. He is committed to academic excellence and advancing research in AI and emerging technologies. With a passion for student development, he actively mentors learners toward innovative problem-solving. His contributions strengthen the growth of AI education and interdisciplinary advancements in engineering.

K. Vijayalakshmi is an Assistant Professor at SRM Institute of Science and Technology, Kattankulathur. Her academic and research interests include Machine learning, intelligent systems, and emerging technologies. She is dedicated to teaching, mentoring students, and contributing to the advancement of her department. She actively engages in research and academic development to promote innovation in higher education.

V. Kalyanasundaram is an Assistant Professor at SRM Institute of Science and Technology Kattankulathur, Chennai. He is a actively researches in power systems and intelligent transportation systems, with multiple publications and citations. Prior to his current role, he worked on academic and research initiatives in his field, contributing to both teaching and scholarly activity.

Manish Kumar is presently working in Department of Biochemistry, J.C. Bose Institute of Life Sciences, Bundelkhand University, Jhansi, Uttar Pradesh, India from 2016 to till date. He has teaching experience more than 9 years and research experience more than 14 years. He did his B.Sc. (H) Biochemistry and M.Sc. Biochemistry from Bundelkhand University, Jhansi (U.P.) and M.Phil. Biochemistry from Singhania University, Jhunjhunu (Rajasthan). He did his doctoral degree Ph.D. Biochemistry from Sam Higginbottom University of Agriculture, Technology and Sciences, Prayagraj (U.P.), India. He received best thesis research award for Ph.D. Biochemistry from UPKAR's foundation, Lucknow. He has Co-supervised 10 PG students. He has been published many research papers and review papers in national and international journals with good Impact Factor and NAAS rating. He is life member of many international and national bodies. He has expertise in biochemistry, enzymology, proteomics and medical biochemistry.

Rocky Kumar is Assistant Professor in Faculty of Computer Science and Engineering at Poornima University Jaipur. He is BCA, MCA and pursuing M Tech.

Abhiram Kumar is a distinguished scientist with over a decade of research experience in leading academic institutions and pharmaceutical R&D organizations, including CSIR-CDRI, Alembic Pharmaceuticals Ltd., and Ipca Laboratories Ltd. He is the author of multiple books, book chapters, research articles, and review papers. A passionate scientist, he is an active member of the Association of Pharmaceutical Teachers of India (APTI), All-India Research Scholars Association (AIRSA), and Vijnana Bharati (VIBHA). Throughout his academic journey, Dr. Kumar has played a pivotal role in numerous high-impact collaborative projects funded by esteemed organizations such as ICMR, CSIR, and DBT. His research expertise spans cancer therapeutics, advanced drug delivery systems, molecular pharmacology, and preclinical animal studies, driving innovation in the fight against cancer.

Anto Gracious L. A. is a faculty member in the Department of Computer Science and Engineering at R.M.K. College of Engineering and Technology, Tamil Nadu. He is dedicated to teaching and mentoring students in emerging areas of computer science and technology. His academic interests include software engineering, data analytics, and innovative teaching methodologies.

Anita M is a faculty member at the SRM Institute of Science and Technology, Kattankulathur, Tamil Nadu, India. She is actively involved in teaching, research, and academic development in the field of computer science and engineering. Her research interests include artificial intelligence, data analytics, and emerging computing technologies, and she is dedicated to advancing innovation and excellence in engineering education.

Ezhilvendan M is a faculty member at Panimalar Engineering College, Tamil Nadu. He is committed to fostering academic excellence and innovation among engineering students. His professional interests include computer science, software engineering, and technology-driven education.

Nalini M is an academic professional in the Department of Cyber Security at S A Engineering College, Chennai. Her research interests include cybersecurity, artificial intelligence, and machine learning applications. She has contributed to scholarly publications and actively works on advancing secure and intelligent computing systems.

Vijayalakshmi G V Mahesh received her BE in Electronics and Communication Engineering from Bangalore University, India in 1999, and M.Tech in Digital Communication and Networking from Visvesvaraya Technological University in 2005 and the Ph.D. degree from the Vellore Institute of Technology, Vellore, India. Currently she is working as an Associate Professor at BMS Institute of Technology and Management, Bangalore, India. She has been in academics for over 19 years and has published her research in various reputed journals and conferences. Her research interests include Machine Learning, Image Processing, Pattern Recognition and Deep learning, Affective computing. She has Memberships in Professional Bodies such as ISTE and IEI.

Renu Mishra is presently working as an Associate Professor at Sharda University, Greater Noida, U.P. She obtained Ph.D. from the School of IT, Rajiv Gandhi Proudyogiki Vishwavidyalaya (RGPV), Bhopal, India. She holds an M-Tech (Honors) from RGPV, Bhopal (M.P.), and a B-Tech from Uttar Pradesh Technical University (UPTU), Lucknow, U.P. With 18 years of teaching experience, Dr. Renu's research interests include Blockchain Technology, Mobile Ad-hoc Networks, Computer Networks, Wireless Sensor Networks, Cryptography, and Network Security. She has also made significant contributions to the academic community by publishing various books and book chapters.

N. Sivakumar is a Post-Doctoral Fellow in Federated Learning from the University of South Florida, USA, and currently serves as an Associate Professor in the Department of Computer Engineering at Marwadi University, Gujarat, India. He holds a Ph.D. in Computer Science from Bharathidasan University, Tamil Nadu, with a specialization in hybrid algorithmic approaches for fraud detection in online transactions. With over 13 years of combined experience in research and teaching, his expertise spans Federated Learning, Data Science, Machine Learning, Cloud Computing, and Programming. He is guiding Ph.D. and Post-Doctoral scholars in the areas of multimodal federated learning for brain tumor classification, FL optimization for medical data, and privacy-preserving distributed AI. Dr. Sivakumar is an active researcher with several Q1/Q2 journal publications and SCOPUS-indexed conference papers. He is also an inventor with five published Indian patents and an author of the book Artificial Intelligence with Machine Learning. His work has received Best Paper Awards, and he has served as a Session Chair and Track Chair in international conferences, including IEEE-sponsored events. He is a Microsoft Certified Azure Data Scientist Associate, AWS Certified Educator, and ORACLE-certified database professional.

Kumar Pranav Narayan is a distinguished academic and researcher in the Department of Biological Sciences at BITS Pilani, Hyderabad Campus, where he has been serving since December 2008. His extensive work spans cancer therapeutics, agricultural microbiology, and bio-based innovations, reflecting a commitment to both scientific advancement and societal impact

Mamta Narwaria is an assistant professor in the School of Computing Science & Engineering, Sharda University, Greater Noida. She has more than 20 years of academic experience. She is Post graduate in Computer Science & Engineering from Rajiv Gandhi Prodyougiki University, Bhopal, India. She has broad research interests in AI/ML,Federated learning, Image processing. She has published 05 papers in reputed conferences and refereed journals.

P. Rashmi is an Assistant Professor at Velammal Engineering College, dedicated to academic excellence and student development. Her teaching and research interests span emerging technologies and innovative learning methodologies. She actively contributes to institutional growth through scholarly activities, mentorship, and professional engagement.

B. Prabha is an Assistant Professor (Grade I) in the Department of Artificial Intelligence and Machine Learning at Panimalar Engineering College, Chennai. She is dedicated to advancing knowledge in AI and ML through innovative teaching and research. Passionate about guiding students, she actively contributes to academic growth and technological development. Her work reflects a commitment to excellence in engineering education and emerging technologies.

Nandita Pradhan is the head of the ECE and CRT department. She did B.E. in Electronics and Telecommunication Engineering in 1985, M.E. in Communication Systems from the Govt. Engineering College, Jabalpur in 1991 and Ph.D. in Electronics and Communication Engineering from AKTU, Lucknow. She began her professional journey as a Lecturer in the Department of ECE at Government Engineering College, Jabalpur in 1985 and served in that position until 1997.

Currently she worked as a Head of the Department of ECE in United Group of Institutions, Prayagraj, India.

Siva Subramanian R. is an Associate Professor in the Department of Computer Science and Engineering at R.M.K. College of Engineering & Technology, Puduvoyal, Tamil Nadu. He is actively engaged in research mentoring and teaching undergraduate and postgraduate students in AI-driven domains.

Ravikumar RN is an experienced educator with 12+ years of teaching in Computer Science across India and overseas. Currently an Assistant Professor at Marwadi University, Rajkot, he specializes in Advanced Java, Machine Learning, Recommendation Systems, Optimization Techniques, and AI. He completed PhD at Amity University, Jaipur, 2025. He completed M.E. in Computer Science Engineering from VMRF University (2013), and B.E. from Anna University (2010). He holds a Mini MBA from the Swiss eLearning Institute (2015). He is dedicated to research, professional development, and advancing education through technology.

Anjali Rai is an academic professional with a Master's degree in Pharmacology (M.Pharm) and a Bachelor's in Pharmacy (B.Pharm). Her academic and research interests include clinical research, pharmacovigilance, cancer biology, and molecular pharmacology. She has contributed to Scopus-indexed review articles and book chapters in the biomedical sciences. Anjali is dedicated to advancing research-driven education and is actively involved in student mentorship and academic development initiatives.

Aarthi S is a dedicated Assistant Professor in Computer Science with over six years of teaching experience. She is currently pursuing her Ph.D. at Marwadi University, Gujarat, and holds an M.Phil. and MCA from Bharathiar University, Tamil Nadu. Throughout her academic career, Aarthi has specialized in programming languages such as C++, Java, and C, and contributed to research in recommender systems and AI-related data privacy. She has published journal articles, contributed to book chapters, and actively participated in technical workshops. Known for her student mentorship and engagement in academic events, Aarthi remains committed to fostering innovation and excellence in computer science education.

Anusha S is an Assistant Professor in the Department of Information Technology at Easwari Engineering College, Chennai. She is passionate about imparting technical knowledge and guiding students in the field of IT. Her areas of interest include data science, artificial intelligence, and software development.

S. Kavitha Bharathi is an Associate Professor in the Department of Computer Technology at Kongu Engineering College, Perundurai, Tamil Nadu, India. She holds a Ph.D. in Computer Science and Engineering from Anna University, Chennai, and has over 21 years of academic experience. Her teaching and research interests include Grid and Mobile Grid Computing, Internet of Things (IoT), Data Structures, Operating Systems, and Artificial Intelligence applications. She has published several research papers in reputed international journals and conferences, focusing on areas such as resource allocation, fault-tolerant models, and machine learning-based prediction systems. Dr. Kavitha Bharathi actively contributes to curriculum development, student mentoring, and collaborative research projects with industry partners. Through her academic leadership and technical expertise, she continues to promote innovation and applied research in emerging computing domains at Kongu Engineering College.

Sujitha S serves as an Assistant Professor in the Department of Information Technology at Kongu Engineering College, Erode, India. Her academic and professional interests span core areas of information technology, including software development, data-driven applications, and emerging digital technologies. She is actively engaged in undergraduate teaching, curriculum delivery, and student mentoring, guiding learners in both course projects and co-curricular technical activities. In addition to her teaching responsibilities, she contributes to departmental initiatives, workshops, and value-added programmes aimed at enhancing students' technical and professional skills. She regularly participates in faculty development programmes, seminars, and conferences to stay abreast of current trends in IT and to integrate them into her teaching-learning practices.

K. Saravanan is an Associate Professor in the Department of Electrical and Electronics Engineering at SRM Institute of Science and Technology, Kattankulathur. He holds a B.Tech in Mechatronics Engineering and a Ph.D. in Electrical Engineering, specializing in power electronics and electric drives. He is actively involved in teaching, research, and publications related to power electronics, contributing to both academic and industry-focused projects.

Somesh Kumar Sharma is an Associate Professor in the Department of Mechanical Engineering at the National Institute of Technology (NIT) Hamirpur, India. He earned his Ph.D. from the Indian Institute of Science, Bangalore, in 2009. Dr. Sharma specializes in Production and Industrial Engineering, with a focus on Computer Integrated Manufacturing, Operations, and Quality Management. His research encompasses areas such as Industrial Engineering and Management. In his administrative roles, Dr. Sharma has served as the Head of the Training and

Placement Cell at NIT Hamirpur. He has also held the position of Associate Dean (Resource Generation & Industrialization) at the institute. Throughout his academic career, Dr. Sharma has contributed significantly to his field, authoring numerous research papers and book chapters. His work has garnered over 200 citations, reflecting his impact in the areas of Industrial Engineering and Management.

Chin-Shiuh Shieh received the M.S. degree in electrical engineering from the National Taiwan University, Taipei, Taiwan, in 1991, and the Ph.D. degree from the Department of Computer Science and Engineering, National Sun Yat-sen University, Kaohsiung, Taiwan, in 2009. He joined the Faculty of the Department of Electronic Engineering, National Kaohsiung University of Science and Technology, Kaohsiung, in August 1991, where he is currently a professor. His research interests include Information Security, Computational Intelligence, Artificial Intelligence, Computer Networking.

Lokesh Kumar Shrivastav is an Assistant Professor in the Department of Computer Science at Atma Ram Sanatan Dharma College (University of Delhi), where he has taught since January 2011. He earned his Ph.D. in Computer Science & Engineering from USICT, GGSIPU, and holds an M.Tech (CSE), an MCA and a B.Sc.(H) in Physics; he is UGC-NET and GATE qualified. His research spans quantum computing, machine learning (including neural networks), numerical optimization, and mathematical modelling of infectious-disease dynamics and prediction. He teaches courses such as Numerical Optimization, Machine Learning, Mathematics for Computing and Data Analytics using Python. Dr. Shrivastav has published peer-reviewed work on COVID-19 modelling and financial time-series prediction and maintains a Google Scholar profile.

Prabhat Chandra Shrivastava is an Assistant Professor in the Department of Electronics and Communication at the University of Allahabad, Prayagraj. He completed his Ph.D. and M.Tech from MNNIT Allahabad and his B.E. from RGPV Bhopal. He has more than 10 years of experience in teaching and research. His main areas of work include Architecture of Digital Filter Design, Digital Signal Processing, Artificial Intelligence (AI), Machine Learning (ML), and Smart Agriculture. He has worked on useful projects like predicting plant health using AI, analyzing ECG signals for heart problems, and recognizing Indian classical music with deep learning. Dr. Shrivastava has published many research papers in international journals and conferences. He has also qualified the GATE exam five times and received fellowships during his M.Tech and Ph.D. studies. He is passionate about using technology to solve real-life problems in agriculture.

Shivang Shukla is an Assistant Professor in the Department of Pharmacy at Jagannath University, Jaipur. He holds an M.Pharm in Pharmacology and is actively involved in teaching and research. His core interests include cancer biology, drug delivery, microbiology, and nanotechnology. He has authored several Scopus-indexed review articles and book chapters in these areas. Shivang also mentors' students in research and skill enhancement and contributes to organizing academic and career-oriented programs within the department.

Sitesh Kumar Singh is an accomplished academic and Chartered Engineer with a Ph.D. in Civil Engineering and over nine years of diverse experience in education, research, consultancy, and educational administration. Currently serving as an Assistant Professor in the Department of Civil and Environmental Engineering at the National University of Science and Technology, Muscat, Sultanate of Oman, he specializes in Civil and Transportation Engineering, Project Management, and Data Analytics. Dr. Singh holds multiple advanced degrees including M.Tech, B.Tech, PGDIPR, PGDUPDL, and an PGD in Education. His career encompasses curriculum development, research supervision, quality assurance, and strategic academic planning. He has supervised over 40 postgraduate and 2 doctoral students, contributing to academic leadership through workshops, conferences, and editorial roles, and has published over 40 research articles, several patents, and authored books and book chapters.

Vikram Singh serves as the Principal of SIRDA Polytechnic in Sundernagar, Himachal Pradesh. He is associated with the Department of Mechanical Engineering at the institution. In his message as Principal, Er. Singh emphasizes the significance of technical education as the backbone of a nation's development and highlights SIRDA's eco-friendly campus equipped with state-of-the-art infrastructure. For direct communication, he can be reached via mobile at 9682170561 or email at vikram_phdme@nith.ac.in.

Anish T. P is an faculty member in the Department of Computer Science and Engineering at R.M.K College of Engineering and Technology, Puduvoyal. His academic interests focus on advancing knowledge and innovation in computing technologies. He is committed to guiding students and contributing to impactful research in the field of Computer Science.

Hemant K. Upadhyay (Ph.D. pursuing) is an accomplished academic leader with over 20 years of experience in higher education, research, and administration. Currently serving as Assistant Professor at Poornima University, Jaipur, he holds B.Tech, M.Tech, and MBA degrees. His research focuses on multi-objective optimization and decision-making models for electric vehicle adoption in India. He has published extensively in SCIE/Scopus-indexed journals, authored international book chapters, and presented papers at global conferences. With eight patents, numerous awards, and memberships in leading professional societies, he is dedicated to advancing sustainable engineering and fostering academic excellence.

Kalai Priya V is an Assistant Professor in the Department of Computer Science and Engineering at RMK College of Engineering and Technology, Tamil Nadu, India. She is passionate about teaching and research in areas such as artificial intelligence, data science, and software engineering. With a strong academic background and commitment to student development, she actively contributes to academic excellence and innovation in the field of computer science.

S. Vidyasagar is an Assistant Professor in the EEE Department at SRM Institute of Science and Technology, Kattankulathur. He is dedicated to quality teaching, student mentorship, and continuous academic contribution through scholarly activities.

Shiferw Garoma Wayessa is expert in Road and Transport Engineering domain. He is working as Senior Lecturer at Department of Civil Engineering, College of Engineering and Technology, Wollega University, Ethiopia. He have published several articles in journals of repute.

Index

Symbols

A

B

C

D

E

F

I

L

M

N

P

Q

R

S

T

U

V

W

www.ingramcontent.com/pod-product-compliance
Lightning Source LLC
LaVergne TN
LVHW081317300926
849161LV00008B/1792
* 9 7 9 8 3 3 7 3 3 1 4 6 1 *